Springer Proceedings in Mathematics & Statistics

Volume 196

Springer Proceedings in Mathematics & Statistics

This book series features volumes composed of selected contributions from workshops and conferences in all areas of current research in mathematics and statistics, including operation research and optimization. In addition to an overall evaluation of the interest, scientific quality, and timeliness of each proposal at the hands of the publisher, individual contributions are all refereed to the high quality standards of leading journals in the field. Thus, this series provides the research community with well-edited, authoritative reports on developments in the most exciting areas of mathematical and statistical research today.

More information about this series at http://www.springer.com/series/10533

L. Andries van der Ark • Marie Wiberg
Steven A. Culpepper • Jeffrey A. Douglas
Wen-Chung Wang
Editors

Quantitative Psychology

The 81st Annual Meeting of the Psychometric Society, Asheville, North Carolina, 2016

Springer

Editors
L. Andries van der Ark
Research Institute for Child Development and Education
University of Amsterdam
Amsterdam, The Netherlands

Steven A. Culpepper
Department of Statistics
University of Illinois at Urbana-Champaign
Champaign, IL, USA

Wen-Chung Wang
Department of Psychology
The Educational University of Hong Kong
Hong Kong, China

Marie Wiberg
Department of Statistics, USBE
Umeå University, Umeå, Sweden

Jeffrey A. Douglas
Department of Statistics
University of Illinois at Urbana-Champaign
Champaign, IL, USA

ISSN 2194-1009 ISSN 2194-1017 (electronic)
Springer Proceedings in Mathematics & Statistics
ISBN 978-3-319-56293-3 ISBN 978-3-319-56294-0 (eBook)
DOI 10.1007/978-3-319-56294-0

Library of Congress Control Number: 2017940525

Printed on acid-free paper

This Springer imprint is published by Springer Nature
The registered company is Springer International Publishing AG
The registered company address is: Gewerbestrasse 11, 6330 Cham, Switzerland

Preface

This volume represents presentations given at the 81st annual meeting of the Psychometric Society in Asheville, North Carolina, during July 11–15, 2016. The meeting, organized by the University of North Carolina at Greensboro, was one of the largest Psychometric Society meetings in the United States, both in terms of participants and number of presentations. It attracted 415 participants, with 204 papers being presented, along with 95 poster presentations, 3 pre-conference workshops, 3 keynote presentations, 6 invited presentations, 2 career-award presentations, a debate, 2 dissertation-award winners, 9 symposia, a trivial-pursuit lunch, and *Psychometrika*'s 80th anniversary celebration.

Since the 77th meeting in Lincoln, Nebraska, Springer publishes the proceedings volume from the annual meeting of the Psychometric Society so as to allow presenters to quickly make their ideas available to the wider research community while still undergoing a thorough review process. The first four volumes of the meetings in Lincoln, Arnhem, Madison, and Beijing were received successfully, and we expect a successful reception of these proceedings too.

We asked authors to use their presentation at the meeting as the basis of their chapters, possibly extended with new ideas or additional information. The result is a selection of 36 state-of-the-art chapters addressing a diverse set of topics, including item response theory, equating, classical test theory, factor analysis, structural equation modeling, dual scaling, multidimensional scaling, power analysis, cognitive diagnostic models, and multilevel models.

Amsterdam	L. Andries van der Ark
Umeå	Marie Wiberg
Urbana-Champaign, IL	Steven A. Culpepper
Urbana-Champaign, IL	Jeffrey A. Douglas
Hong Kong	Wen-Chung Wang

Contents

New Results on an Improved Parallel EM Algorithm for Estimating Generalized Latent Variable Models

Matthias von Davier

Abstract The second generation of a parallel algorithm for generalized latent variable models, including MIRT models and extensions, on the basis of the general diagnostic model (GDM) is presented. This new development further improves the performance of the parallel-E parallel-M algorithm presented in an earlier report by means of additional computational improvements that produce even larger gains in performance. The additional gain achieved by this second-generation parallel algorithm reaches factor 20 for several of the examples reported with a sixfold gain based on the first generation. The estimation of a multidimensional IRT model for large-scale data may show a larger reduction in runtime compared to a multiple-group model which has a structure that is more conducive to parallel processing of the E-step. Multiple population models can be arranged such that the parallelism directly exploits the ability to estimate multiple latent variable distributions separately in independent threads of the algorithm.

Keywords Parallel EM-algorithm • MIRT • Diagnostic modeling • Estimation • Latent variable modeling

1 Introduction

This chapter reports on the second generation of a parallel algorithm for generalized latent variable models on the basis of the general diagnostic model (von Davier 2005, 2008, 2014). This new development further improves the performance of the parallel-E parallel-M algorithm presented in an earlier report (von Davier 2016) by means of additional computational improvements that produce even larger gains in performance. The additional gain achieved by this second-generation parallel

This work was partially completed while the author was at the Educational Testing Service.

M. von Davier (✉)
National Board of Medical Examiners, 3750 Market Street, Philadelphia, PA, 19104-3102, USA
e-mail: mvondavier@nbme.org

L.A. van der Ark et al. (eds.), *Quantitative Psychology*, Springer Proceedings in Mathematics & Statistics 196, DOI 10.1007/978-3-319-56294-0_1

algorithm reaches factor 20 for several of the examples were reported with a sixfold gain based on the first generation. The estimation of a multidimensional IRT model for large-scale data may show a larger reduction in runtime compared to a multiple-group model which has a structure that is more conducive to parallel processing of the E-step. Multiple population models can be arranged such that the parallelism directly exploits the ability to estimate multiple latent variable distributions separately in independent threads of the algorithm.

This development allows estimation of advanced psychometric models for very large datasets in a matter of seconds or minutes, rather than hours. Unlike methods that rely on simplifications of the likelihood equations that are only available for a specific set of constrained problems such as bifactor models, the approach presented here is applicable to all types of multidimensional latent variable models, including multidimensional models, multigroup and mixture models, as well as growth curve and growth mixture models.

Parallel processing is now available in a number of compilers and hence found its way into software packages such as LatentGold, Mplus, and FlexMirt. While these packages allow users to utilize one or multiple cores, their documentation is somewhat limited. In the present report, the approach to parallelism is detailed at the level of algorithmic description, and the types of gains are exemplified based on a range of hardware platforms that are typically available as workstations or servers. Moreover, the software presented here is available for research purposes on all major operating systems, in particular, on Linux, Microsoft Windows, and Apple OS X platforms.

2 A General Latent Variable Model

The general latent variable model used in this evaluation of an improved algorithm for parallel processing is based on the general diagnostic model (GDM) (von Davier 2005). This family of models contains a large class of well-known psychometric approaches as special cases, including IRT, MIRT, latent class models, HYBRID models, and mixture models (von Davier 2008), as well as models for longitudinal data (von Davier et al. 2011) and several diagnostic models (von Davier 2014, 2016).

The probability of a correct item response $X = 1$ by a respondent from a population $C = c$ and with skill attribute pattern $a = (a_1, \ldots, a_k)$ on item i can be written as

$$P(X = 1|i, a, c) = \frac{\exp\left(\beta_{\text{ic}} + \sum_{k=1}^{K} \gamma_{\text{ick}} h(q_{\text{ik}}, a_k)\right)}{1 + \exp\left(\beta_{\text{ic}} + \sum_{k=1}^{K} \gamma_{\text{ick}} h(q_{\text{ik}}, a_k)\right)} \tag{1}$$

This is the general model introduced by von Davier (2005). The q_{ik} are indicator variables for $i = 1, \ldots, I$ and $k = 1, \ldots, K$ and are provided as an input. These Q-matrix entries q_{ik} describe which of the skill attributes is required for which item.

Note that Eq. (1) also contains a population indicator c, which makes it suitable both for multiple-group and mixture distribution models (von Davier and Rost 1995; von Davier and Yamamoto 2004; von Davier 2005, 2008; von Davier and Rost 2016). While the general model given in Eq. (1) served as the basis for the formal specification of the log-linear cognitive diagnostic model (L-CDM) (Henson, Templin and Willse 2009) and other developments for binary skill attributes and data, von Davier (2005, 2008) utilized the general form to derive the linear or partial-credit GDM:

$$P(X = x|i, a, c) = \frac{\exp\left(\beta_{\text{ixc}} + \sum_{k=1}^{K} x\gamma_{\text{ick}} h\left(q_{\text{ik}}, a_k\right)\right)}{1 + \sum_{i}^{m}\left(\beta_{\text{iyc}} + \sum_{k=1}^{K} y\gamma_{\text{ick}} h\left(q_{\text{ik}}, a_k\right)\right)} \tag{2}$$

Note that this leads to a model that contains located latent class models, multiple classification latent class models, IRT models, and multidimensional IRT models, as well as a compensatory version of the reparameterized unified model, as special cases (von Davier 2005). In addition, the linear GDM as well as the general family is suitable for binary, polytomous ordinal, and mixed-format item response data.

One common application of generalized latent variable models is the use for confirmatory analysis. In this case, a Q-matrix provides the required loading pattern, and constraints on the skill attribute space provide the structure of the model. One example is what is commonly called a multi-trait-multi-method model, in which each observed indicator variable is cross classified with respect to two different sets of latent variables. In the examples analyzed for this report, a seven-dimensional model of this type is analyzed, which contains two loadings for each item, one for a set of four latent variables (subdomains) and one for a set of three variables (processes). Figure 1 provides an illustration of this model used as example.

3 Method

The EM algorithm (Dempster et al. 1977) is one of the most frequently used approaches for estimating latent variable models (e.g., McLachlan and Krishnan 1997). The name of the algorithm stems from the alternating, iterative repetition of two steps, the E (expectation) step and the M (maximization) step. The estimation of generalized latent variable models using the EM algorithm requires the estimation of expected values for all required sufficient statistics of the structural parameters of the measurement model as well as the estimation of latent variable distributions in one or more populations. In the M-step, the expected values serve as sufficient statistics for the maximization of parameters. Parallel implementations of the EM algorithm have been used in image processing, in particular in Gaussian mixture modeling for some time (Cui et al. 2014; Cui et al. 2010; Das et al. 2007; Lopez de Teruel et al. 1999). In contrast to Gaussian mixtures, certain latent variable models require computationally more costly calculations in the M-step as well.

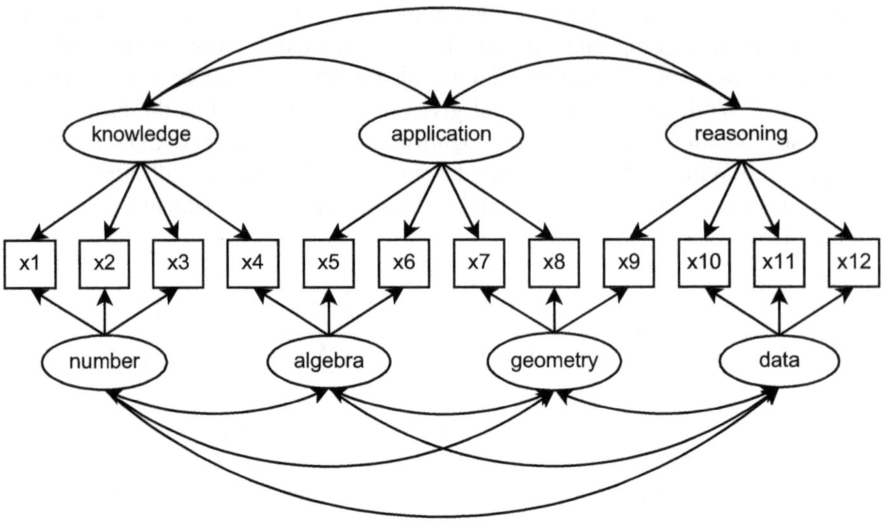

Fig. 1 Confirmatory multidimensional IRT model with seven dimensions. Three of the variables describe processing skills, and four variables describe subdomains of mathematics. While the figure uses on only 12 items, the real data example contains 214 items in a balanced incomplete block design administered to approx. 8000 students

While parallelizing the E-step is straightforward in terms of distribution of the work, the aggregation of the partial results obtained in distributed ways separately by each core is again a potentially costly calculation or aggregation process. A new algorithm was developed based on the first generation parallel-E parallel-M algorithm described in von Davier (2016). This new algorithm can be described as a parallel-E parallel-M algorithm with tiling-based aggregation of results. This new approach is based on three different phases of parallel execution of the necessary calculations:

1. Parallel E-step: Distributed calculation of expected counts for sufficient statistics
2. Tiling: Rearranging distributed latent variable space and parallel aggregation
3. Parallel M-Step: ML-estimation of parameters based on aggregated counts

Gains are largest for the parallelism introduced in part (1) that concerns the E-step by conducting estimation of expected counts separately in subsamples distributed across cores. The smallest gains are obtained by the conversion of the M-step to parallel execution in part (3). The aggregation step that follows the E-step in part (2) provides somewhat more advantages than the parallel M-step, either in the form of a multiple-group approach where aggregation can be completely avoided, or in the form of tiling, where the latent variable space aggregation is rearranged so that it can take place in parallel as well. Shared memory allocation of all latent variable distributions and rearranging the direction of aggregation are crucial in the process. More details about the different approaches utilized in version 1 can be

Table 1 Summary of example analyses used in the comparisons

Case	Scales	Model	Groups	Items	Sample	QPT	Total	Ncat
A	1	2PL/GPCM	312	293	1,614,281	21	21	2–4
B	1	2PL/GPCM	283	133	1,803,599	21	21	2–4
C	7	MTMM	1	214	7377	3	2187	3
D	2	MIRT	1	175	5763	15	225	3
E	2	MIRT	1	175	5763	31	961	3
F	NA	LCA	54	54	246,112	1	1	6
G	5	MIRT	1	150	2026	5	3125	2

The examples cover a wide range of latent variable models from IRT to MIRT, confirmatory models, and latent class models. The items are mixed format, and their number varies from 54 to 293; the number of respondents varies from 2026 to 1.8 million

found in von Davier (2016). The tiling process resulting in an improved ability to utilize parallelism in aggregation is detailed in Gan et al. (2009).

4 Data

Table 1 shows an overview of the test cases. All test cases reported here are based on sequential and parallel versions, except two additional ones, that were only run on the fastest hardware platform, and only in parallel mode, since sequential mode or running these on a laptop would take unacceptably long periods of time. The test cases are from typical applications of generalized latent variable models, ranging from IRT, to classification of respondents by means of a latent class analysis, to MIRT applications with 2, 5, and 7 dimensions, and finally multiple population IRT for linking in large-scale international assessments with approximately 300 populations and 2,000,000 test takers distributed across these populations.

5 Results

Table 2 shows the results for a (somewhat older) Dual-CPU 12-Core Intel Xeon workstation running at 3.46 GHZ per core. These are given for the sequential algorithm, running only on a single core, as well as for parallel-E parallel-M version 1 and the improved version 2 of the PE-PM algorithm, running on all available cores.

Table 3 shows the results for a 4-CPU AMD Opteron (Piledriver architecture) server with 64/32 cores, running at 2.6 GHZ per core. This architecture offers 64 integer arithmetic cores, with each CPU offering 16 integer units that share eight floating point units. In this sense, we see a performance that is more reflective of 32 FPUs, but with some added capacity for caching and pre-fetching and integer processing. The measures are given for the sequential algorithm, running only on a

Table 2 Results of the comparison of parallel-E parallel-M versions 1 and 2 on a 12-Core Xeon workstation as well as the sequentially executed algorithm

		Parallel V1		Parallel V2		
Case	Iterations	Likelihood	Sec.	Sec.	Speedup	Sec.
A	126	−14,547,731.24	1356	168	807%	153
B	112	−14,639,728.20	1127	117	963%	96
C	165	−125,200.51	2465	314	785%	343
D	76	−14,468,510.90	44	11	400%	11
E	86	−14,468,485.63	1155	163	708%	145
F	1028	−1,234,570.30	7444	1039	716%	964
G	277	−130,786.39	2499	949	263%	726

Table 3 Results of the comparison of parallel-E parallel-M versions 1 and 2 on a 64/32-Core AMD Piledriver server as well as the sequentially executed algorithm

		Parallel V1		Parallel V2		
Case	Iterations	Likelihood	Sec.	Sec.	Speedup	Sec.
A	126	−14,547,731.24	2074	256	810%	114
B	112	−14,639,728.20	1553	185	839%	90
C	165	−125,200.51	6131	889	689%	300
D	76	−14,468,510.90	116	21	552%	5
E	86	−14,468,485.63	1945	150	1296%	116
F	1028	−1,234,570.30	6427	377	1704%	227
G	277	−130,786.39	6771	287	2359%	563

single core, as well as for parallel-E parallel-M version 1 and the improved version 2 of the PE-PM algorithm, running on all available cores.

These results show that a gain in the order of 800% for a 12 core workstation and in the order of 2000% for a 32/64 core 4-CPU server is well within reach. The examples provided here show also that for most cases, the version 2 of the parallel algorithm that uses tiling reduction performs for most cases at a much higher level than version 1. Unlike algorithms that either utilize reduction of dimensionality (Gibbons and Hedeker 1992; Rijmen et al. 2014; Cai 2010, 2013), the algorithm presented here is a general-purpose solution for speeding up calculations and can be applied to any latent variable model available through this family of models (von Davier and Rost 2016) to speed up estimation substantially.

6 Discussion

Massive gains in processing speed can be realized by using the parallel-E parallel-M algorithm with tile reduction (PEPM-TR) for estimating generalized latent variable models. The present paper shows that gains in the order of 2000% in processing speed are not uncommon. That is, according to Amdahl's (1967) law, the percent parallel processing with 32 cores is at a level of

$$P = \left(1 - \frac{1}{G}\right)\left(\frac{C}{C-1}\right)$$

see von Davier (2016). For $G = 20$ and $C = 64$, we obtain a value of $P = 0.965$ or a level of parallism of 96.5% for this algorithm. For gains around 800% obtained with the 12 core hardware, we obtain a very similar estimate of a level of 95.5% parallelism. This is a gain that allows using all available data in almost any psychometric analysis. Recent analyses of the combined database of the first five PISA data collections were conducted with almost two million students in more than 300 populations (approximately 60 countries or country/language groups participating on average across 5 cycles) and up to 300 items. The analysis with an IRT model of this very large dataset takes about 2–3 min on the workstation and about 90 s on the server hardware tested here. Multidimensional models for this type of massive databases are easily within reach and can be estimated in less than an hour. This enables a much more rigorous quality control and allows analysts to rerun and to obtain results based on more stringent convergence criteria, resulting in more accurate estimates.

References

G.M. Amdahl, Validity of the single processor approach to achieving large-scale computing capabilities. AFIPS Conf. Proc. **30**, 483–485. doi:10.1145/1465482.1465560

L. Cai, Metropolis–Hastings Robbins–Monro algorithm for confirmatory item factor analysis. J. Educ. Behav. Stat. **35**, 307–335 (2010)

L. Cai, *flexMIRT: A Numerical Engine for Flexible Multilevel Multidimensional Item Analysis and Test Scoring (Version 2.0) [Computer software]* (Vector Psychometric Group, Chapel Hill, NC, 2013)

H. Cui, A. Tumanov, J. Wei, L. Xu, W. Dai, J. Haber-Kucharsky, E. Xing, in *Proceedings of the ACM Symposium on Cloud Computing*. Exploiting iterativeness for parallel ML computations (ACM, New York, NY, 2014), pp. 1–14

H. Cui, X. Wei, M. Dai, Parallel implementation of expectation-maximization for fast convergence (2010). Retrieved from http://users.ece.cmu.edu/~hengganc/archive/report/final.pdf

A.S. Das, M. Datar, A. Garg, S. Rajaram, in *Proceedings of the 16th International Conference on World Wide Web, WWW 07*. Google news personalization: scalable online collaborative filtering (ACM, New York, NY, 2007), pp. 271–280. doi:10.1145/1242572.1242610

A.P. Dempster, N.M. Laird, D.B. Rubin, Maximum likelihood estimation from incomplete data via the EM algorithm. J. R. Stat. Soc. Ser. B **39**, 1–38 (1977)

G. Gan, X. Wang, J. Manzano, G.R. Gao, in *Evolving OpenMP in an Age of Extreme Parallelism: 5th International Workshop on OpenMP, IWOMP 2009 Dresden, Germany, June 3–5, 2009*, ed. by M. S. Muller, B. R. de Supinski, B. M. Chapman. Tile reduction: the first step towards tile aware parallelization in OpenMP (Springer, Berlin, Heidelberg, 2009). doi:10.1007/978-3-642-02303-3_12

R.D. Gibbons, D. Hedeker, Full-information item bi-factor analysis. Psychometrika, **57**, 423–436 (1992)

R. Henson, J. Templin, J. Willse, Defining a family of cognitive diagnosis models using log linear models with latent variables. Psychometrika, **74**, 191–210 (2009)

P.E. Lopez de Teruel, J.M. Garcia, M. Acacio, O. Canovas, P-EDR: an algorithm for parallel implementation of Parzen density estimation from uncertain observations, in *Proceedings of 13th International Parallel Processing symposium and 10th Symposium on Parallel and Distributed Processing (IPPS/SPDP)* (1999). http://ditec.um.es/~jmgarcia/papers/P-EDR.pdf

G. McLachlan, T. Krishnan, The EM Algorithm and Its Extensions (Wiley, New York, 1997)

F. Rijmen, M. Jeon, S. Rabe-Hesketh, M. von Davier, A third order item response theory model for modeling the effects of domains and subdomains in large-scale educational assessment surveys. J. Educ. Behav. Stat. **38**, 32–60 (2014)

M. von Davier, *A General Diagnostic Model Applied to Language Testing Data (Research Report No. RR-05-16)* (Educational Testing Service, Princeton, NJ, 2005). doi:10.1002/j.2333-8504.2005.tb01993.x

M. von Davier, A general diagnostic model applied to language testing data. Br. J. Math. Stat. Psychol. **61**, 287–307 (2008)

M. von Davier, *The Log-Linear Cognitive Diagnostic Model (LCDM) as a Special Case of the General Diagnostic Model (GDM) (Research Report No. RR-14-40)* (Educational Testing Service, Princeton, NJ, 2014). doi:10.1002/ets2.12043

M. von Davier, *High-Performance Psychometrics: The Parallel-E Parallel-M Algorithm for Generalized Latent Variable Models.* ETS Research Report ETS-RR-16-34 (2016)

M. von Davier, J. Rost, Polytomous mixed rasch models, in *Rasch Models: Foundations, Recent Developments and Applications,* ed. by G.H. Fischer, I.W. Molenaar (Springer, New York, 1995), pp. 371–379

M. von Davier, J. Rost, in *Handbook of Item Response Theory*, 2nd edn., ed. by W. van der Linden. Logistic mixture-distribution response models, vol 1 (CRC Press, Boca Raton, FL, 2016), pp. 393–406

M. von Davier, K. Yamamoto, Partially observed mixtures of IRT models: an extension of the generalized partial credit model. Appl. Psychol. Meas. **28**(6), 389–406 (2004)

M. von Davier, X. Xu, C.H. Carstensen, Measuring growth in a longitudinal large scale assessment with a general latent variable model. Psychometrika **76**, 318 (2011). doi:10.1007/s11336-011-9202-z

Properties of Second-Order Exponential Models as Multidimensional Response Models

Carolyn J. Anderson and Hsiu-Ting Yu

Abstract Second-order exponential (SOE) models have been proposed as item response models (e.g., Anderson et al., J. Educ. Behav. Stat. 35:422–452, 2010; Anderson, J. Classif. 30:276–303, 2013. doi: 10.1007/s00357-00357-013-9131-x; Hessen, Psychometrika 77:693–709, 2012. doi:10.1007/s11336-012-9277-1 Holland, Psychometrika 55:5–18, 1990); however, the philosophical and theoretical underpinnings of the SOE models differ from those of standard item response theory models. Although presented as reexpressions of item response theory models (Holland, Psychometrika 55:5–18, 1990), which are reflective models, the SOE models are formative measurement models. We extend Anderson and Yu (Psychometrika 72:5–23, 2007) who studied unidimensional models for dichotomous items to multidimensional models for dichotomous and polytomous items. The properties of the models for multiple latent variables are studied theoretically and empirically. Even though there are mathematical differences between the second-order exponential models and multidimensional item response theory (MIRT) models, the SOE models behave very much like standard MIRT models and in some cases better than MIRT models.

Keywords Dutch Identity • Log-multiplicative association models • Formative models • Reflective models • Composite indicators • Skew normal • Bi-variate exponential

1 Introduction

Philosophical, theoretical, and empirical differences between second-order exponential (SOE) models and multidimensional item response theory (MIRT) models exist; however, these differences that have not been fully discussed nor

C.J. Anderson (✉)
University of Illinois at Urbana-Champaign, Champaign, IL, USA
e-mail: cja@illinois.edu; http://faculty.education.illinois.edu/cja/homepage

H.-T. Yu
National Chengchi University, Taipei City, Taiwan
e-mail: hsiutingyu@gmail.com

L.A. van der Ark et al. (eds.), *Quantitative Psychology*, Springer Proceedings in Mathematics & Statistics 196, DOI 10.1007/978-3-319-56294-0_2

widely recognized in the literature on SOE models are derived based on the Dutch Identity (Holland 1990; Hessen 2012). Equivalent to SOE models, log-multiplicative association (LMA) models were derived as latent variable models from statistical graphical models (Anderson and Vermunt 2000), as well as from item response models using rest scores in lieu of the latent variables (Anderson and Yu 2007; Anderson et al. 2010). Anderson and Yu (2007) studied unidimensional LMA models for dichotomous data. The LMA models are formative measurement models, and they are item response models in their own right. A better understanding of the properties of LMA models as item response models leads to implications regarding the use and performance of LMA models for analyzing response data. The LMA models have a number of advantages, including maximum likelihood estimation does not require an assumption for the marginal distribution of the latent variables and the models can be fit directly to response patterns using Newton-Raphson. The goal of this paper is to extend Anderson and Yu (2007) to study the properties of multidimensional LMA (or equivalently SOE) models for dichotomous and polytomous items.

Holland (1990) proposed and used the Dutch Identity to derive SOE models for data based on underlying uni- and multidimensional IRT models for dichotomous items. The SOE models are equivalent to LMA models, which are special cases of a log-linear model with two-way interactions. Hessen (2012) extended the Dutch Identity to polytomous items and derived an LMA model; however, he focused on models analogous to the partial credit model (i.e., models in the Rasch family), even though his extension of the Dutch Identity is more general. For models in the Rasch family, category scores are set to fixed values (e.g., consecutive integers). Hessen (2012) mentioned that the category scores could be treated as parameters and estimated. We treat category scores as parameters that are estimated. We extend and generalize the results in Anderson and Yu (2007) and Hessen (2012) to the case of multidimensional models for dichotomous and polytomous items. We highlight the philosophical, theoretical, and empirical differences between LMA and MIRT models.

In the first section of this paper, we discuss the philosophical and theoretical differences between standard MIRT and LMA models. In the following two sections, two properties of LMA are theoretically and empirically studied: the downward collapsibility of LMA models and the effect of different marginal distributions of the latent variables on the models' performance. We conclude with a discussion the potential uses of LMA models in measurement contexts.

2 Reflective and Formative Models

The differences between reflective and formative latent variable models have been discussed by Markus and Borsboom (2013), Bollen and Bauldry (2011), and others. Our intent here is to show the philosophical differences between LMA and MIRT models and how they lead to different mathematical models.

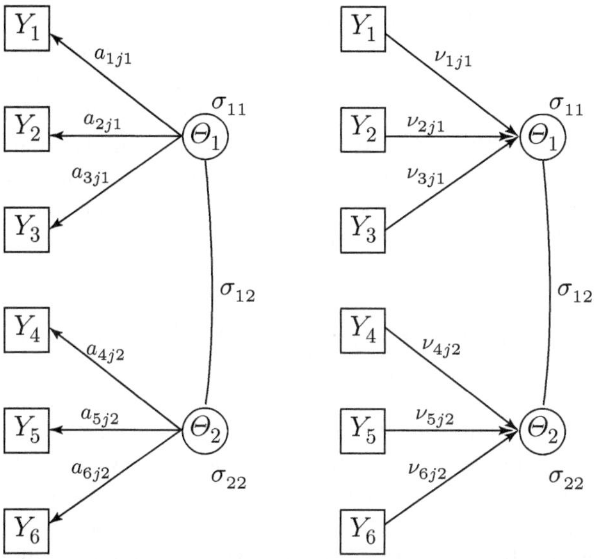

Fig. 1 Graphs corresponding to reflective (*left*) and formative (*right*) models for six items and two latent continuous variables

A reflective model posits that latent variables are prior to behavior, and the latent variables are conceived of as existing whether they are measured or not. A reflective model is illustrated by the graph on the left in Fig. 1. The values on the latent variables lead to observed responses; therefore, behavior indicates or reflects a person's value on the unobserved quantity. A change in the value of a latent variable causes a change in the response behavior. The items are *effect indicators* (Bollen and Bauldry 2011).

To algebraically take into account the directional nature of the relationship between $\boldsymbol{\theta}$ and $\boldsymbol{y}$, models are developed by writing the joint distribution of $\boldsymbol{\theta}$ and $\boldsymbol{y}$ as $f(\boldsymbol{y}, \boldsymbol{\theta}) = f(\boldsymbol{y}|\boldsymbol{\theta})f(\boldsymbol{\theta})$. For a MIRT model, the marginal distribution of the latent variables $f(\boldsymbol{\theta})$ is typically assumed to be multivariate normal, and the distribution for the responses conditional on the latent variables $f(\boldsymbol{y}|\boldsymbol{\theta})$ is a product of multinomial logistic regression models. The model for responses to items is found by numerically integrating over the latent variables; that is, the probability of response pattern $\boldsymbol{y}$ is

$$P(\boldsymbol{y}) = \int_{\theta_1} \cdots \int_{\theta_M} \prod_{i=1}^{I} \frac{\exp[\beta_{ij} + \sum_m \alpha_{ijm}\theta_m]}{\sum_h \exp[\beta_{ih} + \sum_m \alpha_{ihm}\theta_m]} f(\boldsymbol{\theta})d(\boldsymbol{\theta}), \tag{1}$$

where β_{ij} is a location parameter for response option j of item i, and α_{ijm} is the slope parameter for response option j of item i on latent variable θ_m.

In a formative model, the direction of the relationship between $\boldsymbol{\theta}$ and $\boldsymbol{y}$ is reversed relative to the reflective model. A graph representing a formative model is illustrated on the right in Fig. 1. Items define and give meaning to latent variables. The items are *composite indicators* because $\boldsymbol{\theta}$ are composites of the values of the items (Bollen and Bauldry 2011). The joint distribution of $\boldsymbol{y}$ and $\boldsymbol{\theta}$ is found by first specifying the distribution for $f(\boldsymbol{y})$ and then the distribution for $f(\boldsymbol{\theta}|\boldsymbol{y})$; that is, $f(\boldsymbol{y},\boldsymbol{\theta}) = f(\boldsymbol{\theta}|\boldsymbol{y})f(\boldsymbol{y})$. Assuming that $f(\boldsymbol{y})$ is multinomial and $f(\boldsymbol{\theta}|\boldsymbol{y})$ is a homogeneous conditional, Gaussian distribution leads to an LMA model for the probabilities of observed response patterns $\boldsymbol{y}$ (Anderson and Vermunt 2000; Anderson et al. 2010). The model for data is

$$P(\boldsymbol{y}) = \exp\left[\lambda + \sum_{i=1}^{I}\lambda_{ij} + \sum_{i}\sum_{k\geq i}\sum_{m}\sum_{m'\neq m}\sigma_{mm'}\nu_{ijm}\nu_{kjm'}\right], \tag{2}$$

where λ ensures probabilities sum to 1 over response patterns, λ_{ij} is the marginal effect term for response option j to item i, ν_{ijm} is the category scale value for response j to item i on latent variable m, and $\sigma_{mm'}$ is a within response pattern variance or covariance of the latent variable(s). The λ_{ij}s and ν_{ijm}s in (2) are analogous to the β_{ij}s and α_{ijm}s, respectively, in (1). Based on the LMA model, the conditional means of the latent variables given $\boldsymbol{y}$ equal:

$$E(\theta_m|\boldsymbol{y}) = \sum_{m'=1}^{M}\sigma_{mm'}\left(\sum_{i=1}^{I}\nu_{ijm'}\right). \tag{3}$$

Models (1) and (2) are very general models. In this paper, we study the case where each item is directly related to one and only one latent variable, that is, $\nu_{ijm} \neq 0$ and $\alpha_{ijm} \neq 0$ for one and only one m. We expect that the results we find will be the same for more complex models, but we leave this for future study.

The MIRT model given in (1) is not only philosophically different but mathematically different from the LMA model given in (2).

3 Downward Collapsibility of LMA Models

If an item is dropped from data generated from a MIRT model, the data excluding the item still follow a MIRT model and theoretically yield the same estimates of item parameters for the remaining items. If an item is dropped from (or added to) an LMA model, the resulting model is a different model with different parameter estimates. We theoretically and empirically study the effect on LMA model parameter estimates when dropping an item from data (i.e., collapse data over an item). In the first section, we consider the case when data are generated from an LMA model (not collapsible), and in the second section, we consider the case when data are generated from a MIRT model (downward collapsible).

3.1 LMA-Generated Data

Suppose that item 1 is directly related to θ_1 and it is dropped from the data. Let $\boldsymbol{y}_{-1}$ indicate the data excluding item 1. Rather than (3), the conditional means for θ_1 and θ_m are

$$E(\theta_1|\boldsymbol{y}_{-1}) = \sigma_{11}\sum_{i\neq 1}\nu_{ij1} + \sum_{m>1}\left(\sigma_{1m}\sum_{k}\nu_{kjm}\right) + \sigma_{11}\sum_{j}\nu_{1j1}P(Y_1 = j|\boldsymbol{y}_{-1})$$

and

$$E(\theta_m|\boldsymbol{y}_{-1}) = \sigma_{1m}\sum_{i\neq 1}\nu_{ij1} + \sum_{m'>1}\left(\sigma_{mm'}\sum_{k}\nu_{kjm'}\right) + \sigma_{1m}\sum_{j}\nu_{1j1}P(Y_1 = j|\boldsymbol{y}_{-1}),$$

respectively. The last term in each of these equations for the conditional means is unobserved and equals the expected biases of the means due to dropping item 1.

Dropping an item that is directly related to θ_1 changes the conditional variances of θ_1 and any θ_m directly related to θ_1 (i.e., $\sigma_{1m} \neq 0$). In particular, the conditional variances after collapsing over item 1 are

$$\mathrm{var}(\theta_1|\boldsymbol{y}_{-1}) = \sigma_{11} + \sigma_{11}^2\left(\sum_{j}\nu_{ij1}^2P(Y_1 = j|\boldsymbol{y}_{-1}) - \left(\sum_{j}\nu_{ij1}P(Y_1 = j|\boldsymbol{y}_{-1})\right)^2\right),$$

and

$$\mathrm{var}(\theta_m|\boldsymbol{y}_{-1}) = \sigma_{mm} + \sigma_{1m}^2\left(\sum_{j}\nu_{ij1}^2P(Y_1 = j|\boldsymbol{y}_{-1}) - \left(\sum_{j}\nu_{ij1}P(Y_1 = j|\boldsymbol{y}_{-1})\right)^2\right).$$

The conditional variances will increase for larger values of σ_{11} and σ_{1m}. The change of $\mathrm{var}(\theta_m|\boldsymbol{y}_{-1})$ is smaller than that for $\mathrm{var}(\theta_1|\boldsymbol{y}_{-1})$ because $\sigma_{1m}^2 \leq \sigma_{11}^2$. Regardless of the value of σ_{11} and σ_{1m}, the conditional means and variances are affected the most when an item with the largest values of ν_{ij1} is dropped, and they are least affected when the item with the smallest values of ν_{ij1} is dropped.

Our interest is in the theoretical behavior of the LMA models; therefore, $P(\boldsymbol{y})$s were computed from an LMA (six items, three response options per item), so the LMA model fits the data perfectly. The size of the scale values for an item was measured by $\sum_j \nu_{ijm}^2$. Two additional data sets were created by collapsing over the item with the smallest value and the largest value of $\sum_j \nu_{ijm}^2$. The item with the weakest relationship to a θ_m should have the smallest effect on the results, and collapsing over the item with the strongest relationship to a θ_m should have the largest effect.

Throughout this paper, maximum likelihood estimation was used to estimate parameters of LMA and MIRT models. The LMA models were fit to data using SAS$^{\delta}$ PROC NLP (version 9.4, SAS Institute Inc. 2015). The MIRT models were fit to data using *flexMIRT* (Houts and Cai 2013) assuming bivariate (multivariate) normality.[1]

In terms of goodness of fit, the likelihood ratio goodness-of-fit statistic (G^2) is used as an index but is not compared to a χ^2 distribution because there is no sampling variability. As a second index, we used the dissimilarity index:

$$D = \sum_{y} \frac{|P(\boldsymbol{y}) - \hat{P}(\boldsymbol{y})|}{2},$$

where the sum is over all response patterns, $P(\boldsymbol{y})$ is the probability of response pattern $\boldsymbol{y}$, and $\hat{P}(\boldsymbol{y})$ is the fitted value of the probability of response pattern $\boldsymbol{y}$ from a model. The index D is interpretable as the proportion of data that would have to be moved from one response pattern to another for the model to fit perfectly (Agresti 2013).

Any misfit of the LMA model fit to the six items is due to numerical inaccuracy in the data generation and/or model estimation. The LMA model fits the probabilities of response patterns for the six items nearly perfectly. When collapsing over the weak item, the parameter estimates and goodness-of-fit statistics of the LMA model were nearly identical to those when the model was fit to all six items. Specifically, collapsing over the weak item had a smaller impact on the goodness of fit than collapsing over the strong item (i.e., $G^2 = 0.0000$ versus $G^2 = 0.0002$ and $D = 0.0002$ versus $D = 0.0049$). All of the LMA models fit the probabilities better than all of the MIRT models.

When collapsing over the strong item, there were noticeable differences between the estimated parameters from the LMA model fit to those used to generate the data. The variance of θ_m increased the most when the item dropped is the strong item. Specifically, when the strong item is dropped, the variance of the latent variable to which it is connected goes from 0.87 to 1.66, but when the weak item is dropped, the variance of the latent variable to which it is connected goes from 0.77 to 0.88. As predicted, both $\hat{\sigma}_{11}$ and $\hat{\sigma}_{22}$ increased when collapsing over either the weak or strong item. The change in both variances occurs because when we collapse over an item related to, say θ_1, leads to less information to estimate the latent variable θ_2, which increases uncertainty (i.e., larger σ_{22}).

[1]Files containing code and data that reproduce all analyses can be downloaded from http://faculty.education.illinois.edu/cja/homepage.

3.2 MIRT-Generated Data

If θ_1 and θ_2 were discrete, then we could collapse over, for example, item 1 and expect $\hat{\nu}_{ijm}$ for $i \neq 1$ to remain the same. Since for the LMA models $\hat{\boldsymbol{\theta}}$ equals a weighted sum of category scores, $\hat{\boldsymbol{\theta}}$ is empirically discrete and LMA models might be collapsible. When data are generated from a model that implies collapsibility, whether LMA scale values are affected by dropping items is an open question. Since MIRT models imply collapsibility, probabilities were generated from a two-dimensional MIRT model with $\boldsymbol{\theta} \sim MVN(\boldsymbol{\mu} = (0, 0), \rho = 0.5)$ for eight items where items 1–4 were related to θ_1, and items 5–8 were related to θ_2. The generated probabilities were collapsed over one item at a time until there were only four items remaining. We alternated collapsing over an item related to θ_1 and one related to θ_2.

Since LMA models are formative measurement models, we are primarily interested in the $\hat{\nu}_{ijm}$s, which are used to compute estimates of the conditional means of the latent variables (i.e., $\hat{\boldsymbol{E}}(\theta_m|y)$). The scale values $\hat{\nu}_{ijm}$ were essentially unaffected by collapsing the data. When data were collapsed over an item, both of the $\hat{\sigma}_{mm}$s increased. When the first item was dropped, which was related to θ_2, the increase of $\hat{\sigma}_{22}$ was greater than that for $\hat{\sigma}_{11}$. When the second item was dropped, which was related to θ_1, the increase in $\hat{\sigma}_{11}$ was greater than that for $\hat{\sigma}_{22}$. This pattern continued until there are only four items remaining.

In sum, if data are generated from a MIRT model, which collapsibility, then the LMA model yields nearly the same $\hat{\nu}_{ijm}$s when items are dropped. Conversely, we can consider adding items. If the data come from a model that implies collapsibility and then when adding items (assuming that the added items are related to the underlying latent variable(s)), the $\hat{\nu}_{ijm}$s are not expected to change, and $\hat{\sigma}_{mm}$s are expected to be smaller.

4 Different Marginal Distributions

A property often given as an advantage of LMA models is that a marginal distribution of the latent variables is a mixture of normals, which can take on many different shapes. The goal of this section is to determine whether and when an LMA model may perform well in terms of goodness of fit and parameter recovery and compare LMA model performance with a corresponding MIRT model.

In this study, we generated probabilities for response patterns by numerically integrating out the latent variables from a MIRT model assuming one of four different underlying distributions. The multivariate normal (MVN) was chosen because this is the typical assumption made when fitting a MIRT model. The multivariate skew normal was chosen because the MVN is a special case of the skew normal, and there has been some interest in using the skew normal as an alternative to the normal distribution (Azevedo et al. 2011; Casabianca and Junker 2016; Lee 2002). Marshall-Olkin bivariate exponential distribution (Mardia et al. 1979) was chosen because some variables in the data that we often analyze are very skewed. Lastly, a mixture of two multivariate normal distributions was chosen to mimic a situation where individuals have opposite attitudes or views. This also reflects a situation where there is an important group variable that has not been included in the model, and there is differential item functioning.

As the number of items increases to ∞, $\sigma_{mm} \longrightarrow 0$, an LMA model will yield the actual marginal distribution of the latent variables. The behavior of LMA models for short tests or subscales is less certain; therefore, we empirically examine the behavior of the models when fit to generated data for short tests. Probabilities of response patterns were generated using the MIRT model in (1) for $M = 2$ latent variables and $I = 4$ or 6 items with $J = 2$, 3, or 4 response options. For the multivariate normal distribution, we also fit models to data with $M = 3$ and $I = 6$ items.

Both the MIRT and the LMA models were fit to all of the data sets. Albeit naive, when the distribution generating the data is not normal, the MIRT models were fit to data assuming multivariate normality. Although not reported here, two additional models were fit as baseline models: the log-linear model of independence and the homogeneous (all two-way interaction) log-linear model. The probabilities of response patterns were multiplied by 1,000,000 to retain more decimal places and accuracy. Besides the dissimilarity index D, a second measure of goodness of fit is reported for the models: the percent of association accounted for by a model,

$$\text{Percent association} = \frac{G^2_{independence} - G^2_{model}}{G^2_{independence}} \times 100\%,$$

where the likelihood ratio statistic G^2 from the independence model is a measure of the amount of association in the data.

To examine parameter recovery, we used the correlation between the parameters used to generate the data and the estimated parameters from the LMA and MIRT models. Given our focus on LMA models, we are primarily interested in the estimation of the ν_{ijm} parameters. The marginal effect terms λ_{ij} generally are viewed as nuisance parameters from an LMA model framework, but the correlations for marginal terms are reported for the sake of completeness.

The results for different numbers of items and response options are all very similar; therefore, we only report the result for one case (i.e., six items, three response options, and two latent variables). Goodness-of-fit statistics are reported in Table 1, and correlations between estimated parameters and those used to generate the data are reported in Table 2.

When data were generated using the bivariate normal distribution ($\boldsymbol{\mu} = \mathbf{0}$, $\rho = 0.5$), the MIRT model should fit perfectly. Any misfit is due to numerical inaccuracy in generating the probabilities and/or estimating the model. The MIRT models essentially fit perfectly; however, the goodness-of-fit indices for the LMA models are just shy of perfect. When data were generated using a skew normal (i.e., $\boldsymbol{\mu} = \mathbf{0}$, $\rho = 0.75$, and shape parameters 2 and 3) or a bivariate exponential distribution (i.e.,

Table 1 Goodness-of-fit statistics for LMA and MIRT models fit to date generated from a MIRT model with different underlying distributions for $f(\boldsymbol{\theta})$

	Dissimilarity		Percent association	
Underlying distribution	LMA	MIRT	LMA	MIRT
Bivariate normal	0.0016	0.0002	99.99	100.00
Skew normal	0.0268	0.0268	97.14	97.16
Bivariate exponential	0.0127	0.0129	98.44	98.39
Mixture of normals	0.0346	0.0708	99.43	96.05

Table 2 Correlations between LMA and MIRT model parameter estimates and parameters used generated MIRT model probabilities for different $f(\boldsymbol{\theta})$s

	LMA	MIRT	LMA	MIRT
Underlying distribution	$r(\alpha_{ijm}, \hat{\nu}_{ijm})$	$r(\alpha_{ijm}, \hat{a}_{ijm})$	$r(\beta_{ij}, \hat{\lambda}_{ij})$	$r(\beta_{ijm}, \hat{b}_{ij})$
Bivariate normal	0.9980	1.0000	0.9839	1.0000
Skew normal	0.9950	0.9962	0.8361	0.7506
Bivariate exponential	0.9326	0.9077	0.8257	0.7894
Mixture of normals	0.9971	0.9430	0.9665	0.9428

$f(\boldsymbol{\theta}) = \exp(-1.0\theta_1 - 0.5\theta_2 - 0.2\max(\theta_1, \theta_2))\kappa$ where κ normalized the function), the LMA and MIRT models both provide good representations of the data, and there are no systematic differences in terms of which model fits the data better. When data were generated from the mixture of two normals (i.e., $\boldsymbol{\mu}_1 = (-2, -2)'$, $\boldsymbol{\mu}_2 = (2, 2)'$, $\rho = 0.5$, and mixing weight of 0.5), the LMA models clearly fit the data better than the MIRT models.

More differences between the models' performance were found in terms of parameter recovery. When data were from the bivariate normal, MIRT parameters are perfectly correlated with those used to generate the data; however, the LMA parameters were just short of perfect. For the skew normal, the correlations between the α_{ijm}s used to generate the data and the estimated ν_{ijm}s parameters from the LMA models were about the same as the corresponding correlations of parameters from the MIRT models; however, the correlations for the β_{ijm}s were much larger for the LMA model than the MIRT model. For the exponential and mixture of normal distributions, the correlations for the estimated ν_{ijm}s and λ_{ij}s from the LMA models were considerably larger than those for parameters from the MIRT models.

5 Discussion

The LMA models and standard MIRT models were shown to be philosophically and mathematically different models; however, they share some important properties. For short tests, the LMA models performed nearly as well as standard MIRT models when the underlying distribution of the latent variables is multivariate normal, and the LMA and MIRT models empirically perform equally well when the underlying distribution is skew normal. With the skew normal, the goodness of fit is about the same for both the LMA and MIRT models; however, the estimation of the β_{ij}s parameters had lower correlations with the parameters used to generate the data than the LMA model parameters. The LMA models perform better than MIRT models in terms of goodness of model fit to data and parameter recovery when data arise from an LMA model and when $f(\boldsymbol{\theta})$ follows either a bivariate exponential distribution or a mixture of two normal distributions.

The LMA models are more flexible than discussed in this paper. The LMA models can include covariates for the latent variables, the marginal effect terms (i.e., the λ_{ij}), and the conditional variances and covariances of the latent variables (Anderson 2013). The models also permit various restrictions on parameters, including equality, ordinal, partially ordinal, linear transformations, and/or any desired transformation (Anderson 2013). The LMA models can also represent more complex latent variable structures than those studied in this paper, such as those where items "load" on multiple correlated or uncorrelated latent variables (e.g., bi-factor models). Since the assumptions and theory are the same, we expect the same results for more complex models such as those that we found for the simpler models reported in this paper.

Our focus was on short tests because these are cases where LMA and MIRT models may differ. Although we used common commercial software (SAS) to fit the LMA models to data, one bottleneck to more widespread applications of LMA models is a limitation to the size of the problem that can be handled. The size of the cross-classification of items (i.e., number of response patterns) increases exponentially when adding items and/or categories per item. When scores are input, the pseudo-likelihood method given in Anderson et al. (2007) works well and can be implemented in any program that fits conditional multinomial logistic models. Recently Paek (2016), Paek and Anderson (2017) proposed a solution to the more general problem where scores are estimated. In simulations, Paek (2016) showed that the algorithm yields nearly identical parameter estimates as MLE of LMA models for short tests and that the algorithm recovers parameters used to simulate the data in longer tests (i.e., 20 and 50 items). The more general algorithm also can be implemented in any software program that fits conditional multinomial logistic regression models.

We do not advocate that LMA models replace MIRT models because they are philosophically and theoretically different measurement models. The LMA models actually may be complimentary to applications of MIRT models. Suppose a researcher desires a reflective model but does not know what marginal distribution of the latent variable(s) should be used when fitting a MIRT model to data. The LMA models can be used to estimate the marginal distribution of the latent variables, which could confirm or suggest a distribution to be used when fitting the MIRT model to data.

The empirical studies in this paper imply that one cannot conclusively determine whether the model should be formative or reflective. Whether one performs better than the other is an empirical question. The choice between using an LMA model or a MIRT model for a particular case depends on a researcher's conceptualization of the latent variable.

References

A. Agresti, *Categorical Data Analysis*, 3rd edn. (Wiley, New York, 2013)

C.J. Anderson, Multidimensional item response theory models with collateral information as Poisson regression models. J. Classif. **30**, 276–303 (2013). doi: 10.1007/s00357-00357-013-9131-x

C.J. Anderson, J.K. Vermunt, Log-multiplicative association models as latent variable models for nominal and/or ordinal data. Sociol. Methodol. **30**, 81–121 (2000)

C.J. Anderson, H.-T. Yu, Log-multiplicative association models as item response models. Psychometrika **72**, 5–23 (2007)

C.J. Anderson, Z. Li, J.K. Vermunt, Estimation of models in a Rasch family for polytomous items and multiple latent variables. J. Stat. Softw. **20** (2007). http://www.jstatsoft.org/v20/i06/v20i06.pdf

C.J. Anderson, J.V. Verkuilen, B.L. Peyton, Modeling polytomous item responses using simultaneously estimated multinomial logistic regression models. J. Educ. Behav. Stat. **35**, 422–452 (2010)

C.L.N. Azevedo, H. Bolfarine, D.F. Andrade, Bayesian inference for a skew-normal IRT model under the centred parameterization. Comput. Stat. Data Anal. **55**, 353–365 (2011)

K.A. Bollen, S. Bauldry, Three C's in measurement models: causal indicators, composite indicators, and covariates. Psychol. Methods **16**, 265–284 (2011)

J.M. Casabianca, B.W. Junker, Multivariate normal distribution, in *Handbook of Item Response Theory, Volume Two: Statistical Tools*, ed. by W.J. van der Linden (Talyor & Fransics/CRC Press, Boca Raton, 2016), pp. 35–46

D.J. Hessen, Fitting and testing conditional multinomial partial credit models. Psychometrika **77**, 693–709 (2012). doi:10.1007/s11336-012-9277-1

P.H. Holland, The Dutch identity: a new tool for the study of item response models. Psychometrika **55**, 5–18 (1990)

C.R. Houts, L. Cai, *flexMIRT: Flexible Multilevel Multidimensional Item Analysis and Test Scoring* (Vector Psychometric Group, LLC, Chapel Hill, 2013)

J. Lee, Multidimensional item response theory: an investigation of interaction effects between factors on item parameter recovery using Markov Chain Monte Carlo. Unpublished Doctoral Dissertation, Michigan State University (2002)

K.V. Mardia, J.M. Kent, J.M. Bibby, *Multivariate Analysis* (Academic, Orlando, 1979)

K.A. Markus, D. Borsboom, *Frontiers of Test Validity Theory: Measurement, Causation and Meaning* (Routlege, New York, 2013)

Y. Paek, Pseudo-likelihood estimation of multidimensional polytomous item response theory models. Unpublished Doctoral Dissertation, University of Illinois at Urbana-Champaign (2016)

Y. Paek, C.J. Anderson, Pseudo-likelihood estimation of multidimensional response models: polytomous and dichotomous items, in *The 81st Annual Meeting of the Psychometric Society, Asheville, NC*, ed. by L.A. van der Ark, S.A. Culpepper, J.A. Douglas, W.C. Wang, M. Wiberg (Springer, New York, 2017)

SAS Institute Inc., *Statistical Analysis System*, version 9.4 (SAS Institute, Cary, 2015)

Pseudo-likelihood Estimation of Multidimensional Response Models: Polytomous and Dichotomous Items

Youngshil Paek and Carolyn J. Anderson

Abstract Log-multiplicative association (LMA) models have been proposed as uni- and multidimensional item response models for dichotomous and/or polytomous items. A problem that prevents more widespread use of LMA models is that current estimation methods for moderate to large problems are computationally prohibitive. As a special case of a log-linear model, maximum likelihood estimation (MLE) of LMA models requires iteratively computing fitted values for all possible response patterns, the number of which increases exponentially as the number of items and/or response options per item increases. Anderson et al. (J. Stat. Softw. 20, 2007, doi:10.18637/jss.v020.i06) used pseudo-likelihood estimation for linear-by-linear models, which are special cases of LMA models, but in their proposal, the category scores are fixed to specific values. The solution presented here extends pseudo-likelihood estimation to more general LMA models where category scores are estimated. Our simulation studies show that parameter estimates from the new algorithm are nearly identical to parameter estimates from MLE, work for large numbers of items, are insensitive to starting values, and converge in a small number of iterations.

Keywords Log-multiplicative association models • Log linear-by-linear models • Second-order exponential models • Multidimensional item response theory • Formative measurement models

1 Introduction

Log-multiplicative association (LMA) models have been proposed as uni- and multidimensional item response models for dichotomous and/or polytomous items (Anderson et al. 2010; Holland 1990; Hessen 2012, and others). They are formative measurement models (Anderson and Yu 2017) that do not require an assumption for the marginal distribution of the latent variables. Although maximum likelihood

Y. Paek (✉) • C.J. Anderson
University of Illinois at Urbana-Champaign, Champaign, IL, USA
e-mail: ypaek2@illinois.edu; http://faculty.education.illinois.edu/cja/homepage

L.A. van der Ark et al. (eds.), *Quantitative Psychology*, Springer Proceedings in Mathematics & Statistics 196, DOI 10.1007/978-3-319-56294-0_3

estimation can be accomplished for small numbers of items, the estimation of LMA models for moderate to large problems is computationally prohibitive because fitted values for all possible response patterns must be iteratively computed. The number of response patterns increases exponentially as the number of items and/or response options per item increases. Pseudo-likelihood estimation (PLE) was proposed by Anderson et al. (2010) for log linear-by-linear models which are special cases of LMA models where the category scores (e.g., slopes for the latent variables) are set to fixed values at input. We extend the pseudo-likelihood approach to general LMA models where category scores are treated as parameters and are estimated. This method works for large numbers of items and response options.

One of the most widely used programs for estimating LMA models is ℓ_{EM} (Vermunt 1997), which used quasi- or unidimensional Newton-Raphson. With ℓ_{EM} we were able to fit an LMA model to 12 binary items (i.e., $2^{12} = 4096$ response patterns). LMA models can also be fit using analytic derivatives and a Newton-Raphson algorithm as implemented in SAS® procedure NLP (SAS Institute Inc. 2015). Using SAS, the largest problem that we successfully fit had seven items each with five response categories (i.e., $5^7 = 78,125$ response patterns). Adding a single item increased the number of response patterns to $390,625$, and estimation became problematic. Ten items with five response categories per item (i.e., $9,765,625$ response patterns) are beyond the capability of current estimation methods.

Pseudo-likelihood estimation simplifies estimation of large complex models by maximizing the product of likelihoods of a set of conditional models based on the complex model. The method, first proposed by Besag (1974), has been used to solve estimation problems in a number of different settings (Huwang and Huwang 2002; Geys et al. 1999; Liang and Yu 2003; Johnson and Riezler 2002; Strauss and Ikeda 1990; Wasserman and Pattison 1990; Molenberghs and Verbeke 2005). The original uses of PLE to estimate parameters of Rasch models were limited to unidimensional models for pairs of binary items (Arnold and Strauss 1991; Zwinderman 1995). Smit (2000) extended the use of PLE to a set of dichotomous items and studied the quality of the estimates relative to other standard estimation methods. Pseudo-likelihood estimation (Anderson et al. 2007) of LMA models was developed to handle only the special case, when category scale values are assumed and set to fixed values. The estimation method and algorithm that we propose use pseudo-likelihood estimation but add a step for estimating the category scores.

PLE parameter estimates are asymptotically normal and consistent (Geys et al. 1999; Aerts et al. 2002), which is important for forming confidence intervals and hypothesis testing. Other advantages of PLE are that it is fast and stable, and implementation is straightforward.

The structure of the paper is as follows. In the first section, LMA models are presented in a form that is key to our algorithm. In the second section, we discuss pseudo-likelihood estimation and present our algorithm. In the subsequent sections, we present the results of simulation studies showing that the new step for estimation of category scores works (i.e., one latent variable) and simulation studies showing that the method works for multidimensional models. We conclude with a discussion and possible extensions of the algorithm.

2 Log-Multiplicative Association Models

LMA models are log-linear models that include all two-way interactions, but the interaction terms are replaced by products of pairs of category scores and an association parameter. Let $\boldsymbol{y}$ represent a response pattern (i.e., a cell in a cross-classification of I items), $i = 1, \ldots, I$ be an index for items, and $j = 1, \ldots, J$ the index for response options. Items can have different numbers of responses, but to keep notation simple, we will not put subscripts on j or J. Furthermore, let $m = 1, \ldots, M$ be the index for latent variables. The LMA model for the probability of response pattern $\boldsymbol{y}$ is

$$P(\boldsymbol{y}) = \exp\left[\lambda + \sum_{i=1}^{I} \lambda_{ij} + \sum_{i}\sum_{k \geq i}\sum_{m}\sum_{m' \neq m} \sigma_{mm'} \nu_{ijm} \nu_{kjm'}\right], \tag{1}$$

where λ ensures the probabilities sum to 1 over all response patterns, λ_{ij} is the marginal effect terms for item i and response option j, ν_{ijm} is the category score for item i and response option j on latent variable m, and $\sigma_{mm'}$ is the association parameter.

To derive (1) as an item response model, four assumptions are necessary:

1. The distribution of $\boldsymbol{y}$ is multinomial.
2. The responses to variables are independent given the latent variables.
3. The distribution of the latent variables is conditional homogeneous Gaussian.
4. Logits of responses are linear functions of the latent variables.

Details of the derivation using statistical graphical models[1] are given by Anderson and Vermunt (2000), and details using a traditional item response theory perspective are given by Holland (1990) and Hessen (2012). Both derivations yield (1); however, the fact that the underlying model is a formative measurement model was given little attention until Anderson and Yu (2007, 2017).

Although there are no latent variables in (1), the parameters in (1) represent the moments of the distribution of the latent variables conditional on response patterns. The distribution of the latent variable(s) within a response pattern is a multivariate normal, where the mean $\mu_{\theta m}$ depends on the response pattern (i.e., $\boldsymbol{y}$). In particular, the mean for the mth latent variable equals $\mu_{\theta m} = \sum_{m'}^{M} \sigma_{mm'} \sum_{i}^{I} \nu_{ijm'}$. The conditional covariance is $\sigma_{mm'}$.

The key to the algorithm is the model implied by (1) for the conditional probability that $y_i = j$ (response option is j for item i), which is

$$P(y_i = j | \boldsymbol{y}_{-i}) = \frac{\exp[\lambda_{ij} + \sum_m \nu_{ijm}(\sum_{m'} \sigma_{mm'} \sum_{k \neq i} \nu_{k\ell m'})]}{\sum_{h=1}^{J} \exp[\lambda_{ih} + \sum_m \nu_{ihm}(\sum_{m'} \sigma_{mm'} \sum_{k \neq i} \nu_{k\ell m'})]}, \tag{2}$$

[1]The assumption regarding logits is not made in the graphical modeling derivation.

where $\boldsymbol{y}_{-i}$ represents the responses to all items except item i. Note that the quantity $\sum_m \sigma_{mm'} \sum_{k \neq i} \nu_{k\ell m'}$ is an estimate of the value of the latent variable m based on responses to all items except item i. Furthermore,

$$\mu_{\boldsymbol{\theta} m} = \sum_{m'} \sigma_{mm'} \sum_{i} \nu_{ijm'} \approx \sum_{m'} \sigma_{mm'} \sum_{k \neq i} \nu_{k\ell m'}.$$

This approximation is expected to be closer as the number of items increases. The term on the right (i.e., $\sum_{m'} \sigma_{mm'} \sum_{k \neq i} \nu_{k\ell m'}$) is a "rest-score," which is a test total minus the score for the item being studied.

Equation (2) is an item response function for a multidimensional item response model with an estimate of the latent variable. The interpretation of the parameters analogous to traditional IRT parameters is that ν_{ijm} is the slope or discrimination parameter and λ_{ij} is the location parameter.

We can derive the conditional multinomial logistic regression model given by (2) from an LMA given by (1); however, the reverse is also true. Given a set of I equations, one for each item, of the same form as (2), the set uniquely implies (1) (Anderson and Yu 2007; Anderson et al. 2010). This is important for our estimation algorithm.

3 Pseudo-Likelihood Estimation

3.1 *Method*

In PLE, the sum of the logarithms of the conditional likelihoods, which is the pseudo-likelihood function (i.e., $\sum_i \ln(f(y_i | \boldsymbol{y}_{-i}))$), is maximized. If the category scores ν_{ijm} are known, maximizing the pseudo-likelihood function is done by fitting a single conditional multinomial logistic regression model to "stacked" data using MLE (Anderson et al. 2010). In other words, if we vertically concatenate equations defined in (2) over items, then the explanatory or predictor variables equal $\nu_{ijm} \sum_{k \neq i} \nu_{k\ell m'}$, which are input as fixed values (e.g., $\nu_{ijm} = j$ for $j = 1, \ldots, J$). The parameters that are estimated are the λ_{ij}s and $\sigma_{mm'}$s. However, this method works only for the log-linear-by-linear models and does not estimate the ν_{ijm}s.

We extend the PLE algorithm given in Anderson et al. (2010), to provide estimates of ν_{ijm}s. Estimating category scale values requires fitting (2) to each item, one at a time, to get updated estimates of the ν_{ijm}s for $j = 1, \ldots, J$ (fixed item i). These new estimates are then used in computing the value of the predictor variable for the next item. This is repeated until ν_{ijm}s for all items have been updated. For multidimensional models, after updating ν_{ijm}s for all items, a "stacked" regression as discussed previously is performed to get new estimates of $\sigma_{mm'}$ and λ_{ij}. Note that we are maximizing the pseudo-likelihood function by fitting logistic models to data using MLE.

3.2 The Algorithm

The algorithm only requires data manipulation and iteratively fitting of conditional multinomial logistic regressions by MLE. For this study, we wrote a set of SAS® macros implementing the algorithm. The macros along with examples are available at http://faculty.education.illinois.edu/cja/homepage.

Given a set of arbitrary starting values, the algorithm proceeds as follows:

1. Update $\hat{\nu}_{ijm}$ and $\hat{\lambda}_{ij}$
   (a) For item i, use MLE to fit (2) to the data using the *current* values of $\hat{\nu}_{kjm}$ $(k \neq i)$ and $\hat{\sigma}_{mm'}$.
   (b) Repeat step 1(a) until the $\hat{\nu}_{ijm}$s have been updated for all items.
2. Update $\hat{\sigma}_{mm'}$ and $\hat{\lambda}_{ij}$
   (a) Compute $\hat{\nu}_{ijm} \sum_{k \neq i} \hat{\nu}_{k\ell m'}$ using the current estimates.
   (b) Fit a conditional multinomial logistic regression model to the "stacked" data using MLE where $\hat{\nu}_{ijm} \sum_{k \neq i} \hat{\nu}_{k\ell m'}$ is the predictor variable.
3. Repeat Step(s) until the algorithm converges.

Convergence can be assessed in a number of ways. These include (i) no change in the value of the maximum of the likelihood for the stacked regression, (ii) no changes in the values of the maximums of the likelihood for each of the items, (iii) no changes in the parameter estimates, and (iv) the estimated values of $\hat{\lambda}_{ij}$ equal the same values in both Steps 1 and 2. If one of the convergence criteria is satisfied, then all will be satisfied.

For this algorithm to converge, identification constraints need to be imposed on the parameters of the LMA model. For LMA models, these are setting a location for the parameters and scale of the latent variables. For location, one possibility is to use zero-sum constraints, that is, $\sum_j \lambda_{ij} = 0$ and $\sum_j \nu_{ijm} = 0$, which can be achieved by using effect coding. Another possibility is to set one value to a constant, for example, $\lambda_{i1} = 0$ and $\nu_{i1m} = 0$, which can be achieved by dummy coding. One additional constraint is required to set the scale for each latent variable. The scale can be set by setting $\sum_j \nu_{ijm}^2 = 1$ for just one item for each latent variable or by fixing σ_{mm} to a constant. To avoid nonlinear constraints, we used $\sigma_{mm} = 1$. If alternative scaling is desired, the parameters can be linearly transformed after convergence.

This algorithm is modular in the sense that parameters of unidimensional models where scale values are estimated can be obtained using just Steps 1 and 3, log-linear-by-linear models can be obtained using Steps 2 and 3, and multidimensional models with estimated scale values can be obtained using Steps 1, 2, and 3.

The global convergence of the algorithm given in Anderson et al. (2010) is guaranteed provided that the maximum of the likelihood of the logistic regression model fit to the "stacked" data is achieved. Whether the new step of estimating category scale values (i.e., the ν_{ijm}s) combined with the stacked regression finds a global maximum is an open question. No evidence of a local maximum was found in on our simulation studies, which are reported below.

4 Simulation Studies

Step 1 of the algorithm is new; therefore, the first set of simulations examines the performance for unidimensional models, which only requires Steps 1 and 3. The second set of simulations examines the performance for multidimensional models that require all three steps of the algorithm. Each simulation condition was replicated 30 times, and the parameter estimates from the replications were averaged. The averaged parameter estimates were used to compute the evaluation criteria assessing the performance of PLE.

For all simulation studies, data were simulated using the MIRT model:

$$P(\mathbf{y}) = \prod_{i=1}^{I} \frac{\exp[\beta_{ij} + \sum_m \alpha_{ijm}\theta_m]}{\sum_h \exp[\beta_{ih} + \sum_m \alpha_{ihm}\theta_m]}, \tag{3}$$

where β_{ij} is a location parameter for response option j of item i, and α_{ijm} is the slope parameter for response option j of item i on latent variable θ_m. The distribution for θ_m was (multivariate) normal with mean equal to $\mathbf{0}$ and correlations set to 0.50. Values for item parameters were drawn from the following distributions: $\alpha_{ijm} \sim N(0.1, 1)$ and $\beta_{ij} \sim N(0, 1)$.

Theoretically, when data are simulated from a unidimensional item response theory model [i.e., (3) where $M = 1$] using a normal distribution for the latent variable, correlations between the parameters used to simulate the data and the parameter estimates from an LMA model should be very close to 1.000 (Anderson and Yu 2017). Large correlations should be found even though the IRT and the LMA models are mathematically different. When the marginal distribution of θ is not normal (e.g., exponential, skew normal, mixture of normals), the LMA models perform as well as or better than analogous IRT models in terms of parameter recovery and goodness-of-fit model. This is true when the IRT models are estimated assuming a marginal normal distribution for the latent variable. Estimating an LMA does not require an assumption for the marginal distribution, but rather the marginal distribution for the latent variable will be a mixture of as many normal distributions as there are response patterns. This mixture can closely approximate a normal distribution, especially for large numbers of response patterns, and it can approximate many other distributions as well. The same results are true for multidimensional models.

4.1 Unidimensional Models

4.1.1 Small Numbers of Items

The first simulation study was designed to test Step 1 and to determine how similar PLE parameter estimates of an LMA are to the MLE parameter estimates. To test Step 1, data sets with small numbers of items (i.e., 4 and 6) were simulated using

Table 1 Root-mean-square differences between parameter estimates from MLE and PLE for small numbers of items

		2 categories			3 categories			5 categories		
Parameter		$N = 200$	500	1000	200	500	1000	200	500	1000
$M = 1$ *latent variable*										
ν_{ijm}	4 items	0.001	0.001	0.000	0.016	0.011	0.007	0.014	0.007	0.006
	6 items	0.002	0.001	0.001	0.019	0.013	0.008	0.012	0.008	0.014
λ_{ij}	4 items	0.000	0.000	0.000	0.022	0.011	0.009	0.016	0.009	0.007
	6 items	0.000	0.000	0.000	0.035	0.022	0.012	0.017	0.013	0.029
$M = 2$ *latent variables*										
ν_{ijm}	4 items	0.003	0.001	0.000	0.038	0.018	0.013	0.058	0.027	0.010
	6 items	0.011	0.006	0.001	0.035	0.024	0.009	0.021	0.011	0.006
λ_{ij}	4 items	0.000	0.000	0.000	0.012	0.004	0.003	0.028	0.008	0.003
	6 items	0.001	0.000	0.000	0.042	0.037	0.010	0.014	0.006	0.004
$M = 3$ *latent variables*										
ν_{ijm}	6 items	0.019	0.006	0.000	0.057	0.019	0.018	0.067	0.014	0.009
λ_{ij}	6 items	0.001	0.003	0.003	0.029	0.010	0.008	0.024	0.006	0.003

the model in (3), varying the number of response categories (i.e., 2, 3, and 5) and sample size (i.e., 200, 500, and 1000). LMA models were fit to all data sets.

Since the numbers of items in this simulation study are small, we can fit LMA models by MLE, which we can compare with the parameter estimates obtained from PLE. Correlation coefficients were calculated to assess the similarity of the parameter estimates from PLE and MLE of the LMA models. All results for the unidimensional models with small numbers of items are essentially the same regardless of the sample size, the number of response options, the number of items, and the parameter types (i.e., λ_{ij} and ν_{ij}). The correlations are all between 0.999 and 1.000. As further evidence that the algorithm is working, the top of Table 1 contains the root-mean-square differences between the MLE and PLE estimates for the different design conditions. For the unidimensional models with small numbers of items, the root-mean-square differences range from 0.000 to 0.035.

4.1.2 Large Numbers of Items

The second set of simulations of unidimensional models was designed to assess how well the PLE algorithm performs with large numbers of items. For this simulation study, data sets with 20 and 50 items were simulated using the MIRT model given by (3), varying the number of response categories (i.e., 2, 3, and 5) and sample size (i.e., 200, 500, and 1000).

In the top section of Table 2, correlations between PLE estimates of parameters and those used to simulate the data for each sample size, number of categories (response options), parameter types, and number of items are reported. We averaged the correlations over items and categories to simplify the interpretation. The esti-

Table 2 For large numbers of items, correlations between parameter estimates from PLE and those used to simulate the data

		2 categories			3 categories			5 categories		
Parameter	I	$N = 200$	500	1000	200	500	1000	200	500	1000
$M = 1$ *latent variable*										
ν_{ijm}	20	0.979	0.989	0.986	0.994	0.994	0.995	0.992	0.994	0.995
	50	0.987	0.994	0.995	0.976	0.980	0.982	0.973	0.975	0.994
λ_{ij}	20	0.996	0.999	0.999	0.986	0.986	0.987	0.994	0.997	0.998
	50	0.998	0.999	0.999	0.987	0.990	0.991	0.992	0.993	0.990
$M = 2$ *latent variables*										
ν_{ijm}	20	0.989	0.994	0.998	0.994	0.996	0.996	0.992	0.994	0.994
	50	0.989	0.997	0.998	0.988	0.992	0.992	0.994	0.992	0.992
λ_{ij}	20	0.957	0.957	0.957	0.992	0.992	0.992	0.975	0.976	0.974
	50	0.998	0.998	0.998	0.972	0.973	0.974	0.983	0.985	0.898
$M = 3$ *latent variables*										
ν_{ijm}	20	0.987	0.996	0.997	0.993	0.996	0.997	0.992	0.994	0.995
	50	0.974	0.993	0.998	0.991	0.992	0.993	0.994	0.993	0.995
λ_{ij}	20	0.992	0.986	0.985	0.995	0.995	0.995	0.977	0.978	0.978
	50	0.995	0.996	0.995	0.964	0.964	0.964	0.987	0.988	0.990
$M = 4$ *latent variables*										
ν_{ijm}	20	0.973	0.991	0.997	0.992	0.996	0.996	0.991	0.995	0.995
	50	0.989	0.992	0.997	0.993	0.996	0.996	0.994	0.996	0.996
λ_{ij}	20	0.978	0.975	0.972	0.973	0.976	0.974	0.976	0.978	0.978
	50	0.983	0.983	0.985	0.971	0.972	0.973	0.980	0.982	0.982

mates from PLE were highly correlated with the parameters used to simulate the data. Those for ν_{ijm} ranged from 0.973 to 0.995 and those for λ_{ij} ranged from 0.986 to 0.999. There does not seem to be any regular pattern in terms of the conditions of the study design.

In sum, for all conditions, our PLE algorithm for unidimensional models yielded very high correlations between the estimates and the parameters used to simulate the data, which lends support that the new step works well, both with small and large numbers of items.

4.2 *Multidimensional Models*

4.2.1 Small Numbers of Items

The set of simulations for multidimensional models examines the similarity of the parameter estimates from PLE and MLE. The number of items and response categories and sample size were the same as those for unidimensional models with small numbers of items. Since the simulation study was intended for multidimensional models, data were simulated using two correlated latent variables, where each item is related to one and only one latent variable. We also simulated data for a three-dimensional model for six items, where there are two items per latent variable.

All results for multidimensional models are essentially the same over number of latent variables, sample size, number of response options, number of items, and parameter types (λ_{ij} and ν_{ijm}). The correlations were all equal to 1.000, which gives very strong support for our algorithm. Table 1 contains root-mean-square differences between the PLE and MLE parameter estimates. The values range from 0.000 to 0.057 for the ν_{ijm} and from 0.000 to 0.042 for the λ_{ij}s, which indicate the parameter estimates are very close in absolute terms. The PLE and MLE values seem to be closer for lower numbers of latent variables.

In sum, these results support that combining Steps 1 and 2 of the algorithm works very well for multidimensional models.

4.2.2 Large Numbers of Items

This set of simulations was conducted to illustrate that the PLE algorithm works for large numbers of items with multiple latent variables. The simulation design was the same as the one for unidimensional models with 20 and 50 items; however, we used the same multidimensional model where two latent variables are correlated each item is related to only one latent variable. We also added simulations for both 3 and 4 dimensional models.

Table 2 contains the correlations between PLE parameter estimates and those used to simulate the data for different numbers of latent variables, items, sample size, and response options. For the multidimensional models, the correlations for ν_{ijm} range from 0.973 to 0.998, and those for λ_{ij} range from 0.957 to 0.998. No noticeable patterns among the correlations were found based on the factors of the study design.

5 Discussion

The results of our simulation studies show that the parameter estimates from the proposed PLE algorithm are nearly equivalent to the parameters from MLE of LMA models. Furthermore, the PLE algorithm works for large numbers of items and converges in a small number of iterations (around 15 for polytomous items and 30 for dichotomous items). Although the details were not reported, the convergence of the algorithm was not sensitive to starting values. The algorithm could be implemented in any program that fits conditional multinomial logistic regression models. The algorithm and its implementation in SAS are flexible enough to be modified to permit covariates and constraints on parameters. We expect that modifications to the algorithm are possible for more complex LMA models and that they would converge so long as necessary identification constraints for the LMA model are imposed on the parameters.

Although our interest in LMA models is as formative measurement models, the algorithm would work in other cases where models equivalent to LMA models need to be estimated.

References

M. Aerts, H. Geys, G. Molenberghs, L.M. Ryan, *Topics in Modelling of Clustered Data* (Chapman & Hall/CRC, New York, 2002)

C.J. Anderson, Z. Li, J.K. Vermunt, Estimation of models in a Rasch family for polytomous items and multiple latent variables. J. Stat. Softw. **20** (2007) doi:10.18637/jss.v020.i06

C.J. Anderson, J.V. Verkuilen, B.L. Peyton, Modeling polytomous item responses using simultaneously estimated multinomial logistic regression models. J. Educ. Behav. Stat. **35**, 422–452 (2010)

C.J. Anderson, J.K. Vermunt, Log-multiplicative association models as latent variable models for nominal and/or ordinal data. Sociol. Methodol. **30**, 81–121 (2000)

C.J. Anderson, H.-T. Yu, Log-multiplicative association models as item response models. Psychometrika **72**, 5–23 (2007)

C.J. Anderson, H.-T. Yu, Properties of second-order exponential models as multidimensional item response models, in *Quantitative Psychology – The 81st Annual Meeting of the Psychometric Society, Asheville, NC* (Springer, New York, 2017, forthcoming)

B.C. Arnold, D. Strauss, Pseudolikelihood estimation: some examples. Sankhya **53**, 233–243 (1991)

J.E. Besag, Spatial interaction and the statistical analysis of lattice systems. J. R. Stat. Soc. **36**, 192–236 (1974)

H. Geys, G. Molenberghs, L. Ryan, Pseudolikelihood modeling of multivariate outcomes in developmental toxicology. J. R. Stat. Soc. **94**, 734–745 (1999)

D.J. Hessen, Fitting and testing conditional multinomial partial credit models. Psychometrika **77**, 693–709 (2012)

P.H. Holland, The Dutch identity: a new tool for the study of item response models. Psychometrika **55**, 5–18 (1990)

L. Huwang, J.T. Huwang, Prediction and confidence intervals for nonlinear measurement error models. Stat. Probab. Lett. **58**, 355–362 (2002)

M. Johnson, S. Riezler, Statistical models of syntax learning and use. Cogn. Sci. **26**, 239–253 (2002)

G. Liang, B. Yu, Maximum pseudo likelihood estimation in network tomography. IEEE Trans. Signal Process. **51**, 2043–2053 (2003)

G. Molenberghs, G. Verbeke, *Models for Discrete Longitudinal Data* (Springer, New York, 2005)

SAS Institute Inc., *Statistical Analysis System*, version 9.4. (SAS Institute, Cary, 2015)

A. Smit, H. Kelderman, Pseudolikelihood estimation of the Rasch model. J. Outcome Meas. **4**, 513–523 (2000)

D. Strauss, M. Ikeda, Pseudolikelihood estimation for social networks. J. Am. Stat. Assoc. **85**, 204–212 (1990)

J.K. Vermunt, *ℓEM: A General Program for the Analysis of Categorical Data* [Computer software manual] (Tilburg, The Netherlands, 1997). Program and manual retrieved from http://members.home.nl/jeroenvermunt

S. Wasserman, P. Pattison, Logit models and logistics regressions for social networks: I. An introduction to Markov graphs and p*. Psychometrika **61**, 401–425 (1990)

A.H. Zwinderman, Pairwise parameter estimation in Rasch Models. Appl. Psychol. Meas. **19**, 369–375 (1995)

Fitting Graded Response Models to Data with Non-Normal Latent Traits

Tzu-Chun Kuo and Yanyan Sheng

Abstract Fitting item response theory (IRT) models often relies on the assumption of a normal distribution for the person latent trait(s). Violating the assumption of normality may bias the estimates of IRT item and person parameters, especially when sample sizes are not large. In practice, the actual distribution for person parameters may not always be normal, and hence it is important to understand how IRT models perform under such situations. This study focuses on the performance of the multi-unidimensional graded response model using a Hasting-within-Gibbs procedure. The results of this study provide a general guideline for estimating the multi-unidimensional graded response model under the investigated conditions where the latent traits may not assume a normal distribution.

Keywords Polytomous item response theory • Multi-unidimensional graded response models • Hastings-within-Gibbs • Non-normal distributions

1 Introduction

Polytomous item response theory (IRT; Lord 1980) models are applicable for tests with items involving more than two response categories. Polytomous responses include nominal and ordinal responses. Ordinal polytomous responses, such as Likert scale items (Likert 1932), are broadly used in many fields, including education, psychology, and marketing. This study focuses on the graded response model (GRM; Samejima 1969), the most widely used IRT model for polytomous response data (e.g., Ferero and Maydeu-Olivares 2009; Rubio et al. 2007).

T.-C. Kuo (✉)
American Institute for Research, 1000 Thomas Jefferson Street, NorthWest, WA 20007, USA
e-mail: tkuo@air.org

Y. Sheng
Department of Counseling, Quantitative Methods, and Special Education, Southern Illinois University, Carbondale, IL 62901, USA
e-mail: ysheng@siu.edu

L.A. van der Ark et al. (eds.), *Quantitative Psychology*, Springer Proceedings in Mathematics & Statistics 196, DOI 10.1007/978-3-319-56294-0_4

In many circumstances, multidimensional IRT (MIRT; Reckase 1997, 2009) models are adopted when distinct multiple traits are involved in producing the manifest responses for an item. A special case of the MIRT model applies to the situation where the instrument consists of several subscales with each measuring one latent trait, such as the Minnesota Multiphasic Personality Inventory (MMPI; Buchanan 1994). In the IRT literature, such a model is called the *multi-unidimensional* (Sheng and Wikle 2007) or the *simple structure* MIRT (McDonald 1999) model and is the major focus of the study.

The multi-unidimensional GRM applies to situations where a K-item instrument consists of m subscales or dimensions, each containing k_v polytomous response items that measure one latent dimension. With a probit link, the probability that the ith ($i = 1, 2, \ldots N$) person contains a Likert scale response with c categories ($c = 1, 2, \ldots, C_j$) for the jth ($j = 1, 2, \ldots K$) item is defined as

$$P(Y_{vij} = c|\theta_{vi}, \alpha_{vj}, \boldsymbol{\delta}_j) = \Phi(\alpha_{vj}\theta_{vi} - \delta_{j,c-1}) - \Phi(\alpha_{vj}\theta_{vi} - \delta_{j,c})$$
$$= \int_{\delta_{j,c-1}}^{\delta_{j,c}} \phi(z; \alpha_{vj}\theta_{vi})dz, \qquad (1)$$

where $\Phi(\cdot)$ and $\phi(\cdot)$ are the standard normal CDF and PDF, respectively, z is a standard normal variate, α_{vj} and θ_{vi} denote the item discrimination and the person's latent trait in the vth dimension ($v = 1, 2, \ldots, m$), and $\delta_{j,c}$ denotes the item threshold parameter for the cth response category of item j (Samejima 1969), the latter of which satisfies

$$-\infty = \delta_{j,0} < \delta_{j,1} < \ldots < \delta_{j,C_j-1} < \delta_{j,C_j} = \infty. \qquad (2)$$

From a theoretical perspective, latent trait distributions in the IRT literature are often assumed to be normal. Therefore, some common estimation methods, such as marginal maximum likelihood and Bayesian techniques, are developed assuming normal latent traits. However, in some psychological instruments, such as depression and anxiety tests, the population latent traits may follow a non-normal distribution. Research has shown that violating the assumption of normality may bias the estimates of IRT item and latent trait parameters (e.g., Sass et al. 2008; Reise and Revicki 2014). In the literature, studies have been conducted to investigate item and person parameter recovery in estimating unidimensional dichotomous (e.g., Kirisci et al. 2001; Sass et al. 2008) and unidimensional multi-group dichotomous (e.g., Santo et al. 2013) models, where the latent trait follows a non-normal distribution. However, little has been conducted to investigate parameter recovery in estimating multidimensional polytomous models in this regard.

In view of the above, this study focuses on investigating parameter recovery of estimating multi-unidimensional GRMs when latent traits are either normal or non-normal. Specifically, different distributions of person parameters are adopted, and the performances of estimating item and person parameters using Hastings-within-Gibbs (HwG; Kuo and Sheng 2015) are compared. The remainder of the paper is outlined as follows. In Sect. 2, the HwG estimation is introduced. The simulation study is described and the results are discussed in Sect. 3. Finally, the conclusion for this study is summarized in Sect. 4.

2 Hastings-Within-Gibbs Estimation Procedure

For the past two decades, fully Bayesian has gained an increased popularity due to improved computational efficiency. There are two types of fundamental mechanisms among the Markov chain Monte Carlo (MCMC) algorithm: Gibbs sampling (Geman and Geman 1984) and Metropolis-Hastings (MH; Hastings 1970; Metropolis and Ulam 1949). Gibbs sampling is adopted in situations when the full conditional distribution of each parameter can be derived in closed form. If any of the full conditional distribution is not in an obtainable form, MH can be used via choosing a proposal or candidate distribution by the current value of the parameters. Then a proposal value is generated from the proposal distribution and accepted in the Markov chain with a certain amount of probability.

Hastings-within-Gibbs (HwG) is a form of the hybrid between Gibbs sampling and MH and has proved to be useful for complicated IRT models, such as GRMs. In the literature, Albert and Chib (1993) proposed a Gibbs sampler for the unidimensional GRM model. Cowels (1996) proposed a HwG procedure by using an MH step within the Gibbs sampler developed by Albert and Chib (1993) for sampling the threshold parameters to improve mixing and to accelerate convergence. Kuo and Sheng (2015) extended Cowles' approach to the more general multi-unidimensional GRM.

3 Simulation Study

To investigate parameter recovery of the HwG procedure in situations when latent traits are not normal, a Monte Carlo simulation study was carried out where tests with two subscales were considered so that the first half measured one latent trait (θ_1) and the second half measured the other (θ_2).

3.1 Simulated Data

In the study, three factors were manipulated: sample size (N), test length (K), and intertrait correlation (ρ). The choice of N, K, and ρ was based on previous studies with similar models. For example, when investigating multidimensional GRMs, Fu et al. (2010) adopted $N = 500, 1000$, $K = 10, 20, 30$, $\rho = 0.1, 0.3, 0.5, 0.7, 0.9$ for dichotomous items and $N = 1000$, $K = 20$, $\rho = 0.2, 0.4, 0.6, 0.8$ for polytomous items involving three categories. Working with dichotomous multi-unidimensional models, Sheng (2008) adopted $N = 1000$, $K = 18$, $\rho = 0.2, 0.5, 0.8$ in the simulation studies, while Sheng and Headrick (2012) adopted $N = 1000$, $K = 10$, $\rho = 0.2, 0.4, 0.6$. Wollack et al. (2002) conducted simulation studies with nominal response models, and they observed that parameter recovery was improved by increasing the test length from 10 to 30 items but that increasing the test length from 20 to 30 items did not produce a noticeable difference. Consequently, with our study, N polytomous responses ($N = 500, 1000$) to K items ($K = 20, 40$) were generated according to the multi-unidimensional GRM, where the population correlation between the two latent traits (ρ) was set to be 0.2, 0.5, or 0.8. Each item was set to be measured on a Likert scale with three categories so that two threshold parameters were estimated for each item. The item discrimination parameters $\boldsymbol{\alpha}_v$ were generated randomly from uniform distributions so that $\alpha_{vj} \sim U(0, 2)$. The threshold parameters δ_{j1} and δ_{j2} were sorted values based on those randomly generated from a standard normal distribution, i.e., $\delta_{j1} = \min(X_1, X_2)$ and $\delta_{j2} = \max(X_1, X_2)$, where $X_1, X_2 \sim N(0, 1)$.

The person parameters of the first dimension ($\boldsymbol{\theta}_1$) and the second dimension ($\boldsymbol{\theta}_2$) were generated based on the Method of Percentile (MOP; Koran et al. 2015) Power Method transformation. The MOP transformation was developed to generate multivariate distributions with specified values of median, interdecile ranges, left-right tail-weight ratios (a skewness function) and tail-weight factors (a kurtosis function) for each distribution, and the pairwise correlations.

To generate $\boldsymbol{\theta}_1$ and $\boldsymbol{\theta}_2$ using the MOP transformation, $\boldsymbol{\theta}_1$ were generated from a standard normal distribution, and $\boldsymbol{\theta}_2$ were generated from one of the following four distributions: (1) skewness = 0, kurtosis = 0 (Dist. 1), (2) skewness = 0, kurtosis = 25 (Dist. 2), (3) skewness = 2, kurtosis = 7 (Dist. 3), and (4) skewness = 3, kurtosis = 21 (Dist. 4). The correlation between $\boldsymbol{\theta}_1$ and $\boldsymbol{\theta}_2$ (i.e., the true intertrait correlation, ρ) was set to be $0.2, 0.5$, or 0.8. Note that the skewness and kurtosis considered in each of the four distributions are conventional values and they can be transferred to left-right tail-weight ratios and tail-weight factors in order to implement the MOP transformation technique (see Koran et al. 2015).

Harwell et al. (1996) suggested that a minimum of 25 replications for Monte Carlo studies in IRT-based research is needed in order to obtain a better accuracy. Therefore, this study carried out 25 replications for each scenario, where root-mean-squared differences (RMSDs) and bias were used to evaluate the recovery of each item parameter. Let π denote the true value of a parameter (e.g., α_{vj} or $\delta_{j,c}$) and $\hat{\pi}_r$ is the estimate in the rth replication ($r = 1, \ldots, R$). The RMSD is defined as

$$RMSD_\pi = \sqrt{\frac{\sum_{r=1}^{R}(\hat{\pi}_r - \pi)^2}{R}}, \tag{3}$$

and the bias is defined as

$$bias_\pi = \frac{\sum_{r=1}^{R}(\hat{\pi}_r - \pi)}{R}. \tag{4}$$

The 10% trimmed means of these measures were calculated across items to provide summary statistics.

3.2 Results

Tables 1, 2, 3, and 4 display the results of the simulation study under the twelve test situations. The results indicated that the HwG procedure had an overall better estimation when $\boldsymbol{\theta}_2$ followed a normal distribution. The non-normality of $\boldsymbol{\theta}_2$ affected the accuracy of estimating $\boldsymbol{\alpha}_2$. Specifically, distributions 2–4 had overall larger RMSDs of $\boldsymbol{\alpha}_2$ than distribution 1 (normal). $\boldsymbol{\alpha}_1$ had similar RMSDs across these four distributions when $\rho = 0.2$ or 0.5. However, the non-normality of $\boldsymbol{\theta}_2$ had more influence on estimating $\boldsymbol{\alpha}_1$ when the two dimensions were highly correlated (i.e., $\rho = 0.8$). On the other hand, the performance of estimating $\boldsymbol{\delta}$ was affected more by skewness than kurtosis. Specifically, even though distribution 2 had the heaviest kurtosis, its RMSDs for estimating $\boldsymbol{\delta}$ were smaller than those from skewed distributions (i.e., distributions 3 and 4). The estimation of ρ was sensitive to both skewness and kurtosis. Distributions 2–4 had larger RMSDs in estimating ρ than distribution 1. A further comparison of its RMSDs under the four distributions indicated that they were similar when $\rho = 0.2$ but became more different when the actual correlation was higher (i.e., 0.5 or 0.8).

Posterior estimates for the person parameters ($\boldsymbol{\theta}_1$ and $\boldsymbol{\theta}_2$) were also obtained and correlated with their corresponding true values. Tables 1, 2, 3, and 4 summarize all the correlation results, where $r(\boldsymbol{\theta}_1, \hat{\boldsymbol{\theta}}_1)$ and $r(\boldsymbol{\theta}_2, \hat{\boldsymbol{\theta}}_2)$ represent the correlations between the posterior estimates ($\hat{\boldsymbol{\theta}}$) and their corresponding true values ($\boldsymbol{\theta}$) for dimensions 1 and 2, respectively. The results indicate that $\boldsymbol{\theta}_1$ was estimated fairly well due to the satisfaction of normality assumption. On the other hand, the estimation of $\boldsymbol{\theta}_2$ was affected by kurtosis more than skewness, as distribution 2 had an overall lower $r(\boldsymbol{\theta}_2, \hat{\boldsymbol{\theta}}_2)$ than distribution 3 (less kurtotic but more skewed). However, extreme skewed distributions (i.e., distribution 4) had an overall lower $r(\boldsymbol{\theta}_2, \hat{\boldsymbol{\theta}}_2)$ than distributions 2 and 3. In addition, a comparison of $K = 40$ and $K = 20$ for the same sample size conditions (i.e., Table 2 vs. Table 1 and Table 4 vs. Table 3) indicates that the former had consistently larger $r(\boldsymbol{\theta}_2, \hat{\boldsymbol{\theta}}_2)$ values than the latter. This suggests that the accuracy of estimating $\boldsymbol{\theta}_2$ improved with the increase in test length regardless of its distribution.

Further, it is found that an increase of sample size can improve the accuracy of estimating model parameters. For example, with the test length of $K = 20$, the RMSDs of estimating $\boldsymbol{\alpha}$, $\boldsymbol{\delta}$, and ρ when $N = 1000$ were in general smaller than those when $N = 500$, especially when the true intertrait correlation was higher. One shall note that when $\rho = 0.2$, larger sample sizes helped reduce the RMSDs of $\boldsymbol{\alpha}_2$ when $\boldsymbol{\theta}_2$ was non-normal. This is however not observed with $\rho = 0.5$ or 0.8. In terms of estimating $\boldsymbol{\theta}$, larger sample size tended to increase the accuracy of estimating $\boldsymbol{\theta}_1$. This pattern is only observed when estimating $\boldsymbol{\theta}_2$ in distributions 2 and 3 when $\rho < 0.8$.

Table 1 Average RMSD and bias (italic values) for estimating $\boldsymbol{\alpha}$, $\boldsymbol{\delta}$, and ρ when $N = 500$, $K = 20$

	True $\rho = 0.2$				True $\rho = 0.5$				True $\rho = 0.8$			
$\boldsymbol{\theta}_2$	Dist. 1	Dist. 2	Dist. 3	Dist. 4	Dist. 1	Dist. 2	Dist. 3	Dist. 4	Dist. 1	Dist. 2	Dist. 3	Dist. 4
$\boldsymbol{\alpha}_1$	0.1248	0.1337	0.1264	0.1304	0.1242	0.1346	0.1248	0.1310	0.1191	0.1378	0.1395	0.1346
	0.0597	*0.0708*	*0.0657*	*0.0707*	*0.0607*	*0.0739*	*0.0680*	*0.0724*	*0.0556*	*0.0873*	*0.0896*	*0.0863*
$\boldsymbol{\alpha}_2$	0.1088	0.1653	0.1659	0.1664	0.1177	0.1485	0.1468	0.1488	0.1261	0.1538	0.1527	0.1543
	0.0036	*0.0061*	*0.0041*	*0.0048*	*0.0454*	*0.0344*	*0.0308*	*0.0332*	*0.0431*	*0.0589*	*0.0594*	*0.0601*
$\boldsymbol{\delta}_1$	0.1139	0.1217	0.1223	0.1296	0.1444	0.1483	0.1506	0.1558	0.1032	0.1002	0.1122	0.1138
	−0.0988	*−0.0787*	*−0.0775*	*−0.0849*	*−0.0716*	*−0.0815*	*−0.0856*	*−0.0890*	*−0.0623*	*−0.0604*	*−0.0679*	*−0.0656*
$\boldsymbol{\delta}_2$	0.1124	0.1154	0.1193	0.1263	0.1256	0.1373	0.1436	0.1473	0.1041	0.1012	0.1026	0.1051
	−0.0543	*−0.0587*	*−0.0569*	*−0.0656*	*−0.0574*	*−0.0620*	*−0.0651*	*−0.0706*	*−0.0296*	*−0.0352*	*−0.0391*	*−0.0398*
ρ_{12}	0.0026	0.0033	0.0030	0.0031	0.0029	0.0091	0.0079	0.0082	0.0005	0.0118	0.0116	0.0116
	0.0120	*0.0265*	*0.0248*	*0.0256*	*0.0166*	*0.0794*	*0.0762*	*0.0772*	*0.0037*	*0.1066*	*0.1055*	*0.1059*
$r(\boldsymbol{\theta}_1, \hat{\boldsymbol{\theta}}_1)$	0.9160	0.9016	0.9090	0.9045	0.9100	0.9058	0.9141	0.9096	0.9394	0.9336	0.9336	0.9336
$r(\boldsymbol{\theta}_2, \hat{\boldsymbol{\theta}}_2)$	0.9170	0.7875	0.8420	0.7878	0.9148	0.7914	0.8436	0.7912	0.9290	0.8190	0.8477	0.8097

Table 2 Average RMSD and bias (italic values) for estimating $\boldsymbol{\alpha}$, $\boldsymbol{\delta}$, and ρ when $N = 500$, $K = 40$

	True $\rho = 0.2$				True $\rho = 0.5$				True $\rho = 0.8$			
$\boldsymbol{\theta}_2$	Dist. 1	Dist. 2	Dist. 3	Dist. 4	Dist. 1	Dist. 2	Dist. 3	Dist. 4	Dist. 1	Dist. 2	Dist. 3	Dist. 4
$\boldsymbol{\alpha}_1$	0.1109	0.1214	0.1187	0.1204	0.2191	0.5224	0.4451	0.5294	0.1084	0.1446	0.1452	0.1478
	0.0670	*0.0462*	*0.0467*	*0.0478*	*0.1142*	*0.1249*	*0.1165*	*0.1312*	*0.0590*	*0.0801*	*0.0806*	*0.0828*
$\boldsymbol{\alpha}_2$	0.1076	0.1088	0.1103	0.1094	0.1516	0.1365	0.1372	0.1336	0.0992	0.1261	0.1272	0.1289
	0.0357	*0.0333*	*0.0369*	*0.0352*	*0.0883*	*0.0575*	*0.0655*	*0.0580*	*0.0299*	*0.0678*	*0.0692*	*0.0696*
$\boldsymbol{\delta}_1$	0.1256	0.1333	0.1347	0.1382	0.1703	0.2246	0.2333	0.2228	0.1213	0.1235	0.1281	0.1228
	−0.0417	*−0.0549*	*−0.0559*	*−0.0569*	*−0.1217*	*−0.1610*	*−0.1696*	*−0.1608*	*−0.0414*	*−0.0419*	*−0.0532*	*−0.0502*
$\boldsymbol{\delta}_2$	0.1133	0.1259	0.1314	0.1264	0.1577	0.2125	0.2210	0.2110	0.1104	0.1152	0.1188	0.1161
	−0.0302	*−0.0365*	*−0.0369*	*−0.0370*	*−0.1066*	*−0.1463*	*−0.1542*	*−0.1456*	*−0.0277*	*−0.0233*	*−0.0331*	*−0.0298*
ρ_{12}	0.0036	0.0169	0.0157	0.0174	0.0014	0.0076	0.0075	0.0073	0.0007	0.0116	0.0118	0.0118
	0.0092	*0.0506*	*0.0505*	*0.0524*	*0.0019*	*0.0845*	*0.0834*	*0.0825*	*−0.0072*	*0.1074*	*0.1082*	*0.1079*
$r(\boldsymbol{\theta}_1, \hat{\boldsymbol{\theta}}_1)$	0.9224	0.9150	0.9157	0.9145	0.9325	0.9430	0.9431	0.9436	0.9467	0.9458	0.9448	0.9451
$r(\boldsymbol{\theta}_2, \hat{\boldsymbol{\theta}}_2)$	0.9422	0.8221	0.8518	0.8114	0.9448	0.8492	0.8690	0.8372	0.9409	0.8307	0.8508	0.8285

Table 3 Average RMSD and bias (italic values) for estimating $\boldsymbol{\alpha}$, $\boldsymbol{\delta}$, and ρ when $N = 1000$, $K = 20$

	True $\rho = 0.2$				True $\rho = 0.5$				True $\rho = 0.8$			
$\boldsymbol{\theta}_2$	Dist. 1	Dist. 2	Dist. 3	Dist. 4	Dist. 1	Dist. 2	Dist. 3	Dist. 4	Dist. 1	Dist. 2	Dist. 3	Dist. 4
$\boldsymbol{\alpha}_1$	0.0680	0.0724	0.0739	0.0724	0.0720	0.0751	0.0770	0.0755	0.0709	0.0850	0.0854	0.0842
	0.0188	*0.0213*	*0.0220*	*0.0207*	*0.0193*	*0.0200*	*0.0195*	*0.0190*	*0.0193*	*0.0426*	*0.0423*	*0.0419*
$\boldsymbol{\alpha}_2$	0.0753	0.1506	0.1499	0.1492	0.0832	0.1469	0.1451	0.1442	0.0705	0.1511	0.1503	0.1499
	0.0176	*0.0052*	*0.0060*	*0.0077*	*0.0277*	*0.0121*	*0.0158*	*0.0168*	*0.0219*	*−0.0006*	*−0.0004*	*0.0001*
$\boldsymbol{\delta}_1$	0.0582	0.0612	0.0628	0.0632	0.0625	0.0668	0.0697	0.0694	0.0698	0.0703	0.0725	0.0717
	−0.0128	*−0.0125*	*−0.0138*	*−0.0154*	*−0.0233*	*−0.0250*	*−0.0236*	*−0.0267*	*−0.0259*	*−0.0288*	*−0.0288*	*−0.0308*
$\boldsymbol{\delta}_2$	0.0659	0.0660	0.0679	0.0666	0.0618	0.0677	0.0689	0.0685	0.0705	0.0702	0.0714	0.0705
	−0.0011	*−0.0089*	*−0.0099*	*−0.0100*	*−0.0112*	*−0.0179*	*−0.0161*	*−0.0190*	*−0.0131*	*−0.0232*	*−0.0228*	*−0.0249*
ρ_{12}	0.0011	0.0013	0.0013	0.0013	0.0010	0.0043	0.0044	0.0044	0.0004	0.0112	0.0112	0.0113
	−0.0070	*0.0251*	*0.0252*	*0.0253*	*−0.0123*	*0.0601*	*0.0606*	*0.0608*	*−0.0130*	*0.1056*	*0.1054*	*0.1058*
$r(\boldsymbol{\theta}_1, \hat{\boldsymbol{\theta}}_1)$	0.9148	0.9139	0.9137	0.9139	0.9294	0.9200	0.9198	0.9200	0.9487	0.9426	0.9424	0.9426
$r(\boldsymbol{\theta}_2, \hat{\boldsymbol{\theta}}_2)$	0.9111	0.7875	0.8362	0.7811	0.9235	0.8024	0.8469	0.7958	0.9320	0.8199	0.8489	0.8181

Table 4 Average RMSD and bias (italic values) for estimating $\boldsymbol{\alpha}$, $\boldsymbol{\delta}$, and ρ when $N = 1000$, $K = 40$

	True $\rho = 0.2$				True $\rho = 0.5$				True $\rho = 0.8$			
$\boldsymbol{\theta}_2$	Dist. 1	Dist. 2	Dist. 3	Dist. 4	Dist. 1	Dist. 2	Dist. 3	Dist. 4	Dist. 1	Dist. 2	Dist. 3	Dist. 4
$\boldsymbol{\alpha}_1$	0.0582	0.0673	0.0668	0.0666	0.0674	0.0686	0.0679	0.0679	0.0678	0.0905	0.0897	0.0906
	0.0239	*0.0286*	*0.0288*	*0.0281*	*0.0243*	*0.0318*	*0.0317*	*0.0314*	*0.0249*	*0.0585*	*0.0582*	*0.0582*
$\boldsymbol{\alpha}_2$	0.0737	0.0740	0.0737	0.0742	0.0737	0.0765	0.0761	0.0764	0.0725	0.1060	0.1060	0.1055
	0.0218	*0.0230*	*0.0225*	*0.0229*	*0.0288*	*0.0273*	*0.0269*	*0.0274*	*0.0292*	*0.0626*	*0.0619*	*0.0620*
$\boldsymbol{\delta}_1$	0.0718	0.0800	0.0795	0.0808	0.0929	0.0931	0.0932	0.0942	0.0937	0.1102	0.1024	0.1067
	−0.0435	*−0.0437*	*−0.0426*	*−0.0449*	*−0.0549*	*−0.0592*	*−0.0542*	*−0.0605*	*−0.0599*	*−0.0752*	*−0.0662*	*−0.0722*
$\boldsymbol{\delta}_2$	0.0689	0.0761	0.0756	0.0762	0.0815	0.0864	0.0891	0.0889	0.0889	0.0933	0.0968	0.1010
	−0.0357	*−0.0361*	*−0.0352*	*−0.0376*	*−0.0485*	*−0.0499*	*−0.0453*	*−0.0521*	*−0.0514*	*−0.0652*	*−0.0571*	*−0.0635*
ρ_{12}	0.0015	0.0015	0.0015	0.0015	0.0009	0.0055	0.0055	0.0056	0.0002	0.0125	0.0125	0.0125
	−0.0008	*0.0273*	*0.0275*	*0.0275*	*−0.0004*	*0.0708*	*0.0708*	*0.0711*	*−0.0021*	*0.1115*	*0.1115*	*0.1116*
$r(\boldsymbol{\theta}_1, \hat{\boldsymbol{\theta}}_1)$	0.9537	0.9532	0.9533	0.9530	0.9575	0.9545	0.9544	0.9542	0.9575	0.9624	0.9624	0.9621
$r(\boldsymbol{\theta}_2, \hat{\boldsymbol{\theta}}_2)$	0.9527	0.8539	0.8769	0.8338	0.9516	0.8567	0.8771	0.8388	0.9570	0.8609	0.8721	0.8508

4 Conclusion and Discussion

In general, with the use of Monte Carlo simulations, this study demonstrates that departure from normal distributions for the latent traits in the multi-unidimensional GRM does affect the accuracy of its parameter recovery. This is in line with findings from previous studies with unidimensional IRT models (e.g., Sass et al. 2008; Reise and Revicki 2014). Specifically, what we found in our study are that skewed distributions would affect more on the accuracy in estimating the item step parameters and that kurtotic distributions affect the estimation of person parameters. In situations where not all latent traits are normally distributed (such as what was considered in the simulation study), the non-normal shape associated with a few latent traits would affect the estimation of parameters in other dimensions when the intertrait correlation is moderate to high. As non-normal latent trait distributions are common in many polytomous response items, and examples of such instruments include mental tests, business satisfaction, cross-cultural differences, etc., one needs to be aware of the shapes of latent trait distributions before fitting the model to actual data. However, such information may not always be available in practice. It is hence important to find alternate solutions, such as using a more robust estimation method or a non-normal prior distribution. In addition, this study shows that increased sample size and/or test length can help improve the estimation of the multi-unidimensional GRM parameters. This finding not only confirms results from previous studies dealing with normal latent trait(s) (e.g., Linacre 2002; Sheng 2010; Kuo and Sheng 2015; Wollack et al. 2002) but also extends to situations where the latent traits are not normal. One may consider reducing the effect of non-normality by increasing sample size/test length under the non-normal conditions. The minimum number of persons/items necessary to reach a desired level of accuracy can be an interesting study that requires further investigation.

This study focuses on Likert scale items involving three scales, and therefore two threshold parameters need to be estimated for each item. Further study can evaluate the estimation of these procedures using items with more than three scales or with different numbers of scales. In addition, this study investigates the effects of non-normal latent traits using the HwG estimation method. Further study can include other estimation techniques, such as marginal maximum likelihood (Bock and Aitkin 1981) and Metropolis-Hastings Robbins-Monro (Cai 2010a,b). Lastly, the simulation study adopted 25 replications due to the computational expense of the MCMC procedures. Further studies can consider more replications to achieve a better accuracy.

References

J.H. Albert, S. Chib, Bayesian analysis of binary and polychotomous response data. J. Am. Stat. Assoc. **88**, 669–679 (1993)

R.D. Bock, M. Aitkin, Marginal maximum likelihood estimation of item parameters: applications of an EM algorithm. Psychometrika **46**, 443–459 (1981)

R.D. Buchanan, The development of the Minnesota multiphasic personality inventory. J. Hist. Behav. Sci. **30**, 148–161 (1994)
L. Cai, High-dimensional exploratory item factor analysis by a Metropolis-Hastings Robbins-Monro algorithm. Psychometrika **75**, 33–57 (2010a)
L. Cai, Metropolis-Hastings Robbins-Monro algorithm for confirmatory item factor analysis. J. Educ. Behav. Stat. **35**, 307–335 (2010b)
M.K. Cowels, Accelerating Monte Carlo Markov chain convergence for cumulative-link generalized linear models. Stat. Comput. **6**, 101–111 (1996)
C.G. Ferero, A. Maydeu-Olivares, Estimation of IRT graded response models: limited versus full information methods. Psychol. Methods **14**, 275–299 (2009)
Z.H. Fu, J. Tao, N.Z. Shi, Bayesian estimation of the multidimensional graded response model with nonignorable missing data. J. Stat. Comput. Simul. **80**, 1237–1252 (2010)
S. Geman, D. Geman, Stochastic relaxation, Gibbs distribution and the Bayesian restoration of images. IEEE Trans. Pattern Anal. Mach. Intell. **6**, 721–741 (1984)
M. Harwell, C.A. Stone, T.-C. Hsu, L. Kirisci, Monte carlo studies in item response theory. Appl. Psychol. Meas. **20**, 101–125 (1996)
W. Hastings, Monte Carlo sampling methods using Markov chains and their applications. Biometrika **57**, 97–109 (1970)
L. Kirisci, T. Hsu, L. Yu, Robustness of item parameter estimation programs to assumptions of unidimensionality and normality. Appl. Psychol. Meas. **25**, 146–162 (2001)
J. Koran, T.C. Headrick, T.-C. Kuo, Simulating univariate and multivariate nonnormal distributions through the method of percentiles. Multivar. Behav. Res. **50**, 1–17 (2015)
T.C. Kuo, Y. Sheng, Bayesian estimation of a multi-unidimensional graded response IRT model. Behaviormetrika **42**, 79–94 (2015)
R. Likert, A technique for the measurement of attitudes. Arch. Psychol. **22**, 5–55 (1932)
J.M. Linacre, Optimizing rating scale category effectiveness. J. Appl. Meas. **3**, 85–106 (2002)
F.M. Lord, *Applications of Item Response Theory to Practical Testing Problems* (Lawrence Erlbaum, Hillsdale, 1980)
R.P. McDonald, *Test Theory: A Unified Approach* (Lawrence Erlbaum, Mahwah, 1999)
N. Metropolis, S. Ulam, The Monte Carlo method. J. Am. Stat. Assoc. **44**, 335–341 (1949)
M. Reckase, The past and future of multidimensional item response theory. Appl. Psychol. Meas. **21**, 25–36 (1997)
M. Reckase, *Multidimensional Item Response Theory* (Springer, New York, 2009)
S. Reise, D. Revicki, *Handbook of Item Response Theory Modeling: Applications to Typical Performance Assessment*. Multivariate Applications Book Series (Routledge, New York, 2014)
V.J. Rubio, D. Aguado, P.M. Hontangas, J.M. Hernandez, Psychometric properties of an emotional adjustment measure. Eur. J. Psychol. Assess. **23**, 39–46 (2007)
F. Samejima, Estimation of latent ability using a response pattern of graded scores. Psychometrika **35**, 139–139 (1969)
J.R.S. Santo, C.L.N. Azevedo, H. Bolfarine, A multiple group item response theory model with centred skew normal latent trait distributions under a bayesian framework. J. Appl. Stat. **40**, 2129–2149 (2013)
D.A. Sass, T.A. Schmitt, C.M. Walker, Estimating non-normal latent trait distributions within item response theory using true and estimated item parameters. Appl. Meas. Educ. **21**, 65–88 (2008)
Y. Sheng, A sensitivity analysis of gibbs sampling for 3PNO IRT models: effects of prior specifspecific on parameter estimates. Behaviormetrika **37**, 87–110 (2010)
Y. Sheng, T.C. Headrick, A Gibbs sampler for the multidimensional item response model. ISRN Appl. Math. **2012**, 14pp. (2012)
Y. Sheng, C.K. Wikle, Comparing multiunidimensional and unidimensional item response theroy models. Educ. Psychol. Meas. **67**, 899–919 (2007)
Y. Sheng, A MATLAB package for Markov chain Monte Carlo with a multi-unidimensional IRT model. J. Stat. Softw. **28**, 1–20 (2008)
J.A. Wollack, D.M. Bolt, A.S. Cohen, Y.S. Lee, Recovery of item parameters in the nominal response model: a comparison of marginal maximum likelihood estimation and Markov chain Monte Carlo estimation. Appl. Psychol. Meas. **26**, 339–352 (2002)

An Extension of Rudner-Based Consistency and Accuracy Indices for Multidimensional Item Response Theory

Wenyi Wang, Lihong Song, and Shuliang Ding

Abstract Although the field of multidimensional item response theory (MIRT) has enjoyed tremendous growth over recent years, solutions to some problems remain to be studied. One case in point is the estimate of classification accuracy and consistency indices. There have been a few research studies focusing on these indices based on total scores under MIRT. The purposes of this study are to extend Rudner-based index for MIRT under complex decision rules and to compare it with the Guo-based index and the Lee-based index. The Rudner-based index assumes that an ability estimation error follows a multivariate normal distribution around each examinee's ability estimate, and a simple Monte Carlo method is used to estimate accuracy and consistency indices. The simulation results showed that the Rudner-based index worked well under various conditions. Finally, conclusions are described along with thoughts for future research.

Keywords Classification consistency • Classification accuracy • Multidimensional item response theory • Decision rule

1 Introduction

For criterion-referenced tests, classification consistency and accuracy are important indicators to evaluate the reliability and validity of classification results. Numerous procedures have been proposed to estimate these indices in the framework of

W. Wang
College of Computer Information Engineering, Jiangxi Normal University,
99 Ziyang Road, Nanchang, Jiangxi, People's Republic of China
e-mail: wenyiwang2009@gmail.com

L. Song (✉)
Elementary Education College, Jiangxi Normal University, 99 Ziyang Road, Nanchang, Jiangxi, People's Republic of China
e-mail: viviansong1981@163.com

S. Ding
School of Computer and Information Engineering, Jiangxi Normal University, 99 Ziyang Ave., Nanchang, Jiangxi 330022, China
e-mail: ding06026@163.com

L.A. van der Ark et al. (eds.), *Quantitative Psychology*, Springer Proceedings in Mathematics & Statistics 196, DOI 10.1007/978-3-319-56294-0_5

unidimensional item response theory (UIRT) (Huynh 1990; Lathrop and Cheng 2013; Lee 2010; Rudner 2001, 2005; Schulz et al. 1999; Wang et al. 2000; Wyse and Hao 2012). Some of these were based on total scores, while others on latent trait estimates (Lathrop and Cheng 2013). The Lee approach (Lee 2010) belongs to the former, whereas the Rudner approach (Rudner 2001, 2005) and its extension, the Guo approach (Guo 2006), fall into the latter category.

Multidimensional item response theory (MIRT) has been devoted to models that include more than one latent trait to account for the multidimensional nature of complex constructs. MIRT has been successfully employed to analyze many criterion-referenced tests, which are multidimensional to some degree (Bolt and Lall 2003; Chang and Wang 2016; Debeer et al. 2014; Makransky et al. 2012; Rijmen et al. 2014; Yao and Boughton 2007; Zhang 2012). For example, the overall construct of mathematics in IEA's Trends in International Mathematics and Science Study encompassed four content domains: number, algebra, geometry, and data and chance.

Although MIRT has enjoyed tremendous growth, solutions to some problems remain to be studied. One case in point is the estimate of classification accuracy and consistency indices under different decision rules. There have been a few research studies on estimating these indices based on total scores under MIRT (Grima and Yao 2011; LaFond 2014; Yao 2016). It is inflexible to estimate accuracy and consistency indices derived from total scores if a correct one-to-one mapping cannot be established between the decision rules based on latent ability and total scores. For example, if a compensatory model was used to analyze a between-item multidimensional test (Adams et al. 1997) that consists of items measuring more than one content domain with different slope parameters, we often cannot establish a one-to-one mapping between total scores and a decision rule based on a composite latent ability score.

The current paper addresses this issue whenever the decision rules aligned either on latent ability scale or on total score scale. For one of two indices based on latent ability described above, the Guo-based index has been extended to MIRT under complex decision rules for the single administration of a test (Wang et al. 2016). The purposes of this study are to extend the Rudner-based index for MIRT and to compare it with the Guo-based and Lee-based indices. The Rudner-based index assumed that an ability estimation error follows a multivariate normal distribution around the examinee's ability estimate, and Monte Carlo simulation can be easily used to estimate the accuracy and consistency indices. The rest of this article proceeds as follows. Section 2 starts with a review of a MIRT model, decision rule, the Guo-based index, and the Lee-based index. Section 3 introduces test information and the Rudner-based index. Section 4 provides a simulation study and explains the simulation results. Finally, conclusions and suggestions are described in Sect. 5.

2 Model and Methods

2.1 A Multidimensional Graded Response Model

A multidimensional graded response model (MGRM) is a generalization of the unidimensional graded response model, and it uses response function that has the normal ogive or logistic function (Cai 2010; Reckase 2009). The parameterization of this model given here considers the lowest score on item j to be 0 and the highest score to be K_j. Let $\boldsymbol{\theta} = (\theta_1, \theta_2, \ldots, \theta_D)'$ denote a vector of ability with the number of dimensions of D, $\boldsymbol{\alpha}_j$ be a vector of slope parameter related to item discrimination parameter of item j, β_{jk} be a threshold parameter related to item difficulty with which an examinee will reach the kth step of item j, $j = 1,2, \ldots, J$, and y_j is an item response to item j. Given an examinee with abilities in the $\boldsymbol{\theta}$-vector, his probability of successfully performing the work in step k or more advanced steps in answering an item j can be written as (Cai 2010)

$$P_{jk}^{*}(\boldsymbol{\theta}) = P\left(y_j \geq k|\boldsymbol{\theta}, \boldsymbol{\alpha}_j, \boldsymbol{\beta}_j\right) = \frac{1}{1 + \exp\left(\beta_{jk} - \boldsymbol{\alpha}_j'\boldsymbol{\theta}\right)}, \quad (1)$$

where $k = 0,1, \ldots, K_j$ with $P_{j0}^{*}(\boldsymbol{\theta}) = 1$ and $P_{jK_j}^{*}(\boldsymbol{\theta}) = 0$.

The probability of receiving a specific score k is the difference between the probability of successfully performing the work for step k or more advanced steps and that of the work for $k + 1$ or more advanced steps. Then the probability that an examinee will receive a score of k is

$$P_{jk}(\boldsymbol{\theta}) = P\left(y_j = k|\boldsymbol{\theta}, \boldsymbol{\alpha}_j, \boldsymbol{\beta}_j\right) = P_{jk}^{*}(\boldsymbol{\theta}) - P_{j,k+1}^{*}(\boldsymbol{\theta}). \quad (2)$$

Assuming local or conditional independence of the responses given the $\boldsymbol{\theta}$-vector, the likelihood function of the observed data $\mathbf{y}_i$ is

$$L(\mathbf{y}_i|\boldsymbol{\theta}_i, \boldsymbol{\alpha}, \boldsymbol{\beta}) = \prod_{j=1}^{J} \prod_{k=0}^{K_j} P\left(y_{ij} = k|\boldsymbol{\theta}_i, \boldsymbol{\alpha}_j, \boldsymbol{\beta}_j\right)^{1_{(y_{ij}=k)}}, \quad (3)$$

where an indicator function is defined as

$$1_{(y_{ij}=k)} = \begin{cases} 1 & \text{if } y_{ij} = k \\ 0 & \textit{otherwise}. \end{cases} \quad (4)$$

A number of computer programs have been developed for estimating item parameters in the MGRM, such as BMIRT (Yao 2003), IRTPRO program (Cai et al. 2011), MIRT package for the R environment (Chalmers 2012), and so on. For a test data set with a particular sample size N, given an already calibrated set of item parameters, the ability vector could be estimated either via (weighted) maximum likelihood estimation (MLE) or using Bayesian methods such as maximum a posteriori (MAP) estimation or expected a posteriori (EAP) estimation (Wang 2015). In this article, the ability estimate for examinee i, denoted by $\hat{\boldsymbol{\theta}}_i$, was obtained from the MLE.

2.2 Decision Rule for Multidimensional Latent Ability

Decision rules were designed to increase the reliability and validity of the resulting decisions (Douglas and Mislevy 2010). For more details about complex decision rules, please refer to Douglas and Mislevy (2010). For example, an international student will qualify for admission to a China's university as a graduate student if the student has reached a passing grade on China's college entrance examination and the minimum required level of Chinese Proficiency Test (HSK).

For the above example, without loss of generality, we assume that one needs to estimate an accuracy index on a total score scale, and this is achieved via the following decision regions:

$$R_{c0} = \left\{\boldsymbol{\theta} : \tau_{(c-1)0} < \tau\left(\boldsymbol{\theta}\right) < \tau_{c0}\right\}, \tag{5}$$

where $c = 1, 2, \ldots, C$, τ_{c0} is a cut score on the total score scale, $-\infty < \tau_{00} < \tau_{10} < \ldots < \tau_{C0} = +\infty$, and $\tau\left(\hat{\boldsymbol{\theta}}_i\right)$ denote the expected summed score in Eq. (11).

If the accuracy index of each dimension needed to be estimated, it can be achieved via the following decision regions:

$$R_{cd} = \left\{\boldsymbol{\theta} : \tau_{(c-1)d} \leq \theta_d < \tau_{cd}\right\}, \tag{6}$$

where $c = 1, 2, \ldots, C$, $d = 1, 2, \ldots, D$, τ_{ck} is a cut score of the dth dimension, and $-\infty < \tau_{0d} < \tau_{1d} < \ldots < \tau_{Cd} = +\infty$.

If a compensatory rule was applied on a composite score scale, the decision regions are defined as follows:

$$R_{c(D+1)} = \left\{\boldsymbol{\theta} : \tau_{(c-1)(D+1)} \leq \sum_{d=1}^{D} w_k \theta_k < \tau_{c(D+1)}\right\}, \tag{7}$$

where $c = 1, 2, \ldots, C$, $\tau_{(c-1)(D+1)}$ is a cut score on the composite score scale, and w_d is a weight on the dth dimension.

2.3 Lee-Based Indices

First, we briefly describe the accuracy and consistency indices for total scores using MIRT model (Grima and Yao 2011; Yao 2016), which are based on the Lee approach (Lee 2010). Let us assume now that the test scores on one test form are used to classify examinees into C categories defined by cutoff scores $\tau_{00}, \tau_{10}, \ldots, \tau_{C0}$. That is, examinees with an observed score greater than or equal to $\tau_{(c-1)0}$ and less than τ_{c0} are assigned to the cth category.

Next, we will first present some formulas. Let $x = \sum_{j=1}^{J} y_j$ be a particular realization of the total score X for an examinee, and $P\left(X = x|\hat{\boldsymbol{\theta}}_i\right)$ be a conditional distribution of X given an ability estimate $\hat{\boldsymbol{\theta}}_i$. Supposing that item and ability parameters are estimated, because of the MIRT's assumption of conditional independence of the responses given the $\boldsymbol{\theta}$-vector, the conditional probability of total score x located in the cth category can be written as

$$\hat{p}_{ic} = \sum_{y_1, y_2, \ldots, y_J : \tau_{(c-1)0} \leq \sum_{j=1}^{J} y_j < \tau_{c0}} \prod_{j=1}^{J} P_{jy_j}\left(\hat{\boldsymbol{\theta}}_i\right), \tag{8}$$

where $c = 1, 2, \ldots, C$ and $P_{jy_j}(\boldsymbol{\theta})$ is defined by Eq. (2). The conditional distribution of X was approximated by using Monte Carlo simulation.

Given the conditional probability of scoring in each performance category, the conditional classification consistency index $\hat{\phi}_i$ is defined as the probability that an examinee with $\hat{\boldsymbol{\theta}}_i$ is classified into the same category in two independent administrations of parallel forms of a test, and it can be written as (Wyse and Hao 2012)

$$\hat{\phi}_i = \sum_{c=1}^{C} \left(\hat{p}_{ic} * \hat{p}_{ic}\right). \tag{9}$$

The conditional classification consistency index quantifies classification consistency for different levels of $\boldsymbol{\theta}$. The marginal classification consistency index $\hat{\phi}$ is given by

$$\hat{\phi} = \frac{\sum_{i=1}^{N} \hat{\phi}_i}{N} = \frac{\sum_{i=1}^{N} \sum_{c=1}^{C} \left(\hat{p}_{ic} * \hat{p}_{ic}\right)}{N}. \tag{10}$$

Now, supposing we have a set of true cut scores on the summed-score metric, $\tau_{00}, \tau_{10}, \ldots, \tau_{C0}$, we need to determine the "true" category of each examinee with $\hat{\boldsymbol{\theta}}_i$ or $\tau\left(\hat{\boldsymbol{\theta}}_i\right)$ (i.e., expected summed score). The expected summed score for an examinee with ability $\hat{\boldsymbol{\theta}}_i$ is obtained by

$$\tau\left(\hat{\boldsymbol{\theta}}_i\right) = \sum_{j=1}^{J} \sum_{k=0}^{K_j} k P_{jk}\left(\hat{\boldsymbol{\theta}}_i\right). \tag{11}$$

Then by comparing $\tau\left(\hat{\boldsymbol{\theta}}_i\right)$ with true cut scores, we know the "true" classification of $\tau\left(\hat{\boldsymbol{\theta}}_i\right)$ of examinee i. If $\tau\left(\hat{\boldsymbol{\theta}}_i\right) \in \left[\tau_{(c-1)0}, \tau_{c0}\right)$ is satisfied for a particular category, then the cth category is assumed as the "true" category of the examinee. Let $\hat{w}_{ic}$ be the flag of the performance-level category, meaning that $\tau\left(\hat{\boldsymbol{\theta}}_i\right)$ is classified into category c, $\hat{w}_{ic} = 1$, and 0 otherwise. The conditional classification accuracy index $\hat{\gamma}_i$ is defined as the probability that an examinee with $\hat{\boldsymbol{\theta}}_i$ is classified into the "true" category assuming known cut scores on a single test, and it can be written as

$$\hat{\gamma}_i = \hat{p}_{ic}, for\ \tau\left(\hat{\boldsymbol{\theta}}_i\right) \in \left[\tau_{(c-1)0}, \tau_{c0}\right). \tag{12}$$

The marginal classification accuracy index γ is given by (Wyse and Hao 2012)

$$\hat{\gamma} = \frac{\sum_{i=1}^{N} \hat{\gamma}_i}{N} = \frac{\sum_{i=1}^{N} \sum_{c=1}^{C} \left(\hat{p}_{ic}^{*} \hat{w}_{ic}\right)}{N}. \tag{13}$$

2.4 Guo-Based Indices

In this section, the Guo approach is described to estimate the consistency and accuracy indices for multidimensional latent ability. This approach is computationally easy and can be directly adapted to MIRT because it is closely tied to the normalized likelihood.

Guo (2006) defined classification accuracy index as the percentage of agreement between the observed and expected proportions of examinees in each of the categories under the UIRT framework. Next, we will introduce an extension of the Guo approach. The extension was proposed by Wang et al. (2016) and is suitable for estimating the consistency and accuracy indices for complex decision rules in MIRT. Supposing the $\boldsymbol{\theta}$ space can be partitioned into C separate decision regions, $R_1, R_2, \ldots, R_C$, corresponding to the various categories, we can determine the true category of each examinee with $\boldsymbol{\theta}$. In other words, a decision rule is a function from the $\boldsymbol{\theta}$ space into the set of categories. From the idea of the Guo approach, the expected probability of scoring in any particular category can be obtained using the likelihood functions as

$$\hat{p}_{ic} = p_i\left(R_c\right) = \frac{\int_{R_c} L\left(\mathbf{y}_i|\boldsymbol{\theta}, \boldsymbol{\alpha}, \boldsymbol{\beta}\right) d\boldsymbol{\theta}}{\sum_{c=1}^{C} \int_{R_c} L\left(\mathbf{y}_i|\boldsymbol{\theta}, \boldsymbol{\alpha}, \boldsymbol{\beta}\right) d\boldsymbol{\theta}}, \tag{14}$$

where $L(\mathbf{y}_i| \boldsymbol{\theta}, \boldsymbol{\alpha}, \boldsymbol{\beta})$ is defined by Eq. (3), and $c = 1, 2, \ldots, C$.

Classification consistency provides a measure of the proportion of examinees who would be classified into the same category on parallel replications of the same test. The classification consistency index can be expressed as

$$\hat{\phi} = \frac{\sum_{i=1}^{N} \sum_{c=1}^{C} \left(\hat{p}_{ic}^{*} \hat{p}_{ic}\right)}{N}. \tag{15}$$

As discussed above, the entry $\hat{w}_{ic}$ is 1 if an examinee's ability estimate $\hat{\boldsymbol{\theta}}_i$ is classified into category c, and 0 otherwise. Then the classification accuracy index can be written as

$$\hat{\gamma} = \frac{\sum_{i=1}^{N} \sum_{c=1}^{C} \left(\hat{p}_{ic} * \hat{w}_{ic}\right)}{N}. \tag{16}$$

3 Rudner-Based Indices

3.1 Fisher Information Under the MGRM

Within MIRT, test information is used to evaluate the measurement precision for the ability estimate. For example, the asymptotic variance of the MLE can be approximated by the inverse of test information (Wang 2015). A definition of item information (Ackerman 1994; Yao and Schwarz 2006) is the following:

$$I_j(\boldsymbol{\theta}) = -E\left[\frac{\partial^2 \log P_j(\boldsymbol{\theta})}{\partial\boldsymbol{\theta}\,\partial\boldsymbol{\theta}^T}\right], \tag{17}$$

where $P_j(\boldsymbol{\theta}) = \prod_{k=0}^{K_j} P_{jk}(\boldsymbol{\theta})^{1(y_j=k)}$. When it was applied to MGRM, the diagonal elements of $I_j(\boldsymbol{\theta})$ become

$$\begin{aligned}\left[I_j(\boldsymbol{\theta})\right]_{dd} &= \sum_{k=0}^{K_j} \frac{1}{P_{jk}(\boldsymbol{\theta})}\left[\frac{\partial P_{jk}(\boldsymbol{\theta})}{\partial\theta_d}\right]^2 \\ &= \sum_{k=0}^{K_j} \frac{\left[a_d P_{jk}^*(\boldsymbol{\theta}_i)\left(1-P_{jk}^*(\boldsymbol{\theta}_i)\right) - a_d P_{jk+1}^*(\boldsymbol{\theta}_i)\left(1-P_{jk+1}^*\right)\right]^2}{P_{jk}^*(\boldsymbol{\theta}_i) - P_{jk+1}^*(\boldsymbol{\theta}_i)},\end{aligned} \tag{18}$$

where $d = 1, 2, \ldots, D$. Note that this formula was originally shown by Chang (1996), Reckase (2009), and Samejima (1969) in the unidimensional case. The nondiagonal elements of $I_j(\boldsymbol{\theta})$ are

$$\left[I_j(\boldsymbol{\theta})\right]_{dd'} = \sum_{k=0}^{K_j} \frac{1}{P_{jk}(\boldsymbol{\theta})}\left(\frac{\partial P_{jk}(\boldsymbol{\theta})}{\partial\theta_d}\right)\left(\frac{\partial P_{jk}(\boldsymbol{\theta})}{\partial\theta_{d'}}\right), \tag{19}$$

where $d, d' = 1, 2, \ldots, D$ and $d \neq d'$. The diagonal and nondiagonal elements of $I_j(\boldsymbol{\theta})$ can be expressed by a unified formula as

$$I_j(\boldsymbol{\theta}) = \left\{\sum_{k=0}^{K_j} \frac{\left[P_{jk}^*(\boldsymbol{\theta}_i)\left(1-P_{jk}^*(\boldsymbol{\theta}_i)\right) - P_{jk+1}^*(\boldsymbol{\theta}_i)\left(1-P_{jk+1}^*\right)\right]^2}{P_{jk}^*(\boldsymbol{\theta}_i) - P_{jk+1}^*(\boldsymbol{\theta}_i)}\right\} \begin{bmatrix} a_{j1}^2 & a_{j1}a_{j2} & \vdots & a_{j1}a_{jD} \\ a_{j2}a_{j1} & a_{j2}^2 & \vdots & a_{j2}a_{jD} \\ \cdots & \cdots & \ddots & \cdots \\ a_{jD}a_{j1} & a_{jD}a_{j2} & \vdots & a_{jD}^2 \end{bmatrix}. \tag{20}$$

Since the Fisher information is additive (Chang 1996) for the local independent assumption, the test information is the sum of item information functions at point $\boldsymbol{\theta}$

$$I^{(J)}(\boldsymbol{\theta}) = \sum_{j=1}^{J} I_j(\boldsymbol{\theta}). \tag{21}$$

3.2 Rudner-Based Indices

In this section, the Rudner approach was extended to estimate the consistency and accuracy indices for MIRT. The computation of this approach is relatively easy by assuming that the estimated ability is distributed asymptotically according to a multivariate normal distribution $N(\boldsymbol{\theta}, \boldsymbol{\Sigma})$. Here, the asymptotic covariance matrix of the MLE can be approximated by the inverse of the test information matrix (Chang 1996; Chang and Stout 1993; Wang 2015).

Similar to the notation in the definition of the Guo-based index, the Rudner-based accuracy index, which is the expected probability of an examinee's true ability in any particular category, can be obtained using multivariate normal distribution as

$$\hat{p}_{ic} = \int_{R_c} \frac{1}{(2\pi)^{d/2}\left|I^{(J)}\left(\hat{\boldsymbol{\theta}}_i\right)\right|^{-1/2}} \exp\left[-\frac{1}{2}\left(\boldsymbol{\theta} - \hat{\boldsymbol{\theta}}_i\right)^T I^{(J)}\left(\hat{\boldsymbol{\theta}}_i\right)\left(\boldsymbol{\theta} - \hat{\boldsymbol{\theta}}_i\right)\right] d\boldsymbol{\theta}, \tag{22}$$

where $I^{(J)}\left(\hat{\boldsymbol{\theta}}_i\right)$ is the determinant of the test information matrix. The integrations involved in Eq. (22) can be approximated by using Monte Carlo simulation.

The Rudner-based consistency index is the expected probability of examinees who would be classified into the same category under independence administration, and it can be expressed as

$$\hat{\phi} = \frac{\sum_{i=1}^{N}\sum_{c=1}^{C}\left(\hat{p}_{ic} * \hat{p}_{ic}\right)}{N}. \tag{23}$$

By contrast, as discussed above, the entry $\hat{w}_{ic}$ is 1 if an examinee's ability estimate $\hat{\boldsymbol{\theta}}_i$ is classified into category c, and 0 otherwise. Then the classification accuracy index can be written as

$$\hat{\gamma} = \frac{\sum_{i=1}^{N}\sum_{c=1}^{C}\left(\hat{p}_{ic} * \hat{w}_{ic}\right)}{N}. \tag{24}$$

3.3 Theoretical Analysis of Lee-, Rudner-, and Guo-Based Indices

Next, we briefly analyze the above three indices. Firstly, all of three indices need to compute the values of large sums of expressions in Eq. (8) or complex integrals in Eqs. (14) and (22). Monte Carlo simulation provides a powerful method for generating random numbers to compute these values. The Rudner's method assumes that the ability estimate follows a multivariate normal distribution. Because the method of generating multivariate normal random vectors in estimating Rudner-based index is usually much easier than Metropolis-Hastings and Gibbs sampling in estimating the Guo- and Lee-based indices. Thus, the Rudner's method is expected to perform best. Secondly, both the Guo- and Lee-based accuracy indices are expected to yield similar performance, because we have proved that they have the same theoretical value under certain conditions in the following theorem.

Theorem 1 *If the prior distribution of ability is non-informative and the decision rule can establish the unique correspondence between the set of ability space* $\Theta = \{\Theta_c, c = 1, 2, \ldots, C\}$ *and the set of true scores* $T = \{T_c, c = 1, 2, \ldots, C\}$, *then both the Guo- and Lee-based accuracy indices have the same theoretical value.*

Proof Let $g(\boldsymbol{\theta})$ be a true distribution of ability. Let γ_{Lee} and $\gamma_{\boldsymbol{\theta}}(c)$ be the marginal and conditional classification accuracy index which were given by Lee (2010). The following equations are derived:

$$\begin{aligned}\gamma_{Lee} &= \int_{\boldsymbol{\theta}\in\Theta} \gamma(\boldsymbol{\theta})\, g(\boldsymbol{\theta})\, d\boldsymbol{\theta} = \sum_{c=1}^{C} \int_{\boldsymbol{\theta}\in\Theta_c} \gamma_{\boldsymbol{\theta}}(c) g(\boldsymbol{\theta})\, d\boldsymbol{\theta} = \sum_{c=1}^{C} P(x \in T_c, \boldsymbol{\theta} \in \Theta_c) \\ &= \sum_{c=1}^{C} \sum_{\mathbf{y}_i:\sum_j y_{ij}\in T_c} P(Y=\mathbf{y}_i, \boldsymbol{\theta} \in \Theta_c) = \sum_{c=1}^{C} \sum_{\mathbf{y}_i:\sum_j y_{ij}\in T_c} P(Y=\mathbf{y}_i)\, p_i(\Theta_c),\end{aligned} \tag{25}$$

where the third equation is satisfied because the events $x \in T_c$ and $\boldsymbol{\theta} \in \Theta_c$ imply the correct classification; the last equation is satisfied because a uniform prior distribution of ability is employed, and the factor of $p_i(\Theta_c)$ in the last term is defined by Eq. (14).

4 Simulation Study

Given that the classification consistency and accuracy indices based on the Rudner approach are new to multidimensional latent ability, an important question is whether the Rudner-based indices can accurately estimate their true values. The true accuracy indices were computed as the method proposed by Lathrop and Cheng (2013). The true/simulated accuracy was the proportion of examinees whose $\hat{\boldsymbol{\theta}}$ or x was classified in the same category as their simulated $\boldsymbol{\theta}$. Similarly, the true/simulated consistency was the proportion of examinees whose $\hat{\boldsymbol{\theta}}$ or x on two parallel tests were classified in the same category. This was also called a test-retest consistency rate.

4.1 Simulation Design

A simulation study following the MGRM was conducted. The dimensions were initialized to 1, 2, and 4, respectively. In a two- or four-factor model, three levels of correlation between pairs of dimensions, $\rho = 0.00$, $\rho = 0.50$, and $\rho = 0.80$ were considered. The sample consisted of either $N = 1000$ or $N = 3000$ examinees. The sample size of $N = 1000$ was chosen as the lower bound (Yao and Boughton 2007). The ability vectors were generated from multivariate normal distributions with an appropriately sized mean vector of 0 and covariance matrix Σ, where the diagonal elements of Σ were all 1 and the off-diagonal elements were given by the correlation for the associated condition.

Test length for the one-, two, and four-factor model could be either 10 or 20, either 15 or 30, and either 30 or 60. In order to balance the information of the domains or dimensions (Yao 2012, 2014), the number of items for each dimension (Kroehne et al. 2014; Yao 2012, 2014) was constrained. For example, in the two-factor model, the constraints for a 15-item test are such that two five-item sets each loaded exclusively on one of the two dimensions and the remaining five items loaded on both of the two dimensions. There are ten items measuring each dimension. The above simulation conditions have been often used in the literature (Wang 2015; Wang and Nydick 2015; Yao and Boughton 2007).

The fully crossed design yielded 14 conditions for each sample size, where each condition was replicated ten times to estimate an averaged simulated consistency. Item parameters were fixed across all replications. They were originally described and used by Cai (2010) (Table 1 in Sect. 4.4) with two dimensions and ten three-category items. Considering the above constraints, these item parameters were used to construct six tests. For example, the slope parameters for the first and third or second and fourth dimensions are between 1.6 and 2.6 or 1.1 and 2.6.

The three decision rules were shown in Table 1. Let's take decision rule A as an example: when a test had ten items and each item was scored against three ordered categories, the two cut scores of 10 and 16 were used to classify examinees into three categories. Note that Monte Carlo method can be used to tackle intractable summations or high-dimensional integrals. Therefore, Monte Carlo method with Monte Carlo sample size of 3000 was employed to estimate Eqs. (8), (14) and (22), based on one of our previous studies (Wang et al. 2016).

It should be noted that for decision rules A and B, we can compute an expected summed score by using Eq. 11 given an examinee's ability estimate. Then we know under which category of the examinee should be classified. But for decision rules

Table 1 Three decision rules

Decision rule	Scale of decision rule	Cut scores
A	Total score scale	50% and 80% of perfect score
B	The θ_k total score scale	50% and 80% of perfect score
C	An equally weighted composite of the θ_s	0 and 0.75 of composite score

C, we often cannot establish a one-to-one mapping between total scores and the equally weighted composites of latent ability because of test items with different slope parameters.

4.2 Results

Due to similar results, we only presented the results with the sample size of 3000. Tables 2, 3, and 4 display the simulated and estimated classification accuracy for the three decision rules under 14 conditions for $N = 3000$ and Monte Carlo sample size = 3000. The results suggest that:

(a) The Rudner-based indices worked well because their values matched closely with the simulated accuracy rates.
(b) The difference between the three simulates and estimates tended to increase when the number of dimensions increased.
(c) The difference between the three simulates tended to become trivial when the test length increased.
(d) For decision rule A or B, the difference between the Lee- and Guo-based accuracy indices was very small.
(e) For decision rule B, different methods provided similar magnitude of accuracy when the test length was the same, regardless of the number of dimensions.
(f) For decision rule C, the Guo- and Rudner-based indices had similar performance across various conditions.

If we look at Table 3, for one dimension, the simulated accuracy of total score was found be smaller than the simulated accuracy of estimated ability, which was consistent with the previous research (Lathrop and Cheng 2013). For four

Table 2 Simulated and estimated classification accuracy for decision rule A

			Simulated			Estimated		
Dimension	Correlation	Test length	Guo	Lee	Rudner	Guo	Lee	Rudner
1	NA	10	0.820	0.813	0.820	0.833	0.825	0.821
		20	0.867	0.862	0.867	0.877	0.871	0.874
2	0.0	15	0.849	0.844	0.849	0.859	0.855	0.843
		30	0.870	0.868	0.870	0.900	0.896	0.896
	0.5	15	0.860	0.857	0.860	0.875	0.868	0.849
		30	0.866	0.866	0.866	0.907	0.902	0.899
	0.8	15	0.864	0.863	0.864	0.880	0.872	0.849
		30	0.864	0.861	0.864	0.913	0.908	0.902
4	0.0	30	0.867	0.870	0.867	0.873	0.876	0.868
		60	0.876	0.881	0.876	0.906	0.914	0.915
	0.5	30	0.859	0.865	0.859	0.895	0.895	0.883
		60	0.873	0.877	0.873	0.924	0.928	0.926
	0.8	30	0.864	0.871	0.864	0.910	0.908	0.886
		60	0.887	0.887	0.887	0.932	0.933	0.929

Table 3 Simulated and estimated classification accuracy of one dimension for decision rule B

			Simulated			Estimated		
Dimension	Correlation	Test length	Guo	Lee	Rudner	Guo	Lee	Rudner
2	0.0	15	0.827	0.828	0.827	0.841	0.840	0.825
		30	0.862	0.862	0.862	0.885	0.885	0.882
	0.5	15	0.837	0.837	0.837	0.856	0.852	0.830
		30	0.858	0.857	0.858	0.892	0.890	0.885
	0.8	15	0.841	0.841	0.841	0.860	0.856	0.832
		30	0.854	0.853	0.854	0.900	0.897	0.890
4	0.0	30	0.817	0.820	0.817	0.826	0.839	0.827
		60	0.850	0.851	0.850	0.863	0.879	0.878
	0.5	30	0.823	0.832	0.823	0.844	0.851	0.835
		60	0.847	0.852	0.847	0.882	0.890	0.887
	0.8	30	0.828	0.836	0.828	0.856	0.859	0.837
		60	0.856	0.859	0.856	0.890	0.894	0.890

Table 4 Simulated and estimated classification accuracy for decision rule C

			Simulated		Estimated	
Dimension	Correlation	Test length	Guo	Rudner	Guo	Rudner
1	NA	10	0.811	0.811	0.822	0.805
		20	0.859	0.859	0.869	0.863
2	0.0	15	0.835	0.835	0.846	0.827
		30	0.863	0.863	0.888	0.885
	0.5	15	0.854	0.854	0.871	0.845
		30	0.874	0.874	0.905	0.898
	0.8	15	0.864	0.864	0.878	0.852
		30	0.873	0.873	0.914	0.905
4	0.0	30	0.848	0.848	0.840	0.837
		60	0.855	0.855	0.885	0.894
	0.5	30	0.863	0.863	0.883	0.875
		60	0.874	0.874	0.914	0.919
	0.8	30	0.877	0.877	0.907	0.895
		60	0.896	0.896	0.929	0.930

dimensions, however, the simulated accuracy of total score was relatively larger than the simulated accuracy of estimated ability.

For clearly indicating the difference between the simulated and estimated accuracy, Table 5 provides summaries of the bias, absolute error (ABS), and root mean square error (RMSE) of the classification accuracy estimates. The results indicate that among the three methods, the Rudner's method typically had the lowest bias, the lowest ABS, and the lowest RMSE across all conditions, while the Guo's and Lee's methods were both fairly comparable. In addition, all of the three methods overestimated the simulated classification accuracy.

Table 5 Error of estimation for three accuracy indices

Decision rule	Dimension	Indices	BIAS	ABS	RMSE
Decision rule A	ALL	Guo	0.0282	0.0282	0.0323
		Lee	0.0262	0.0262	0.0304
		Rudner	**0.0181**	**0.0227**	**0.0277**
Decision rule B	Dimension 1	Guo	0.0228	0.0228	0.0251
		Lee	0.0231	0.0231	0.0253
		Rudner	**0.0147**	**0.0174**	**0.0216**
	Dimension 2	Guo	0.0239	0.0239	0.0253
		Lee	0.0225	0.0225	0.0246
		Rudner	**0.0132**	**0.0188**	**0.0218**
	Dimension 3	Guo	0.0235	0.0235	0.0280
		Lee	0.0280	0.0280	0.0303
		Rudner	**0.0231**	**0.0231**	**0.0274**
	Dimension 4	Guo	0.0268	0.0268	0.0301
		Lee	0.0296	0.0296	0.0313
		Rudner	**0.0241**	**0.0241**	**0.0278**
Decision rule C	ALL	Guo	0.0216	0.0228	0.0253
		Rudner	**0.0131**	**0.0199**	**0.0236**

Note. The lowest BIAS, ABS, RMSE in each condition are in boldface type. Dimension 1 belongs to one-, two-, and four-factor model; Dimension 2 belongs to two- and four-factor model; Dimensions 3 and 4 belong to four-factor model

For the simulated and estimated consistency of the three decision rules, the results (were not shown here) suggest that:

(a) The Lee-based index was better than the Rudner- and Guo-based indices because the Guo-based indices often exceeded the simulated indices, while the Rudner-based indices did not.
(b) The difference between the three simulates and estimates tended to become trivial when the test length increased.
(c) For decision rule C, the Guo- and Rudner-based indices had similar performance across various conditions.

5 Discussions

Based on previous studies (Grima and Yao 2011; Guo 2006; Lathrop and Cheng 2013; Rudner 2001, 2005; Wyse and Hao 2012; Yao 2016), the Rudner-based consistency and accuracy indices have been adapted for MIRT in this paper, and their performance was evaluated under the MGRM through the simulation study. The simulation results show that:

(a) The Rudner-based indices worked very well because their values matched closely with the test-retest consistency rates or the true accuracy rates.

(b) For the decision rule based on total scores, the difference between the Lee- and Guo-based accuracy indices was very small and had a very similar trend as test length increased, which completely conformed to the theoretical results.
(c) The findings show somewhat difference between the three indices, but the difference tended to be small with increasing test length.
(d) The Lee-based indices could not be applicable for a decision rule based on a composite latent ability score when a one-to-one mapping cannot be established between total scores and the decision rule, but the flexible Guo- and Rudner-based indices can be used in this case and tended to perform similarly.

Several directions are described based on the current research. First, it is worthy to study how to choose a better decision rule for making more valid decision closely related to improve teaching and learning. Second, they might be useful for developing an item selection algorithm in adaptive tests since item selection is the most important procedure in adaptive testing (Chang and Wang 2016). Third, the construction of their confidence intervals needed further investigation in future. Fourth, they should be applied to many other MIRT models. Finally, we should consider making estimate of consistency and accuracy based on the Rudner- and Guo- based indices where the asymptotic posterior covariance matrix of ability estimate was obtained by Bayesian method (Wang 2015). In addition, as a remark, if the slope parameters were very diverse, the total score method might perform poorly. In this study, the slope parameters for one dimension are between 1.6 and 2.6 or 1.1 and 2.6, which were not very diverse. Replicating this study with more diverse slope parameters would be an important area for future research.

Acknowledgments This research is supported by the National Natural Science Foundation of China (Grant Nos. 31500909, 31360237, and 31160203), the Key Project of National Education Science "Twelfth Five Year Plan" of Ministry of Education of China (Grant No. DHA150285), the National Social Science Fund of China (Grant No. 16BYY096), the Humanities and Social Sciences Research Foundation of Ministry of Education of China (Grant Nos. 13YJC880060 and 12YJA740057), the National Natural Science Foundation of Jiangxi Province (Grant No. 20161BAB212044), Jiangxi Education Science Foundation (Grant No. 13YB032), the Science and Technology Research Foundation of Education Department of Jiangxi Province (Grant No. GJJ13207), the China Scholarship Council (CSC No. 201509470001), and the Youth Growth Fund and the Doctoral Starting up Foundation of Jiangxi Normal University. The authors would like to thank Prof. Hua-Hua Chang for his kind support and Prof. Wen-Chung Wang for his valuable comments.

References

T.A. Ackerman, Full-information factor analysis for polytomous item responses. Appl. Psychol. Meas. **18**(3), 257–275 (1994)

R.J. Adams, M. Wilson, W.-C. Wang, The multidimensional random coefficients multinomial logit model. Appl. Psychol. Meas. **21**(1), 1–23 (1997)

D.M. Bolt, V.F. Lall, Estimation of compensatory and noncompensatory multidimensional item response models using Markov chain Monte Carlo. Appl. Psychol. Meas. **27**(6), 395–414 (2003)

L. Cai, High-dimensional exploratory item factor analysis by a metropolis–Hastings Robbins–Monro algorithm. Psychometrika **75**(1), 33–57 (2010)

L. Cai, D. Thissen, S.H.C. du Toit, *IRTPRO: Flexible, Multidimensional, Multiple Categorical IRT Modeling [Computer software]* (Scientific Software International, Lincolnwood, IL, 2011)

R.P. Chalmers, Mirt: a multidimensional item response theory package for the R environment. J. Stat. Softw. **48**(6), 1–29 (2012)

H.-H. Chang, The asymptotic posterior normality of the latent trait for polytomous IRT models. Psychometrika **61**(3), 445–463 (1996)

H.-H. Chang, W. Stout, The asymptotic posterior normality of the latent trait in an IRT model. Psychometrika **58**(1), 37–52 (1993)

H.-H. Chang, W. Wang, Internet plus measurement and evaluation: a new way for adaptive learning. J. Jiangxi Norm. Univ. (Nat. Sci.) **40**(5), 441–455 (2016)

D. Debeer, J. Buchholz, J. Hartig, R. Janssen, Student, school, and country differences in sustained test-taking effort in the 2009 PISA reading assessment. J. Educ. Behav. Stat. **39**(6), 502–523 (2014)

K.M. Douglas, R.J. Mislevy, Estimating classification accuracy for complex decision rules based on multiple scores. J. Educ. Behav. Stat. **35**(3), 280–306 (2010)

A. Grima, L. H. Yao, in Classification consistency and accuracy for test of mixed item types: unidimensional versus multidimensional IRT procedures. Paper presented at the annual meeting of National Council on Measurement in Education, New Orleans, LA, (2011)

F. Guo, Expected classification accuracy using the latent distribution. Pract. Assessm. Res. Eval. **11**(6), 1–6 (2006)

H. Huynh, Computation and statistical inference for decision consistency indexes based on the Rasch model. J. Educ. Stat. **15**(4), 353–368 (1990)

U. Kroehne, F. Goldhammer, I. Partchev, Constrained multidimensional adaptive testing without intermixing items from different dimensions. Psychol. Test Assess. Modeling **56**(4), 348–367 (2014)

L.J. LaFond, *Decision Consistency and Accuracy Indices for the Bifactor and Testlet Response Theory Models Detecting Heterogeneity in Logistic Regression Models* (University of Iowa, Iowa City, IA, 2014.) (Unpublished doctoral dissertation)

Q.N. Lathrop, Y. Cheng, Two approaches to estimation of classification accuracy rate under item response theory. Appl. Psychol. Meas. **37**(3), 226–241 (2013)

W.-C. Lee, Classification consistency and accuracy for complex assessments using item response theory. J. Educ. Meas. **47**(1), 1–17 (2010)

G. Makransky, E.L. Mortensen, C.A.W. Glas, Improving personality facet scores with multidimensional computer adaptive testing: an illustration with the Neo Pi-R. Assessment **20**(1), 3–13 (2012)

M.D. Reckase, *Multidimensional Item Response Theory* (Springer, New York, NY, 2009)

F. Rijmen, M. Jeon, M. von Davier, S. Rabe-Hesketh, A third-order item response theory model for modeling the effects of domains and subdomains in large-scale educational assessment surveys. J. Educ. Behav. Stat. **39**(4), 235–256 (2014)

L.M. Rudner, Computing the expected proportions of misclassified examinees. Pract. Assess. Res. Eval. **7**(14), 1–8 (2001)

L.M. Rudner, Expected classification accuracy. Pract. Assess. Res. Eval. **10**(13), 1–4 (2005)

F. Samejima, *Estimation of Latent Ability Using a Response Pattern of Graded Scores (Psychometric Monograph No. 17)* (Psychometric Society, Richmond, VA, 1969)

E.M. Schulz, M.J. Kolen, W.A. Nicewander, A rationale for defining achievement levels using IRT-estimated domain scores. Appl. Psychol. Meas. **23**(4), 347–362 (1999)

C. Wang, On latent trait estimation in multidimensional compensatory item response models. Psychometrika **80**(2), 428–449 (2015)

C. Wang, S. Nydick, Comparing two algorithms for calibrating the restricted non-compensatory multidimensional IRT model. Appl. Psychol. Meas. **39**(2), 119–134 (2015)

T. Wang, M.J. Kolen, D.J. Harris, Psychometric properties of scale scores and performance levels for performance assessments using polytomous IRT. J. Educ. Meas. **37**(2), 141–162 (2000)

W. Wang, L. Song, S. Ding, Y. Meng, in *Quantitative Psychology Research: The 80th Annual Meeting of the Psychometric Society*, ed. by L. A. van der Ark, D. M. Bolt, W.-C. Wang, J. A. Douglas, M. Wiberg. Estimating classification accuracy and consistency indices for multidimensional latent ability (Springer International Publishing, Cham, 2016), pp. 89–103

A.E. Wyse, S. Hao, An evaluation of item response theory classification accuracy and consistency indices. Appl. Psychol. Meas. **36**(7), 602–624 (2012)

L. Yao, *BMIRT: Bayesian Multivariate Item Response Theory [Computer Software]* (CTB/McGraw-Hill, Monterey, 2003)

L. Yao, Multidimensional CAT item selection methods for domain scores and composite scores: theory and applications. Psychometrika **77**(3), 495–523 (2012)

L. Yao, Multidimensional CAT item selection methods for domain scores and composite scores with item exposure control and content constraints. J. Educ. Meas. **51**(1), 18–38 (2014)

L. Yao, The BMIRT toolkit. Retrieved March 1, 2016., from http://www.bmirt.com/media/f5abb5352d553d5fffff807cffffd524.pdf

L. Yao, K.A. Boughton, A multidimensional item response modeling approach for improving subscale proficiency estimation and classification. Appl. Psychol. Meas. **31**(2), 1–23 (2007)

L. Yao, R.D. Schwarz, A multidimensional partial credit model with associated item and test statistics: an application to mixed-format tests. Appl. Psychol. Meas. **30**(6), 469–492 (2006)

J. Zhang, Calibration of response data using MIRT models with simple and mixed structures. Appl. Psychol. Meas. **36**(5), 375–398 (2012)

Supporting Diagnostic Inferences Using Significance Tests for Subtest Scores

William Lorié

Abstract Users of content-heterogeneous assessments based on unidimensional trait models often request information about examinee strengths and weaknesses in specific subareas. This is commonly called diagnostic information, and a standard way of providing it is by computing and reporting subscores. However, in the many cases where subscores fail to provide reliable information sufficiently independent of the total score, they cannot support claims about subarea strengths and weaknesses relative to total score expectations. These kinds of claims are referred to here as *diagnostic inferences*. This paper introduces a method to support diagnostic inferences for assessment programs developed and maintained using item response theory (IRT). The method establishes null and alternative hypotheses for the number correct on subsets of items or subtests. Statistical significance testing is then conducted to determine the strength of the statistical evidence in favor of a diagnostic inference. If the subtest score is modeled as a Poisson binomial distribution with probabilities set to those expected by the IRT model conditional on fixed item parameters and person scores, then a determination can be made, by individual or groups, whether and which diagnostic inferences are supported. This paper presents results of power computations showing the subtest lengths generally required for supporting diagnostic inferences under different conditions and effect sizes.

Keywords Item response theory • Subscores • Diagnostic inferences • Statistical power • Poisson binomial

1 Introduction

Users of assessments based on unidimensional trait models often call for more information, commonly called diagnostic information, about student strengths and weaknesses in specific subareas. Several researchers have questioned a standard response to this call—reporting subscores—arguing that subscores seldom carry

W. Lorié (✉)
Capital Metrics, LLC, Washington, DC 20037, USA
e-mail: williamlorie@alumni.stanford.edu

L.A. van der Ark et al. (eds.), *Quantitative Psychology*, Springer Proceedings in Mathematics & Statistics 196, DOI 10.1007/978-3-319-56294-0_6

reliable information sufficiently independent of the total score to support justifiable claims about strengths and weaknesses. A *diagnostic inference* is defined in this paper as a claim regarding a student's performance in a subarea, after conditioning on what is known and reported about his or her overall performance.

This paper (1) presents a conceptual framework for supporting diagnostic inferences; (2) shows how testing for diagnostic inferences can be conducted using subtest responses, regardless of the combination of items a student has taken; and (3) reports the numbers of items required to detect diagnostic information for different combinations of overall score and strength (or weakness) effect magnitude.

2 Diagnostic Information and Subscores

Diagnostic information refers to claims about student knowledge, skill, or ability with respect to subareas of tests assessing heterogeneous content. Such tests are typically scaled using unidimensional models when tests for dimensionality (e.g., the DIMTEST procedure, Nandakumar and Stout 1993; Stout 1987; Stout et al. 2001; Stout et al. 1992) reveal that the collection of items is essentially unidimensional. However, even when tests are sufficiently unidimensional to report one total score, test users often request information supporting diagnostic inferences, and testing programs report subscores to satisfy this need. These scores are subject to the same technical requirements of all scores, such as having adequate reliability and sufficient distinctiveness from other scores, including each other (Standards 1.13–1.15, American Educational Research Association et al. 2014).

Several researchers (Haberman 2008; Monaghan 2006; Sinharay 2010), however, have noted a lack of evidence of adequate reliability or distinctiveness for subscores on many tests. Sinharay (2010), for example, showed that out of 25 testing programs reporting between two and seven subscores, only nine had at least one subscore providing added value above and beyond the total score. Sinharay (2010) notes that subscores "have to consist of at least 20 items and have to be sufficiently distinct from each other to have any hope of adding value" (p.169), placing a great price—in terms of testing cost and time—on subscore-based diagnostic information.

3 A Conceptual Framework for Supporting Diagnostic Inferences

This paper presents a hypothesis testing approach to investigating and supporting diagnostic inferences on tests designed and built using item response theory (IRT). For this method, item and person parameters set the conditions for a null hypothesis against which one may test alternative hypotheses about an individual's performance on items in a subarea of the tested domain.

The approach is based on the idea that, per the theory, a test furnishes all the information it is designed to provide via the total score, but that when a new individual or group takes the test, their performance in subareas may deviate from what is expected by theory, and occasionally such deviations may be sufficiently extreme to justify a claim about relative strength or weakness in one or more subareas.

Typically, there is a moderate-to-strong positive correlation between overall performance and performance in a subarea because the two measures share items and because the skills they measure are conceptually related. Thus, reporting on performance in a subarea can appear to add little information beyond what is conveyed in the total score. But, subareas vary in difficulty and can cover instructionally distinct content, and to that extent, diagnostic inferences might provide meaningful, unique information to test score users. In addition, some students may follow patterns of content coverage, exhibit study habits, or otherwise manifest strengths and weaknesses that deviate from the norm under which a unidimensional scale is constructed and applied to all. The perspective of this paper is that a subtest score or subscore conveys new information for a student if that subscore deviates sufficiently from what would have been predicted by the student's reported total score.

3.1 Hypothesis Testing at the Individual Level

This paper adopts a hypothesis testing approach at the level of the individual to assess the evidence for a relative strength or weakness in a subarea. The null hypothesis is that a person's test performance in a subarea is as expected, based on their total score. Several alternative hypotheses are possible for a specific person *j* and subarea *s*:

1*js*. The performance of *j* on *s* is better than expected by chance.
2*js*. The performance of *j* on *s* is worse than expected by chance.
3*js*. The performance of *j* on *s* is outside of the range expected by chance.

Rendering a diagnostic inference that person *j* has a relative strength in subarea *s* is equivalent to formally testing for and finding statistically significant evidence in support of alternative hypothesis (1*js*). Likewise, a diagnostic inference that *j* has a relative weakness in *s* is supported by evidence against the null and for hypothesis (2*js*). The first two alternatives correspond to one-tailed tests in the same way that hypothesis (3*js*) corresponds to a two-tailed test.

Including other individuals and/or other subareas extends the set of alternative hypotheses about subarea performance. Assume all individuals belong to group *g*. One important set of alternative hypotheses relates to the group level:

1_s. The performance of *g* on *s* is better than expected by chance.
2_s. The performance of *g* on *s* is worse than expected by chance.

3_s. The performance of g on s is outside of the range expected by chance.

These alternatives will be considered in connection to results of power computations presented later.

Two analogues to alternative hypotheses for the ANOVA omnibus tests across treatment conditions, formulated here for individuals and groups, are noteworthy:

3_j. The performance of j on at least one subarea is outside of the range expected by chance.

3. The performance of g on at least one subarea is outside of the range expected by chance.

The well-known multiple comparisons problem applies to these alternative hypotheses, when investigators finding a positive result wish to identify the specific subarea(s) for which an individual or group is exhibiting atypical performance, and the direction of the effect—i.e., whether it is a relative strength or a relative weakness. Controlling the type I error rate (usually denoted α) for (3*j*) and (3) is relevant when there are many subareas. If a hypothesis test is run for each subarea individually—i.e., if a test for alternative hypothesis (3*js*) is run on each subarea *s*—the likelihood of one being flagged in the absence of any real underlying strength or weakness is the type I error rate for hypothesis (3*js*). Thus, it is important to interpret results of diagnostic inferences considering precisely which alternatives were hypothesized.

The central question for diagnostic inferences from the perspective of a student or educator is whether that student is relatively strong or weak in each area, or if their performance is otherwise typical. Accordingly, our focus here is on alternative hypotheses (1*js*)–(3*js*).

3.2 Formulating the Null and Alternative Hypotheses

Unidimensional dichotomous IRT models and their multidimensional IRT (MIRT) extensions assume that responses to items are independent after conditioning on ability, and thus, the count of correct responses to a subset of dichotomous items (a subtest) follows a Poisson binomial distribution with parameters given by the probabilities of each item's being correctly responded to by a person at a given level of the latent trait. This property has also been demonstrated in González et al. (2016).

More specifically, the counts of correct responses n_{si} by a person i in a subarea s with J total dichotomous items, indexed by j, follow a Poisson binomial distribution, with the trial probabilities $p_{ij} = (p_{i1}, \ldots, p_{iJ})$ given by the theoretical probabilities of correct responses on the items, conditional on person parameters:

$$n_{si} \sim PBD\left(p_{i1}, \ldots, p_{iJ}\right),$$

where

$$p_{ij} = P\left(X_{ij} = 1|\theta_i, \gamma_j\right), \quad (1)$$

and PBD denotes the Poisson binomial distribution, X_{ij} is zero (if the response of person i to item j is incorrect) or 1 (if correct), θ_i is the (set of) latent trait parameter(s) for person i, and θ_j is the (set of) item parameter(s) for item j.

In practice, the parameters for items are usually estimated from field-testing and treated as known when estimating person parameters. The probabilities are then generated from the measurement model in Eq. (1), with estimates in place of the parameters.

The paradigm of statistical hypothesis testing can thus be brought to bear on the task of supporting diagnostic inferences. Formally, one can test the hypothesis that the count of correct responses on a subtest is greater (or less) than expected, given a significance level α. This furnishes direct evidence for relative strength or weakness in that subarea.

4 Testing for Diagnostic Inferences in Practice

This section and the next address the power of the Poisson binomial test for providing evidence of relative strength or weakness in a subarea. Testing for diagnostic inferences can be practically conducted in large-scale assessment settings, as illustrated here for a statewide testing program.

4.1 Data Sparseness and Possible Diagnostic Inferences

A statewide grade 6 mathematics computerized adaptive testing (CAT) program administers 45 items per student in four broad reporting categories, which are further subdivided into 36 learning objectives or subareas. Each item is classified into one of these subareas.

When a student takes the test, each item is drawn from a large pool of operational items, per the program's rules for meeting the test blueprint during a test session, algorithms for minimizing error in estimating the student's total score, and other constraints. Thus, and typical of a large-scale CAT, any given administration of the assessment yields a very sparse student-by-item matrix, with approximately 95% of the data "missing," by design.

This sparseness is not considered a problem in operational testing; in fact, it is a sought-after benefit because it allows for the control of item exposure and increases a testing program's longevity. However, it complicates psychometric reanalysis of operational data because less information is available to estimate reliability, (re)calibrate items, estimate person parameters, etc. It is more difficult to investigate

Table 1 Possible diagnostic inferences for different statistical power profiles

Power profile		Possible diagnostic inferences				
Power to detect relative strength?	Power to detect relative weakness?	An area of relative strength	An area of relative weakness	Not an area of relative strength	Not an area of relative weakness	Neither an area of relative strength nor weakness
No	No	No	No	No	No	No
Yes	No	Yes	No	Yes	No	No
No	Yes	No	Yes	No	Yes	No
Yes	Yes	Yes	Yes	No	No	Yes

whether any new, proposed subscores add value above and beyond what is furnished by the total score, because subscores with identical intended interpretations would have to be calculated from different sets of items.

These complications arising from data sparseness are not an issue for hypothesis testing to support diagnostic inferences, because the question of whether an individual took sufficient items to support a diagnostic inference can be answered person by person. By computing the conditional probabilities of the most extreme responses (all incorrect or all correct) and comparing these to the threshold of statistical significance (α), researchers can determine whether there is sufficient power to detect an effect. For each student and subarea, if there is power to detect a relative strength of any magnitude, then a statistical test is conducted with the alternative ($1js$). If there is power to detect a relative strength, then a test is conducted to evaluate alternative ($2js$) against the null. If there is sufficient power for both tests, then failure to reject the null in both cases is evidence for alternative ($3js$).

Whether there is sufficient power to detect an effect for each alternative hypothesis determines the diagnostic inferences possible for that subarea and student (Table 1).

Reports can state the status of diagnostic inference testing and results for each student and subarea. The set of possible flags should include, at minimum, two elements: “an area of relative strength” and “an area of relative weakness.” There is an important distinction to be made (and communicated to report audiences) between a subarea where a student exhibits typical performance and one where there simply is no basis for making a claim about relative strength, relative weakness, or lack thereof (the first row of Table 1). Moreover, although the distinction can be difficult to communicate, performance that is “typical” because tests of relative strength and weakness failed to prove otherwise—in contrast to performance in a subarea for which there were few items to make such a claim—is a meaningful one that can also be conveyed to report audiences.

4.2 *Computation of Statistical Power Profiles and Diagnostic Inferences*

To show the feasibility of practical application of these hypothesis tests for diagnostic inferences, results and computation time are reported for assessing power to detect and generating detection results on 36 subareas for approximately[1] 65,000 students taking the assessment in 2015.

Second, testing for the group-level alternatives (1_s)–(3_s) by grouping students into schools enhances power considerably. Fortunately, as will be shown, more extensive Poisson binomial calculations did not significantly affect computation time.

4.2.1 Many Subareas and Students, Individual Level

To generate the results needed to produce diagnostic inferences for the above-referenced program, the author wrote *R* code (version 3.2.3, R Core Team 2015) to read student response and item parameter data and compute power-to-detect and statistical significance flag results for all students and subareas. Poisson binomial probability masses were computed with the R package "poibin" (Hong 2013b).

For any given student, there was sufficient power to detect a relative weakness for very few of the subareas (on average 5.2 subareas out of 36), since students take very few items in any given subarea. Summarizing the student-by-subarea item exposure count matrix reveals that the most likely number of exposures to any subarea for any student is just one (56% of the elements). The next most likely value is two (30%), followed by zero (11%) and three (3%). In just 16 cases was a student assessed on a subarea four times. With such low item exposures, it is very difficult to reach the power-to-detect threshold.

Despite such low by-standard item exposures, there was sufficient power for testing for relative weakness on at least one subarea for over 80% of the tested group. Thus, if their subtest scores were low enough to merit the inference, it was possible to report at least one positive diagnostic inference ("an area of relative weakness") for these students. A negative diagnostic inference ("not an area of relative weakness") could also be reported for a student on any subarea for which there was sufficient power to test for relative weakness, but where no relative weakness was found (formally, no rejection of the null). At least one positive diagnostic inference of relative weakness was flagged for 17% of those students for whom there was sufficient power to test for relative weakness on at least one subarea.

[1]Approximate numbers of students and schools are reported here to preserve anonymity of the data source.

Power for detecting a relative strength was lower than that for detecting a relative weakness, because the conditional probabilities for correct responses for the tested population tended to be higher than 0.5. A MacBook Pro with a 2.5 GHz Intel Core i7 processor and 8 GB in two 1333 MHz DDR3 memory modules, running *R* 3.2.3 GUI 1.66 Mavericks build (7060) (R Core Team 2015), took 306 s to complete the calculations.

4.2.2 Many Subareas and Students, Group Level

The analysis was repeated, this time after grouping students into schools (*N* approximately 650). Each school in the data file contained between 1 and 424 records of students taking the test, with a mean of 99.8. About 5% of schools had five or fewer records.

For any given school, there was sufficient power to detect a relative weakness for almost every subarea (on average 34.8 subareas out of 36). There was sufficient power to detect a relative strength for a by-school average of 32.9 of the subareas.

Each subarea was flagged for relative weakness in fewer than ten schools to over half of the schools, depending on the subarea. Subareas were flagged for relative strength in less than ten schools to about 70% of the schools, again depending on the subarea. The above-referenced computer took 366 s to complete the calculations.

The next section incorporates effect magnitude, as well as other considerations, in estimating a rule of thumb for the numbers of items needed per subarea to detect a relative strength or weakness with a lower bound on its magnitude.

5 Subtest Length Requirements for Diagnostic Inferences

Simulation-based research was conducted to identify the subtest length required for supporting diagnostic inferences. Statistical power was estimated as a function of θ, θ_c (the student's true ability on the class of items in the subtest), and the number of items in the subtest. Three values of θ were chosen (0, ± 1 SD units), and six different effects, or deviations of θ_c from θ (± 1, ± 2, ± 5 SD units). The three effect magnitudes are termed "moderate," "large," and "very large" solely for this study. Where effects are positive, power was estimated for hypothesis tests concerning relative strength. Where they are negative, power was estimated for hypothesis tests concerning relative weakness.

The largest effect magnitudes were chosen to approximate cases in which a student's responses to the subtest are best characterized under a mastery framework, in which correct responses for true non-masters in the subarea are "lucky guesses" and incorrect responses from true masters are "careless slips." As such, the "very large" effect size should be such that it does not depend on θ, and this is better obtained with a deviation of ± 5 SD units than with deviations of ± 3 or ± 4 SD units.

Table 2 Experimental conditions classified by location of the null and effect size

	Effect size		
Location of null	Moderate	Large	Very large
Same side of mean as effect direction	$\theta = 1$ and $\theta_c = 2$; or $\theta = -1$ and $\theta_c = -2$	$\theta = 1$ and $\theta_c = 3$; or $\theta = -1$ and $\theta_c = -3$	$\theta = 1$ and $\theta_c = 6$; or $\theta = -1$ and $\theta_c = -6$
Mean	$\theta = 0$ and $\theta_c = \pm 1$	$\theta = 0$ and $\theta_c = \pm 2$	$\theta = 0$ and $\theta_c = \pm 5$
Opposite side of mean as effect direction	$\theta = 1$ and $\theta_c = 0$; or $\theta = -1$ and $\theta_c = 0$	$\theta = 1$ and $\theta_c = -1$; or $\theta = -1$ and $\theta_c = 1$	$\theta = 1$ and $\theta_c = -4$; or $\theta = -1$ and $\theta_c = 4$

Subtest length varied from 1 to 60 items. Response probabilities were computed based on the 1-parameter logistic (1PL) IRT model, with difficulty parameters drawn from a standard normal distribution.

For each of the $3 \times 6 \times 60 = 1080$ θ by θ_c by subtest length conditions, a response vector was randomly generated using the 1PL-predicted probabilities based on θ_c, a number correct was computed, and the appropriate tail probability was calculated for a Poisson binomial distribution under the (null) item success probabilities determined by θ. Any tail probability less than or equal to α (set to 0.05) indicated a null rejection. This was replicated 50,000 times for each condition. Poisson binomial probability cumulative distribution functions and point masses were computed with the R package "poibin" (Hong 2013b).

Inspection of the resulting power plots confirmed symmetry in those pairs of θ by θ_c conditions where the direction of the effect and its relation to the sign of θ are the same. An example is detecting a moderate relative weakness for a student one standard deviation below the mean and detecting a moderate relative strength for a student one standard deviation above the mean. Results for pairs such as this were averaged and treated as one condition, resulting in an effective replication count of 100,000 per condition. In the example just cited, this combined condition can be described as detecting a moderate effect in the direction of the same side of the mean as the (null) θ.

The nine combined conditions are described in Table 2. They are ordered left to right and top to bottom by decreasing hypothesized item requirements. Detecting larger effect sizes requires fewer cases than detecting smaller effect sizes. Thus, detecting very large effects should require the fewest items. As for the vertical ordering, detecting an effect directed toward the same side of the mean as the null hypothesis (e.g., detecting a relative weakness for a student who is already performing below the mean) should require more items than detecting an effect directed in the opposite direction (e.g., detecting a relative strength for the same student). Detecting effects for students at the mean might occupy an intermediate position with respect to item requirements.

The ordering of hypothesized item requirements bears out in the results of the simulation study. Table 3 displays the numbers of items required to detect an effect

Table 3 Subtest length required and estimated power for nine experimental conditions

Location of null	Effect size		
	Moderate	Large	Very large
Same side of mean as effect direction	49 (0.800)	19 (0.825)	10 (0.826)
Mean	36 (0.802)	12 (0.815)	6 (0.905)
Opposite side of mean as effect direction	36 (0.804)	10 (0.813)	4 (0.932)

for these nine scenarios, with the power threshold set to 0.8. Estimates of the power at the item counts are in parentheses. The power estimates for one item less than the shown item counts are all smaller than 0.800 (to three decimal places).

Three observations can readily be made about item requirements. First, diagnostic inferences are best supported when there is a very large effect. To detect these effects, between four and six items are needed, but ten are required to detect "very large" strengths for a student already performing above average or to detect "very large" weaknesses for one below average. Second, large effects—on the order to 2 SDs—can be detected with ten items but only when testing for relative strength in students performing below average or relative weakness in those performing well overall. Third, reliable detection of moderate (1 SD) or weaker effects requires more items than are typically included in subtests.

6 Conclusions

Users of assessments based on unidimensional trait models often want more information about student strengths and weaknesses in specific subareas, commonly called diagnostic information. Several researchers have questioned a standard method of providing this information—reporting subscores—arguing that subscores seldom carry reliable information sufficiently independent of the total score, or each other, to support justifiable claims about strengths and weaknesses.

This paper presented a conceptual framework for supporting diagnostic inferences grounded in statistical hypothesis testing and standard IRT assumptions. A Poisson binomial model provides the probabilities of observing extreme values of number correct in a subarea, conditional on the null hypothesis item response probabilities determined by estimates of the total score. Testing for diagnostic inferences can be conducted in practice at the individual or group level and directly on subtest responses regardless of the combination of items a student has taken. Previously prohibitive computations such as those needed to obtain Poisson binomial probability mass functions are now feasible with advances in computing power and algorithmic efficiency (see, e.g., Barrett and Gray 2014; Hong 2013a). This means that whether there is sufficient information in a student's or a group's responses to support a diagnostic inference is a matter that can be determined separately for each student or group. This research shows that in IRT contexts, reliable total scores can be leveraged in many cases to meet test users' diagnostic information needs directly.

The numbers of items required for diagnostic inferences is consistent with other findings on subtest length requirements for reliable subscores (e.g., Sinharay 2010). Unsurprisingly, the size of the effect to be detected has a strong influence on item requirements, but so does the direction of the effect with respect to the location of the null hypothesis, with effects harder to detect when the null is on the same side of the mean as the effect direction and easier to detect when the null and the effect direction are on opposite sides of the mean. Item requirements become to item *exposure* requirements in the context of detecting effects for groups rather than individuals.

Diagnostic inferences through hypothesis testing is particularly well-suited to contexts in which validating proposed subscores may not be practical due to data matrix sparseness or possible group differences in subscale structure.

References

American Educational Research Association, American Psychological Association, National Council on Measurement in Education, *Standards for Educational and Psychological Testing* (American Educational Research Association, Washington, DC, 2014)

B.E. Barrett, J.B. Gray, Efficient computation for the Poisson binomial distribution. Comput. Stat. **29**(6), 1469–1479 (2014). doi:10.1007/s00180-014-0501-6

J. González, M. Wiberg, A.A. von Davier, A note on the Poisson's binomial distribution in item response theory. Appl. Psychol. Meas. **40**(4), 302–310 (2016). doi:10.1177/0146621616629380

S.J. Haberman, When can subscores have value? J. Educ. Behav. Stat. **33**(2), 204–229 (2008)

Y. Hong, On computing the distribution function for the Poisson binomial distribution. Comput. Stat. Data Anal. **59**, 41–51 (2013a). doi:10.1016/j.csda.2012.10.006

Y. Hong, *Poibin: The Poisson Binomial Distribution (Version R package version 1.2)* (2013b). Retrieved from http://CRAN.R-project.org/package=poibin

W. Monaghan, *The Facts About Subscores*. Educational testing service (2006)

R. Nandakumar, W. Stout, Refinements of Stout's procedure for assessing latent trait unidimensionality. J. Educ. Stat. **18**(1), 41 (1993). doi:10.2307/1165182

R Core Team, *R: A Language and Environment for Statistical Computing* (R Foundation for Statistical Computing, Vienna, 2015.) Retrieved from https://www.R-project.org/

S. Sinharay, How often do subscores have added value? Results from operational and simulated data. J. Educ. Meas. **47**(2), 150–174 (2010)

W.F. Stout, A nonparametric approach for assessing latent trait dimensionality. Psychometrika **52**, 589–617 (1987)

W. Stout, R. Nandakumar, B. Junker, H.-H. Chang, D. Steidinger, DIMTEST: a Fortran program for assessing dimensionality of binary item responses. Appl. Psychol. Meas. **16**(3), 236–236 (1992). doi:10.1177/014662169201600303

W. Stout, A.G. Froelich, F. Gao, Using resampling methods to produce an improved DIMTEST procedure, in *Essays on Item Response Theory*, ed. by ed. by A. Boomsma, M. A. J. van Duijn, T. A. B. Snijders, vol. 157, (Springer, New York, NY, 2001), pp. 357–375. Retrieved from http://link.springer.com/10.1007/978-1-4613-0169-1_19

A Comparison of Two MCMC Algorithms for the 2PL IRT Model

Meng-I Chang and Yanyan Sheng

Abstract Markov chain Monte Carlo (MCMC) techniques have become popular for estimating item response theory (IRT) models. The current development of MCMC includes two major algorithms: Gibbs sampling and the No-U-Turn sampler (NUTS), which can be implemented in two specialized software packages JAGS and Stan, respectively. This study focused on comparing these two algorithms in estimating the two-parameter logistic (2PL) IRT model where different prior specifications for the discrimination parameter were considered. Results suggest that Gibbs sampling performed similarly to the NUTS under most of the conditions considered. In addition, both algorithms recovered model parameters with a similar precision except for small sample size situations. Findings from this study also shed light on the use of the two MCMC algorithms with more complicated IRT models.

Keywords Item response theory • Markov chain Monte Carlo • Gibbs sampling • No-U-Turn sampler

1 Introduction

Item response theory (IRT; Lord 1980) is a measurement theory used in educational and psychological measurements (e.g., achievement tests, rating scales, and inventories) that investigates a mathematical relationship between individuals' abilities (or other mental traits) and item responses. It is based on the idea that the probability of a correct response to an item is a mathematical function of person and item parameters (Hemker et al. 1995). Fully Bayesian estimation can be used to estimate model parameters by summarizing the posterior distribution. For decades during which IRT has been developed, fully Bayesian was not computationally practical for models with a large number of parameters such as IRT models. Modern computational technology and the development in Markov chain Monte Carlo (MCMC; e.g., Metropolis et al. 1953) algorithms, however, have changed

M.-I. Chang (✉) • Y. Sheng
Department of Counseling, Quantitative Methods, and Special Education, Southern Illinois University, Carbondale, IL, 62901, USA
e-mail: mengi@siu.edu

L.A. van der Ark et al. (eds.), *Quantitative Psychology*, Springer Proceedings in Mathematics & Statistics 196, DOI 10.1007/978-3-319-56294-0_7

that (Béguin and Glas 2001; Bolt and Lall 2003; Bradlow et al. 1999; Chib and Greenberg 1995; de la Torre et al. 2006; Fox and Glas 2001; Johnson and Sinharay 2005; Patz and Junker 1999).

1.1 Markov Chain Monte Carlo (MCMC)

MCMC methods are a class of algorithms that can be used to simulate draws from the posterior distribution. The concept of MCMC methods is to generate samples from a probability distribution via constructing a Markov chain that has the desired distribution as its stationary distribution. In MCMC, the quality of the sample improves as a function of the number of steps. After a number of steps, the state of the chain is then used as a sample of the desired distribution. MCMC methods have been proved useful in practically all aspects of Bayesian inference, such as parameter estimation and model comparisons. They can be applied in situations (e.g., with small sample sizes) where the maximum likelihood methods are difficult to implement. The samples produced by the MCMC procedure can also be used for conducting model fit diagnosis, model selection, and model-based prediction.

1.2 Gibbs Sampling

Among MCMC algorithms, Gibbs sampling (Geman and Geman 1984) is one of the simpler random walk algorithms. The idea of the random walk method is that at each step, the direction of the proposed move is random. If the relative probability of the proposed position is more than that of the current position, then the proposed move is always accepted. If the relative probability of the proposed position, however, is less than that of the current position, the acceptance of the proposed move is by chance. Due to the randomness, if the process were started over again, then the movement would certainly be different. Regardless of the specific movement, in the long run the relative frequency of visits will be close to the target distribution.

Gibbs sampling is applicable when the joint posterior distribution is not explicitly known, but the conditional posterior distribution of each parameter is known. The process of a Gibbs sampler is to obtain the joint posterior distribution by iteratively generating a random sample from the full conditional distribution for each parameter. The problem of the random walk algorithms, however, is that they may need too much time to reach convergence to the target distribution for complicated models with many parameters. These methods tend to explore parameter space via inefficient random walks (Neal 1993).

1.3 No-U-Turn Sampler (NUTS)

Other MCMC algorithms such as No-U-Turn sampler (NUTS; Hoffman and Gelman 2014) try to avoid the random walk behavior by introducing an auxiliary momentum vector and implementing Hamiltonian dynamics so the potential energy function is the target density. Basically, NUTS generates a proposal in a way similar to rolling a small marble on a hilly surface (the posterior distribution). The marble is given a random velocity and can move for several discrete steps in that direction. The movement follows the laws of physics, so the marble gains kinetic energy when it falls down the hill and earns potential energy when it climbs back up the hill. In this manner, a proposal is generated that can be a great distance from the original starting point. The proposed point is then accepted or rejected according to the Metropolis rule. NUTS utilizes a recursive algorithm to construct a set of possible candidate points that cross a wide strip of the target distribution, stopping automatically when it starts to double back and retrace its steps.

One of the primary challenges in implementing MCMC algorithms such as Gibbs sampling and NUTS is the availability of accessible software. This issue, however, can be resolved via two emerging computer programs: JAGS (Plummer 2003) and Stan (Stan Development Team 2016) developed for implementing Gibbs sampling and NUTS, respectively.

1.4 Two-Parameter Logistic IRT Model

In this study, the main focus is on the two-parameter logistic (2PL) IRT model (Birnbaum 1968), and it is defined as

$$P\left(Y_{ij} = 1|\theta_i,\ a_j, b_j\right) = \frac{1}{1 + \exp\left[-a_j\left(\theta_i - b_j\right)\right]}, \tag{1}$$

where Y_{ij} is the probability that the ith individual responds to the jth item correctly ($Y_{ij} = 1$) or incorrectly ($Y_{ij} = 0$), θ_i is the latent ability for subject i, a_j is the discrimination parameter, and b_j is the difficulty parameter for item j.

Prior research has been conducted on the development and application of IRT models under the fully Bayesian framework using Gibbs sampling (e.g., Albert 1992; Baker 1998; Ghosh et al. 2000; Sheng 2010; Sheng and Wikle 2007) as well as NUTS (e.g., Caughey and Warshaw 2014; Copelovitch et al. 2015). Recently, Grant et al. (2016) fitted the Rasch model (Rasch 1960) using both Gibbs sampling and NUTS, which were implemented in JAGS and Stan, respectively. They noted the memory problems of using JAGS for huge number of items. Their study, however, only focused on the computation speed and scalability, and the results showed that NUTS performed better than Gibbs sampling. To date, no study has compared these algorithms in estimating IRT models. Therefore, the purpose of this study is to investigate their performance in parameter recovery of the 2PL IRT model where different sample sizes, test lengths, and prior specifications for the discrimination parameter are concerned.

2 Methods

This study was conducted by using the computer program R (R Core Team 2016). Data were generated for the 2PL IRT model as defined in (1). Test length (K) was manipulated to be 10 and 20 items and sample size (N) to be 100, 500, and 1000 examinees. Model parameters were generated such that $\theta_i \sim N(0, 1)$, $a_j \sim \text{lognormal}(0, 0.5)$, and $b_j \sim N(0, 1)$. For the MCMC procedures, normal priors were assumed for both θ_i and b_j such that $\theta_i \sim N(0, 1)$ and $b_j \sim N(0, 1)$. Three prior specifications were considered for a_j such that (1) $a_j \sim \text{lognormal}(0, 0.5)$, which is commonly used in BILOG-MG (Zimowski et al. 2003); (2) $a_j \sim N_{(0,\infty)}(0, 1)$, which is another common way to specify the discrimination parameter (Sahu 2002; Sheng 2008; Spiegelhalter et al. 2003); and via a transformation to α_j so that a_j is assumed to be exp (α_j) and having a standard normal prior for α_j. Any real value exponentiated will be positive. Therefore, with the third prior specification, Eq. (1) can be written as

$$\text{logit}\left(p_{ij}\right) = \exp\left(\alpha_j\right)\left(\theta_i - b_j\right), \tag{2}$$

where $\alpha_j \sim N(0, 1)$. Gibbs sampling and NUTS were implemented to each simulated data set via the use of JAGS and Stan where the burn-in stage used in JAGS or warm-up stage used in Stan was set to be 3000 iterations followed by four chains with 5000 iterations. For both algorithms, the initial values for the model parameters were set the same. Convergence was evaluated using the Gelman-Rubin R statistic (Gelman and Rubin 1992). For each simulated condition, ten replications were conducted to avoid erroneous results in estimation due to sampling error. The accuracy of parameter estimates was evaluated using *bias* and the root mean square error (*RMSE*), which are defined as

$$bias_\pi = \frac{\sum_{j=1}^{n}\left(\widehat{\pi}_j - \pi_j\right)}{n}, \tag{3}$$

$$RMSE_\pi = \sqrt{\frac{\sum_{j=1}^{n}\left(\widehat{\pi}_j - \pi_j\right)^2}{n}}, \tag{4}$$

where π is the true value of an item parameter (e.g., a_j or b_j), $\widehat{\pi}$ is the estimated value of that parameter in the kth replication using either Gibbs sampling or NUTS, and n is the total number of replications. In addition, the root mean square of difference (*RMSD*) was used to assess the consistence between parameter estimates of competing algorithms (Jurich and Goodman 2009) and is defined as

$$RMSD_{\pi} = \sqrt{\frac{\sum_{j=1}^{n} \left(\widehat{\pi}_{j,Gibbs\,sampling} - \widehat{\pi}_{j,NUTS}\right)^2}{n}}, \tag{5}$$

where $\widehat{\pi}_{j,Gibbs\,sampling}$ and $\widehat{\pi}_{j,NUTS}$ are the estimates of any parameter estimated via the use of Gibbs sampling and NUTS, respectively, and n is as defined in (3) and (4). These measures were averaged over items to provide summary information.

3 Results

For the 2PL IRT model, $\widehat{R}$ is less than 1.1 for each model parameter under all test conditions, suggesting that convergence appears satisfactory for both algorithms.

The results of *bias*, *RMSE*, and *RMSD* averaged across items from Gibbs sampling and NUTS for recovering the discrimination and difficulty parameters are summarized in Tables 1, 2, and 3.

The results indicate that Gibbs sampling performs comparably to NUTS under most conditions tested. Both algorithms recover true item parameters with similar precision as *RMSE*s are virtually identical except that they tend to be larger with the maximum value of 0.632 for the condition where the prior distribution for a_j is lognormal using Gibbs sampling than NUTS when sample size is small (i.e., $N = 100$) (see Table 1). When given adequate sample size and sufficient number of items (e.g., $N = 1000$ and $K = 10$), discrimination parameter estimates become more stable with the maximum value of *RMSE* equal to 0.141 (see Table 3). In addition, *bias* is close to zero for all conditions, indicating that both algorithms

Table 1 Average Bias, RMSE, and RMSD for recovering item parameters in the 2PL IRT model when $N = 100$

			Gibbs sampling		NUTS		
K	Prior for a_j	Parameters	*Bias*	*RMSE*	*Bias*	*RMSE*	*RMSD*
10	1	*a*	0.070	0.472	−0.003	0.324	0.206
		b	−0.028	0.263	−0.160	0.252	0.051
	2	*a*	−0.041	0.335	−0.042	0.335	0.006
		b	−0.027	0.253	−0.027	0.254	0.005
	3	*a*	−0.025	0.404	−0.026	0.401	0.009
		b	−0.026	0.261	−0.029	0.261	0.005
20	1	*a*	0.139	0.632	−0.030	0.283	0.428
		b	−0.027	0.285	−0.019	0.264	0.060
	2	*a*	−0.107	0.299	−0.108	0.300	0.005
		b	−0.026	0.274	−0.026	0.276	0.005
	3	*a*	−0.022	0.433	−0.025	0.432	0.009
		b	−0.026	0.279	−0.026	0.281	0.006

Note. Prior 1, a_j~lognormal(0, 0.5); prior 2, $a_j \sim N_{(0,\infty)}$ (0, 1); prior 3, $a_j = \exp(\alpha_j)$, $\alpha_j \sim N(0, 1)$

Table 2 Average Bias, RMSE, and RMSD for recovering item parameters in the 2PL IRT model when $N = 500$

			Gibbs sampling		NUTS		
K	Prior for a_j	Parameters	*Bias*	*RMSE*	*Bias*	*RMSE*	*RMSD*
10	1	a	0.049	0.225	0.030	0.199	0.077
		b	−0.048	0.181	−0.044	0.166	0.042
	2	a	0.015	0.197	0.012	0.196	0.005
		b	−0.049	0.181	−0.050	0.181	0.003
	3	a	0.027	0.216	0.025	0.216	0.006
		b	−0.048	0.179	−0.048	0.180	0.005
20	1	a	0.051	0.190	0.035	0.160	0.051
		b	−0.002	0.176	0.012	0.163	0.032
	2	a	0.020	0.159	0.019	0.158	0.003
		b	−0.003	0.172	−0.004	0.170	0.003
	3	a	0.034	0.178	0.035	0.179	0.004
		b	−0.004	0.173	−0.003	0.173	0.004

Note. Prior 1, $a_j \sim$lognormal(0, 0.5); prior 2, $a_j \sim N_{(0,\infty)}$ (0, 1); prior 3, $a_j = \exp(\alpha_j)$, $\alpha_j \sim N(0, 1)$

Table 3 Average Bias, RMSE, and RMSD for recovering item parameters in the 2PL IRT model when $N = 1000$

			Gibbs sampling		NUTS		
K	Prior for a_j	Parameters	*Bias*	*RMSE*	*Bias*	*RMSE*	*RMSD*
10	1	a	−0.001	0.141	−0.0003	0.132	0.025
		b	0.016	0.133	0.020	0.129	0.018
	2	a	−0.007	0.135	−0.007	0.134	0.007
		b	0.017	0.139	0.018	0.132	0.007
	3	a	−0.008	0.139	−0.008	0.130	0.004
		b	0.018	0.133	0.017	0.187	0.003
20	1	a	−0.006	0.105	−0.017	0.098	0.036
		b	−0.014	0.078	−0.015	0.083	0.017
	2	a	−0.028	0.100	−0.030	0.101	0.003
		b	−0.016	0.081	−0.016	0.081	0.003
	3	a	−0.015	0.101	−0.017	0.101	0.003
		b	−0.012	0.077	−0.013	0.080	0.002

Note. Prior 1, $a_j \sim$lognormal(0, 0.5); prior 2, $a_j \sim N_{(0,\infty)}$ (0, 1); prior 3, $a_j = \exp(\alpha_j)$, $\alpha_j \sim N(0, 1)$

estimate parameters with little bias. When sample size increases, the *RMSE*s for the discrimination parameter and difficulty parameter tend to decrease with both Gibbs sampling and NUTS. This pattern, however, is observed with *bias* only for the discrimination parameter. When test length increases, the *RMSE*s for the discrimination parameter and difficulty parameter appear to decrease using either algorithm when $N \geq 500$. This pattern, however, is not directly observed with *bias*. *RMSD*s between Gibbs sampling and NUTS are approximately zero for nearly all conditions except for the lognormal prior and small sample size ($N = 100$) condition where the maximum value is 0.428. Low *RMSD*s suggest that when estimation errors are made, the direction of these errors is consistent across algorithms. With Gibbs sampling, the truncated normal prior for the discrimination parameter recovers better than the other two prior specifications across all test length and sample size conditions especially when $N = 100$ and $K = 20$ (see Table 1). However, with NUTS, the lognormal prior for the discrimination parameter results in a better estimation than the other two approaches when sample size is small (i.e., $N = 100$) (see Table 1).

4 Conclusions

Overall, both algorithms recover true item parameters in a consistent manner. Also, estimation error and bias reduce as samples size and test format conditions (e.g., sample sizes are not extremely small or too big) are sufficient for IRT estimation. More importantly for this study, the two algorithms produce nearly identical estimates across most conditions except for the lognormal condition. Given the results, it is suggested that when dealing with sample sizes such as $N < 100$, NUTS should be adopted when a lognormal prior is assumed for the discrimination parameter for the 2PL IRT model. The results also provide some sense of assurance that decisions about which algorithm to use should be made on considerations other than accuracy in estimation (e.g., budget, user-friendliness, institutional availability, need for customization). It is, however, noted that results of this study were based on ten replications due to the computational expense of the MCMC algorithm. For example, the computation time of implementing Gibbs sampling in JAGS to data with $N = 1000$ and $K = 20$ was about 72 min to complete four chains with 5000 iterations on a computer with 2.5 GHz Core i5 and 8G memory. For the same data size and number of iterations, NUTS via the use of Stan took about 41 min for each replication. Given the small number of iterations, and given that Harwell et al. (1996) suggested a minimum of 25 replications for Monte Carlo studies in typical IRT-based research, *bias*, *RMSE*, and *RMSD* values presented in Tables 1 and 2 need to be verified with further studies before one can generalize the results to similar conditions.

Simulation studies often demonstrate performance under ideal situations. In this case, the true IRT model was known, and fit can be assumed nearly perfectly. Future studies may use these IRT programs to fit the 2PL IRT model to real data. Other test format conditions should be also explored. This would include expanding current simulation conditions to compare other MCMC algorithms (e.g., Metropolis-Hastings and Hastings-within-Gibbs) or other estimation methods (e.g., marginal maximum likelihood) and compare the two algorithms on models that have more latent dimensions (e.g., multidimensional IRT models), item parameters (e.g., three-parameter logistic IRT model), or response categories (e.g., polytomous IRT models). Other prior specifications for model parameters a_j, b_j, and θ_i should be also considered.

References

J.H. Albert, Bayesian estimation of normal ogive item response curves using Gibbs sampling. J. Educ. Behav. Stat. **17**(3), 251–269 (1992). doi:10.3102/10769986017003251

F.B. Baker, An investigation of the item parameter recovery characteristics of a Gibbs sampling procedure. Appl. Psychol. Meas. **22**(2), 153–169 (1998). doi:10.1177/01466216980222005

A.A. Béguin, C.A.W. Glas, MCMC estimation and some model-fit analysis of multidimensional IRT models. Psychometrika **66**(4), 541–561 (2001). doi:10.1007/BF02296195

A. Birnbaum, in *Statistical Theories of Mental Test Scores*, ed. by F. M. Lord, M. R. Novick. Some latent trait models and their use in inferring an examinee's ability (Addison-Wesley, Reading, MA, 1968), pp. 453–479

D.M. Bolt, V.F. Lall, Estimation of compensatory and noncompensatory multidimensional item response models using Markov chain Monte Carlo. Appl. Psychol. Meas. **27**(6), 395–414 (2003). doi:10.1177/0146621603258350

E.T. Bradlow, H. Wainer, X.H. Wang, A Bayesian random effects model for testlets. Psychometrika **64**(2), 153–168 (1999). doi:10.1007/BF02294533

D. Caughey, C. Warshaw, in *Dynamic Representation in the American States, 1960–2012*. American Political Science Association 2014 Annual Meeting Paper

S. Chib, E. Greenberg, Understanding the Metropolis-Hastings algorithm. Am. Stat. **49**(4), 327–335 (1995)

M. Copelovitch, C. Gandrud, M. Hallerberg, in *Financial Regulatory Transparency, International Institutions, and Borrowing Costs*. The Political Economy of International Organizations Annual Conference, University of Utah, Salt Lake City, Utah, vol. 3, (2015), p. 2015, http://wp.peio.me/wp-content/uploads/PEIO8/ Copelovitch, Gandrud, Hallerberg

J. de la Torre, S. Stark, O.S. Chernyshenko, Markov chain Monte Carlo estimation of item parameters for the generalized graded unfolding model. Appl. Psychol. Meas. **30**(3), 216–232 (2006). doi:10.1177/0146621605282772

J.P. Fox, C.A. Glas, Bayesian estimation of a multilevel IRT model using Gibbs sampling. Psychometrika **66**(2), 271–288 (2001). doi:10.1007/BF02294839

A. Gelman, D.B. Rubin, Inference from iterative simulation using multiple sequences. Stat. Sci. **7**, 457–472 (1992)

S. Geman, D. Geman, Stochastic relaxation, Gibbs distributions, and the Bayesian restoration of images. IEEE Trans. Pattern Anal. Mach. Intell. **6**, 721–741 (1984). doi:10.1109/TPAMI.1984.4767596

M. Ghosh, A. Ghosh, M.H. Chen, A. Agresti, Noninformative priors for one-parameter item response models. J. Stat. Plan. Inference **88**(1), 99–115 (2000). doi:10.1016/S0378-3758(99)00201-3

R.L. Grant, D.C. Furr, B. Carpenter, A. Gelman, Fitting Bayesian item response models in Stata and Stan. Preprint arXiv:1601.03443 (2016), https://arxiv.org/abs/1601.03443

M. Harwell, C.A. Stone, T.C. Hsu, L. Kirisci, Monte Carlo studies in item response theory. Appl. Psychol. Meas. **20**(2), 101–125 (1996). doi:10.1177/014662169602000201

B.T. Hemker, K. Sijtsma, I.W. Molenaar, Selection of unidimensional scales from a multidimensional item bank in the polytomous Mokken IRT model. Appl. Psychol. Meas. **19**(4), 337–352 (1995). doi:10.1177/014662169501900404

M.D. Hoffman, A. Gelman, The No-U-turn sampler: adaptively setting path lengths in Hamiltonian Monte Carlo. J. Mach. Learn. Res. **15**(1), 1593–1623 (2014). https://arxiv.org/abs/1111.4246

M.S. Johnson, S. Sinharay, Calibration of polytomous item families using Bayesian hierarchical modeling. Appl. Psychol. Meas. **29**(5), 369–400 (2005). doi:10.1177/0146621605276675

D. Jurich, J. Goodman, in *A Comparison of IRT Parameter Recovery in Mixed Format Examinations Using PARSCALE and ICL*. Annual Meeting of Northeastern Educational Research Association: 21–23 October 2009

F.M. Lord, *Applications of Item Response Theory to Practical Testing Problems* (L. Erlbaum Associates, Hillsdale, NJ, 1980), p. 1980

N. Metropolis, A.W. Rosenbluth, M.N. Rosenbluth, A.H. Teller, E. Teller, Equation of state calculations by fast computing machines. J. Chem. Phys. **21**(6), 1087–1092 (1953). doi:10.1063/1.1699114

R. Neal, *Probabilistic Inference Using Markov Chain Monte Carlo Methods*. Technical Report CRG-TR-93-1, Department of Computer Science, University of Toronto, Toronto, 1993

R.J. Patz, B.W. Junker, Applications and extensions of MCMC in IRT: multiple item types, missing data, and rated responses. J. Educ. Behav. Stat. **24**(4), 342–366 (1999). doi:10.3102/10769986024004342

M. Plummer, in *JAGS: A Program for Analysis of Bayesian Graphical Models Using Gibbs Sampling*. Proceedings of the 3rd International Workshop on Distributed Statistical Computing, vol. 124 (2003), p. 125. http://mcmc-jags.sourceforge.net

R Core Team, *R: A Language and Environment for Statistical Computing* (R Foundation for Statistical Computing, Vienna, 2016.) https://www.R-project.org/

G. Rasch, *Probabilistic Models for Some Intelligence and Attainment Tests* (University of Chicago Press, Chicago, IL, 1960)

S.K. Sahu, Bayesian estimation and model choice in item response models. J. Stat. Comput. Simul. **72**(3), 217–232 (2002). doi:10.1080/00949650212387

Y. Sheng, A MATLAB package for Markov chain Monte Carlo with a multi-unidimensional IRT model. J. Stat. Softw. **28**, 1–20 (2008)

Y. Sheng, A sensitivity analysis of Gibbs sampling for 3PNO IRT models: effects of prior specifications on parameter estimates. Behaviormetrika **37**(2), 87–110 (2010). doi:10.2333/bhmk.37.87

Y. Sheng, C.K. Wikle, Comparing multiunidimensional and unidimensional item response theory models. Educ. Psychol. Meas. **67**(6), 899–919 (2007). doi:10.1177/0013164406296977

D. Spiegelhalter, A. Thomas, N. Best, D. Lunn, WinBUGS version 1.4 user manual. MRC Biostatistics Unit (2003), http://www.mrc-bsu.cam.ac.uk/bugs/

Stan Development Team, Stan modeling language users guide and reference manual, version 2.12.0 (2016), http://mc-stan.org/

M. Zimowski, E. Muraki, R.J. Mislevy, R.D. Bock, *BILOG-MG 3: Item Analysis and Test Scoring with Binary Logistic Models [Computer Software]* (Scientific Software, Chicago, IL, 2003) [illegible] Metropolis, A.W., Rosenbluth, M.N., Rosenbluth, A.H., Teller, E.: Teller, Equation of state calculations by fast computing machines. J. Chem. Phys. 21(6), 1087–1092 (1953). doi:10.1063/1.1699114

Neal, R.: Probabilistic Inference Using Markov Chain Monte Carlo Methods. Technical Report CRG-TR-93-1, Department of Computer Science, University of Toronto, Toronto (1993)

[illegible]

Similar DIFs: Differential Item Functioning and Factorial Invariance for Scales with Seven ("Plus or Minus Two") Response Alternatives

David Thissen

Abstract Measurement invariance in the factor analytic tradition and a lack of differential item functioning (DIF) in item response theory (IRT) are essentially the same. However, the two types of analysis have rarely been cast in exactly parallel form. Either categorical (IRT) or continuous linear/confirmatory factor analysis (CFA) models may usefully be applied to item response data with seven (or so) response alternatives. This chapter is intended to clarify some of issues involved in the application of CFA to DIF detection, by using the CFA procedures slightly differently than they are used in the analysis of factorial invariance. Summaries are provided of DIF detection using parametric IRT models and the conventional evaluation of factorial invariance. The evaluation of factorial invariance is reformulated to make it more like DIF detection, for the context of item analysis. An empirical illustration is provided using both parametric IRT DIF detection procedures and a parallel version of the evaluation of factorial invariance using CFA models.

Keywords Item response theory • Measurement invariance • DIF

1 Introduction

It has long been established that measurement invariance, as it is known in the factor analytic tradition, and a lack of differential item functioning (DIF), as described in the item response theory (IRT) literature, are essentially the same conceptually (Meredith and Millsap 1992; Meredith 1993). However, due to the different questions that gave rise to the ideas of DIF and factorial invariance, the two types of analysis have rarely been cast in parallel form. DIF analysis originated in educational measurement, with the goal of detecting and removing "biased" items, to reduce the overall bias against some demographic group(s) (Lord 1977, 1980). It has become standard practice to include DIF detection in item analysis for any kind

D. Thissen (✉)
Department of Psychology and Neuroscience, The University of North Carolina at Chapel Hill, 235 E. Cameron Avenue, Chapel Hill, NC 27599, USA
e-mail: dthissen@email.unc.edu

L.A. van der Ark et al. (eds.), *Quantitative Psychology*, Springer Proceedings in Mathematics & Statistics 196, DOI 10.1007/978-3-319-56294-0_8

of test construction (Edwards and Edelen 2009). The origins of factorial invariance involved the theoretical question of whether sampling a subgroup from a population would affect the factor structure (Meredith 1954).

Meredith and Millsap (1992) and Meredith (1993) drew together ideas of factorial invariance, measurement invariance, and "item bias" (now DIF), but still with the factor analytic style of a division into weak, strong, and strict factorial invariance, without the concentration on item-by-item analysis that is the hallmark of DIF detection in test theory. CFA is conventionally used to provide the parameter estimation and statistical tests upon which the analysis of factorial invariance is based.

For tests or scales using Likert-type response scales with seven (plus or minus a couple) alternatives, cases can be made for the use of either IRT-based (categorical) or CFA-based (linear, continuous) models (Rhemtulla et al. 2012). If the goal is item analysis, and specifically DIF detection, modifications of conventional factor analytic approaches are useful to make the analysis more like what is usually done in test theory. The procedures involved are closely related in many respects to those suggested by Raykov et al. (2013) in a recent article on factorial invariance, which borrows from methods commonly used in DIF analysis. But that work by Raykov et al. is not focused on DIF analysis. This chapter is intended to clarify some of the issues involved in the use of linear factor models in DIF detection.

2 Conventional IRT and DIF

2.1 The Graded Response IRT Model

The graded response IRT model (Samejima 1969, 1997) describes the probability of each item response as a function of a set of item parameters and θ, the latent variable measured by the scale, as follows: The conditional probability, or *trace line*, of graded response $u = 1, 2, \ldots, m$ as a function of the latent variable being measured (θ) is

$$T_{ui}(\theta) = T^*_{ui}(\theta) - T^*_{u+1i}(\theta) \tag{1}$$

in which $T^*_{ui}(\theta)$ is a curve tracing the probability of a response in category u or higher: $T^*_1(\theta) = 1$, $T^*_{m+1}(\theta) = 0$, and for $u = 2, 3, \ldots, m$

$$T^*_{ui}(\theta) = \frac{1}{1 + \exp(-[a_i\theta + c_{ui}])} = \frac{1}{1 + \exp(-a_i[\theta - b_{ui}])}\,. \tag{2}$$

For computational convenience, the graded response model is usually fitted in the slope-intercept form in the center of Eq. (2), with a_i as the slope or discrimination parameter and c_{ui} as an intercept parameter for each response u to item i. A more interpretable form [rightmost in Eq. (2)] has threshold parameters $b_{ui} = -c_{ui}/a_i$. The thresholds are the values on the θ scale at which a respondent has a 50% chance of responding in category u or higher.

2.2 DIF Analysis with the Graded Response IRT Model

DIF analysis rests on the idea, most succinctly stated by Lord (1980, p. 212), "If …an item has a different item response function for one group than for another, it is clear that the item is biased." In parametric IRT, there is a one-to-one relation between an item response function and the item's parameters, so the question of whether "an item has a different item response function for one group than for another" is answered with a statistical test of the equality of the item's parameters for one group and those for the other. Such a statistical test has two elements: One is some mechanism to determine, and "correct for," whatever overall differences exist between the groups in the latent variable measured by the test or scale. That is usually done by designating some set of items as the *anchor*, by analogy with the anchor in test linking designs. While the best way to designate the anchor is on some theoretical grounds (Thissen et al. 1993), most often as a practical matter the anchor comprises all of the other items on the test or scale.

The second element is the statistical test itself; in this chapter, we use the likelihood ratio (L.R.) test; the L.R. G_i^2 test for item i is computed as

$$\begin{aligned} G_i^2 =& (-2\text{loglikelihood}[\text{Model : item } i \text{ parametersequal}]) \\ &- (-2\text{loglikelihood}[\text{Model : item } i \text{ parametersfree}])\,. \end{aligned} \tag{3}$$

The loglikelihood in the first term in Eq. (3) is for an IRT model in which all of the item parameters are constrained to be equal for the two groups, and the second term is the loglikelihood for a model in which the item parameter estimates for the *studied* item are free to differ between the two groups, but the item parameters for the anchor items (usually, all of the other items) are still constrained equal between the two groups. The latter constraint provides the basis for the estimation of the parameters of the population distribution, usually the mean and variance, for the *focal group* relative to a standard normal distribution for the *reference group* that defines the scale of the latent variable (Thissen et al. 1988, 1993). Under the null hypothesis of no DIF, G_i^2 is distributed as χ^2 with degrees of freedom equal to the difference in the number of free parameters between the two models, which is equal to the number of parameters for the item in one group. The procedure can be extended to provide separate statistical tests for the equality of subsets of parameters, like the slope (a) parameters and separately the thresholds (b) (or the intercept (c) parameters given equal slope parameters).

Because there are many statistical tests performed when all items on a scale are checked for DIF, some multiple comparisons procedure is useful to provide protection against excessive Type I errors. In the analyses in subsequent sections, we use the Benjamini–Hochberg procedure (Benjamini and Hochberg 1995) to set the *false discovery rate* (FDR) at 0.05 across I overall DIF tests. The FDR is defined as "the expected proportion of errors among the rejected hypotheses" (Benjamini and Hochberg 1995, p. 290); this is a less-stringent criterion than the older standard, the *family-wise error rate*, that required control of "the probability of committing any type I error in families of comparisons under simultaneous consideration" (Benjamini and Hochberg 1995, p. 289). The sequential testing algorithm used here to control FDR involves sorting the p-values (in this case, for the I overall DIF tests) in decreasing order. Then the smallest p-value is compared to the Bonferroni standard, α/I; the largest p-value is compared to α, and intermediate p-values in the sort order are compared to a linearly interpolated sequence between α/I and α. The hypotheses for the largest obtained p-value that exceeds its comparison value, and those with all smaller p-values, are significant. Williams et al. (1999) provide a detailed introduction to the use of the Benjamini–Hochberg procedure in educational measurement, where it is commonly used (e.g., as the standard multiple comparisons procedure for the National Assessment of Educational Progress). Edwards and Edelen (2009) illustrate the use of the Benjamini–Hochberg procedure in DIF analysis, and Thissen et al. (2002) provide a tutorial on the quick and easy computation of the required values.

In our application of the Benjamini–Hochberg procedure in DIF analysis, we apply the multiple comparisons procedure to the I overall DIF tests for I items. Then for items for which the overall test shows significant DIF, we test parameter subsets at standard α levels.

2.2.1 An Example: Eight Items from the Bem Femininity Scale

The illustration makes use of responses to eight items from the femininity scale of the Bem Sex-Role Inventory (Bem 1974). The data are from the 1985–1988 entries for a test battery that included the scale, archived by the UNC Dataverse (2009). The original femininity scale comprised 20 items; subsequently, a ten-item short form was developed (Bem 1981). Neither the original nor the short form provides a good illustration of the concepts discussed here. To create the example used in this chapter, six items were extracted from the short form, and two items that exhibit strong DIF were added, to make a compact eight-item scale within which DIF can be detected. The response scale was 1–7, with only the endpoints labeled as "never or almost never true" for 1 and "always or almost always true" for 7. The respondents were 199 men and 200 women who were undergraduate students. We investigate DIF between men and women.

As is usually the case with as many as seven response alternatives, and a sample size (within each group) around 200, there are no responses in some of the seven categories for some items. To test the hypothesis that an item's parameters are the same in two groups, the item must have the same set of parameters in each group. The graded response model's intercept or threshold parameters can only be

Table 1 IRT DIF detection results; significant tests with the p-values evaluated using the Benjamini–Hochberg procedure are bold

Item	Item stem	L.R. G^2	*df*	p
2	**Yielding**	**29.0**	**6**	**<0.001**
11	Affectionate	5.0	6	0.537
20	**Feminine**	**345.1**	**5**	**<0.001**
23	**Sympathetic**	**17.8**	**7**	**0.013**
29	Understanding	9.4	5	0.093
32	Compassionate	7.0	6	0.322
35	Eager to soothe hurt feelings	2.2	7	0.945
56	Loves children	1.4	7	0.986

estimated between response categories for which there are observed responses. For that reason, categories must be collapsed in one or both groups until the item has non-zero response counts in all of the (collapsed) responses in both groups. For this example, three items have responses (in both groups) in all seven categories; three items (2, 11, and 32) had categories 1–2 collapsed to yield six response categories, item 20 had both categories 1–2 and 6–7 combined, and item 29 had categories 1-2-3 collapsed into one.

We test each item for DIF with all other items as the anchor. Estimation of the IRT parameters and computation of the loglikelihood for the L.R. tests were done using the IRTPRO software (Cai et al. 2011). A summary of the results is in Table 1. Three items exhibit significant DIF (with the p-values evaluated using the Benjamini–Hochberg procedure): "Yielding," "Feminine," and "Sympathetic." Partitioning the DIF statistics into components attributable to differences between groups in slopes and differences between sets of intercepts given that the slopes are equal, we find that "Yielding" exhibits both slope (sometimes called *nonuniform*) DIF ($G^2(1) = 10.7, p = 0.001$) and intercept/threshold (sometimes called *uniform*) DIF ($G^2(5) = 18.3$, $p = 0.003$) . The same is true for "Feminine" (for a: $G^2(1) = 9.6, p = 0.002$; for $c|a$: $G^2(4) = 335.5, p < 0.001$). For "Sympathetic," only the test for the intercepts / thresholds is significant ($G^2(6) = 16.7, p = 0.002$). All of these results are easier to understand with graphical displays, as suggested by Steinberg and Thissen (2006). We will examine graphical displays of these results jointly with the CFA results in a subsequent section.

3 The Factor Analysis Model and DIF

3.1 The Linear Model

When applied to Likert-type item response data, linear factor analysis models the responses $u = 1, 2, \ldots, m$ for item i as a linear function of the latent factor score f, the factor-analytic equivalent of θ in IRT models:

$$u_i = \alpha_i + \lambda_i f + \epsilon_i \,. \tag{4}$$

The latent variable f is usually defined to be $N(0, 1)$, so α_i is the intercept (and the mean of the observed responses); λ_i is the factor loading, or the regression coefficient of u_i on f; and ϵ_i is random error, defined by its distribution, $N(0, \sigma_i^2)$; σ_i^2 is a third parameter for each item. While it cannot be strictly true that the discrete categorical responses $u = 1, 2, \ldots, m$ are linearly dependent on a continuous latent variable, simulation studies have shown that for about seven (or more) response categories, the linear model can yield results as useful as categorical models (Rhemtulla et al. 2012).

3.2 Measurement Invariance and DIF

Measurement invariance is nearly certainly implied by factorial invariance (Meredith 1993). So to elucidate the similarities and differences between IRT and factor analytic approaches to DIF/measurement invariance, we consider the analysis of factorial invariance between groups.

The starting point for the analysis of factorial invariance is a test of *configural* or *pattern* invariance: Do the two groups have the same general factor structure? In the absence of configural invariance, further tests of parameter equality usually make little sense. For some data, configural invariance is a real question. However, in the context of DIF analysis, the variables are usually the items on a single measurement instrument that is assumed to be essentially unidimensional, so configural invariance is the assumed unidimensional model.

Given configural invariance, factorial invariance analysis proceeds in steps based on categories of model parameters first and variables (items) within those categories second. First comes the test of *weak* or *metric* invariance, testing the equality of factor loadings between groups. There can be partial weak (or metric) invariance, if only some of the factor loadings differ between groups. Second, for variables (items) for which the loadings are constrained equal, *strong* or *scalar* invariance is tested by constraining the intercept parameters to be equal; partial scalar invariance is also a possibility. Finally, *strict* or *residual variance* invariance can be tested by constraining the unique variances to be equal for the two groups (Meredith 1993).

In contrast, in the DIF literature, the focus is on the items (variables) first and the parameters of the model second. The items are divided into an anchor set (items that are assumed to have no DIF) and a studied item (or items). DIF is tested one item at a time, with between-group differences for all of its parameters evaluated simultaneously, then perhaps separately (by classes of parameters) within each item.

Figure 1 illustrates the difference between DIF and factorial invariance analyses. The important thing to notice is that DIF analysis works with items as the superordinate classification and then parameters within items, while the analysis of factorial invariance examines classes of model parameters first and then items within parameter sets.

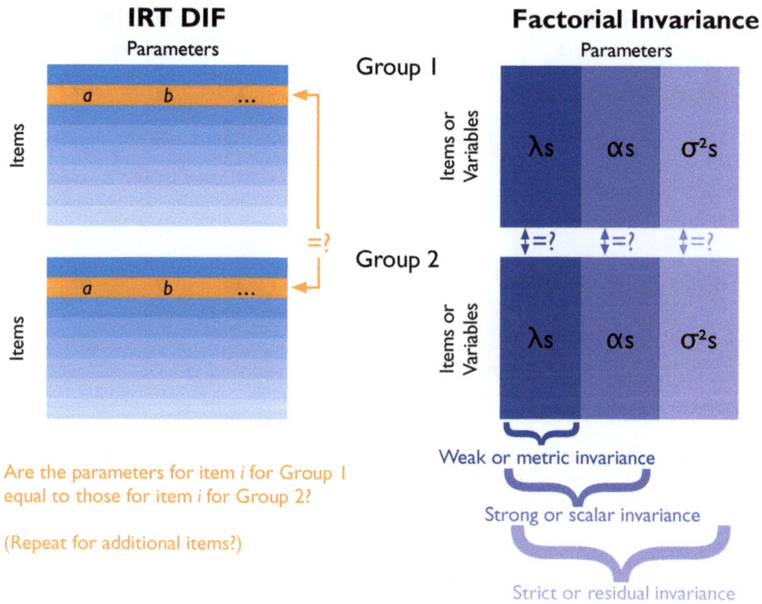

Fig. 1 *Left panel*: Schematic depiction of IRT DIF analysis, checking the equality of the parameters for a single item (*orange*) between groups using the other items (*cyan*) as the anchor. *Right panel*: Schematic depiction of analysis of factorial invariance, checking first equality of factor loadings between groups, then constraining intercepts equal, and then making specific variances equal

Figure 2 shows a schematic depiction of IRT DIF analysis, checking the equality of the parameters for a single item between groups using the other items as the anchor, translated (in the right panel) into the parallel procedure using the CFA model. We use this idea to repeat the analysis of the eight-item subset of the Bem femininity scale with a linear model.

3.2.1 The Example Continued: Eight Items from the Bem Femininity Scale

We make use of the same responses to eight items that were used previously to illustrate IRT DIF analysis, now with the linear model and maximum likelihood estimation as implemented in the *lavaan* software (Rosseel 2012). We again use L.R. G^2 tests for each item with all other items as anchor. These statistical tests are based on the difference between the values of the loglikelihood for two models for each item, with and without equality constraints for all three-item parameters (λ_i, α_i, σ_i^2) for each item.

A summary of the results is in Table 2. In parallel with the IRT analyses, the same three items exhibit significant DIF: “Yielding,” “Feminine,” and “Sympathetic.”

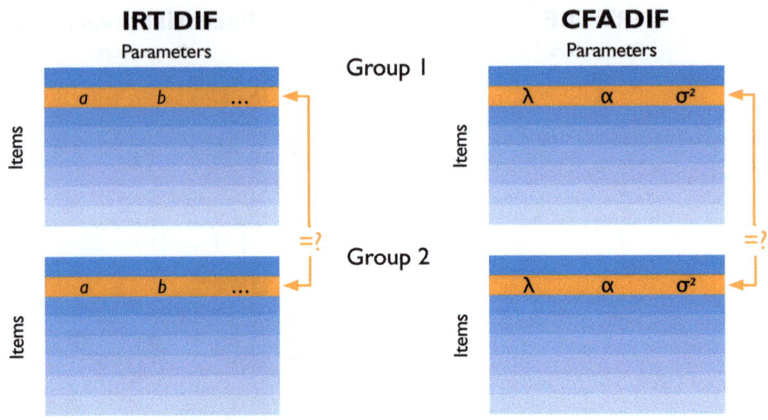

Fig. 2 *Left panel*: Schematic depiction of IRT DIF analysis, checking the equality of the parameters for a single item (orange) between groups using the other items (*cyan*) as the anchor. *Right panel*: Parallel depiction of CFA DIF analysis, checking equality of each item's parameter set

Table 2 Linear CFA DIF detection results; significant tests with the p-values evaluated using the Benjamini–Hochberg procedure are bold

Item	Item stem	All other items as anchor		
		L.R. G^2	*df*	*p*
2	**Yielding**	**15.1**	**3**	**0.002**
11	Affectionate	0.6	3	0.899
20	**Feminine**	**478.0**	**3**	**<0.001**
23	**Sympathetic**	**10.1**	**3**	**0.018**
29	Understanding	6.8	3	0.079
32	Compassionate	6.9	3	0.074
35	Eager to soothe hurt feelings	5.2	3	0.159
56	Loves children	0.9	3	0.836

Partitioning the DIF statistics into components attributable to differences between the three parameters separately for the two groups for items with significant DIF, we find: "Yielding" exhibits a significant difference between groups only in λ_i ($G^2(1) = 9.7, p = 0.002$). "Feminine" has significant differences for all three parameters (for λ_i: $G^2(1) = 12.2, p < 0.001$; for α_i: $G^2(1) = 447.3, p < 0.001$; for σ_i^2: $G^2(1) = 18.5, p < 0.001$). For "Sympathetic," only the test for the intercept is significant ($G^2(1) = 10.2, p = 0.002$).

Figure 3 shows the expected score curves as functions of the latent variable, and the modeled conditional distributions of the responses at f or θ values of -2, 0, and 2, for the three items that exhibit DIF with the linear (CFA) results on the left and the IRT DIF results on the right.

For "Yielding," the results from both analyses can be stated succinctly: The slope differs between men (blue lines) and women (magenta lines); "Yielding" is a "good"

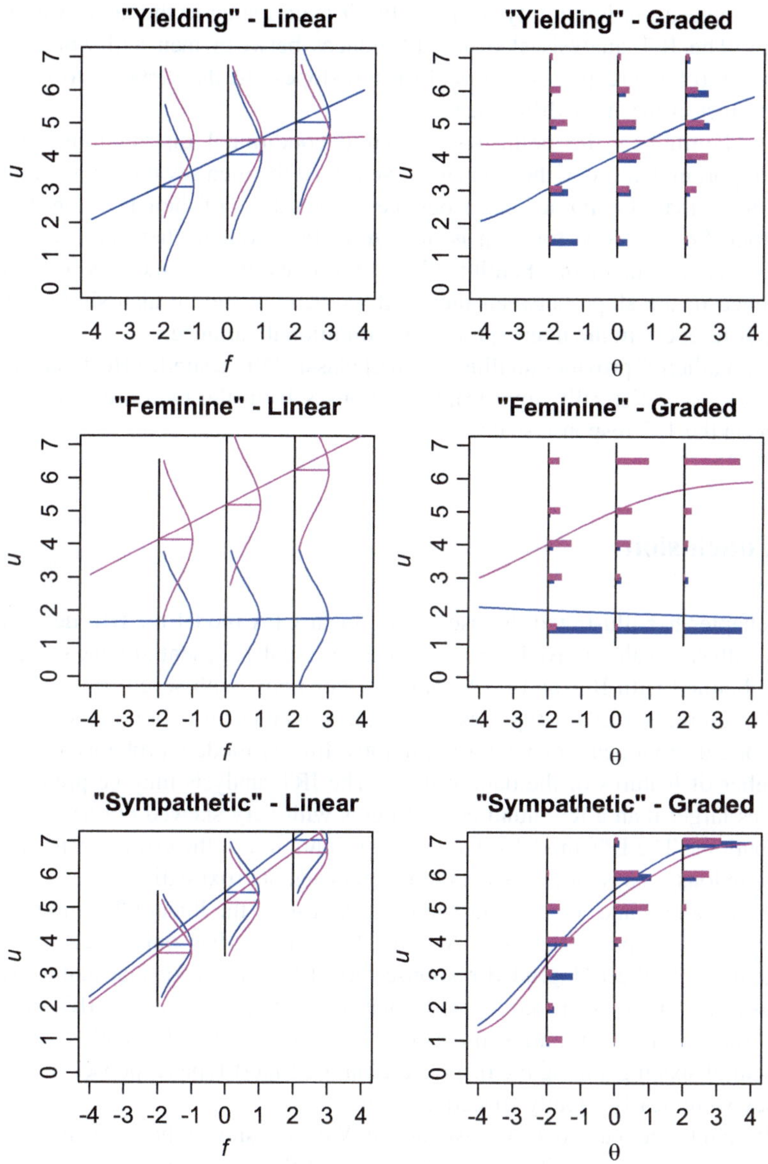

Fig. 3 Linear CFA models (*left*) and IRT models (*right*) for the three DIF items. Each *graphic* shows the expected score curve (a *straight line* for the linear models) on the 1–7 response scale as a function of the latent variable measured by the other items, with modeled response distributions at *f* or θ values of −2, 0, and 2. The models for the men are shown in *blue* and those for the women in *magenta*

item for measuring femininity for men, but it is unrelated to the latent variable for women. The IRT analysis also has differences between men and women in the thresholds (or intercepts, conditional on the slope) but that appears to have been a side effect of the slope difference.

For "Feminine," the point-mass discrete conditional distributions make the most important feature of the data clear: Nearly all the men say it is "never or almost never true" that "Feminine" describes them, regardless of their level on the latent construct. So the slope for men is near zero. For women, there is a relationship between endorsement of "Feminine" and the construct, so there is a significant difference in the slope between men and women. There are also significant, and large, differences in the intercept and the conditional variance.

"Sympathetic" provides an illustration of classic DIF: a subtle effect that is barely detectable as statistically significant. Responses from the men are a uniform 0.4 higher on the 1–7 response scale.

4 Conclusion

This chapter has illustrated the use of the linear CFA model for DIF detection for items with seven alternative Likert-type response scales. Essentially the same results were obtained with IRT and CFA analyses. For items with seven (plus or minus a couple) alternative Likert-type response scales, whether it is preferable to use IRT (categorical, nonlinear) or CFA (continuous, linear) models probably depends on a number of features of the data at hand. The IRT analysis may be preferable for samples larger than a few hundred and items with very skewed observed response distributions. The IRT model's discrete representation of the error distribution is in some obvious senses more correct than the normal approximation.

On the other hand, for smaller samples, the use of the linear CFA model avoids the need to collapse categories. And the CFA model has only three parameters per item where the IRT graded response model has seven for that many response categories; it is not clear that the additional four parameters for the IRT model are worth their cost in estimation precision. Those parameters basically "adjust for" differential spacing among the responses, but the Likert-type responses may "act as if" they were nearly equally spaced numbers.

Ultimately the data analyst must choose. We have shown that basically the same results can be obtained either way, if the use of the linear model tests hypotheses in the same order as has been used in IRT approaches to DIF detection. As we noted earlier, the procedures involved are closely related in many respects to those suggested by Raykov et al. (2013) in a recent article on factorial invariance, which borrows from methods commonly used in DIF analysis, including the use of the Benjamini–Hochberg procedure to adjust for multiple comparisons among groups for many items. However, the work by Raykov et al. was not focused on DIF analysis; this chapter has intended to present the use of the linear CFA model from the perspective of IRT.

References

S. Bem, The measurement of psychological androgyny. J. Consult. Clin. Psychol. **42**, 155–162 (1974)
S. Bem, *A Manual for the Bem Sex Role Inventory* (Consulting Psychologist Press, Palo Alto, 1981)
Y. Benjamini, Y. Hochberg, Controlling the false discovery rate: a practical and powerful approach to multiple testing. J. R. Stat. Soc. Ser. B **57**, 289–300 (1995)
L. Cai, D. Thissen, S. du Toit, *IRTPRO Version 2: Flexible, Multidimensional, Multiple Categorical IRT Modeling [Computer Software Manual]* (Scientific Software International, Chicago, 2011)
M. Edwards, M. Edelen, Special topics in item response theory, in *The Sage Handbook of Quantitative Methods in Psychology*, ed. by R. Millsap, A. Maydeu-Olivares (Sage, London, 2009), pp. 178–198
F. Lord, A study of item bias, using item characteristic curve theory, in *Basic Problems in Cross-Cultural Psychology*, ed. by Y.H. Poortinga (Swets and Zeitlinger, Amsterdam, 1977), pp. 19–29
F. Lord, *Applications of Item Response Theory to Practical Testing Problems* (Lawrence Erlbaum Associates, Hillsdale, 1980)
W. Meredith, Notes on factorial invariance. Psychometrika **29**, 177–185 (1954)
W. Meredith, Measurement invariance, factor analysis, and factorial invariance. Pyschometrika **58**, 525–543 (1993)
W. Meredith, R. Millsap, The misuse of manifest variables in the detection of measurement bias. Psychometrika **57**, 289–311 (1992)
T. Raykov, G. Marcoulides, R. Millsap, Factorial invariance in multiple populations: a multiple testing procedure. Educ. Psychol. Meas. **73**, 713–727 (2013)
M. Rhemtulla, P. Brosseau-Liard, V. Savalei, When can categorical variables be treated as continuous? A comparison of robust continuous and categorical sem estimation methods under suboptimal conditions. Psychol. Methods **17**, 354–373 (2012)
Y. Rosseel, lavaan: an R package for structural equation modeling. J. Stat. Softw. **48**, 1–36 (2012)
F. Samejima, Estimation of latent ability using a response pattern of graded scores. Psychom. Monogr. **18**, 1–100 (1969)
F. Samejima, Graded response model, in *Handbook of Modern Item Response Theory*, ed. by W. van der Linden, R.K. Hambleton (Springer, New York, 1997), pp. 85–100
L. Steinberg, D. Thissen, Using effect sizes for research reporting: examples using item response theory to analyze differential item functioning. Psychol. Methods **11**, 402–415 (2006)
D. Thissen, L. Steinberg, H. Wainer, Use of item response theory in the study of group differences in trace lines, in *Test Validity*, ed. by H. Wainer, H. Braun (Erlbaum, Hillsdale, 1988), pp. 147–169
D. Thissen, L. Steinberg, H. Wainer, Detection of differential item functioning using the parameters of item response models, in *Differential Item Functioning*, ed. by P. Holland, H. Wainer (Lawrence Erlbaum Associates, Hillsdale, 1993), pp. 67–113
D. Thissen, L. Steinberg, D. Kuang, Quick and easy implementation of the Benjamini-Hochberg procedure for controlling the false positive rate in multiple comparisons. J. Educ. Behav. Stat. **27**(1), 77–83 (2002)
UNC Dataverse, Bem sex-role inventory [Data files and code books] (2009). Retrieved from http://hdl.handle.net/1902.29/CAPS-BEM
V. Williams, L. Jones, J. Tukey, Controlling error in multiple comparisons, with examples from state-to-state differences in educational achievement. J. Educ. Behav. Stat. **24**, 42–69 (1999)

Finally! A Valid Test of Configural Invariance Using Permutation in Multigroup CFA

Terrence D. Jorgensen, Benjamin A. Kite, Po-Yi Chen, and Stephen D. Short

Abstract In multigroup factor analysis, configural measurement invariance is accepted as tenable when researchers either (a) fail to reject the null hypothesis of exact fit using a χ^2 test or (b) conclude that a model fits approximately well enough, according to one or more alternative fit indices (AFIs). These criteria fail for two reasons. First, the test of perfect fit confounds model fit with group equivalence, so rejecting the null hypothesis of perfect fit does not imply that the null hypothesis of configural invariance should be rejected. Second, treating common rules of thumb as critical values for judging approximate fit yields inconsistent results across conditions because fixed cutoffs ignore sampling variability of AFIs. As a solution, we propose replacing χ^2 and fixed AFI cutoffs with permutation tests. Iterative permutation of group assignment yields an empirical distribution of any fit measure under the null hypothesis of invariance. Simulations show the permutation test of configural invariance controls Type I error rates better than χ^2 or AFIs when a model has parsimony error (i.e., negligible misspecification) but the factor structure is equivalent across groups (i.e., the null hypothesis is true).

Keywords Measurement equivalence • Configural invariance • Permutation • Multiple group confirmatory factor analysis

1 Introduction

The assumption of measurement equivalence/invariance (ME/I) is required to draw inferences about how latent constructs might differ across different contexts, such as different occasions or populations. Configural ME/I, in particular, must

T.D. Jorgensen (✉)
University of Amsterdam, Amsterdam, The Netherlands
e-mail: T.D.Jorgensen@uva.nl

B.A. Kite • P.-Y. Chen
University of Kansas, Lawrence, KS 66045, USA

S.D. Short
College of Charleston, Charleston, SC 29424, USA

L.A. van der Ark et al. (eds.), *Quantitative Psychology*, Springer Proceedings in Mathematics & Statistics 196, DOI 10.1007/978-3-319-56294-0_9

be implicitly assumed before measurement parameters can be compared across contexts, whose equality is required before comparing common-factor parameters across contexts. Multigroup confirmatory factor analysis (CFA) is one of the most common frameworks used to test ME/I across groups (Vandenberg and Lance 2000), and multigroup models provided the only avenue for testing whether factor structure is configured equivalently across groups.

We first describe the current recommended best practices for testing configural invariance, as well as their limitations. We then propose a permutation randomization test of configural invariance across groups. We present Monte Carlo simulation studies to compare the power and Type I error rates of the permutation method to other methods.

1.1 Assessing Configural Invariance

To test for configural invariance (i.e., the same form), researchers fit a model with identical factor structure across groups, but allow all freely estimated measurement-model parameters (factor loadings, intercepts, and residual variances) to differ between groups (except scale-identification constraints). The likelihood ratio test (LRT or χ^2 statistic of exact fit) is used to judge whether the configural invariance model is an acceptable baseline model before constraining measurement parameters (Byrne et al. 1989). If the test is not significant at the specified α level, the analyst fails to reject the null hypothesis (H_0) that the configural model fits well and proceeds to test equality of item parameters (e.g., factor loadings, intercepts or thresholds, residual variances) across groups.

The LRT confounds two sources of model misfit (Cudeck and Henly 1991; MacCallum 2003): estimation discrepancy (due to sampling error) and approximation discrepancy (due to a lack of correspondence between the population and analysis models). Because configural ME/I is assessed by testing the absolute fit of the configural model, an LRT for a multigroup model further confounds two sources of approximation discrepancy. The overall lack of correspondence between the population and analysis models could theoretically be partitioned into (a) differences among the groups' true population models and (b) discrepancies between each group's population and analysis models. It is possible (perhaps even probable) that an analysis model corresponds only approximately to the groups' population models (Byrne et al. 1989), yet the analysis model may be equally (in)appropriate for each group. Although overall model fit is certainly important to assess in conjunction with tests of ME/I, the H_0 of configural invariance is only concerned with group equivalence, so the LRT does not truly provide a test of configural invariance.

Large sample sizes make the LRT sensitive even to minute differences in model form, which would have little or no practical consequence on parameter estimates. Many researchers would prefer to use an alternative fit index (AFI) to assess the approximate fit of the configural model. Putnick and Bornstein (2016) found that only 17% of ME/I tests are decided by the LRT alone, whereas 46% also involve at least one AFI, and 34% are decided using AFIs alone. The comparative fit index (CFI) (Bentler 1990) was reported for 73.2% of ME/I tests, making it the most popular AFI in this context. This chapter will focus mainly on CFI, but the root mean square error of approximation (RMSEA) (Steiger and Lind 1980) is also very popular. AFIs are functions of overall discrepancies between observed and model-implied sample moments, so using them to assess configural invariance would confound group equivalence with overall misfit, just like the LRT. However, we discuss their additional limitations below.

Most AFIs do not have known sampling distributions,[1] so evaluating the fit of a configural model involves some subjective decisions (e.g., which fit indices to use, what values indicate acceptable fit). Sometimes there are conflicting recommendations based on different criteria. For example, Bentler and Bonett (1980) suggested CFI > 0.90 indicates good fit, yet Hu and Bentler (1999) recommended CFI > 0.95 as a stricter criterion. Browne and Cudeck (1992) suggested RMSEA < 0.05 indicates close fit, RMSEA < 0.08 indicates reasonable fit, and RMSEA > 0.10 indices poor fit (RMSEA between 0.08 and 0.10 indicates mediocre fit) (MacCallum et al. 1996), yet Hu and Bentler recommended RMSEA < 0.06 as a stricter criterion. According to an October 2016 Google Scholar search, Hu and Bentler's criteria seem to be more widely applied (35,474 citations) than Bentler and Bonett's (13,815 citations) or Browne and Cudeck's (2843 citations).

The problem with using fixed cutoffs, even as mere rules of thumb, is that they ignore conditions specific to the study, such as sample size (and by implication, sampling error), number of groups, sample size ratios, number (and pattern) of indicators and factors, etc. Fixed cutoffs can also lead to the apparent paradox that larger samples yield lower power, which occurs when the AFI cutoff is more extreme than the population-level AFI[2] (Marsh et al. 2004).

1.2 A Permutation Randomization Test of Configural Invariance

When a theoretical distribution of a statistic is unavailable for null-hypothesis significance testing, it is possible for researchers to use a resampling method to create an empirical sampling distribution from their observed data. Rodgers (1999) provided a useful taxonomy of resampling methods. One flexible method that can be used to create an empirical approximation of a sampling distribution is the permutation randomization test. If a method of resampling the data can be conceived such that a H_0 is known to be true (in the permutation distribution),

---

[1]A notable exception is RMSEA. See an excellent discussion by Kenny et al. (2015).

[2]Population-level AFIs can be obtained by fitting the analysis model to the population moments or can be estimated from the average AFI across Monte Carlo samples.

then reference distributions can be empirically approximated for statistics whose sampling distributions are unknown or intractable.

The logic of the permutation test is related to the use of random assignment in experimental designs. Random assignment of subjects to two (or more) groups will average out any between-group differences, so that on average, group mean differences would be zero, resulting in two (or more) comparable groups before administering different treatments. Due to sampling fluctuation, observed differences will not be exactly zero after any single random assignment, but differences will be zero on average across replications of random assignment. Capitalizing on this effect of randomization, when a set of observed outcome scores (*Y*) is randomly (re)assigned to the two different observed groups (natural[3] or experimental), any existing between-group differences would be zero, on average.

A simple example of a permutation test is to compare two group means. The grouping variable (*G*) can be resampled without replacement and paired with values on the dependent variable (*Y*). The resulting randomization is a single permutation (reordering) of the data. Because H_0: $\mu_1 - \mu_2 = 0$ is true (i.e., the groups do not systematically differ in a permuted data set), the calculated *t* value is one observation from a theoretically infinite population of *t* statistics that could be calculated under the H_0 of no group mean difference. Repeating this process 100 times results in a distribution of 100 *t* statistics under H_0, one *t* value from each permutation of the data. As the number of permutations increases, the shape of the empirical distribution of the *t* values will become a closer approximation of the true, but unknown, sampling distribution. Using the empirical approximation of the sampling distribution under H_0, a researcher can calculate a good approximate *p* value by determining the proportion of the permutation distribution that is more extreme than the *t* value calculated from the original, unpermuted data.

We propose a permutation method for testing configural invariance. Randomly permuting group assignment yields resampled data for which the H_0 of group equivalence in model fit is true, even if the model does not fit perfectly. The steps to test configural ME/I are:

1. Fit the hypothesized multiple-group model to the original data, and save the fit measure(s) of interest.
2. Sample *N* values without replacement from the observed grouping-variable vector *G*. The new vector $G_{\text{perm(i)}}$ contains the same values as *G*, but in a new randomly determined order (i.e., $G_{\text{perm(i)}}$ is a permutation of *G*).
3. Assign the *n*th row of original data to the *n*th permuted value from $G_{\text{perm(i)}}$. On average, group differences are removed from this *i*th permuted data set.
4. Fit the same multiple-group model from step 1 to the permuted data, and save the same fit measure(s).

[3]The exchangeability assumption might be violated for natural groups (Hayes 1996), which we bring up in the Discussion.

5. Repeat steps 2–4 I times, resulting in a vector of length I for each fit measure.
6. Make an inference about the observed fit measure by comparing it to the vector of permuted fit measures.

Step 6 can test H_0 in either of two ways, yielding the same decision:

- Calculate the proportion of the vector of permuted fit measures that is more extreme (i.e., worse fit) than the observed fit measure. This is a one-tailed p value that approximates the probability of obtaining a fit measure at least as poor as the observed one, if the H_0 of ME/I for all groups holds true. Reject H_0 if $p < \alpha$.
- Sort the vector of permuted fit measures in ascending order for badness of fit measures like χ^2 or RMSEA or sort in descending order for goodness of fit indices like CFI. Use the $[100 \times (1 - \alpha)]$th percentile as a critical value, and reject H_0 if the observed fit measure is more extreme than the critical value.

Because permutation removes group differences (on average) without altering the structure among the variables in any other way, this method provides a simple framework to test configural ME/I separately from overall model fit. Furthermore, permutation provides empirical sampling distributions of AFIs, which generally have unknown sampling distributions. Researchers using permutation methods would not need to rely on fixed cutoff criteria based on intuition or studies whose simulated conditions might not closely resemble their own data and model(s), such as CFI > 0.90 (Bentler and Bonett 1980) or CFI > 0.95 (Hu and Bentler 1999). As we demonstrate using simulation studies, none of these fixed rules-of-thumb consistently control Type I error rates. In contrast, permutation distributions implicitly take into account the unique qualities of the data and model under consideration.

2 Monte Carlo Simulations

To evaluate the permutation methods proposed in the previous section, we present results from two simulation studies. The first evaluated Type I error rates when H_0 is true and second evaluated power when H_0 is false. In each study, we compared H_0 rejection rates between permutation methods and currently recommended practices under a variety of sample-size and model-size conditions.

We chose conditions based on Meade et al. (2008) ME/I study, which included approximation error in the form of near-zero cross-loadings in the population models that were fixed to zero in the analysis models (i.e., simple structure was only approximately true; see Table 1). When fitting simple structure CFA models to the model-implied population moments, AFIs indicated good model fit using standard conventions (e.g., CFI > 0.98, RMSEA < 0.03).

Based on Meade et al. (2008), we varied sample size (N) in each of two groups across five levels, 100, 200, 400, 800, and 1600 per group. We varied model complexity via number of factors (2 or 4) and number of items per factor (4 or 8), using the same population values for factor loadings as Meade et al. (2008) (see

Table 1 Population factor loadings (Λ matrix)

Item	Factor 1	Factor 2	Factor 3	Factor 4
1	0.68 (0.54)	−0.03	−0.02	−0.11
2	0.76 (0.62)	0.02	−0.03	−0.03
3	0.74 (0.60)	−0.04	−0.03	0.00
4	0.75 (0.61)	0.00	−0.01	0.08
5	0.04	0.76 (0.61)	0.07	0.00
6	−0.06	0.56 (0.41)	−0.03	0.04
7	−0.08	0.75 (0.60)	0.07	0.06
8	−0.02	0.72 (0.57)	0.05	−0.03
9	0.07	−0.01	0.80 (0.65)	0.00
10	−0.01	−0.03	0.58 (0.43)	−0.02
11	−0.04	0.06	0.80 (0.65)	0.03
12	0.04	0.00	0.39 (0.24)	0.05
13	−0.02	−0.02	−0.01	0.65 (0.51)
14	0.00	−0.13	−0.03	0.67 (0.53)
15	0.00	0.03	−0.01	0.59 (0.45)
16	0.00	0.03	0.02	0.67 (0.53)

Note. For conditions with eight indicators per factor, λ_s in parentheses were used as population parameters, and λ_s for items 17–32 were identical to λ_s for items 1–16. Cells with only one value (near zero) are minor discrepancies from simple structure (approximation error)

Table 1). In the population models, we fixed all intercepts to zero, discussed next), factor means to zero, factor variances to one, factor correlations to 0.3, and residual variances to values that would set total item variances to one (i.e., standard normal variables). We simulated 2000 replications in each condition and used $I = 200$ permutations to calculate p values associated with fit measures.

Whereas Meade et al. (2008) simulated configural lack of invariance (LOI) by adding additional factors to group two's population model (resulting in dozens of different population models), we simply changed one, two, three, or four of the zero (or nonsalient) parameters in group two's population model. The first level of configural LOI was to change factor loading λ_{51} from 0.04 to 0.7. The second level was to make the same change to λ_{51} and to add a residual covariance ($\theta_{72} = 0.2$). The third level made the same additions and changed λ_{12} from −0.03 to 0.7, and the fourth level also added another residual covariance ($\theta_{84} = 0.2$). These arbitrary levels of configural LOI served to compare the power of different methods to detect the same lack of correspondence between the groups' population models.

We used R (R Core Team 2016) to generate multivariate normal data and the R package lavaan (version 0.5–20; Rosseel 2012) to fit models to simulated data. Using lavaan's default settings, the scales of the latent factors were identified by fixing the first factor loading to one, and the latent means were fixed to zero.

3 Results

We first used a 5 (N) × 2 (2 or 4 factors) × 2 (4 or 8 items per factor) design, holding LOI constant at zero. Because the population models included minor approximation discrepancy in the form of near-zero cross-loadings, we expected Type I error rates to exceed 5% and for these rates to increase with N. Because fixed cutoffs do not take sampling variability or model complexity into account, we expected results to vary across Ns and model sizes. We expected permutation to yield nominal Type I error rates in all conditions for all fit measures, which would indicate a valid test of configural invariance.

As expected, using the traditional LRT resulted in extremely high Type I error rates. Figure 1 confirms that even in the condition with the smallest N and model, Type I errors were almost 20%, approaching 100% as N increased. For larger Ns, rejection rates matched the expected power using the Satorra and Saris (1985) method, but rejection rates were inflated at smaller N, especially in larger models, due to the small-sample bias of the LRT (Nevitt and Hancock 2004). In contrast, permutation provided nominal Type I error rates across conditions.

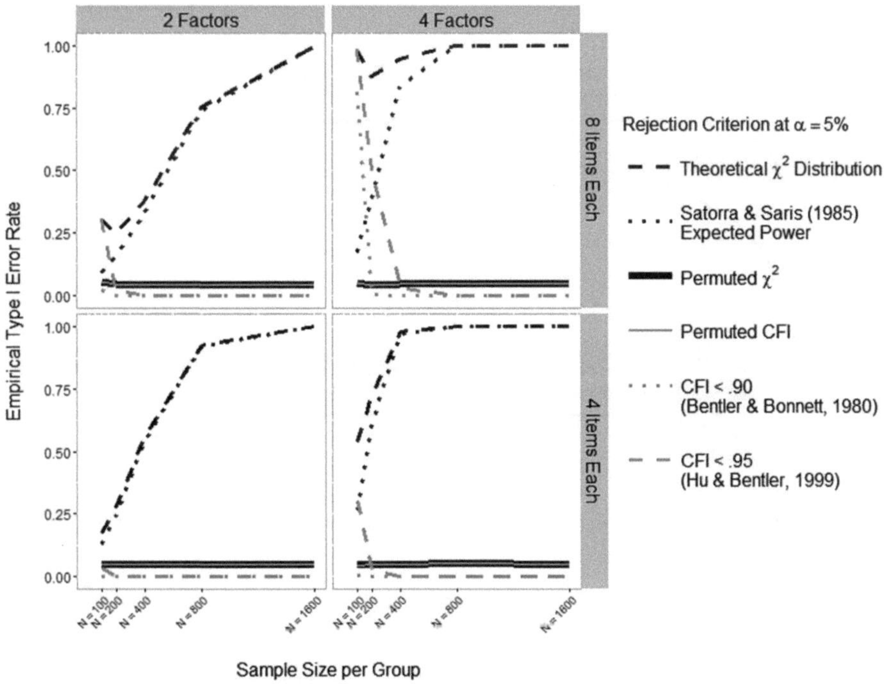

Fig. 1 Observed Type I error rates for LRT, CFI rules of thumb, and permutation tests of configural invariance, as well as expected power of LRT using the Satorra and Saris (1985) method

Using AFIs to assess approximate fit of the configural model only appeared to yield inflated Type I errors under small-N conditions, but that depended heavily on the size of the model and on which rule of thumb was used. Figure 1 shows that larger models yielded more errors at smaller N. Similar results were found for RMSEA guidelines. In contrast, permuting CFI (or any AFI) maintained nominal Type I error rates across all conditions.

We next used a 5 (N) × 4 (LOI) design, holding model complexity constant (4 items for each of 2 factors, the condition in which fixed cutoffs for CFI showed ≤5% Type I errors). We expected permutation to have lower power than the LRT, which already had high rejection rates when H_0 was true. Given that Type I error rates for AFI cutoffs were typically close to zero for this particular population model, we had no specific hypotheses about how their power would compare to power using permutation, but we did expect lower power with increasing N in conditions where population AFIs met guidelines for acceptable fit.

Figure 2 confirms our expectation that the LRT had the highest power to detect LOI, particularly at the lowest level of LOI and the smallest N. But as Fig. 1 shows, the greater power came at the expense of high Type I errors because the LRT tests overall model fit rather than configural invariance alone. Hu and Bentler's (1999) more stringent criterion (CFI > 0.95) yielded power almost as high as the LRT,

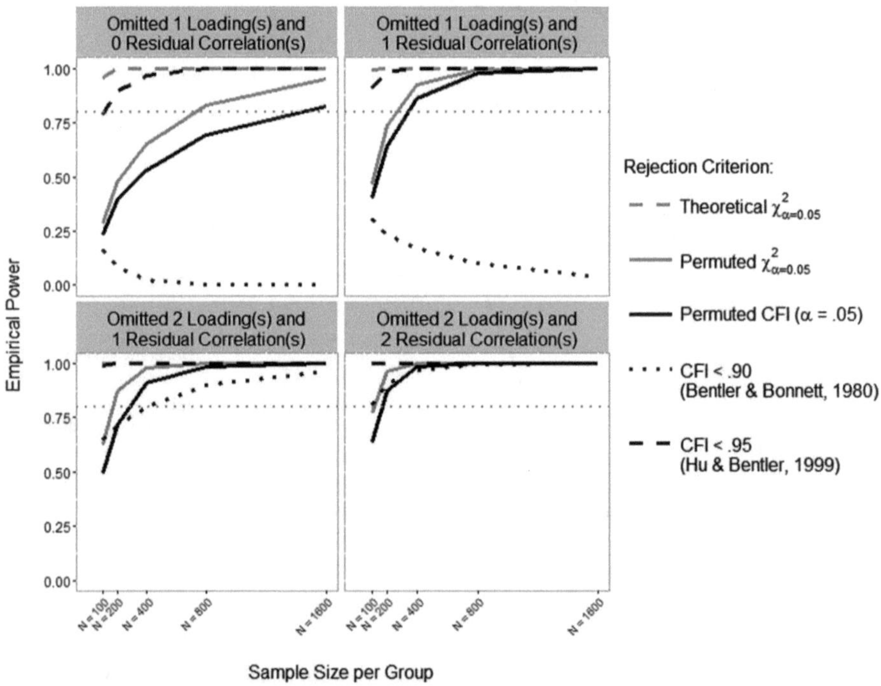

Fig. 2 Power for LRT (gray lines) and CFI (*black lines*) using theoretical (or fixed) vs. permutation-based critical values. The *dotted gray line* indicates 80% power

whereas Bentler and Bonett's (1980) less stringent criterion (CFI > 0.90) yielded lower power that decreased as N increased in conditions where only one or two salient population parameters differed between groups. We found the same pattern of results for RMSEA.

Permutation yielded inadequate power when only a single parameter differed between populations, unless $N \geq 800$ per group. Adequate power to detect greater LOI was achieved at smaller N. The permuted LRT tended to have greater power than permuted CFI, but the discrepancy was small when N and LOI were large. Permuted RMSEA had power similar to the permuted LRT.

4 Discussion

We proposed a permutation randomization framework for using multigroup CFA to test ME/I. We proposed this framework to address some limitations of current best practices. First, the LRT of exact (or equal) fit does not test the correct H_0 of group equivalence for the configural model. Assessing overall model fit confounds any group differences with overall model misspecification. Irrespective of how well a model only approximates a population process, the model may be equally well specified for both groups, in which case the H_0 of group equivalence should not be rejected. Our simulation studies showed that current best practices can lead to highly inflated Type I error rates, even for models with very good approximate fit. Permutation, on the other hand, yields well-controlled Type I error rates even when the model does not fit perfectly, providing the only valid test of configural invariance across groups that we are currently aware of.

Second, most researchers prefer AFIs over the LRT (Putnick and Bornstein 2016) because of the latter's sensitivity to differences that are negligible in practice, which could be thought of as inflated Type I error rates when assessing approximate fit in large samples. However, lack of known distributions for ΔAFIs leads to reliance on rule-of-thumb cutoffs that, as we have shown, lead to inflated Type I error rates in smaller (albeit still large) samples, especially in larger models. Our simulations showed that regardless of which fit measure is preferred, permutation provides well controlled Type I error rates, with power to detect true differences that is comparable to the LRT.

We recommend that applied researchers interested in testing configural invariance use the permutation method, which is implemented in a function called "`permuteMeasEq()`" in the R package `semTools` (semTools Contributors 2016). If the overall fit of the configural model is satisfactory, the permutation method provides a valid test of the H_0 of group equivalence in model form and is currently the only method to do so. Permutation may be particularly valuable in conditions with inflated error rates, such as missing or categorical data, but its utility may be limited by the exchangeability assumption. We encourage further investigation of permutation methods for testing group equivalence, particularly for developing guidelines for modifying individual group models (when configural invariance does not hold) versus making modifications to poorly fitting models simultaneously across groups (when configural invariance does hold).

References

P.M. Bentler, Comparative fit indexes in structural models. Psychol. Bull. **107**(2), 238–246 (1990). doi:10.1037/0033-2909.107.2.238

P.M. Bentler, D.G. Bonett, Significance tests and goodness of fit in the analysis of covariance structures. Psychol. Bull. **88**(3), 588–606 (1980). doi:10.1037/0033-2909.88.3.588

M.W. Browne, R. Cudeck, Alternative ways of assessing model fit. Sociol. Methods Res. **21**, 230–258 (1992). doi:10.1177/0049124192021002005

B.M. Byrne, R.J. Shavelson, B. Muthén, Testing for the equivalence of factor co-variance and mean structures: The issue of partial measurement invariance. Psychol. Bull. **105**(3), 456–466 (1989). doi:10.1037/0033-2909.105.3.456

R. Cudeck, S.J. Henly, Model selection in covariance structures analysis and the "problem" of sample size: a clarification. Psychol. Bull. **109**(3), 512–519 (1991). doi:10.1037//0033-2909.109.3.512

A.F. Hayes, Permutation test is not distribution-free: Testing H0: $\rho = 0$. Psychol. Methods **1**(2), 184–198 (1996). doi:10.1037/1082-989X.1.2.184

L.-T. Hu, P.M. Bentler, Cutoff criteria for fit indexes in covariance structure analysis: conventional criteria versus new alternatives. Struct. Equ. Model. **6**(1), 1–55 (1999). doi:10.1080/10705519909540118

D.A. Kenny, B. Kaniskan, D.B. McCoach, The performance of RMSEA in models with small degrees of freedom. Sociol. Methods Res. **44**(3), 486–507 (2015). doi:10.1177/0049124114543236

R.C. MacCallum, 2001 presidential address: working with imperfect models. Multivar. Behav. Res. **38**(1), 113–139 (2003). doi:10.1207/S15327906MBR3801_5

R.C. MacCallum, M.W. Browne, H.M. Sugawara, Power analysis and determination of sample size for covariance structure modeling. Psychol. Methods **1**(2), 130–149 (1996). doi:10.1037//1082-989X.1.2.130

H.W. Marsh, K.-T. Hau, Z. Wen, In search of golden rules: comment on hypothesis-testing approaches to setting cutoff values for fit indexes and dangers in overgeneralizing Hu and Bentler's (1999) findings. Struct. Equ. Model. **11**(3), 320–341 (2004). doi:10.1207/s15328007sem1103_2

A.W. Meade, E.C. Johnson, P.W. Braddy, Power and sensitivity of alternative fit indices in tests of measurement invariance. J. Appl. Psychol. **93**(3), 568–592 (2008). doi:10.1037/0021-9010.93.3.568

J. Nevitt, G.R. Hancock, Evaluating small sample approaches for model test statistics in structural equation modeling. Multivar. Behav. Res. **39**(3), 439–478 (2004). doi:10.1207/S15327906MBR3903_3

D.L. Putnick, M.H. Bornstein, Measurement invariance conventions and reporting: The state of the art and future directions for psychological research. Dev. Rev. **41**, 71–90 (2016). doi:10.1016/j.dr.2016.06.004

R Core Team, R: a language and environment for statistical computing (version 3.3.0). R Foundation for Statistical Computing (2016). Available via CRAN, https://www.R-project.org/

J.L. Rodgers, The bootstrap, the jackknife, and the randomization test: a sampling taxonomy. Multivar. Behav. Res. **34**(4), 441–456 (1999). doi:10.1207/S15327906MBR3404_2

Y. Rosseel, Lavaan: an R package for structural equation modeling. J. Stat. Softw. **48**(2), 1–36 (2012.) http://www.jstatsoft.org/v48/i02/

A. Satorra, W.E. Saris, Power of the likelihood ratio test in covariance structure analysis. Psychometrika **50**, 83–90 (1985). doi:10.1007/BF02294150

J.H. Steiger, J.C. Lind, Statistically-based tests for the number of common factors. Paper presented at the annual meeting of the Psychometric Society, Iowa City (1980)

semTools Contributors, semTools: useful tools for structural equation modeling (version 0.4–12) (2016). Available via CRAN. https://www.R-project.org/

R.J. Vandenberg, C.E. Lance, A review and synthesis of the measurement invariance literature: Suggestions, practices, and recommendations for organizational research. Organ. Res. Methods **3**(1), 4–70 (2000). doi:10.1177/109442810031002

Outcries of Dual Scaling: The Key Is Duality

Shizuhiko Nishisato

Abstract There are a number of points in the development of dual scaling which have escaped our attention. In my Beijing paper, problems with joint graphical display were discussed to fill the gap of understanding, and the current paper deals with some other points. These two papers can be regarded as a sequel to my paper, entitled "Gleaning in the field of dual scaling," written 20 years ago. Noting that the basic premise of dual scaling lies in duality of exhaustive analysis, we will look at a few more points in this paper. Outcry one is on linear and nonlinear analysis. As is well known, dual scaling is a method for simultaneous regressions of row variates and column variates on data, capturing all linear and nonlinear relations contained in the data. From this point of view, Likert scores, used as scores for data analysis, are far from satisfactory, for it is a strictly linear and data-independent procedure. Outcry two is on our definition of multidimensional quantification space, because the traditional framework needs to be modified so as to satisfy our objective, that is, describing both row and column structure of data in a symmetric comprehensive way. Outcry three is on a logical alternative to problem-plagued joint graphical display, and a recommended alternative is cluster analysis. Finally, outcry four is on the distinction between dual space and total space, leading to the suggestion that simple correspondence analysis fails to provide exhaustive analysis of information in data.

Keywords Coordinates for data • Simultaneous symmetric analysis • Joint graphical display • Doubled space • Cluster analysis • Dual space versus total space

1 Introduction

In 1996, Nishisato presented his presidential address, entitled "Gleaning in the field of dual scaling," in which he identified a number of hidden or unsolved aspects

S. Nishisato (✉)
Professor Emeritus, University of Toronto, 30 Old Mill Road, Suite 308, Toronto, Ontario M8X 0A5, Canada
e-mail: shizuhiko.nishisato@utoronto.ca

L.A. van der Ark et al. (eds.), *Quantitative Psychology*, Springer Proceedings in Mathematics & Statistics 196, DOI 10.1007/978-3-319-56294-0_10

of dual scaling (Nishisato 1996). It is 20 years since then, and one wonders if dual scaling is well understood and if some of the problems raised then have been solved to our satisfaction. Some major problems were discussed in the paper, entitled "Multidimensional joint graphical display of symmetric analysis: Back to the fundamentals" (Nishisato 2016a). The current paper supplements it with further discussion of the problems in quantification theory. Current concerns are with the nature of multidimensional space used in quantification, in particular about the point that we must at least double the dimensionality of the space to accommodate quantified variates, which makes us wonder if we should still pursue joint graphical display or consider an alternative to graphical display. Simple correspondence analysis is known as one of the main realms of quantification theory, and it is dual scaling of the contingency table. The current paper, however, will take us to the point at which we may have to say "farewell" to it. Let us discuss these problems as outcries.

1.1 Outcry 1: Linear and Nonlinear Analysis

This is a well-known aspect of quantification theory, but it seems that the point needs to be reemphasized. Suppose that we collect data on preference of tea under different water temperatures. Each subject is given ten cups of tea, ranging from freezing cold to boiling hot, and is asked to rate the preference of ten cups of tea on the 10-point scale, ranging from the worst to the best. If we use 10-point Likert scales for the temperature and for the preference ratings, the data can be presented as a 10-by-10 contingency table of choice frequencies. Typical analysis of the table using Likert scores without transformation, however, would not capture such a nonlinear relation as might be expected, namely, the preference being at the lowest (least liked) when the tea temperature is boiling hot, followed by freezing cold, then lukewarm, then ordinary cold ice tea, and finally optimally hot tea at the highest (most preferred). There are at least two distinct approaches to this kind of nonlinear relation. The first approach is to predict the preference Likert scores as a nonlinear function of the temperature of tea, indicated by Likert scores. Should we use a quadratic term, a cubic term, interaction terms, or higher order terms? The choice of these is not easy, but we must seek the best possible nonlinear function, and this is, however, not what most investigators would normally do—they do not consider any nonlinear function. Furthermore, what can we do to deal with multidimensional aspects of the data in this nonlinear regression approach? This is not a simple problem. The second approach is via correlation of the Likert scores of the two variables. In this approach, it is well known that Pearsonian correlation captures only linear relations; thus this is not an appropriate way to analyze nonlinear relations. One should realize then that Likert scores are predetermined quantities, independently of the data structure, and without additional operations of nonlinear transformations, one cannot generally expect exhaustive analysis of information in data through Likert scores. In contrast to these two approaches, dual scaling (correspondence analysis, homogeneity analysis, optimal scaling) is a method to find optimal scores for both the temperature and the preference ratings as regressions on the data. In other words, dual scaling is used to transform Likert scores typically nonlinearly so as to make the regression of rows (ten cups of tea) on preference ratings and the regression of columns (ten preference values) on tea simultaneously linear (Hirschfeld 1935). Thus, this is a data-dependent method of scaling row values and column values in the optimal way, the reason why it is also called optimal scaling (Bock 1960), and multidimensional aspects of the data can be handled without problems. In this context, dual scaling is a method of projecting row values to the column values and column values to row values in the symmetric way. The common projection operators are known as singular values, which are also Hirschfeld's simultaneous regression coefficients and Guttman's maximal correlation coefficients between row quantification and column quantification. In terms of multidimensional decomposition, Nishisato (2006) has shown that dual scaling maximizes the Cramér's coefficient (Cramér 1946) and that this coefficient is the sum of the squared nonlinear correlation coefficients of principal components. This indicates how dual scaling deals with multidimensional nonlinear relations in the data.

In summary, Likert scores are predetermined scores, independently of the data, and should be used only for the purpose of data collection. Once data are collected, Likert scores should be subjected to transformation, typically nonlinear, so as to best describe the information in data.

There is a caution on the use of order constraints in analysis. Because the response categories are ordered (e.g., never < sometimes < often < always), one may wish to derive scores for these categories under the order constraint. This may sound reasonable, but one should not even be tempted to impose such an order constraint if the study aims to explore the information in the data, that is, if the research is exploratory. The reason is clear. The order constraint permanently wipes out the possibility of ever finding nonlinear relations in the data (e.g., one's ability to lift a heavy object increases as one gets older to a certain point and then decreases beyond a certain age). Thus, a general advice is not to use the order constraint in exploratory research. Note that there are many studies on ordered categories in quantification theory, but that the above advice should be kept in mind.

1.2 *Outcry 2: Nature of Multidimensional Space for Symmetric Analysis*

Dual scaling is based on the mathematical decomposition of data, called dual relations:

$$\frac{\sum_i^m f_{ij} y_{ik}}{f_{.j}} = \rho_k x_{kj}; \quad \frac{\sum_j^n f_{ij} y_{ki}}{f_{i.}} = \rho_k y_{ik}$$

where f_{ij} is the frequency of cell (*i, j*) of a contingency table, y_{ik} and x_{kj} are weights for row *i* and column *j*, called standard coordinates, of component *k*, $\rho_k x_{kj}$ and $\rho_k y_{ik}$ are the corresponding principal coordinates, and ρ_k is the singular value of component *k*. This is nothing but Hirschfeld's simultaneous linear regressions with the singular value as the regression coefficient, and the singular value is also Guttman's maximal correlation between the rows and the columns and also Nishisato's projection operator for rows onto columns and vice versa. From the last point, we can conclude that the row axis and the column axis for each component are separated by the angle $\theta_k = \cos^{-1}\rho_k$ (Nishisato and Clavel 2003, 2008). This space discrepancy indicates that if we analyze a two-by-two contingency table, we obtain a single component, but the fact of the matter is that dual scaling of this contingency table requires a two-dimensional graph, one for row variables and the other for column variables with the two axes separated by the angle θ. This means that one component of dual scaling outcome requires two dimensions and two components four dimensions. From this point of view, the currently most popular graphical methods used in quantification studies are all problematic. The first two (symmetric and nonsymmetric graphs) are traditional quantification approaches to graphical display (see, e.g., Benzécri et al. 1973; Nishisato 1980, 1994, 2007; Greenacre 1984; Lebart et al. 1984; Gifi 1990; Le Roux and Rouanet 2004; Beh and Lombardo 2014), and the third one (biplot) is a more general and mathematical invention with a variety of graphical choices (see, e.g., Gabriel 1971; Gower and Hand 1996).

1. Symmetric display or French plot: The two sets of principal coordinates, $\rho_k x_{kj}$ and $\rho_k y_{ik}$, are plotted in the same space (i.e., without taking the space discrepancy θ_k into consideration). In other words, a two-dimensional configuration of data points is plotted in a unidimensional graph; similarly, a four-dimensional configuration is plotted in a two-dimensional graph. Thus, unless the singular value ρ is *very close* to 1, the symmetric display does not offer a usable graph (see, for example, the warning by Lebart et al. 1977). Generally speaking, symmetric display is an illogical and obviously wrong graph for the data, but for its simplicity, it has unfortunately become a routine method for graphing quantification results. This practice should immediately be discarded.
2. Nonsymmetric display: This method plots the principal coordinates of one variable and the standard coordinate of the other variable, for example, $\rho_k x_{kj}$ and y_{ik}. This is the projection of x onto the standard space of y. But, the standard coordinates are not the coordinates of the data, but artificially adjusted for the common variance, independently of the data at hand. Thus, projecting data onto these coordinates is not a logical way to describe data, thus making the joint graph not usable. See the demonstration (Nishisato 1996) that the standard coordinates associated with a small singular value are much further from the origin than those associated with a large singular value because the standard coordinates reciprocally compensate the frequencies of data points. One can consider the problem of principal component analysis, in which we start with a linear combination of variables, then find the principal axis, which is defined as the axis on which projections of data have the largest variance. Those projections of data on the principal axis are called principal coordinates. Therefore, principal

coordinates are the coordinates of data in the most informative way. Standard coordinates, on the other hand, do not represent projections of data points unless the singular value is 1.

3. Biplot: Consider the singular-value decomposition of a two-way data, $\boldsymbol{Y}\Delta^{a}\Delta^{1-a}\boldsymbol{X}$, where $\boldsymbol{Y}$ and $\boldsymbol{X}$ are, respectively, matrices of left and right singular vectors of the data matrix, Δ is the diagonal matrix of singular values, and α is bounded by 0 and 1. In biplot, graphical display of both variates are considered for various values of α. Notice, however, that only when α is either 0 or 1, it offers a plot comparable to the above two traditional plots, that is, nonsymmetric display of (2). In introducing coordinate systems for a set of variables, one of the most popular methods is through principal component analysis, where principal coordinates are the projections of data on principal axes. In this regard, principal coordinates represent data structure. It is true that the principal coordinate system is only one way of representing data, and there are an infinite number of coordinates systems, which, however, should be orthogonal transformations of the principal coordinates so long as we want to represent the data structure. Those variates used in biplots are not related to principal coordinates in any imaginable ways, except for one set of variates, Y or X, when α is 0 or 1. From the view of the graphical display in Euclidean space, therefore, it is the current author's personal view that a question mark has to be placed on the use of biplots for exploring data structure.

Considering that each of these popular methods for joint graphical display leaves a serious concern from the viewpoint that we wish to represent data in Euclidean multidimensional space, there seems to be an urgent problem of either finding a better method of graphical display or to give up a graphical display completely and look for a non-graphical way of summarizing the outcome of quantification.

1.3 Outcry 3: From "Graphing Is Believing" to Cluster Analysis

"Graphing is believing" (Nishisato 1997) was an attempt to legitimize joint graphical display of quantification results in Euclidean space. Since then, the author realized that a complete description of data requires a large number of dimensions, more precisely at least twice the dimensions that the traditional joint graphical display deals with. To clarify why we must at least double the dimensionality of space, Nishisato and Clavel (2010) proposed a framework for comprehensive dual scaling with doubled dimensions, and noting this aspect of expanded (doubled) dimensionality for graphical display, the authors proposed the use of cluster analysis as an alternative to the traditional graphical displays.

To illustrate their procedure, let us use an example from Stebbins (1950): 500 seeds of six varieties of barley were planted at six agricultural stations in the United States, and at the harvest time, 500 seeds at each station were randomly chosen

Table 1 Varieties of barley seeds after a number of years at different locations (from Stebbins 1950)

Locations						
	Arlington	Ithaca	St. Paul	Moccasin	Moro	Davis
Barley	(Virginia)	(New York)	(Minnesota)	(Montana)	(Oregon)	(California)
CT[a]	446	57	83	87	6	362
Ha	4	34	305	19	4	34
WS	4	0	4	241	480	65
Ma	1	343	2	21	0	0
Ga	13	9	15	58	0	1
Me	4	0	0	4	0	27

[a]Barley: *CT* Coast & Trebi, *Ha* Hanchen, *WS* White Smyrna, *Ma* Manchuria, *Ga* Gatemi, *Me* Meloy

and sorted into six varieties of barley, and those seeds were again planted in the following year, and at the harvest time, 500 randomly chosen seeds were again classified into six varieties, and so on. This experiment was repeated over a number of years to see if certain varieties of barley will become dominant at particular locations. The numbers of years of these experiments are not uniform but different from station to station. The final counts reported in Stebbins (1950) are summarized in the 6×6 contingency table (Table 1).

A complete dual scaling analysis of this data set is reported in Nishisato (1994), which shows the percentage contributions of the five components are, in the descending order, 38, 33, 25, 3, and 1%, showing the dominance of three components. Following Nishisato and Clavel (2003), the 12×12 super-distance matrix, consisting of the within-set distances of stations (between-station distances), the between-set distances (those between stations and barley varieties), and the within-set distances of barley varieties (between barley varieties), was calculated as given in Table 2.

This 12×12 matrix contains the distance information of all the variables in Euclidean space. Clavel and Nishisato (2008) and Nishisato and Clavel (2008) thoroughly analyzed this table by the hierarchical clustering method and the k-means clustering (see the results in their papers). Nishisato (2012) argued, however, that the investigators would typically be interested in the relations between row variables (stations) and column variables (varieties of barley), not relations within stations or within barley varieties, and therefore proposed that we should analyze only the between-set distance matrix, that is, the "barley varieties"-by-"locations" distance matrix. Although the current example of the between-set distance matrix is 6×6, the number of rows and the number of columns are not always equal; hence the between-set distance matrix is typically rectangular, as opposed to square. In order to deal with a rectangular distance matrix for clustering, Nishisato (2012) proposed a very simple and intuitive method of clustering, called clustering with the p-percentile filter. This method is very simple and does not require a complicated algorithm: calculate the p-percentile distance (the criterion distance) out of the

Table 2 Within-set and between-set distances in five-dimensional space (from Clavel and Nishisato 2008)

CT	0.0	(Symmetric upper portion is omitted)										
Ha	2.1	0.0										
WS	1.9	2.5	0.0									
Ma	2.6	2.9	2.8	0.0								
Ga	1.7	2.1	1.7	2.6	0.0							
Me	1.3	2.6	2.2	3.0	2.2	0.0						
Ar[a]	0.6	2.3	2.1	2.7	1.8	1.5	0.0					
It	2.3	2.7	2.6	0.9	2.4	2.8	2.5	0.0				
SP	2.0	0.9	2.5	2.9	2.1	2.5	2.3	2.7	0.0			
Mo	1.5	2.2	0.9	2.5	1.2	1.9	1.7	2.3	2.2	0.0		
Mo	2.1	2.6	0.7	2.9	1.9	2.4	2.4	2.8	2.7	1.1	0.0	
Da	0.6	2.1	1.8	2.6	1.7	1.3	0.6	2.4	2.1	1.4	2.0	0.0

[a]Locations: *Ar* Arlington, *It* Ithaca, *SP* St. Paul, *Mo* Mocasin, *Mo* Moro, *Da* Davis

Table 3 The 6 × 6 filtered between-set distance matrix, using 22-percentile criterion point

	CT[a]	Ha	WS	Ma	Ga	Me
Ar[b]	0.6	–	–	–	–	–
It	–	–	–	0.9	–	–
SP	–	0.9	–	–	–	–
Moc	–	–	0.9	–	1.2	–
Mor	–	–	0.7	–	–	–
Da	0.6	–	–	–	–	1.3

[a]*CT* Coast & Trebi, *Ha* Hanchen, *WS* White Smyrna, *Ma* Manchuria, *Ga* Gatemi, *Me* Meloy
[b]*Ar* Arlington, *It* Ithaca, *SP* St. Paul, *Mo* Mocasin, *Mo* Moro, *Da* Davis

elements of the between-set distance matrix, discard all distances which are larger than the criterion distance (i.e., variables which are widely separated do not belong to the same cluster), and see what clusters one can see among the remaining distances. The underlying idea is that we are interested only in those variables which are close to one another, thus we might as well discard all irrelevant distances from clustering. This method is simple and depending the value of p one chooses, the cluster can be tight or loose, and two clusters may or may not overlap. See its application in Nishisato (2014).

Let us apply the clustering with the p-percentile filter to the 6 × 6 matrix of the between-set distances, that is, the distance matrix between the six barley varieties and the six locations (see the 6 × 6 part of the left-bottom of the distance matrix). At the current stage of development, the choice of p is arbitrary, and for this example, $p = 22$ percentile was chosen, that is, all distances greater than this were discarded from the original 6 × 6 distance matrix, as shown in Table 3.

From this choice of the cutting point, we can identify the following clusters (Coast & Trebi at Arlington and Davis), (Hanchen at St. Paul), (White Smyrna at Mocasin and Moro), (Manchuria at Ithaca), (Gatemi at Mocasin), and (Meloy at Davis). As we can see, some clusters are overlapping. We can also guess that the overlapping can be eliminated by reducing the percentile point, although this may result in discarding some variables from analysis.

This filtering method is still at its infancy, and there are many studies needed before it can compete with other existing clustering methods, for example, how to determine the optimal p value for a given data set and how to calculate the distance between clusters. Its advantages over other methods are, among others, easiness or simplicity and capability to deal with rectangular matrices unlike some other existing methods. As for the traditional analysis through graphical display, see Nishisato (1994), noting that we must sacrifice much information in graphical display.

1.4 Outcry 4: Limitation of Simple Correspondence Analysis?

Traditionally, the French correspondence analysis identifies simple correspondence analysis and multiple correspondence analysis as two distinct forms of quantification. These "simple" and "multiple" methods correspond to dual scaling of the contingency table and that of multiple-choice data, respectively.

As was described in Nishisato (1980, 2016b), however, the two types of analysis are closely related to each other. Let us reproduce the example from Nishisato (2016b)— in response to the two multiple-choice questions:

Q1: Do you smoke? (yes, no)
Q2: Do you prefer coffee to tea? (yes, not always, no)

The data can be represented in three forms as shown in Table 4.

Table 4 Three forms for representing the information in the contingency table

$\boldsymbol{C} = \begin{bmatrix} 3\ 2\ 1 \\ 1\ 2\ 4 \end{bmatrix}$	$\boldsymbol{F}_a = \begin{bmatrix} 10100 \\ 10100 \\ 10100 \\ 10010 \\ 10010 \\ 10001 \\ 01100 \\ 01010 \\ 01010 \\ 01001 \\ 01001 \\ 01001 \\ 01001 \end{bmatrix}$	$\boldsymbol{F}_b = \begin{bmatrix} 30300 \\ 20020 \\ 10001 \\ 01100 \\ 02020 \\ 04004 \end{bmatrix}$

If we are given one of the three forms, the other two forms can be generated from it. In this regard, the three formats are "equivalent" in some sense. Since the latter two forms yield identical singular values, let us eliminate the second format $\boldsymbol{F}_a$ from our discussion. The remaining formats are data forms respectively for simple correspondence analysis and multiple-correspondence analysis. But, as Nishisato (1980) has shown, the two formats yield different numbers of components. An $m \times n$ table of $\boldsymbol{C}$ yields the "smaller number of m and n minus 1" components, while the corresponding $\boldsymbol{F}_b$ provides "$m + n - 2$" components. Nishisato (2016b) calls twice of the space of $\boldsymbol{C}$ as "dual space" and the space for $\boldsymbol{F}_b$ as "total space." As is clear, when $m = n$, dual space and total space have the same number of dimensions, but when $m \neq n$, the dimensionality of total space is greater than that of dual space. What is the nature of these extra components in total space when $m \neq n$?

The implication of this discrepancy is that simple correspondence analysis, which deals with the format of $\boldsymbol{C}$, fails to capture the total information in $\boldsymbol{F}_b$ when $m \neq n$, which is the format for multiple correspondence analysis. Thus, if we are to analyze the data information exhaustively, the conclusion is that we should always use multiple correspondence analysis, that is, dual scaling of multiple-choice data $\boldsymbol{F}_b$, rather than simple correspondence analysis or dual scaling of contingency tables $\boldsymbol{C}$. Does this mean "Limitation of simple correspondence analysis"?

We can stretch our imagination to the quantification of multimode contingency tables. For example, consider a three-mode data, which can be described as a trilinear decomposition of frequency f_{ijk}. The contingency table format will again restrict the total number of dimensions to the smallest number of categories of the variables minus 1. In this case, one can always represent the data in the response-pattern format with frequencies (e.g., $\boldsymbol{F}_b$), which will most typically yield more components than the corresponding analysis of the three-way contingency table. Can we then abandon simple correspondence analysis completely and always use multiple correspondence analysis? The author's view is "yes, we can."

2 Concluding Remarks

Dual scaling quantifies categorical data in such a way that variates for the rows and those for the columns are determined as simultaneous regressions of them on the data in hand. As such, dual scaling provides the optimal way to explain the data. As is clear from such phrases as simultaneous linear regressions (Hirschfeld 1935), reciprocal averaging (Horst 1935), and dual scaling (Nishisato 1980), the basic premise of dual scaling lies in symmetric analysis of rows and columns of a data matrix. It was clarified in the current paper as well as my Beijing paper that we need to expand the multidimensional space to accommodate both variates. This awareness of expanded space has led to the criticism of the current methods of joint graphical display, leading to the suggestion for an alternative method of graphical display, that is, cluster analysis. In the same context, we were brought back to Nishisato (1980) on the analytical comparisons between the contingency table format and its response-pattern format of the same data. When the number of rows is not equal to the number of columns of the data matrix, the response-pattern format of the same data yields more components than the contingency table format. If we are to pursue exhaustive analysis in the data, therefore, it is recommended that we should analyze the data represented in the response-pattern format rather than the contingency format. Data-dependent quantification, analysis in expanded multidimensional space, and exhaustive analysis using the response-pattern representation of the data are three major messages of the current paper.

Acknowledgment Due to an unexpected circumstance, the author could not present the paper at the conference, but Dr. Pieter M. Kroonenberg kindly offered his help and presented it for the author. His kind help is noted here and much appreciated. Computational assistance of Dr. J.G. Clavel is also appreciated.

References

E.J. Beh, R. Lombardo, *Correspondence Analysis: Theory, Practice and New Strategies* (Wiley, New York, NY, 2014)
J.P. Benzécri et al., *L'analyse Des Données: II. L'analyse Des Correspondances* (Dunod, Paris, 1973)
R.D. Bock, Methods and applications of optimal scaling. *The University of North Carolina Psychometric Laboratory Research Memorandum*, No. 25 (1960)
J.G. Clavel, S. Nishisato, in *New Trends in Psychometrics*, ed. by K. Shigemasu, A. Okada, T. Imaizumi, T. Hoshino. Joint analysis of within-set and between-set distances (Universal Academy Press, Tokyo, 2008), pp. 41–50
H. Cramér, *Mathematical Methods of Statistics* (Princeton University Press, Princeton, NJ, 1946)
K.R. Gabriel, The biplot graphical display of matrices with applications to principal component analysis. *Biometrics**58***, 453–476 (1971)
A. Gifi, *Nonlinear Multivariate Analysis* (Wiley, New York, NY, 1990)
J.C. Gower, D.J. Hand, *Biplots* (Chapman and Hall, London, 1996)
M.J. Greenacre, Theory and Applications of Correspondence Analysis (Academic, London, 1984)
H.O. Hirschfeld, A connection between correlation and contingency. *Camb. Philos. Soc. Proc.***31**, 520–524 (1935)
P. Horst, Measuring complex attitudes. *J. Soc. Psychol.**6***, 369–374 (1935)
B. Le Roux, H. Rouanet, *Geometric Data Analysis: From Correspondence Analysis to Structured Data* (Kluwer, Dordrecht, 2004)
L. Lebart, A. Morineau, N. Tabard, *Tecniques de la Description Statistique: Méthodes et L'ogiciels Pour L'analyse des Grands Tableaux* (Dunod, Paris, 1977)
L. Lebart, A. Morineau, K.M. Warwick, *Multivariate Descriptive Statistical Analysis* (Wiley, New York, NY, 1984)
S. Nishisato, *Analysis of Categorical Data: Dual Scaling and its Applications* (University of Toronto Press, Toronto, ON, 1980)
S. Nishisato, *Elements of Dual Scaling: An Introduction to Practical Data Analysis* (Lawrence Erlbaum, Hillsdale, NJ, 1994)
S. Nishisato, Gleaning in the field of dual scaling. *Psychometrika**1996**(61)*, 559–599 (1996)
S. Nishisato, in *Visualization of Categorical Data*, ed. by J. Blasius, M. J. Greenacre. Graphing is believing: Interpretable graphs for dual scaling (Academic Press, London, 1997), pp. 185–196

S. Nishisato, in *Multiple Correspondence Analysis and Related Methods*, ed. by M. J. Greenacre, J. Blasius. Correlational structure of multiple-choice data as viewed from dual scaling (Chapman and Hall/CRC, Boca Raton, FL, 2006), pp. 161–177

S. Nishisato, *Multidimensional Nonlinear Descriptive Analysis* (Chapman and Hall/CRC, Boca Raton, FL, 2007)

S. Nishisato, in *Challenges at the Interface of Data analysis, Computer Science and Optimization*, ed. by W. Gaul, A. Geyer-Schultz, L. Schmidt-Thiéme, I. Luntz. Quantification theory: reminiscence and a step forward (Springer, Heidelberg, 2012), pp. 109–119

S. Nishisato, Structural representation of categorical data and cluster analysis through filters, in German-Japanese Interchange of Data Analysis results, ed. by W. Gaul, A. Geyer-Schultz, Y. Baby, A. Okada (Springer, Berlin, 2014), pp. 81–91

S. Nishisato, in *Quantitative Psychology Research*, ed. by L. A. van der Ark, D. M. Bolt, W. C. Wang, J. A. Douglas, M. Wiberg. Multidimensional joint graphical display of symmetric analysis: back to the fundamentals (Springer, Berlin, 2016a), pp. 291–298. (Paper presented at the Annual Meeting of the Psychometric Society, Beijing)

S. Nishisato, Quantification theory: dual space and total space. Paper presented at the annual meeting of the Japanese behaviormetric Society, Sapporo, August (in Japanese), 2016b

S. Nishisato, J.G. Clavel, A note on between-set distances in dual scaling and Correspondence analysis. *Behaviormetrika* **30**, 87–98 (2003)

S. Nishisato, J.G. Clavel, in *New Trends in Psychometrics*, ed. by K. Shigemasu, A. Okada, T. Imaizumi, T. Hoshino. Interpreting data in reduced space: a case of what is not what in multidimensional data analysis (Universal Academy Press, Tokyo, 2008), pp. 347–356

S. Nishisato, J.G. Clavel, Total information analysis: comprehensive dual scaling. *Behaviormetrika* **37**, 15–32 (2010)

C.L. Stebbins, *Variation and Evolution* (Columbia University Press, New York, NY, 1950)

The Most Predictable Criterion with Fallible Data

Seock-Ho Kim

Abstract Hotelling's canonical correlation is the Pearson product moment correlation between two weighted linear composites from two sets of variables. The two composites constitute a set of canonical variates, namely, a criterion variate and a predictor variate. Many statistical analyses in psychometrics deal with fallible data that contain measurement errors. A method of obtaining canonical correlations from the true-score covariance matrix is presented and contrasted with Meredith's method for which the disattenuated canonical correlations are obtained from the correlation matrix of fallible data. Illustrations are presented with modified data from two seminal papers.

Keywords Canonical correlation • Disattenuated canonical correlation • Fallible data

1 Introduction

According to Bock (1975), Harold Hotelling (1935, 1936) derived the canonical correlation as a response to a query by Truman Kelley. The query is finding the variate as a linear combination of criterion variables that has the greatest multiple correlation with the variate of a linear combination of predictor variables.

It can be noticed that in fact the canonical correlation derived by Hotelling (1935, 1936) might not be the true answer to the Kelley's original query. Hotelling seemed to have modified the original query to the one he could find the solution. The solution is, of course, the canonical correlation, and the overall procedure to obtain such a correlation is called canonical correlation analysis. Today, virtually every textbook on multivariate statistics includes a chapter on canonical correlation analysis (e.g., Bock 1975; Johnson and Wichern 2007).

Almost all textbooks and reportings of empirical studies (e.g., Cooley and Lohnes 1976) present only Hotelling's (1936) method of canonical correlation analysis. It took about 40 years until Meredith (1964) proposed a viable solution to Kelley's

S.-H. Kim (✉)
The University of Georgia, Athens, GA 30602, USA
e-mail: shkim@uga.edu

L.A. van der Ark et al. (eds.), *Quantitative Psychology*, Springer Proceedings in Mathematics & Statistics 196, DOI 10.1007/978-3-319-56294-0_11

original query. It should be noted that Kelly's data analyzed in Hotelling (1936) are the so-called measurement data for which the concept of attenuated reliability in classical test theory has an important role. The solution to the disattenuated canonical correlation might have not been of any interest to Hotelling, even though the attenuation problem was clearly mentioned in one of the Hotelling's work on the canonical correlation (Hotelling 1936, p. 377).

A summary of canonical correlation analysis is presented subsequently. Basic algebra, computational methods, sampling properties, and statistical testing procedures are included. Canonical correlation analysis with fallible data in the form of the true-score covariance is presented together with examples. Summary remarks and discussion follow in the concluding section.

2 Canonical Correlations

2.1 *Population Canonical Correlations*

Let the population covariance matrix $\boldsymbol{\Sigma}$ of the p criterion variables $\mathbf{y}$ and q predictor variables $\mathbf{x}$ with full rank be partitioned into four submatrices,

$$\boldsymbol{\Sigma} = \begin{bmatrix} \boldsymbol{\Sigma}_{yy} & \boldsymbol{\Sigma}_{yx} \\ \boldsymbol{\Sigma}_{xy} & \boldsymbol{\Sigma}_{xx} \end{bmatrix}. \tag{1}$$

Assume linear combinations having unit variances provide simple summary measures of a set of variables, and let the variate $\eta = \boldsymbol{\alpha}'\mathbf{y}$ be a linear combination of criterion variables and the variate $\xi = \boldsymbol{\beta}'\mathbf{x}$ be a linear combination of predictor variables. Then the correlation between the linear combinations is given by

$$\rho = \frac{\boldsymbol{\alpha}'\boldsymbol{\Sigma}_{yx}\boldsymbol{\beta}}{\sqrt{\boldsymbol{\alpha}'\boldsymbol{\Sigma}_{yy}\boldsymbol{\alpha}}\sqrt{\boldsymbol{\beta}'\boldsymbol{\Sigma}_{xx}\boldsymbol{\beta}}}. \tag{2}$$

Canonical correlation analysis sets forth to find coefficient vectors $\boldsymbol{\alpha}$ and $\boldsymbol{\beta}$ so as to maximize the absolute value of ρ. The weights $\boldsymbol{\alpha}$ and $\boldsymbol{\beta}$ that maximize the absolute value of the correlation can be determined only up to proportionality constants. The proportional constants can be chosen to yield $\boldsymbol{\alpha}'\boldsymbol{\Sigma}_{yy}\boldsymbol{\alpha} = \boldsymbol{\beta}'\boldsymbol{\Sigma}_{xx}\boldsymbol{\beta} = 1$. Using Lagrange multipliers in the constrained maximization with partial derivatives and solving for $\boldsymbol{\alpha}$ and $\boldsymbol{\beta}$, we obtain

$$(\boldsymbol{\Sigma}_{yx}\boldsymbol{\Sigma}_{xx}^{-1}\boldsymbol{\Sigma}_{xy} - \rho^2\boldsymbol{\Sigma}_{yy})\boldsymbol{\alpha} = \mathbf{0} \tag{3}$$

and

$$(\boldsymbol{\Sigma}_{xy}\boldsymbol{\Sigma}_{yy}^{-1}\boldsymbol{\Sigma}_{yx} - \rho^2\boldsymbol{\Sigma}_{xx})\boldsymbol{\beta} = \mathbf{0}. \tag{4}$$

It can be recognized that each of Eqs. (3) and (4) as the eigenvalue-eigenvector problem (Johnson and Wichern 2007, pp. 97–98). The nontrivial solutions of the equations are eigenvectors associated with the eigenvalues of ρ^2 that satisfy

$$|\boldsymbol{\Sigma}_{yx}\boldsymbol{\Sigma}_{xx}^{-1}\boldsymbol{\Sigma}_{xy} - \rho^2\boldsymbol{\Sigma}_{yy}| = 0 \tag{5}$$

and

$$|\boldsymbol{\Sigma}_{xy}\boldsymbol{\Sigma}_{yy}^{-1}\boldsymbol{\Sigma}_{yx} - \rho^2\boldsymbol{\Sigma}_{xx}| = 0. \tag{6}$$

The number of nonzero roots (i.e., eigenvalues) of Eqs. (5) and (6) is determined by rank($\boldsymbol{\Sigma}_{yx}$). Assuming that $\boldsymbol{\Sigma}$ of the two sets of variables $\mathbf{y}$ and $\mathbf{x}$ has full rank, the number of nonzero roots s equals to $\min(p, q)$. The jth canonical correlation ρ_j is attained by the pair of canonical variates,

$$\eta_j = \boldsymbol{\alpha}_j'\mathbf{y} = \mathbf{e}_j'\boldsymbol{\Sigma}_{yy}^{-1/2}\mathbf{y} \tag{7}$$

and

$$\xi_j = \boldsymbol{\beta}_j'\mathbf{x} = \mathbf{f}_j'\boldsymbol{\Sigma}_{xx}^{-1/2}\mathbf{x}, \tag{8}$$

where ρ_j^2 is the jth largest eigenvalue of $\boldsymbol{\Sigma}_{yy}^{-1/2}\boldsymbol{\Sigma}_{yx}\boldsymbol{\Sigma}_{xx}^{-1}\boldsymbol{\Sigma}_{xy}\boldsymbol{\Sigma}_{yy}^{-1/2}$, $\mathbf{e}_j$ is the associated normalized eigenvector, ρ_j^2 is also the jth largest eigenvalue of $\boldsymbol{\Sigma}_{xx}^{-1/2}\boldsymbol{\Sigma}_{xy}\boldsymbol{\Sigma}_{yy}^{-1}\boldsymbol{\Sigma}_{yx}\boldsymbol{\Sigma}_{xx}^{-1/2}$, and $\mathbf{f}_j$ is the associated normalized eigenvector (Johnson and Wichern 2002, pp. 546–547).

2.2 Sample Canonical Correlations

A random sample of N observations on each of the p criterion variables $\mathbf{y}$ and the q predictor variables $\mathbf{x}$ can be assembled to have the sample covariance matrix

$$\mathbf{S} = \begin{bmatrix} \mathbf{S}_{yy} & \mathbf{S}_{yx} \\ \mathbf{S}_{xy} & \mathbf{S}_{xx} \end{bmatrix}. \tag{9}$$

The linear combinations $\hat{\eta} = \mathbf{a}'\mathbf{y}$ and $\hat{\xi} = \mathbf{b}'\mathbf{x}$ have the sample correlation given by

$$r = \frac{\mathbf{a}'\mathbf{S}_{yx}\mathbf{b}}{\sqrt{\mathbf{a}'\mathbf{S}_{yy}\mathbf{a}}\sqrt{\mathbf{b}'\mathbf{S}_{xx}\mathbf{b}}}. \tag{10}$$

Assume that $\mathbf{S}_{yx}$ has a full rank, there are $s = \min(p, q)$ canonical correlations exist. The jth sample canonical correlation is attained by the pair of sample canonical variates,

$$\hat{\eta}_j = \mathbf{a}_j'\mathbf{y} = \hat{\mathbf{e}}_j'\mathbf{S}_{yy}^{-1/2}\mathbf{y} \tag{11}$$

and

$$\hat{\xi}_j = \mathbf{b}_j'\mathbf{x} = \hat{\mathbf{f}}_j'\mathbf{S}_{xx}^{-1/2}\mathbf{x}. \tag{12}$$

The squared canonical correlation r_j^2 is the jth largest eigenvalue of $\mathbf{S}_{yy}^{-1/2}\mathbf{S}_{yx}\mathbf{S}_{xx}^{-1} \times \mathbf{S}_{xy}\mathbf{S}_{yy}^{-1/2}$, and $\hat{\mathbf{e}}_j$ is the associated normalized eigenvector. The squared canonical correlation r_j^2 is also the jth largest eigenvalue of $\mathbf{S}_{xx}^{-1/2}\mathbf{S}_{xy}\mathbf{S}_{yy}^{-1}\mathbf{S}_{yx}\mathbf{S}_{xx}^{-1/2}$, and $\hat{\mathbf{f}}_j$ is the associated normalized eigenvector.

2.3 Statistical Testing of Canonical Correlations

When $\boldsymbol{\Sigma}_{yx} = \mathbf{0}$, then $\boldsymbol{\alpha}'\mathbf{y}$ and $\boldsymbol{\beta}'\mathbf{x}$ have covariance $\boldsymbol{\alpha}'\boldsymbol{\Sigma}_{yx}\boldsymbol{\beta} = 0$ for all vectors $\boldsymbol{\alpha}$ and $\boldsymbol{\beta}$. There are several ways of testing $\boldsymbol{\Sigma}_{yx} = \mathbf{0}$ or equivalently all $\rho_j = 0$ for large samples. Many textbooks and statistical software contain Bartlett's (1947) chi-square test or Rao's (1951) F test.

The likelihood ratio test of the null hypothesis $\boldsymbol{\Sigma}_{yx} = \mathbf{0}$ versus the alternative hypothesis $\boldsymbol{\Sigma}_{yx} \neq \mathbf{0}$ rejects the null for a large value of $-2\log_e \Lambda = -N\log_e \prod_{j=1}^{s}(1 - r_j^2)$ which is distributed as a chi-square random variable with the degrees of freedom of pq, that is, $\chi^2(pq)$. Bartlett (1941) suggests replacing the multiplicative factor N in the likelihood ratio test with the factor $(N - 1) - (p + q + 1)/2$ to improve the chi-square approximation. Note that the composite test of no association can be modified to test the dimensionality of significant relationships between the two sets of variables (Bartlett 1947).

An F test based on Rao's approximation to the distribution of likelihood ratio (Rao 1973, p. 556; Rao 1951) can be used to perform sequentially for the nil of each canonical correlation. The test statistic is

$$R = \frac{1 - \Lambda^{1/n}}{\Lambda^{1/n}}\left(\frac{mn - 2l}{pq}\right), \tag{13}$$

where $m = t - (p + q + 1)/2$, $n = [(p^2q^2 - 4)/(p^2 + q^2 - 5)]^{1/2}$, $l = (pq - 2)/4$, and $t = N - 1$ in the current context following Bartlett (1941). The statistic R is distributed as F with $\nu_1 = pq$ and $\nu_2 = mn - 2l$. For cases where n cannot be defined (e.g., $p = 1$ and $q = 2$), the likelihood ratio test of Bartlett (1941) can be employed. There are other statistical methods to the null hypothesis testing of the canonical correlations (see Bock 1975, pp. 377–378).

3 Canonical Correlations from Fallible Data

When fallible data are used, we can view variables as the observed scores that consist of the true scores and the error scores. Hence, a criterion variable can be expressed as $Y = T_y + E_y$, and a predictor variable as $X = T_x + E_x$. Based on the assumption of the classical test theory model (Gulliksen 1950/1987, pp. 4–11; Lord and Novick 1968, p. 36), the covariance matrix of the p criterion variables $\mathbf{y}$ and q predictor variables $\mathbf{x}$ can be written as

$$\boldsymbol{\Sigma} = \boldsymbol{\Sigma}_T + \boldsymbol{\Sigma}_E \tag{14}$$

or

$$\begin{bmatrix} \boldsymbol{\Sigma}_{yy} & \boldsymbol{\Sigma}_{yx} \\ \boldsymbol{\Sigma}_{xy} & \boldsymbol{\Sigma}_{xx} \end{bmatrix} = \begin{bmatrix} \boldsymbol{\Sigma}_{Tyy} & \boldsymbol{\Sigma}_{yx} \\ \boldsymbol{\Sigma}_{xy} & \boldsymbol{\Sigma}_{Txx} \end{bmatrix} + \begin{bmatrix} \boldsymbol{\Sigma}_{Eyy} & \mathbf{0} \\ \mathbf{0} & \boldsymbol{\Sigma}_{Exx} \end{bmatrix}, \tag{15}$$

where subscripts T and E designate the terms are obtained from the respective true and error scores.

If reliabilities or errors of measurement of the fallible data are known, it is possible to obtain disattenuated canonical correlations. The disattenuated correlation between the linear combinations of the respective criterion and predictor variables is given by

$$\rho_T = \frac{\boldsymbol{\alpha}'_T \boldsymbol{\Sigma}_{yx} \boldsymbol{\beta}_T}{\sqrt{\boldsymbol{\alpha}'_T \boldsymbol{\Sigma}_{Tyy} \boldsymbol{\alpha}_T}\sqrt{\boldsymbol{\beta}'_T \boldsymbol{\Sigma}_{Txx} \boldsymbol{\beta}_T}} \tag{16}$$

with the subscript T that denotes the terms are obtained from the true scores (cf. Meredith 1964, p.57). In Meredith (1964) the observed correlation was decomposed into two parts, one from the true score and the other from the error score.

The disattenuated canonical correlations can be attained by finding and taking square root of the s largest eigenvalues of either $\boldsymbol{\Sigma}_{Tyy}^{-1/2} \boldsymbol{\Sigma}_{yx} \boldsymbol{\Sigma}_{Txx}^{-1} \boldsymbol{\Sigma}_{xy} \boldsymbol{\Sigma}_{Tyy}^{-1/2}$ or $\boldsymbol{\Sigma}_{Txx}^{-1/2} \boldsymbol{\Sigma}_{xy} \boldsymbol{\Sigma}_{Tyy}^{-1} \boldsymbol{\Sigma}_{yx} \boldsymbol{\Sigma}_{Txx}^{-1/2}$. The associated normalized eigenvectors and the coefficient vectors $\boldsymbol{\alpha}$ and $\boldsymbol{\beta}$ can also be obtained.

When data from simple random sampling are used, the Greek letters can be replaced with the corresponding Latin or Roman letters to express respective statistics. Estimates needed to perform canonical correlation analysis can be obtained from the statistics. Note again that the subscript T can be used to emphasize that the terms are from the true score. The equations based on the true and error scores can be easily constructed and are not repeated here.

4 Illustrations

4.1 Hotelling's Data

Hotelling (1936, p. 342) performed canonical correlation analysis of correlation data from 140 seventh-grade schoolchildren who took four tests on reading speed Y_1, reading power Y_2, arithmetic speed X_1, and arithmetic power X_2. The two sample canonical correlations are 0.3945 and 0.0688. Pairs of canonical variates in terms of the standardized variables can be obtained.

Kelley (1928, p. 100) reported means, standard deviations, and reliability coefficients for the four tests analyzed in Hotelling (1936). Portions that contain information about the four tests are presented in Table 1. The sample covariance matrix can be constructed as

$$\mathbf{S} = \begin{bmatrix} \mathbf{S}_{yy} & \mathbf{S}_{yx} \\ \mathbf{S}_{xy} & \mathbf{S}_{xx} \end{bmatrix} = \begin{bmatrix} 421.0704 & 409.9382 & 158.6290 & 20.1655 \\ 409.9382 & 996.6649 & -55.9536 & 34.6776 \\ 158.6290 & -55.9536 & 1027.2025 & 228.3209 \\ 20.1655 & 34.6776 & 228.3209 & 281.2329 \end{bmatrix}. \tag{17}$$

When covariance matrices are used, canonical variates with coefficients on the scales of the respective criterion and predictor variables can be found.

The two sample canonical correlations are, of course, 0.3945 and 0.0688. For r_1, the pair of canonical variates are $\hat{\eta}_1 = 0.0612Y_1 - 0.0325Y_2$ and $\hat{\xi}_1 = 0.0345X_1 - 0.0270X_2$. For r_2, the pair of canonical variates are $\hat{\eta}_2 = 0.0145Y_1 + 0.0249Y_2$ and $\hat{\xi}_2 = -0.0006X_1 + 0.0601X_2$.

As noted in Hotelling (1936, p. 342, footnote), the original book by Kelley (1928, p. 100) contains not only the raw correlations analyzed by Hotelling but also the correlations corrected for attenuation and other summary statistics. Utilizing information on Table 1 the sample true-score covariance matrix of the four variables can be expressed as

$$\mathbf{S}_T = \begin{bmatrix} \mathbf{S}_{Tyy} & \mathbf{S}_{yx} \\ \mathbf{S}_{xy} & \mathbf{S}_{Txx} \end{bmatrix} = \begin{bmatrix} 387.2584 & 409.9382 & 158.6290 & 20.1655 \\ 409.9382 & 891.2178 & -55.9536 & 34.6776 \\ 158.6290 & -55.9536 & 933.0080 & 228.3209 \\ 20.1655 & 34.6776 & 228.3209 & 158.5872 \end{bmatrix}. \tag{18}$$

We want to inquire the disattenuated relationship between reading ability and arithmetic ability indicated by the four fallible tests.

The two sample canonical correlations are 0.5344 and 0.0952. For r_{T1}, the pair of disattenuated canonical variates are $\hat{\eta}_{T1} = -0.0683T_{Y1} + 0.0405T_{Y2}$ and $\hat{\xi}_{T1} = -0.0407T_{X1} + 0.0589T_{X2}$. For r_{T2}, the pair of canonical variates are $\hat{\eta}_{T2} = 0.0192T_{Y1} + 0.0234T_{Y2}$ and $\hat{\xi}_{T2} = 0.0002T_{X1} + 0.0792T_{X2}$.

Table 1 Correlations, means, and standard deviations of four tests for 140 seventh-grade children

Test		Reading Speed	Reading Power	Arithmetic Speed	Arithmetic Power	Mean	Standard Deviation
Reading speed	Y_1	*0.9197*				93.06	20.52
Reading power	Y_2	0.6328	*0.8942*			194.53	31.57
Arithmetic speed	X_1	0.2412	−0.0553	*0.9083*		147.39	32.05
Arithmetic power	X_2	0.0586	0.0655	0.4248	*0.5639*	136.11	16.77

Note. Values along the main diagonal are reliability coefficients

Following Meredith (1964, p. 56) and assuming that all variables are standardized to have the means of zero and the standard deviations of unity, canonical correlation analysis could be performed with the matrix of covariances where the standardized error variances have been removed. The two sample canonical correlations are 0.5344 and 0.0952. Pairs of canonical variates on the scale of the standardized variables can also be obtained (cf. Darlington et al. 1973).

4.2 *Meredith's Data*

Meredith (1964, pp. 63–65) reported results from canonical correlation analysis based on the disattenuated intercorrelations of the 12 subtests of the Wechsler Intelligence Scale for Children (WISC) from 100 boys and 100 girls of years of age 7.5 (Wechsler 1949, p. 10). Both results from the matrix of the usual correlations and from the matrix of the correlations modified to account for the effect of attenuation are reported with the sets of weights (i.e., canonical correlation coefficients) for canonical variates of the WISC. Because Meredith (1964, pp. 63–64) presented the canonical correlation coefficient vectors $\mathbf{a}_j$ and $\mathbf{b}_j$ from the ordinary canonical correlation analysis and the canonical correlation coefficient vectors $\mathbf{a}_{Tj}$ and $\mathbf{b}_{Tj}$ for $j = 1, \ldots, 5$ (although $s = 6$) from the disattenuated canonical correlation analysis, the results need not be repeated here. Table 2 presents the sample true-score covariance matrix and other related summary statistics of the 12 subtests of the WISC from 100 boys and 100 girls of age 7.5 (cf. Wechsler 1949, pp. 10–13). Canonical correlation analysis was performed with the sample true-score covariance matrix in Table 2. The six criterion variables are arbitrarily chosen to be the subtests in the verbal subgroup. The six predictor variables are the subtests in the performance subgroup.

Table 3 contains the canonical correlation coefficient vectors $\mathbf{a}_{Tj}$ and $\mathbf{b}_{Tj}$ for $j = 1, \ldots, 6$ from the sample true-score covariance matrix. The canonical correlations obtained from the disattenuated covariance matrix are 0.971, 0.466, 0.351, 0.297, 0.239, and 0.097. Rao's test statistic was $R = 26.909$; for testing the null hypothesis of all population, canonical correlations are nil. The test statistic is distributed as F with $\nu_1 = 36$ and $\nu_2 = 828.327$ and is statistically significant at the nominal

Table 2 Sample true-score covariance matrix, means, standard deviations, and reliability coefficients of 12 subtests of the WISC from 100 boys and 100 girls of age 7.5

Test		Verbal subgroup						Performance subgroup					
		Y_1	Y_2	Y_3	Y_4	Y_5	Y_6	X_1	X_2	X_3	X_4	X_5	X_6
Information	Y_1	5.5506											
Comprehension	Y_2	3.0044	4.6256										
Arithmetic	Y_3	3.9933	2.3436	4.5927									
Similarities	Y_4	3.9788	2.8224	3.0240	5.1744								
Vocabulary	Y_5	4.1470	3.7128	3.2292	3.2760	5.2052							
Digit span	Y_6	2.6622	2.1924	2.9160	2.4948	3.0186	4.3740						
Picture completion	X_1	1.9488	3.0576	2.1924	2.1168	2.6208	2.4948	4.6256					
Picture arrangement	X_2	3.0276	3.1668	2.9754	3.0856	2.9406	3.4452	2.5984	6.0552				
Block design	X_3	2.6796	2.5088	2.0412	2.2736	2.4024	1.8144	2.1952	3.0044	6.5856			
Object assembly	X_4	2.3490	2.1000	2.3490	2.4360	2.3400	1.7820	2.3520	4.1760	4.4520	5.6700		
Coding A	X_5	2.3374	1.9096	2.6784	1.3020	1.7732	2.2599	1.0416	2.1576	2.2568	2.7900	5.7660	
Mazes	X_6	2.0880	1.9320	1.6200	2.1000	1.7160	1.2960	1.6800	3.1320	4.1160	4.3200	1.7670	7.1100
Mean		10.0	10.0	10.1	9.9	10.1	9.8	10.0	10.1	10.1	9.9	10.1	10.0
Standard deviation		2.9	2.8	2.7	2.8	2.6	2.7	2.8	2.9	2.8	3.0	3.1	3.0
Reliability		0.66	0.59	0.63	0.66	0.77	0.60	0.59	0.72	0.84	0.63	0.60	0.79

Table 3 Canonical correlation coefficient vectors of 12 subtests of the WISC to produce the canonical variates corresponding to the six disattenuated canonical correlations

	Coefficient vector of verbal subgroup					
Test	$\mathbf{a}_{T1}$	$\mathbf{a}_{T2}$	$\mathbf{a}_{T3}$	$\mathbf{a}_{T4}$	$\mathbf{a}_{T5}$	$\mathbf{a}_{T6}$
Information	0.1268	−0.7092	−0.1929	0.5093	−0.1869	−0.1460
Comprehension	0.3730	−0.1102	−0.1544	−0.2266	−0.3876	0.3927
Arithmetic	0.0401	0.3420	0.7139	−0.0308	−0.0189	0.2982
Similarities	−0.0920	0.4249	−0.2777	0.1540	0.3489	0.2537
Vocabulary	−0.2712	0.5716	0.0482	−0.0737	−0.1074	−0.6282
Digit span	0.3493	−0.3340	−0.1260	−0.2150	0.3919	−0.1819
	Coefficient vector of performance subgroup					
Test	$\mathbf{b}_{T1}$	$\mathbf{b}_{T2}$	$\mathbf{b}_{T3}$	$\mathbf{b}_{T4}$	$\mathbf{b}_{T5}$	$\mathbf{b}_{T6}$
Picture completion	0.2148	0.2364	−0.0014	−0.3553	−0.2579	0.0797
Picture arrangement	0.3280	−0.1833	−0.2095	0.0149	0.4197	−0.0727
Block design	0.1222	−0.2104	−0.2502	0.1853	−0.2715	−0.3453
Object assembly	−0.3840	0.6493	0.3403	0.2016	0.1204	−0.0999
Coding A	0.1970	−0.2273	0.3548	0.0161	−0.0936	0.0713
Mazes	0.0407	−0.1006	−0.1410	0.0775	−0.0835	0.4826
Canonical correlation	0.9711	0.4660	0.3506	0.2971	0.2388	0.0969
P	<.0001	<0.0001	<0.0001	0.0003	0.0107	0.1778

Note. P is the observed level of significance

level of 0.01. Subsequent Rao's tests for the second (n.b., actually the testing of the second to the last canonical correlations), third, and fourth canonical correlations are statistically significant at the nominal level of 0.01. The fifth canonical correlation is significant at the nominal level of 0.05. The last Rao's test for the sixth canonical correlation was not statistically significant at the nominal level of 0.05. The observed level of significance for each sequential testing is reported in Table 3.

The usual method of the canonical correlation analysis yielded canonical correlations of 0.680, 0.197, 0.163, 0.116, 0.107, and 0.045. Rao's test statistic was $R = 4.066$; for testing the null hypothesis of all population, canonical correlations are nil. The test statistic is distributed as F with $\nu_1 = 36$ and $\nu_2 = 828.327$ and is statistically significant at the nominal level of 0.01. Except for the first canonical correlation, all the remaining tests of the five canonical correlations are not statistically significant at the nominal level of 0.05.

5 Summary and Discussion

In this paper, a summary of canonical correlation analysis is presented, and the solutions by Hotelling (1936) and Meredith (1964) were reconsidered. A modified solution over Meredith (1964) is proposed with examples.

Although some monographs and journal articles (e.g., Darlington et al. 1973; Thompson (1984)) briefly mentioned the disattenuated canonical correlation by Meredith (1964), the usual textbooks on multivariate statistics do not contain any presentation of canonical correlation analysis in conjunction with fallible data consisted with variables that contain measurement errors.

Van de Geer (1971) presented a rather lucid connection between canonical correlation analysis and latent variable modeling using path diagrams and illustrations. A brief discussion was presented about canonical correlation analysis for fallible data using the correlation matrix of true scores but without mentioning Meredith's (1964) solution (see Van de Geer 1971, pp. 168–169). Kenny (1979) also contained a discussion about canonical correlation analysis, factor analysis, and causal models with unmeasured variables using path diagrams and examples. The discussion did not consider either canonical correlation analysis for fallible data or Meredith's (1964) solution.

Canonical correlation analysis is truly an inclusive multivariate statistical method that subsumes nearly all other well-known linear models and procedures including simple and multiple regression analysis, analysis of variance, discriminant analysis, and chi-square test of independence as special cases. These standard techniques are usually applied without considering measurement errors or fallibility of data. Although several works exist that specifically addressed the effect of measurement errors (e.g., Cochran 1968; Pedhazur 1997, pp. 292–294), finding satisfactory procedures for coping with fallible data seems to be a difficult task. It will be interesting to present illustrations of obtaining canonical correlations from fallible data for these special cases employing the methods presented in this paper because the methods may provide a unified framework for analyzing fallible data.

References

M.S. Bartlett, The statistical significance of canonical correlations. Biometrika **32**, 29–37 (1941)

M.S. Bartlett, Multivariate analysis. Suppl. J. R. Stat. Soc. **9**, 176–197 (1947)

R.D. Bock, *Multivariate Statistical Methods in Behavioral Research* (McGraw-Hill, New York, 1975)

W.G. Cochran, Errors of measurement in statistics. Technometrics **10**, 637–666 (1968)

W.W. Cooley, P.R. Lohnes, *Evaluation Research in Education* (Irvington, New York, 1976)

R.B. Darlington, S.L. Weinberg, H.J. Walberg, Canonical variate analysis and related techniques. Rev. Educ. Res. **43**, 433–454 (1973)

H. Gulliksen, *Theory of Mental Tests* (Lawrence Erlbaum, Hillsdale, 1987) (Original work published 1950)

H. Hotelling, The most predictable criterion. J. Educ. Psychol. **26**, 139–142 (1935)

H. Hotelling, Relations between two sets of variates. Biometrika **28**, 321–377 (1936)

R.A. Johnson, D.W. Wichern, *Applied Multivariate Statistical Analysis*, 5th edn. (Prentice Hall, Upper Saddle River, 2002)

R.A. Johnson, D.W. Wichern, *Applied Multivariate Statistical Analysis*, 6th edn. (Prentice Hall, Upper Saddle River, 2007)

T.L. Kelley, *Crossroads in the Mind of Man: A Study of Differentiable Mental Abilities* (Stanford University Press, Stanford, 1928)

D.A. Kenny, *Correlation and Causality* (Wiley, New York, 1979)
F.M. Lord, M.R. Novick, *Statistical Theories of Mental Test Scores* (Addison-Wesley, Reading, 1968)
W. Meredith, Canonical correlations with fallible data. Psychometrika **29**, 55–65 (1964)
E.J. Pedhazur, *Multiple Regression in Behavioral Research: Explanation and Prediction*, 3rd edn. (Harcourt Brace College Publishers, Fortworth, 1997)
C.R. Rao, An asymptotic expansion of the distribution of Wilks' Λ criterion. Bull. Int. Stat. Inst. **33**, 177–180 (1951)
C.R. Rao, *Linear Statistical Inference and Its Application*, 2nd edn. (Wiley, New York, 1973)
B. Thompson, *Canonical Correlation Analysis: Uses and Interpretation* (Sage, Newbury Park, 1984)
J.P. Van de Geer, *Introduction to Multivariate Analysis for the Social Sciences* (W. H. Freeman, San Francisco, 1971)
D. Wechsler, *Wechsler Intelligence Scale for Children: Manual* (The Psychological Corporation, New York, 1949)

Asymmetric Multidimensional Scaling of Subjective Similarities Among Occupational Categories

Akinori Okada and Takuya Hayashi

Abstract The subjective similarity among ten occupational categories is analyzed by the asymmetric multidimensional scaling based on singular value decomposition. The similarity among occupational categories is obtained by a procedure where the similarity from occupational categories j to k is judged by respondents engaged in occupational category j, and the similarity from occupational categories k to j is judged by respondents engaged in occupational category k. These two similarities are not necessarily equal. This makes it possible to analyze asymmetric relationships of the subjective similarity. The three-dimensional solution disclosed two kinds of asymmetry between two occupational categories which are caused by the difference of the status between two occupational categories.

Keywords Asymmetry • Multidimensional scaling • Occupational category • Similarity • Status of occupation

1 Introduction

The similarity among occupations has been studied as *social relations* perspective, where the distance or the similarity is implicitly assumed to be symmetric, i.e., the similarity from occupations j to k is equal to that from k to j. While some researchers (Laumann and Guttman 1966; Prandy 1990; Rytina 1992; Chan and Goldthorpe 2004) say that the similarity between the occupation of oneself and the occupation of one's parents, relatives, or friends should be treated as asymmetric, the asymmetry of similarities is ignored in their analyses. Some studies derived two configurations separately to represent asymmetric relationships such as husband-wife and father-

A. Okada (✉)
Research Institute, Tama University, 4-1-1 Hijirigaoka Tama-shi, Tokyo 206-0022, Japan
e-mail: okada@rikkyo.ac.jp

T. Hayashi
Faculty Division of Humanities and Social Sciences, Nara Women's University, Kitauoyanishimachi, Nara 630-8506, Japan
e-mail: t-hayashi@cc.nara-wu.ac.jp

L.A. van der Ark et al. (eds.), *Quantitative Psychology*, Springer Proceedings in Mathematics & Statistics 196, DOI 10.1007/978-3-319-56294-0_12

son (Bakker 1993; Kondo 2006), but they did not succeed in representing the asymmetry, because derived two configurations were almost identical.

Aside from the similarity in social relations abovementioned, the subjective similarity is also important in studying social psychological effects or consequences of social relations among social groups. It seems that the status of an occupation of a respondent plays an important role in judging the subjective similarity among occupations (Ikeda 1973; Wegener 1992). We can consider two subjective similarities between two occupations of higher and lower status: (a) one is the subjective similarity from a higher-status occupation to a lower-status one judged by people engaged in a higher-status occupation, and (b) the other is the subjective similarity from a lower-status occupation to a higher-status one judged by people engaged in a lower-status occupation. The difference of the status of two occupations brings about two different effects on the asymmetry of the subjective similarity between higher- and lower-status occupations as described below, i.e., (a) > (b) or (a) < (b).

When people of one group feel a barrier to those of the other group but people of the latter group do not feel the barrier (feel a lower barrier) to those of the former group, this causes the asymmetry of the subjective similarity between two groups. For a person engaged in lower-status occupations, the subjective similarity from lower-status occupations to higher-status ones is small, while for a person engaged in higher-status occupations, the subjective similarity from higher-status occupations to lower-status ones is large (Ikeda 1973, pp. 46–49). It might reflect the situation that only people engaged in the lower-status occupations feel difficulty in getting higher-status occupations, as in a study on aspiration of status attainment that showed ambitions shifted downward over time in lower-status-background people (Hanson 1994). This causes (a) > (b).

On the contrary, people engaged in higher-status occupations have more discriminatory feeling to those engaged in lower-status occupations, and people engaged in lower-status occupations have less discriminatory feeling to those engaged in higher-status occupations (Wegener 1992). People engaged in higher-status occupations exaggerate the difference of higher- and lower-status occupations, which suggests smaller similarity from the higher- to the lower-status occupations. People engaged in lower-status occupations undervalue the difference of higher- and lower-status occupations, which suggests large similarity from the lower- to the higher-status occupations. This causes (a) < (b). Thus, the difference of the status of two occupations brings two contradictory effects on the asymmetry of the subjective similarity between higher- and lower-status occupations.

Laumann (1965), Ikeda (1973), and Laumann and Senter (1976) investigated subjective similarity among occupations, but they have flaws in common. Firstly, the obtained similarity does not represent the asymmetry among occupational categories. Secondly, they regard the similarity as unidimensional, while the similarity in social relations has a multidimensional structure (Chan and Goldthorpe 2004; Laumann and Guttman 1966; Okada and Imaizumi 1997). These studies were not able to represent two contradictory effects on the asymmetry of the subjective similarity and to represent a multidimensional structure. In the present study we analyze subjective similarities among occupational categories, where the similarity from occupational categories j to k was judged by respondents engaged in occupational category j, and the similarity from occupational categories k to j was judged by respondents engaged in occupational category k. These two similarities are not necessarily equal, and the relationship between occupational categories j and k can be asymmetric. The purpose of the present study is to investigate two contradictory effects of the status of occupations on the asymmetry of the subjective similarity and the multidimensional structure of asymmetric relationships among occupations.

2 Data

The data were collected from a survey conducted in 2013. Each respondent judged the similarity from the occupation in which a respondent oneself is engaged to each of the other occupations by a 5-point rating scale (5: Most similar, $\cdots$, 1: Least similar). There were 36 occupations in the survey which were classified into ten occupational categories which correspond to *meso classes* of micro-class scheme (Jonsson et al. 2009). While 2069 respondents judged the similarities from one's occupation to 36 occupations, 2017 responses gave the valid response to their own occupations. The analysis in the present study was done based on ten occupational categories shown in the leftmost column of Table 1, where (a) to (f) are nonmanual and (g) to (j) are manual occupational categories.

The mean of obtained similarities from occupational categories j to k were derived, which resulted in a 10×10 matrix whose (j, k) element represents the mean of obtained similarity from occupational categories j to k. Then the mean of all elements of the matrix was subtracted from each element to normalize the similarity matrix. The normalization gives positive and negative elements of the resulting matrix. This suggests that occupational categories represented in a configuration along Dimension 1 derived by the singular value decomposition (Eckart and Young 1936) can be represented in all four quadrants, which can represent more subtle aspects of relationships among occupational categories along Dimension 1. The similarity matrix shown in Table 1 is asymmetric, because the similarity from occupational categories j to k judged by respondents engaged in occupational category j is not necessarily equal to the similarity from occupational categories k to j judged by respondents engaged in occupational category k.

3 Method and Analysis

In the present study, the asymmetric multidimensional scaling (Okada and Tsurumi 2012) based on singular value decomposition was used to analyze the subjective similarity among ten occupational categories. The asymmetric multidimensional scaling represents asymmetries along each dimension by two terms: the closeness

Table 1 Similarity matrix among ten occupational categories

From	To occupational category									
	(a)	(b)	(c)	(d)	(e)	(f)	(g)	(h)	(i)	(j)
(a) Classical professions	0.536	0.242	−0.114	0.018	−0.062	0.270	−0.048	−0.376	−0.215	−0.082
(b) Managers and officials	0.015	1.016	−0.314	0.035	0.458	0.372	−0.374	−0.355	−0.308	−0.366
(c) Other professions	0.201	−0.090	0.184	0.035	0.103	0.364	−0.289	−0.501	−0.065	0.087
(d) Proprietors	−0.177	0.251	−0.209	1.183	0.694	−0.088	0.168	0.059	−0.015	0.747
(e) Sales	−0.520	−0.176	−0.425	0.051	1.432	0.285	−0.364	−0.351	0.153	−0.110
(f) Clerical	−0.207	−0.006	−0.267	−0.231	0.490	1.260	−0.481	−0.527	0.003	−0.354
(g) Craft	−0.152	−0.021	−0.420	0.283	0.154	0.049	0.484	0.235	0.008	0.165
(h) Lower manual	−0.394	−0.567	−0.431	0.028	0.369	0.324	0.257	0.151	0.343	0.244
(i) Service workers	−0.413	−0.585	−0.255	0.038	0.532	0.376	−0.176	−0.166	0.554	0.016
(j) Primary sector	−0.154	−0.292	−0.304	0.452	0.350	−0.058	0.126	0.211	0.121	2.381

from an occupational category to the other occupational categories which is called the outward tendency of an occupational category and the closeness from the other occupational categories to an occupational category which is called the inward tendency of an occupational category.

The procedure of the asymmetric multidimensional scaling is briefly described. Let **A** be an $n \times n$ matrix of asymmetric similarities among n occupational categories. The (j, k) element of **A** represents the similarity from occupational categories j to k, which is not necessarily equal to the (k, j) element. Based on r ($< n$) largest singular values and corresponding left and right singular vectors, **A** is approximated by

$$\mathbf{A} \simeq \mathbf{XDY}', \tag{1}$$

where **D** is the $r \times r$ diagonal matrix of r largest singular values in descending order at its diagonal elements, **X** is the $n \times r$ matrix whose ith column is the left singular vector corresponding to the ith singular value (normalized so that the length is unity), and **Y** is the $n \times r$ matrix whose ith column is the right singular vector corresponding to ith singular value (normalized so that the length is unity).

The jth element of the ith column of **X**, x_{ji}, represents the closeness from occupational category j to the other occupational categories along Dimension i, because rows of **A** correspond to occupational categories from which similarities to the others were judged. x_{ji} is the outward tendency of occupational category j along Dimension i. The kth element of the ith column of **Y**, y_{ki}, represents the closeness from the other occupational categories to occupational category k along Dimension i, because columns of **A** correspond to occupational categories to which similarities from other occupational categories were judged. y_{ki} is the inward tendency of occupational category k along Dimension i. The (j, k) element of Eq. (1) is represented as $a_{jk} \simeq \sum_{i=1}^{r} d_i x_{ji} y_{ki}$, where d_i is the ith largest singular value. This equation shows that the similarity from occupational categories j to k is approximated by the algebraic sum of the product of the outward tendency of occupational category j (x_{ji}) and the inward tendency of occupational category k along Dimension i (y_{ki}) multiplied by d_i.

As shown in Figs. 1, 2, and 3, the result of the present asymmetric multidimensional scaling is shown by r planar configurations each of which represents the similarity along each of r dimensions. In the planar configuration along Dimension i, the abscissa represents the outward tendency which corresponds to the ith left singular vector, and the ordinate represents the inward tendency which corresponds to the ith right singular vector. In the configuration, $x_{ji}y_{ki}$ means an area (or the area with a negative sign) of a rectangle made by two sides: x_{ji} and y_{ki}, in a plane spanned by the ith left and right singular vectors. x_{ji} and y_{kl} can be negative, and $x_{ji}y_{ki}$ can be negative. Let two points representing occupational categories j and k be in the same quadrant or in two neighboring quadrants. When occupational category k is ahead of the counterclockwise direction than occupational category j is, $x_{ji}y_{ki}$ is larger than $x_{ki}y_{ji}$, suggesting occupational category j is more similar to occupational category k than k is to j. The similarity between two occupational categories along the counterclockwise direction is larger than that along the clockwise direction.

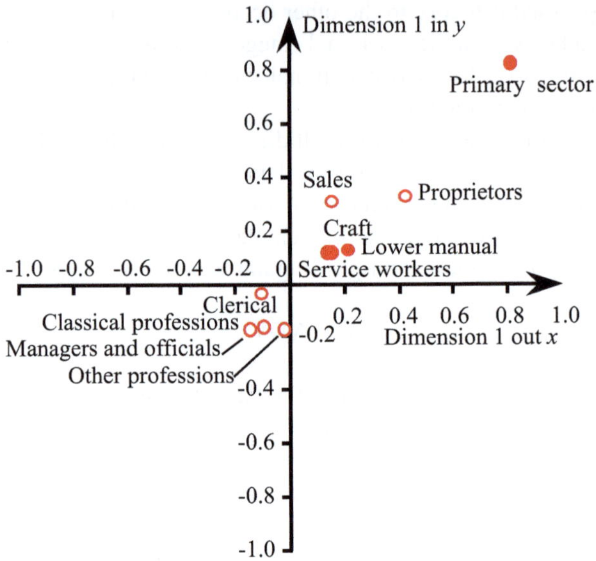

Fig. 1 Configuration along Dimension 1. *Solid circles* represent manual occupational categories, and *open circles* represent nonmanual occupational categories

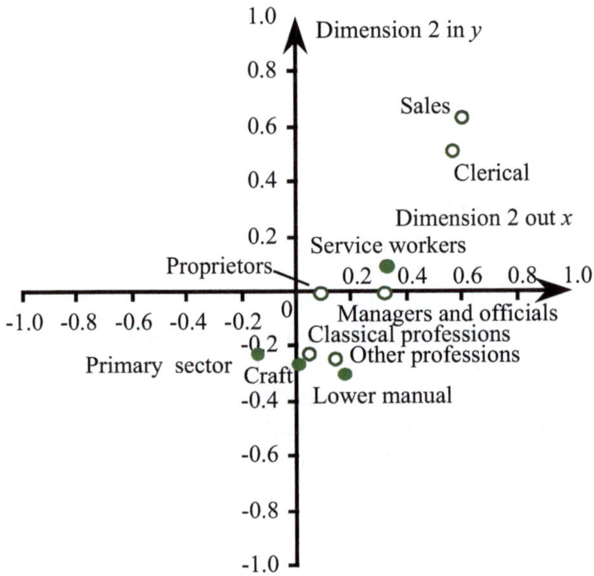

Fig. 2 Configuration along Dimension 2. *Solid circles* represent manual occupational categories, and *open circles* represent nonmanual occupational categories

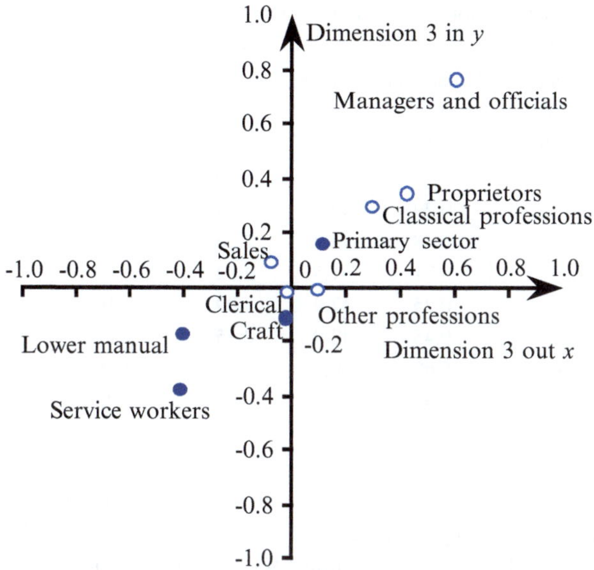

Fig. 3 Configuration along Dimension 3. *Solid circles* represent manual occupational categories, and *open circles* represent nonmanual occupational categories

When two points are in the first and the third quadrants, respectively, $x_{ji}y_{ki}$ and $x_{ki}y_{ji}$ are negative, because either x_{ji} or y_{ki} is negative and either x_{ki} or y_{ji} is negative. This tells that two occupational categories represented in the first and the third quadrants are not similar. When two points are in the second and the fourth quadrants, respectively, $x_{ji}y_{ki}$ and $x_{ki}y_{ji}$ are positive, because x_{ji} and y_{ki} have the same sign and x_{ki} and y_{ji} have the same sign. This tells that two occupational categories represented in the second and the fourth quadrants are similar.

The analysis was done by using a software which was published in association with Okada and Imaizumi (1994). Ten singular values of **A** are 2.940, 2.504, 1.629, 1.311, 0.826, 0.718, 0.467, 0.270, 0.084, and 0.032. The three-dimensional result was chosen as the solution, because the three-dimensional result is easy to interpret but the four-dimensional result is difficult.

4 Results and Discussions

Figure 1 shows the configuration along Dimension 1. The horizontal axis represents the outward tendency of ten occupational categories along Dimension 1, and the vertical axis represents the inward tendency of ten occupational categories along Dimension 1. The asymmetry of similarities among occupational categories represented in Fig. 1 is not large, because points are close to the 45-degree line from the lower left to the upper right direction passing the origin and they are

Table 2 The mean score of four questions along with the rate of female and the index of prestige of occupational categories

Occupational category	Female rate	Prestige	Specialized	Autonomous	Authority	Discretion
(a) Classical professions	0.43	68.6	3.83	2.87	2.96	2.67
(b) Managers and officials	0.16	63.7	3.33	2.88	3.08	2.77
(c) Other professions	0.57	62.8	3.60	2.72	2.84	2.39
(d) Proprietors	0.29	48.6	3.33	3.39	3.39	3.30
(e) Sales	0.48	44.7	2.66	2.37	2.35	2.06
(f) Clerical	0.71	51.5	2.82	2.57	2.48	2.19
(g) Craft	0.25	47.9	3.09	2.39	2.46	1.91
(h) Lower manual	0.45	42.9	2.35	2.17	2.17	1.79
(i) Service workers	0.60	40.9	2.52	2.25	2.25	1.76
(j) Primary sector	0.38	45.6	3.17	3.02	3.02	2.96

either in the first or the third quadrants. In the first quadrant, the similarity to sales from the others is larger than that of the other way around. The configuration along Dimension 1 differentiates primary sector from the others. All four manual occupational categories are in the first quadrant, and four of six nonmanual occupational categories are in the third quadrant.

To interpret the result, we use answers of respondents to four questions and two characteristics (female rate and prestige) shown in Table 2. Four questions are:

- The job I am engaged in needs **specialized** knowledge or skill.
- I can decide the content or pace of my own job **autonomously**.
- I have the **authority** to reflect my opinions on work assignment in workplace as a whole.
- I can decide the time I begin and end work at my **discretion**.

Respondents answered by a 4-point rating scale (4: True, $\cdots$, 1: Not true). The emboldened word is used to represent each question henceforth.

The female rate of each occupational category is obtained by using the data of the Japanese Population Census conducted in 2010 (Ministry of Internal Affairs and Communications, Statistics Bureau. 2015). The prestige reflects the hierarchy of individual social position (Wegener 1992, p. 273), which is based on peoples' evaluation on occupations in a society. The prestige score of each occupation in the present study is assigned according to those measured in the Japanese National Survey of Social Stratification and Mobility (SSM) which was conducted in 1995 (Tsuzuki 1998, Appendix, pp. 231–236).

The mean prestige scores of occupational categories in the first and the third quadrants are 44.8 and 59.0. The mean scores of autonomous, authority, and discretion of those in the first and the third quadrants are 2.48 and 2.69, 2.49 and 2.73, and 2.14 and 2.39, respectively. These figures suggest that the configuration along Dimension 1 classify occupational categories into two groups: one having lower prestige, autonomous, authority, and discretion scores and the other having higher prestige and scores.

Table 3 Mean of female rate and mean scores of autonomous, authority, and discretion for occupational categories in the first, the third, and the fourth quadrants in the configuration along Dimension 2

Female rate		Autonomous		Authority		Discretion	
Q2: –	Q1: 0.62	Q2: –	Q1: 2.44	Q2: –	Q1: 2.40	Q2: –	Q1: 2.06
Q3: 0.38	Q4: 0.41	Q3: 3.02	Q4: 2.69	Q3: 3.02	Q4: 2.77	Q3: 2.96	Q4: 2.40

Figure 2 shows the configuration along Dimension 2. The horizontal and the vertical axes represent the outward tendency and the inward tendency of ten occupational categories along Dimension 2, respectively. The appreciable asymmetry of similarities among occupational categories is represented in Fig. 2, because ten occupational categories are represented in the first, the third, and the fourth quadrants. The similarity from six occupational categories in the fourth quadrant to those in the first quadrant is larger than that of the other way around. The similarity from primary sector in the third quadrant to the six occupational categories is larger than that of the other way around. The similarity between primary sector in the third quadrant and occupational categories in the first quadrant is small.

Table 3 shows the mean of female rate and mean scores of autonomous, authority, and discretion for occupational categories in each of the first (Q1), the third (Q3), and the fourth quadrants (Q4) in the configuration along Dimension 2. Table 3 tells that the female rate increases along the counterclockwise direction from the third to the first quadrants, while mean scores of autonomous, authority, and discretion decrease along the counterclockwise direction from the third to the first quadrants. This is compatible with the fact that occupational categories with higher prestige would have the lower female rate and have higher autonomous, authority, and discretion scores.

Figure 3 shows the configuration along Dimension 3. The horizontal and the vertical axes represent the outward tendency and inward tendency of ten occupational categories along Dimension 3, respectively. The appreciable asymmetry of similarities among occupational categories is represented in Fig. 3, because occupational categories are distributed over all four quadrants. Four occupational categories in the first quadrant and three manual occupational categories in the third quadrant are not similar along Dimension 3. The similarity from four occupational categories in the first quadrant to sales in the second quadrant is larger than that of the other way around. The similarity from sales to three manual occupational categories in the third quadrant is larger than that of the other way around.

Table 4 shows the mean scores of autonomous, authority, and discretion for occupational categories in each of the four quadrants in the configuration along Dimension 3. In Table 4, mean scores of autonomous and authority increase along the counterclockwise direction from the second to the first quadrants. And the mean score of discretion increases along the counterclockwise direction from the third to the first quadrants. On the contrary, these mean scores decrease along the counterclockwise direction in the configuration along Dimension 2. The relationships between these mean scores and the direction (clockwise or counterclockwise) in the configurations along Dimensions 2 and 3 are opposite.

Table 4 Mean scores of autonomous, authority, and discretion for occupational categories in four quadrants in the configuration along Dimension 3

Autonomous		Authority		Discretion	
Q2: 2.37	Q1: 3.04	Q2: 2.35	Q1: 3.12	Q2: 2.06	Q1: 2.91
Q3: 2.40	Q4: 2.72	Q3: 2.38	Q4: 2.84	Q3: 1.99	Q4: 2.39

5 Conclusions

Subjective similarities among occupational categories, where the similarity from occupational categories j to k was judged by respondents engaged in occupational category j and the similarity from occupational categories k to j was judged by respondents engaged in occupational category k (thus, two similarities are not necessarily equal), were analyzed by the asymmetric multidimensional scaling. The three-dimensional solution disclosed three different aspects of relationships among occupational categories which represent the multidimensional structure of asymmetric relationships among occupational categories. Dimension 1 represented almost symmetric relationships among occupational categories and showed two groups of occupational categories; one consists of (four nonmanual) occupational categories having higher prestige as well as higher autonomous, authority, and discretion scores, and the other consists of (four manual and two nonmanual) occupational categories having lower prestige and lower scores.

Dimensions 2 and 3 represent asymmetric relationships among occupational categories. They represent different aspects of asymmetric relationships, respectively, and demonstrate that the difference of the status, especially denoted by autonomous, authority, and discretion, between two occupational categories simultaneously have two different effects on the asymmetry of the subjective similarity. Dimension 2 showed that the subjective similarity from the higher-status to the lower-status occupational categories is larger than that from the lower-status to the higher-status occupational categories. This represents the effect of the status suggested by Ikeda (1973). On the contrary, Dimension 3 showed that the subjective similarity from the lower-status to the higher-status occupational categories is larger than that from the higher-status to the lower-status occupational categories. This represents the effect of the status suggested by Wegener (1992).

Acknowledgements This work was supported by JSPS KAKENHI (Grant-in-Aid for Scientific Research (C)) Grant Number 24530625. The authors would like to express their gratitude to the reviewer who gave us constructive comments. They hope to thank for helpful suggestions given to the earlier version by Hiroshi Inoue. They also wish to their appreciation to Reginald Williams concerning English.

References

B.F.M. Bakker, A new measure of social status for men and women: the social distance scale. Neth. J. Soc. Sci. **29**, 113–129 (1993)

T.W. Chan, J.H. Goldthorpe, Is there a status order in contemporary British society? evidence from the occupational structure of friendship. Eur. Sociol. Rev. **20**, 383–401 (2004)

C. Eckart, G. Young, The approximation of one matrix by another of lower rank. Psychometrika **1**, 211–218 (1936)

S. Hanson, Lost talent: unrealized educational aspirations and expectations among U.S. youths. Sociol. Educ. **67**, 159–183 (1994)

M. Ikeda, Shokugyou hyouka to kaikyuu ishiki (Occupational evaluation and class consciousness), pp. 383–401, (in Japanese) in *Gendai nihon no kaikyuu ishiki (Class Consciousness in Contemporary Japan)*, ed. by S. Yasuda Yuhikaku, Tokyo (1973), pp. 31–58 (in Japanese)

J.O. Jonsson, M. Di Carlo, M.C. Brinton, D.B. Grusky, R. Pollak, Microclass mobility: social reproduction in four countries. Am. J. Sociol. **114**, 977–1036 (2009)

H. Kondo, Idohyou niyoru shokugyouteki chiishakudo no kousei: ordination giho no ouyou (Constructing occupational status scale from mobility tables: an application of ordination methods). Riron to Ouyou (Sociol. Theory and Methods) **21**, 313–332 (2006). (in Japanese)

E.O. Laumann, Subjective social distance and urban occupational stratification. Am. J. Sociol. **71**, 26–36 (1965)

E. Laumann, L. Guttman, The relative associational contiguity of occupations in an urban setting. Am. Sociol. Rev. **31**, 169–178 (1966)

E. Laumann, R. Senter, Subjective social distance, occupational stratification, and forms of status and class consciousness: a cross-national replication and extension. Am. J. Sociol. **81**, 1304–1338 (1976)

Ministry of Internal Affairs and Communications, Statistics Bureau, Detailed sample tabulation, Table10-1. Employed persons 15 years of age and over, by occupation (minor groups), employment status (8 groups) and sex - Japan. Retrieved October 9, 2015, from http://www.e-stat.go.jp/SG1/estat/List.do?bid=000001050829&cycode=0

A. Okada, T. Imaizumi, *Pasokon tajigen shakudo kouseiho (Multidimensional scaling by personal computer)* (Kyoritsu Shuppan, Tokyo, 1994). (in Japanese)

A. Okada, T. Imaizumi, Asymmetric multidimensional scaling of two-mode three-way proximities. J. Classif. **14**, 195–224 (1997)

A. Okada, H. Tsurumi, Asymmetric multidimensional scaling of brand switching among margarine brands. Behaviormetrika **39**, 111–126 (2012)

K. Prandy, The revised cambridge scale of occupations. Sociology **24**, 629–655 (1990)

S. Rytina, Scaling the intergenerational continuity of occupation: is occupational inheritance ascriptive after all? Am. J. Sociol. **97**, 1658–1688 (1992)

K. Tsuzuki, Shokugyou hyouka no kouzou to shokugyou ishin sukoa (The structure of occupational evaluation and occupational prestige scores), Appendix Table. 1995 SSM Survey Study Group, Tokyo (1998), pp. 231–236. (in Japanese)

B. Wegener, Concepts and measurement of prestige. Annu. Rev. Sociol. **18**, 253–280 (1992)

On the Relationship Between Squared Canonical Correlation and Matrix Norm

Kentaro Hayashi, Ke-Hai Yuan, and Lu Liang

Abstract In research on approximating factor analysis (FA) by principal component analysis (PCA), FA loadings and PCA loadings are typically compared using some measure of closeness or distance. Previous studies have used the average squared canonical correlation (ASCC) between the two loading matrices as a measure of closeness. This measure has the advantages of being invariant with respect to sign and column changes, and most conveniently, it is not affected by rotations. However, the drawback of ASCC is that it is hard to intuitively perceive the size of the distance between the (elements of) two loading matrices. Therefore, other measures of difference between matrices such as the Frobenius norm are sometimes preferred. However, then complexities might occur such as the sign changes and the column alignment of the corresponding factors/components as well as rotational indeterminacy. The current study aims to characterize the relationship between the ASCC and a direct measure derived from matrix norms (e.g., Frobenius norm), which facilitates the understanding of the closeness between PCA and FA.

Keywords Factor analysis • Principal component analysis • High dimension • Large p small N

K. Hayashi (✉)
Department of Psychology, University of Hawaii at Manoa, 2530 Dole Street, Sakamaki C400, Honolulu, HI, 96822, USA
e-mail: hayashik@hawaii.edu

K.-H. Yuan
Department of Psychology, University of Notre Dame, 123A Haggar Hall, Notre Dame, IN, 46556, USA
e-mail: kyuan@nd.edu

L. Liang
Department of Psychology, Florida International University, 11200 S.W. 8th Street, Miami, FL, 33199, USA
e-mail: luliang@fiu.edu

L.A. van der Ark et al. (eds.), *Quantitative Psychology*, Springer Proceedings in Mathematics & Statistics 196, DOI 10.1007/978-3-319-56294-0_13

1 Introduction

Principal component analysis (PCA) and factor analysis (FA) are frequently used multivariate statistical methods for data reduction (Anderson 1963; Anderson 2003; Lawley and Maxwell 1971). Oftentimes, PCA is used to approximate FA, and an important research question is under what condition PCA gives a good approximation of FA (Guttman 1956; Bentler and Kano 1990; Krijnen 2006; Schneeweiss and Mathes 1995; Schneeweiss 1997).

Let $\mathbf{\Lambda}^*$ be the matrix of principal component loadings and $\mathbf{\Lambda}$ be the matrix of factor loadings, both with dimension $p \times m$, where $p > m$. Then the difference $\mathbf{\Delta}$ between the two loading matrices can be expressed as $\mathbf{\Delta} = \mathbf{\Lambda}^* - \mathbf{\Lambda}$ or equivalently $\mathbf{\Lambda}^* = \mathbf{\Lambda} + \mathbf{\Delta}$. Here, we use the "original" principal components so that they are uncorrelated, and likewise we assume that the factors are also uncorrelated (orthogonal). Related with the column alignment, we arrange the factors/components in the descending order of the column sum of squares of the loading matrices.

Practically speaking, for the estimates $\widehat{\mathbf{\Lambda}}^*$ and $\widehat{\mathbf{\Lambda}}$, we need to try every different sign change and column alignment that makes $\widehat{\mathbf{\Delta}}$ the smallest in terms of, e.g., the squared matrix norm (Ichikawa and Konishi 1995)—to be discussed below. If we know the true population loading matrices as in a simulation, we can alternatively try every different sign change and column alignment that makes the difference measured by, e.g., the squared matrix norm between the estimated loading matrix and the true population loading matrix the smallest. The sign and column alignments have made the computation of the direct measure of the difference between $\widehat{\mathbf{\Lambda}}^*$ and $\widehat{\mathbf{\Lambda}}$ substantially more complicated and thus less popular. In particular, for each column, there are two sign changes (+ or –). With m factors (m columns), there are 2^m different combinations of sign changes. For column alignments, there are $m!$ kinds of different alignments. Therefore, we must examine a total of $2^m \cdot m!$ different possibilities as for which one leads to the smallest sum of squared differences. Thus, even with a small number of factors and components, the number of comparisons is relatively large, e.g., for $m = 3$, there are a total of $2^m \cdot m! = 2^3 \cdot 3! = 48$ combinations, and as the number of factors and components increases, the number of combinations of different signs and column alignments rapidly increases. For simplicity, we assumed that the issues with sign changes and column alignments between $\mathbf{\Lambda}^*$ and $\mathbf{\Lambda}$ or $\widehat{\mathbf{\Lambda}}^*$ and $\widehat{\mathbf{\Lambda}}$ have been resolved beforehand.

The squared matrix (Frobenius) norm (see, e.g., p. 291 of (Horn and Johnson 1985); p. 165 of (Schott 2005)) of the difference between matrices $\mathbf{\Lambda}^*(p \times m)$ and $\mathbf{\Lambda}(p \times m)$ is given by

$$|| \mathbf{\Lambda}^* - \mathbf{\Lambda} ||^2 = tr\left\{\left(\mathbf{\Lambda}^* - \mathbf{\Lambda}\right)' \left(\mathbf{\Lambda}^* - \mathbf{\Lambda}\right)\right\} = tr\left(\mathbf{\Delta}'\mathbf{\Delta}\right), \tag{1}$$

where $tr(\mathbf{A})$ is the trace (sum of the diagonal elements) of a square matrix $\mathbf{A}$.

If we want to avoid examining all the combinations of different sign changes and column alignments, one recommended treatment would be to employ the squared canonical correlations (SCCs) between the two loading matrices $\mathbf{\Lambda}^*$ and $\mathbf{\Lambda}$, instead of an ordinary matrix norm as a measure of closeness/difference between them (see, e.g., (Schneeweiss and Mathes 1995; Schneeweiss 1997)). The SCCs between $\mathbf{\Lambda}^*$ and $\mathbf{\Lambda}$ are given by the eigenvalues of $(\mathbf{\Lambda}'\mathbf{\Lambda})^{-1}(\mathbf{\Lambda}'\mathbf{\Lambda}^*)(\mathbf{\Lambda}^{*\prime}\mathbf{\Lambda}^*)^{-1}(\mathbf{\Lambda}^{*\prime}\mathbf{\Lambda})$ and are known to be invariant with respect to sign changes and column alignments. Thus, the average squared canonical correlation (ASCC) between matrices $\mathbf{\Lambda}^*$ and $\mathbf{\Lambda}$ is given by $\rho^2(\mathbf{\Lambda},\mathbf{\Lambda}^*) = (1/m)tr\{(\mathbf{\Lambda}'\mathbf{\Lambda})^{-1}(\mathbf{\Lambda}'\mathbf{\Lambda}^*)(\mathbf{\Lambda}^{*\prime}\mathbf{\Lambda}^*)^{-1}(\mathbf{\Lambda}^{*\prime}\mathbf{\Lambda})\}$.

2 Squared Canonical Correlations and Matrix Norms

Our objective is to find the relationship between the matrix of differences $\mathbf{\Delta}$ or the squared matrix norm $||\mathbf{\Delta}||^2 = tr(\mathbf{\Delta}'\mathbf{\Delta})$ and the ASCC $\rho^2(\mathbf{\Lambda},\mathbf{\Lambda}^*)$. Concretely speaking, we aim to express $(\mathbf{\Lambda}'\mathbf{\Lambda})^{-1}(\mathbf{\Lambda}'\mathbf{\Lambda}^*)(\mathbf{\Lambda}^{*\prime}\mathbf{\Lambda}^*)^{-1}(\mathbf{\Lambda}^{*\prime}\mathbf{\Lambda})$ in terms of $\mathbf{\Lambda}$ and $\mathbf{\Delta} = \mathbf{\Lambda}^* - \mathbf{\Lambda}$ and then further examine the properties of ASCC. For such a purpose, first, we express $\mathbf{\Lambda}^{*\prime}\mathbf{\Lambda}^*$ as

$$\mathbf{\Lambda}^{*\prime}\mathbf{\Lambda}^* = (\mathbf{\Lambda} + \mathbf{\Delta})'(\mathbf{\Lambda} + \mathbf{\Delta}) = \mathbf{\Lambda}'\mathbf{\Lambda} + \mathbf{\Pi}, \tag{2}$$

where $\mathbf{\Pi} = \mathbf{\Lambda}'\mathbf{\Delta} + \mathbf{\Delta}'\mathbf{\Lambda} + \mathbf{\Delta}'\mathbf{\Delta}$ is the residual matrix of order $O(\mathbf{\Delta})$ and

$$\left(\mathbf{\Lambda}^{*\prime}\mathbf{\Lambda}^*\right)^{-1} = \{\mathbf{\Lambda}'\mathbf{\Lambda} + \mathbf{\Pi}\}^{-1}. \tag{3}$$

Now, using the matrix identity $(\boldsymbol{A}+\boldsymbol{B})^{-1} = \boldsymbol{A}^{-1}(\boldsymbol{I}+\boldsymbol{B}\boldsymbol{A}^{-1})^{-1}$ and letting $A = \mathbf{\Lambda}'\mathbf{\Lambda}, \boldsymbol{B} = \mathbf{\Pi}$, and $A + B = \mathbf{\Lambda}^{*\prime}\mathbf{\Lambda}^*$, we can rewrite Eq. (3) as

$$\left(\mathbf{\Lambda}^{*\prime}\mathbf{\Lambda}^*\right)^{-1} = \left(\mathbf{\Lambda}'\mathbf{\Lambda}\right)^{-1}\left\{\boldsymbol{I}_m + \mathbf{\Pi}\left(\mathbf{\Lambda}'\mathbf{\Lambda}\right)^{-1}\right\}^{-1}. \tag{4}$$

Let $\boldsymbol{H} = \mathbf{\Pi}(\mathbf{\Lambda}'\mathbf{\Lambda})^{-1}$ and $\boldsymbol{U} = -\boldsymbol{H}$. According to Eq. (5) of (Strang 1988), p. 270, if the absolute value of every eigenvalue of the matrix $\boldsymbol{U}$ is strictly less than 1, the right-hand side of $(\boldsymbol{I} - \boldsymbol{U})^{-1} = \boldsymbol{I} + \boldsymbol{U} + \boldsymbol{U}^2 + \boldsymbol{U}^3 + \ldots$converges. (Note: For $\boldsymbol{I} - \boldsymbol{U}$ to be invertible, we assume that every eigenvalue of $\boldsymbol{U}$ is less than 1. We write this as $\boldsymbol{I} - \boldsymbol{U} > 0$ or $\boldsymbol{I} > \boldsymbol{U}$.) Thus, if the absolute value of every eigenvalue of $\boldsymbol{H}$ is strictly less than 1, we can further rewrite Eq. (4) as

$$\left(\mathbf{\Lambda}^{*\prime}\mathbf{\Lambda}^*\right)^{-1} = \left(\mathbf{\Lambda}'\mathbf{\Lambda}\right)^{-1}\left\{\boldsymbol{I}_m + \sum_{k=1}^{\infty}(-1)^k\left[\mathbf{\Pi}\left(\mathbf{\Lambda}'\mathbf{\Lambda}\right)^{-1}\right]^k\right\}.$$

That is,

$$\left(\mathbf{\Lambda}^{*\prime}\mathbf{\Lambda}^*\right)^{-1} = \left(\mathbf{\Lambda}'\mathbf{\Lambda}\right)^{-1} + \mathbf{\Omega}, \tag{5}$$

where

$$\boldsymbol{\Omega} = \left(\boldsymbol{\Lambda}'\boldsymbol{\Lambda}\right)^{-1} \sum_{k=1}^{\infty} (-1)^k \left[\boldsymbol{\Pi}\left(\boldsymbol{\Lambda}'\boldsymbol{\Lambda}\right)^{-1}\right]^k \tag{6}$$

is the remainder term. (See Appendix 2 for an alternative expression of the remainder term.) By keeping the two leading terms, Eq. (6) becomes

$$\boldsymbol{\Omega} = -\left(\boldsymbol{\Lambda}'\boldsymbol{\Lambda}\right)^{-1}\boldsymbol{\Pi}\left(\boldsymbol{\Lambda}'\boldsymbol{\Lambda}\right)^{-1} + \left(\boldsymbol{\Lambda}'\boldsymbol{\Lambda}\right)^{-1}\left(\boldsymbol{\Lambda}'\boldsymbol{\Delta} + \boldsymbol{\Delta}'\boldsymbol{\Lambda}\right)\left(\boldsymbol{\Lambda}'\boldsymbol{\Lambda}\right)^{-1} \left(\boldsymbol{\Lambda}'\boldsymbol{\Delta} + \boldsymbol{\Delta}'\boldsymbol{\Lambda}\right)\left(\boldsymbol{\Lambda}'\boldsymbol{\Lambda}\right)^{-1} + o\left(\boldsymbol{\Delta}^2\right). \tag{7}$$

Here, to guarantee the convergence of the series in (6), we need to examine whether (or when) the absolute value of every eigenvalue of

$$\boldsymbol{H} = \boldsymbol{\Pi}\left(\boldsymbol{\Lambda}'\boldsymbol{\Lambda}\right)^{-1} = \left(\boldsymbol{\Lambda}'\boldsymbol{\Delta} + \boldsymbol{\Delta}'\boldsymbol{\Lambda} + \boldsymbol{\Delta}'\boldsymbol{\Delta}\right)\left(\boldsymbol{\Lambda}'\boldsymbol{\Lambda}\right)^{-1}$$

is less than 1. An easy algebra shows that the answer is when $0 < \boldsymbol{\Lambda}^{*\prime}\boldsymbol{\Lambda}^* < 2\boldsymbol{\Lambda}'\boldsymbol{\Lambda}$. Thus, Eq. (6) converges if $0 < \boldsymbol{\Lambda}^{*\prime}\boldsymbol{\Lambda}^* < 2\boldsymbol{\Lambda}'\boldsymbol{\Lambda}$.

We next turn to ASCC, which contains the terms $\boldsymbol{\Lambda}'\boldsymbol{\Lambda}^*$ and $\boldsymbol{\Lambda}^{*\prime}\boldsymbol{\Lambda}$. It follows from $\boldsymbol{\Lambda}^* = \boldsymbol{\Lambda} + \boldsymbol{\Delta}$ that

$$\boldsymbol{\Lambda}'\boldsymbol{\Lambda}^* = \boldsymbol{\Lambda}'\left(\boldsymbol{\Lambda} + \boldsymbol{\Delta}\right) = \boldsymbol{\Lambda}'\boldsymbol{\Lambda} + \boldsymbol{\Lambda}'\boldsymbol{\Delta} \tag{8}$$

and

$$\boldsymbol{\Lambda}^{*\prime}\boldsymbol{\Lambda} = \left(\boldsymbol{\Lambda} + \boldsymbol{\Delta}\right)'\boldsymbol{\Lambda} = \boldsymbol{\Lambda}'\boldsymbol{\Lambda} + \boldsymbol{\Delta}'\boldsymbol{\Lambda}. \tag{9}$$

By additional algebraic computation, together with Eqs. (5), (7), (8), and (9), we obtain

$$\left(\boldsymbol{\Lambda}'\boldsymbol{\Lambda}\right)^{-1}\left(\boldsymbol{\Lambda}'\boldsymbol{\Lambda}^*\right)\left(\boldsymbol{\Lambda}^{*\prime}\boldsymbol{\Lambda}^*\right)^{-1}\left(\boldsymbol{\Lambda}^{*\prime}\boldsymbol{\Lambda}\right) = \boldsymbol{I}_m + \boldsymbol{R}, \tag{10}$$

where

$$\boldsymbol{R} = -\left(\boldsymbol{\Lambda}'\boldsymbol{\Lambda}\right)^{-1}\boldsymbol{\Delta}'\left\{\boldsymbol{I}_p - \boldsymbol{\Lambda}\left(\boldsymbol{\Lambda}'\boldsymbol{\Lambda}\right)^{-1}\boldsymbol{\Lambda}'\right\}\boldsymbol{\Delta} + o\left(\boldsymbol{\Delta}^2\right) \tag{11}$$

is the remainder term.

3 The Dominant Term in the Remainder Term

We have just shown that the dominant term in $\boldsymbol{R}$ in Eq. (11) is

$$\boldsymbol{R}\left(\boldsymbol{\Delta}^2\right) = -\left(\boldsymbol{\Lambda}'\boldsymbol{\Lambda}\right)^{-1}\boldsymbol{\Delta}'\left\{\boldsymbol{I}_p - \boldsymbol{\Lambda}\left(\boldsymbol{\Lambda}'\boldsymbol{\Lambda}\right)^{-1}\boldsymbol{\Lambda}'\right\}\boldsymbol{\Delta}. \tag{12}$$

Here, note that the terms of order $O(\boldsymbol{\Delta})$ cancel out and the dominant term is of order $O(\boldsymbol{\Delta}^2)$, which is nonpositive definite since $\boldsymbol{I}_p - \boldsymbol{\Lambda}(\boldsymbol{\Lambda}'\boldsymbol{\Lambda})^{-1}\boldsymbol{\Lambda}'$ is a projection matrix and thus semi-positive definite (i.e., $\boldsymbol{I}_p - \boldsymbol{\Lambda}(\boldsymbol{\Lambda}'\boldsymbol{\Lambda})^{-1}\boldsymbol{\Lambda}' \geq 0$), which is guaranteed from the fact that every projection matrix is an idempotent matrix (Theorem 12.3.4 (6) of (Harville 1997)) and that every idempotent matrix is semi-positive definite, with the eigenvalues being either 1 or 0 (Theorem 10.2 of (Schott 2005)). Also, because $\boldsymbol{\Delta} = \boldsymbol{\Lambda}^* - \boldsymbol{\Lambda}$ appears twice in the dominant term, defining $\boldsymbol{\Delta}$ as $\boldsymbol{\Lambda}^* - \boldsymbol{\Lambda}$ or $\boldsymbol{\Lambda} - \boldsymbol{\Lambda}^*$ does not change the value of $\boldsymbol{R}(\boldsymbol{\Delta}^2)$. Thus, in short, the ASCC $\rho^2(\boldsymbol{\Lambda}, \boldsymbol{\Lambda}^*) = (1/m)tr\{(\boldsymbol{\Lambda}^*\boldsymbol{\Lambda})^{-1}(\boldsymbol{\Lambda}'\boldsymbol{\Lambda}^*)(\boldsymbol{\Lambda}^{*\prime}\boldsymbol{\Lambda}^*)^{-1}(\boldsymbol{\Lambda}^{*\prime}\boldsymbol{\Lambda})\}$ can approximately be expressed as

$$\begin{aligned}
\rho^2\left(\boldsymbol{\Lambda}, \boldsymbol{\Lambda}^*\right) &= 1 + (1/m)\, tr\left\{\boldsymbol{R}\left(\boldsymbol{\Delta}^2\right)\right\} + o\left(\boldsymbol{\Delta}^2\right) \\
&= 1 - (1/m)\, tr\left\{\left(\boldsymbol{\Lambda}'\boldsymbol{\Lambda}\right)^{-1}\boldsymbol{\Delta}'\left[\boldsymbol{I}_p - \boldsymbol{\Lambda}\left(\boldsymbol{\Lambda}'\boldsymbol{\Lambda}\right)^{-1}\boldsymbol{\Lambda}'\right]\boldsymbol{\Delta}\right\} + o\left(\boldsymbol{\Delta}^2\right) \\
&= 1 - (1/m)\, tr\left\{\left[\boldsymbol{I}_p - \boldsymbol{\Lambda}\left(\boldsymbol{\Lambda}'\boldsymbol{\Lambda}\right)^{-1}\boldsymbol{\Lambda}'\right]\left[\boldsymbol{\Delta}\left(\boldsymbol{\Lambda}'\boldsymbol{\Lambda}\right)^{-1}\boldsymbol{\Delta}'\right]\right\} + o\left(\boldsymbol{\Delta}^2\right)
\end{aligned} \tag{13}$$

if $0 < \boldsymbol{\Lambda}^{*\prime}\boldsymbol{\Lambda}^* < 2\boldsymbol{\Lambda}'\boldsymbol{\Lambda}$. Now, applying Claim 13 and Corollary 14 of (Harvey 2011) (see Appendix 1) to $tr\{[\boldsymbol{I}_p - \boldsymbol{\Lambda}(\boldsymbol{\Lambda}'\boldsymbol{\Lambda})^{-1}\boldsymbol{\Lambda}'][\boldsymbol{\Delta}(\boldsymbol{\Lambda}'\boldsymbol{\Lambda})^{-1}\boldsymbol{\Delta}']\}$ in Eq. (13), we can show that

$$\begin{aligned}
tr\left\{\left[\boldsymbol{I}_p - \boldsymbol{\Lambda}\left(\boldsymbol{\Lambda}'\boldsymbol{\Lambda}\right)^{-1}\boldsymbol{\Lambda}'\right]\boldsymbol{\Delta}\left(\boldsymbol{\Lambda}'\boldsymbol{\Lambda}\right)^{-1}\boldsymbol{\Delta}'\right\} &\leq tr\left\{\boldsymbol{I}_p - \boldsymbol{\Lambda}\left(\boldsymbol{\Lambda}'\boldsymbol{\Lambda}\right)^{-1}\boldsymbol{\Lambda}'\right\} \\
&\quad \cdot \|\boldsymbol{\Delta}\left(\boldsymbol{\Lambda}'\boldsymbol{\Lambda}\right)^{-1}\boldsymbol{\Delta}'\|.
\end{aligned} \tag{14}$$

Here, using the well-known formula regarding the trace of an idempotent matrix (Theorem 10.1(d) of (Schott 2005)), $tr(\boldsymbol{I}_p - \boldsymbol{\Lambda}(\boldsymbol{\Lambda}'\boldsymbol{\Lambda})^{-1}\boldsymbol{\Lambda}') = p - m$, it follows that

$$\begin{aligned}
\rho^2\left(\boldsymbol{\Lambda}, \boldsymbol{\Lambda}^*\right) &\geq 1 - \left(\frac{p}{m} - 1\right)\|\boldsymbol{\Delta}\left(\boldsymbol{\Lambda}'\boldsymbol{\Lambda}\right)^{-1}\boldsymbol{\Delta}'\| + o\left(\boldsymbol{\Delta}^2\right) \\
&= 1 - \left(\frac{p}{m} - 1\right)\sqrt{tr\left\{\left[\left(\boldsymbol{\Lambda}'\boldsymbol{\Lambda}\right)^{-1}\boldsymbol{\Delta}'\boldsymbol{\Delta}\right]^2\right\}} + o\left(\boldsymbol{\Delta}^2\right).
\end{aligned} \tag{15}$$

This equation connects $\boldsymbol{\Delta}$ (the difference between the matrix of PC loadings $\boldsymbol{\Lambda}^*$ and the matrix of factor loadings $\boldsymbol{\Lambda}$) and $\rho^2(\boldsymbol{\Lambda}, \boldsymbol{\Lambda}^*)$ (the squared CC between the two matrices).

Now, let us further assume $\mathbf{\Lambda}'\mathbf{\Lambda} = O(p)$ and equivalently, $(\mathbf{\Lambda}'\mathbf{\Lambda})^{-1} = O(p^{-1})$. This is a natural extension of the same assumption introduced in, e.g., (Bentler and Kano 1990) under the one-factor model. Then, we can show the order of $(\frac{p}{m} - 1)\sqrt{tr\left\{\left[(\mathbf{\Lambda}'\mathbf{\Lambda})^{-1}\mathbf{\Delta}'\mathbf{\Delta}\right]^2\right\}}$ to be

$$O\left(\left(\frac{p}{m} - 1\right)\sqrt{tr\left\{\left[(\mathbf{\Lambda}'\mathbf{\Lambda})^{-1}\mathbf{\Delta}'\mathbf{\Delta}\right]^2\right\}}\right) = O\left(\left(\frac{p}{m} - 1\right)p^{-1}\sqrt{tr\left\{\left[\mathbf{\Delta}'\mathbf{\Delta}\right]^2\right\}}\right)$$
$$= O\left(\left(\frac{1}{m} - \frac{1}{p}\right)||\mathbf{\Delta}'\mathbf{\Delta}||\right). \tag{16}$$

In addition, with the common assumption of $m/p = o(1)$ (Guttman 1956), we can further rewrite Eq. (16) as $O(m^{-1}||\mathbf{\Delta}'\mathbf{\Delta}||)$. Therefore,

$$\rho^2\left(\mathbf{\Lambda}, \mathbf{\Lambda}^*\right) \geq 1 - (1/m)\, R\left(\mathbf{\Lambda}, \mathbf{\Lambda}^*\right) \quad \text{with} \quad R\left(\mathbf{\Lambda}, \mathbf{\Lambda}^*\right) = O\left(||\mathbf{\Delta}'\mathbf{\Delta}||\right). \tag{17}$$

Because $||\mathbf{\Delta}'\mathbf{\Delta}|| \leq ||\mathbf{\Delta}||^2$ (see Appendix A.3), we can also show that

$$\rho^2\left(\mathbf{\Lambda}, \mathbf{\Lambda}^*\right) \geq 1 - (1/m)\, R_2\left(\mathbf{\Lambda}, \mathbf{\Lambda}^*\right) \quad \text{with} \quad R_2\left(\mathbf{\Lambda}, \mathbf{\Lambda}^*\right) = O\left(||\mathbf{\Delta}^2\right). \tag{18}$$

Obviously, the lower bound in Eq. (17) is sharper than that in Eq. (18). However, Eq. (18) connects the SCC and the matrix norm more directly.

4 Example

The sample correlation matrix in Table 1 was reproduced from (Emmett 1949). It is computed from a sample of 211 observations with 9 observed variables ($p = 9$). Only the first two eigenvalues of the sample correlation matrix are above 1, so we employed a two-factor/component model ($m = 2$). The FA and PCA loading matrices are listed in Table 2. For this example, the matrix norm (without squaring) was $||\mathbf{\Lambda}^* - \mathbf{\Lambda}|| = ||\mathbf{\Delta}|| = \sqrt{tr\left(\mathbf{\Delta}'\mathbf{\Delta}\right)} = 0.439$, the squared matrix norm was $||\mathbf{\Lambda}^* - \mathbf{\Lambda}||^2 = 0.193$, and ASCC was $\rho^2(\mathbf{\Lambda}, \mathbf{\Lambda}^*) = 0.982$. When we took the trace of the dominant term $\boldsymbol{R}(\mathbf{\Delta}^2)$ (given in Eq. (12)) of the remainder terms, divide it by m to take the average, and add 1 (as in Eq. (13) ignoring the terms with $o(\mathbf{\Delta}^2)$), the value was $1 + (1/m)tr\{\boldsymbol{R}(\mathbf{\Delta}^2)\} = 0.976$, which was very close to the value of ASCC. The value of the lower bound given in Eq. (17) when $O(||\mathbf{\Delta}'\mathbf{\Delta}||)$ was replaced by $||\mathbf{\Delta}'\mathbf{\Delta}||$ was $1 - (1/m)||\mathbf{\Delta}'\mathbf{\Delta}|| = 0.922$, which was not as close to ASCC as the value created from Eq. (13); however, it still gave a relatively good approximation to the ASCC value as a lower bound. The value of the lower bound given in Eq.

Table 1 Sample covariance matrix (Emmett, 1949) with the FA (maximum likelihood) and PCA loading matrices

1.000								
0.523	1.000							
0.395	0.479	1.000						
0.471	0.506	0.355	1.000					
0.346	0.418	0.270	0.691	1.000				
0.426	0.462	0.254	0.791	0.679	1.000			
0.576	0.547	0.452	0.443	0.383	0.372	1.000		
0.434	0.283	0.219	0.285	0.149	0.314	0.385	1.000	
0.639	0.645	0.504	0.505	0.409	0.472	0.680	0.470	1.000

Table 2 Estimated FA and PCA loading matrices (two factors/components)

Factor analysis		Principal component analysis	
F1	F2	PC1	PC2
0.668	0.303	0.749	0.265
0.692	0.237	0.762	0.125
0.500	0.286	0.596	0.303
0.840	−0.322	0.792	−0.453
0.701	−0.318	0.680	−0.565
0.800	−0.372	0.747	−0.512
0.670	0.384	0.754	0.305
0.442	0.245	0.521	0.355
0.775	0 .425	0.832	0.285

(18) when $O(||\mathbf{\Delta}^2)$ was replaced by $||\mathbf{\Delta}||^2$ was $1-(1/m)||\mathbf{\Delta}||^2 = 0.904$, which was slightly lower than when $||\mathbf{\Delta}'\mathbf{\Delta}||$ was used instead of $||\mathbf{\Delta}||^2$.

5 Concluding Remarks

In this work, we obtained some inequalities connecting the SCC and the matrix norm, under the assumption that the column sum(s) of squares of the FA loading matrix are of order p. An obvious concern in any efforts to connect between the SCC and the matrix norm is that the SCC ranges between 0 and 1, while the matrix norm can be any non-negative value. More specifically, as the number of variables p increases, the lower bound for the SCC given in terms of the matrix norm in Eq. (18) becomes pessimistically low. A way to improve the situation would be to define the matrix norm as, for example,

$$|| \mathbf{\Delta} ||_*^2 = \left(\frac{1}{pm}\right) tr\left(\mathbf{\Delta}'\mathbf{\Delta}\right) \tag{19}$$

instead of $||\mathbf{\Delta}||^2 = tr(\mathbf{\Delta}'\mathbf{\Delta})$.

Acknowledgment The authors appreciate the comments of Dr. Jeffrey Douglas who read our manuscript very carefully. Ke-Hai Yuan's work was supported by the National Science Foundation under Grant No. SES-1461355.

Appendix 1: Claim 13 and Corollary 14 by Professor Nick Harvey (University of British Columbia)

Claim 13: Let A, B, and C be Symmetric d × d Matrices Satisfying A ≥ 0 and B ≤ C. Then tr(AB) ≤ tr(AC)

Proof *First consider the case that* $\boldsymbol{A}$ *is rank one, i.e.,* $\boldsymbol{A} = \nu\nu'$ *for some vector* ν. *Then*

$$tr(\boldsymbol{AB}) = tr\left(\boldsymbol{\nu\nu}'\boldsymbol{B}\right) = tr\left(\boldsymbol{\nu}'\boldsymbol{B\nu}\right) = \boldsymbol{\nu}'\boldsymbol{B\nu} \le \boldsymbol{\nu}'\boldsymbol{C\nu} = tr\left(\boldsymbol{\nu}'\boldsymbol{C\nu}\right)$$
$$= tr\left(\boldsymbol{\nu\nu}'\boldsymbol{C}\right) = tr(\boldsymbol{AC}).$$

Now we consider the case where $\boldsymbol{A}$ has arbitrary rank. Let $\boldsymbol{A} = \sum_{i=1}^{d} \lambda_i \boldsymbol{u}_i \boldsymbol{u}_i{}'$. Since we assume that $\boldsymbol{A} \ge 0$ we can set $\boldsymbol{v}_i = \sqrt{\lambda_i}\boldsymbol{u}_i$ and write $\boldsymbol{A} = \sum_{i=1}^{d} \boldsymbol{v}_i \boldsymbol{v}_i{}'$. Then

$$tr(\boldsymbol{AB}) = tr\left(\sum_{i=1}^{d} \boldsymbol{v}_i\boldsymbol{v}_i{}'\boldsymbol{B}\right) = \sum_{i=1}^{d} tr\left(\boldsymbol{v}_i\boldsymbol{v}_i{}'\boldsymbol{B}\right) \le \sum_{i=1}^{d} tr\left(\boldsymbol{v}_i\boldsymbol{v}_i{}'\boldsymbol{C}\right)$$
$$= tr\left(\sum_{i=1}^{d} \boldsymbol{v}_i\boldsymbol{v}_i{}'\boldsymbol{C}\right) = tr(\boldsymbol{AC}).$$

Here the second and third equalities are by linearity of trace, and the inequality follows from the rank one case.

Corollary 14: If A ≥ 0, Then tr (AB) ≤ ||B|| · tr (A)

Proof *Apply Claim 13 with* $\boldsymbol{C} = ||\boldsymbol{B}|| \cdot \boldsymbol{I}$. *We get* $tr(\boldsymbol{AB}) \le tr(\boldsymbol{A} \cdot ||\boldsymbol{B}|| \cdot \boldsymbol{I}) \le ||\boldsymbol{B}|| \cdot tr(\boldsymbol{A} \cdot \boldsymbol{I})$, *by linearity of trace since* $||\boldsymbol{B}||$ *is a scalar.*

Proof for $||\boldsymbol{\Delta}'\boldsymbol{\Delta}|| \le ||\boldsymbol{\Delta}||^2$

By applying Corollary (A.2), $||\boldsymbol{\Delta}'\boldsymbol{\Delta}|| = \{tr(\boldsymbol{\Delta}'\boldsymbol{\Delta}\boldsymbol{\Delta}'\boldsymbol{\Delta})\}^{1/2} \le \{tr(\boldsymbol{\Delta}'\boldsymbol{\Delta})\}^{1/2}$ $||\boldsymbol{\Delta}'\boldsymbol{\Delta}||^{1/2}$. Because $||\boldsymbol{\Delta}'\boldsymbol{\Delta}|| \ge 0$, the same inequality holds when each side is squared. Thus, $||\mathbf{\Delta}'\mathbf{\Delta}||^2 \leq \{tr(\mathbf{\Delta}'\mathbf{\Delta})\}||\mathbf{\Delta}'\mathbf{\Delta}||$. If $||\mathbf{\Delta}'\mathbf{\Delta}|| > 0$, because the inequality still holds after both sides are divided by the same positive number $||\mathbf{\Delta}'\mathbf{\Delta}||$, it follows that $||\mathbf{\Delta}'\mathbf{\Delta}|| \leq \{tr(\mathbf{\Delta}'\mathbf{\Delta})\}$. Finally, by definition, $||\mathbf{\Delta}||^2 = tr(\mathbf{\Delta}'\mathbf{\Delta})$. Therefore, $||\mathbf{\Delta}'\mathbf{\Delta}|| \leq ||\mathbf{\Delta}||^2$. If $||\mathbf{\Delta}'\mathbf{\Delta}|| = 0$, then $\mathbf{\Delta} = 0$, and there exists $||\mathbf{\Delta}'\mathbf{\Delta}|| = ||\mathbf{\Delta}||^2$.

Appendix 2: Alternative Formula for Equation (6)

Alternatively, we can use another well-known identity $(\boldsymbol{A}+\boldsymbol{B})^{-1} = \boldsymbol{A}^{-1} - \boldsymbol{A}^{-1}(\boldsymbol{B}^{-1}+\boldsymbol{A}^{-1})^{-1}\boldsymbol{A}^{-1}$ (e.g., Corollary 1.7.1 of (Harvey 2011–2012)). This identity holds as long as the matrices $\boldsymbol{A}$, $\boldsymbol{B}$, and $\boldsymbol{A}+\boldsymbol{B}$ are all square, non-singular matrices so that the inverses exist. In our context, $\boldsymbol{A} = \mathbf{\Lambda}'\mathbf{\Lambda}$, $\boldsymbol{B} = \mathbf{\Pi} = \mathbf{\Lambda}'\mathbf{\Delta} + \mathbf{\Delta}'\mathbf{\Lambda} + \mathbf{\Delta}'\mathbf{\Delta}$, and $\boldsymbol{A}+\boldsymbol{B} = \mathbf{\Lambda}^{*\prime}\mathbf{\Lambda}^*$, so that

$$(\mathbf{\Lambda}^{*\prime}\mathbf{\Lambda}^*)^{-1} = (\mathbf{\Lambda}'\mathbf{\Lambda})^{-1} - (\mathbf{\Lambda}'\mathbf{\Lambda})^{-1}\left\{\mathbf{\Pi}^{-1} + (\mathbf{\Lambda}'\mathbf{\Lambda})^{-1}\right\}^{-1}(\mathbf{\Lambda}'\mathbf{\Lambda})^{-1}.$$

That is,

$$(\mathbf{\Lambda}^{*\prime}\mathbf{\Lambda}^*)^{-1} = (\mathbf{\Lambda}'\mathbf{\Lambda})^{-1} + \mathbf{\Omega}_A, \tag{20}$$

where $\mathbf{\Omega}_A = -\{(\mathbf{\Lambda}'\mathbf{\Lambda})\mathbf{\Pi}^{-1}(\mathbf{\Lambda}'\mathbf{\Lambda}) + (\mathbf{\Lambda}'\mathbf{\Lambda})\}^{-1}$.

The Eq. (20) connects $(\mathbf{\Lambda}'\mathbf{\Lambda})^{-1}$ and $(\mathbf{\Lambda}^{*\prime}\mathbf{\Lambda}^*)^{-1}$ in an alternative way. The only requirement for Eq. (20) is the existence of the inverse of $\mathbf{\Pi} = \mathbf{\Lambda}'\mathbf{\Delta} + \mathbf{\Delta}'\mathbf{\Lambda} + \mathbf{\Delta}'\mathbf{\Delta}$. If we express $\mathbf{\Pi}$ as $\mathbf{\Pi} = (\mathbf{\Lambda}'\mathbf{\Lambda} + \mathbf{\Lambda}'\mathbf{\Delta} + \mathbf{\Delta}'\mathbf{\Lambda} + \mathbf{\Delta}'\mathbf{\Delta}) - \mathbf{\Lambda}'\mathbf{\Lambda} = \mathbf{\Lambda}^{*\prime}\mathbf{\Lambda}^* - \mathbf{\Lambda}'\mathbf{\Lambda} > 0$, we can easily see that the inverse of $\mathbf{\Pi}$ exists if $\mathbf{\Lambda}'\mathbf{\Lambda} < \mathbf{\Lambda}^{*\prime}\mathbf{\Lambda}^*$. Unfortunately, this condition may be harder to satisfy in practice than Eq. (6).

References

T.W. Anderson, Asymptotic theory for principal component analysis. Ann. Math. Stat. **34**, 122–148 (1963)

T.W. Anderson, *An Introduction to Multivariate Statistical Analysis*, 3rd edn. (Wiley, New York, NY, 2003)

P.M. Bentler, Y. Kano, On the equivalence of factors and components. Multivar. Behav. Res. **25**, 67–74 (1990)

W.G. Emmett, Factor analysis by Lawley's method of maximum likelihood. Br. J. Stat. Psychol. **2**, 90–97 (1949)

L. Guttman, "Best possible" estimates of communalities. Psychometrika **21**, 273–286 (1956)

N. Harvey, 2011–2012. Notes on symmetric matrices. Retrieved from http://www.cs.ubc.ca/~nickhar/W12/NotesMatrices.pdf on 9/1/2016

D.A. Harville, *Matrix Algebra from a Statistician's Perspective* (Springer, New York, NY, 1997)

R.A. Horn, C.R. Johnson, *Matrix Analysis* (Cambridge University Press, Cambridge, 1985)

M. Ichikawa, S. Konishi, Application of the bootstrap method in factor analysis. Psychometrika **60**, 77–93 (1995)

W.P. Krijnen, Convergence of estimates of unique variances in factor analysis, based on the inverse sample covariance matrix. Psychometrika **71**, 193–199 (2006)

D.N. Lawley, A.E. Maxwell, *Factor Analysis as a Statistical Method*, 2nd edn. (American Elsevier, New York, NY, 1971)

H. Schneeweiss, Factors and principal components in the near spherical case. Multivar. Behav. Res. **32**, 375–401 (1997)

H. Schneeweiss, H. Mathes, Factor analysis and principal components. J. Multivar. Anal. **55**, 105–124 (1995)

J.R. Schott, *Matrix Analysis for Statistics*, 2nd edn. (Wiley, New York, NY, 2005)

G. Strang, *Linear Algebra and its Applications*, 3rd edn. (Harcourt Brace Jovanovich, San Diego, CA, 1988)

Breaking Through the Sum Scoring Barrier

James O. Ramsay and Marie Wiberg

Abstract The aim of this paper is to reflect around what would be needed in order to replace sum scoring, including technical advances, communication with both test constructors and examinees, and organizational strategy. Sum scoring are proposed to be replaced by smart scoring and a brief description, and some theoretical support for smart scoring and methods for achieving it are given together with an example from a large-scale assessment test.

Keywords Smart scoring • Technical advances • Item impact function

1 Introduction

Test scoring by counting the number of correct answers is still, after more than half a century of psychometric effort, the most common method for estimating ability. However, this practice, which we call "sum scoring," is inefficient because it ignores variation in the amount of information the response to an item provides over both items and examinees. "Smart scoring," on the other hand, allows for an interaction between ability and item effectiveness in a way that we detail in Sect. 2. For example, Ramsay and Wiberg (2017) demonstrate improvements in root mean squared error (RMSE) for ability estimation of 6% for a 36-item carefully designed National Assessment of Educational Progress (NAEP) history test and 13% for a 100-item university classroom test.

Although these improvements in the score accuracy may not impress an individual examinee, when aggregated over the many millions of students assessed each spring in a country such as the USA, these would be an invaluable improvement in educational technology and one that is available for practically no cost. Similar progress in, say, breast cancer mortality would merit a Nobel Prize, and an

J.O. Ramsay (✉)
Department of Psychology, McGill University, Montreal, QC, Canada
e-mail: james.ramsay@mcgill.ca; ramsay@psych.mcgill.ca

M. Wiberg
Department of Statistics, USBE, Umeå University, Umeå, Sweden
e-mail: marie.wiberg@umu.se

L.A. van der Ark et al. (eds.), *Quantitative Psychology*, Springer Proceedings in Mathematics & Statistics 196, DOI 10.1007/978-3-319-56294-0_14

investment firm achieving this amount of increase in return would astonish Wall Street. Moreover, they show that the improvement in RMSE for high-end examinees scoring at 90% or more would be even greater, promising an invaluable benefit for the college admission process.

But sum scoring is simple to understand, is easy to execute, and has the image of being "fair" in some sense. Its supporters would argue that it works well enough to have adequately served the needs of schools and colleges and that the percentage improvements possible for more sophisticated approaches are too trivial to be worth the effort involved or could be achieved more simply by adding a few more items to the test. In short, replacing sum scoring by what we call "smart scoring" is apt to be a large challenge in statistical and social engineering, and the task must not be underestimated.

We first reflect on what would be needed in order to replace sum scoring, including technical advances, communication strategies for both test constructors and examinees, and organizational issues. The final part of the paper contains a brief description and some theoretical support for smart scoring and methods for achieving it.

2 What Replacing Sum Scoring Would Require

We need to work on a scale that is already familiar in order to sidestep the problem of interpreting numbers on the whole real line that are employed in most variants of item response theory. The percentage interval [0,100] is easy to use, to understand, and to interpret both for test takers and for the test constructors.

The score distribution is a central issue. Sum scoring offers no control over the score distribution, but we often need the capacity to define the score distribution, such as one which approaches a preassigned distribution target set by an educational institution or which preserves the score distribution of previous test administrations in multi-administration testing contexts. We consider some theoretical issues around manipulating score distribution in Sect. 3.

To spread the use of smart scoring, we need to develop application programs that any teacher, administrator, or test taker can use to analyze test data. An application must have a version for handheld platforms such as smartphones, as well as for tablets and laptops, and be web based as well. A test scoring application needs to be fast enough that it can score 1000 test takers in tens of seconds. Covariates such as gender, language groups, age, and so on should be potentially a part of the analysis in order to detect, in the usual case, undesirable contributors to differential item functioning (DIF).

Test designers and administrators will want to see graphical displays of the performance of test items, and, where appropriate, examinees should have access to displays of confidence intervals as well as best estimates for their performance, along with indications of what each response contributed to their result. Storage and printing of these displays should be seamless.

Applications would need to access exam data in standard data formats, including ASCII, Microsoft Excel, and Apple Numbers, as well as formats used by automatic answer sheet processing hardware. The test administrator would want extensive editing capacity so as to add or delete items, score subsets of items, and modify text in the examination itself. Test composition utilities should be a part of such a package. Application displays should be easily portable to various types of printing, storage, and display resources.

The application must be accompanied by a manual that is comprehensible to the widest possible range of users, with extensive illustrations and user templates. The manual would need to be published in all of the world's major languages since testing is a multinational industry. The scientific community would undoubtedly be challenged to replace some of its more arcane jargon by friendlier options. Folks on the street do not use "item" to refer to an exam question, for example, and "item characteristic curve" would surely not survive this editing process.

Data simulation would play a large role in using this software. Someone considering the use of smart smoothing would want to see its benefits for a prospective or existing exam. A simulation tool would enable the simulation and analysis of up to a thousand replicates of simulated exams which match its design and examinee characteristics. The simulation application should give a variety of accuracy assessment displays as well as show the implications of such data design elements as number of examinees and items. A portfolio of previous successful and credible applications must be assembled and be readily available to prospective users of smart scoring.

These exam scoring resources would have to be available at a cost sufficient to support the expenditures required for distribution, possible research and development, risk protection, etc. Ideally, the distributor would be an entity acting in the public interest and might be involved, for example, in a partnership with an existing nonprofit testing agency and the Psychometric Society. Test scoring services might be an outlet for these exam processing utilities.

3 The Smart Scoring Equation

Smart scoring is briefly summarized here but is given in more detail in Ramsay and Wiberg (2017). Using $i, i = 1, \ldots, n$ to index items and $j = 1, \ldots, N$ to index examinees, let $P_i(\theta), i = 1, \ldots, n$, be the proportion test takers with ability θ who answer item i correctly, which we might call the item's "profile." Let the item function $W_i(\theta)$ be the corresponding log-odds ratio

$$W_i(\theta) = \log\left(\frac{P_i(\theta)}{1 - P_i(\theta)}\right), \tag{1}$$

perhaps called the "performance" of the item.

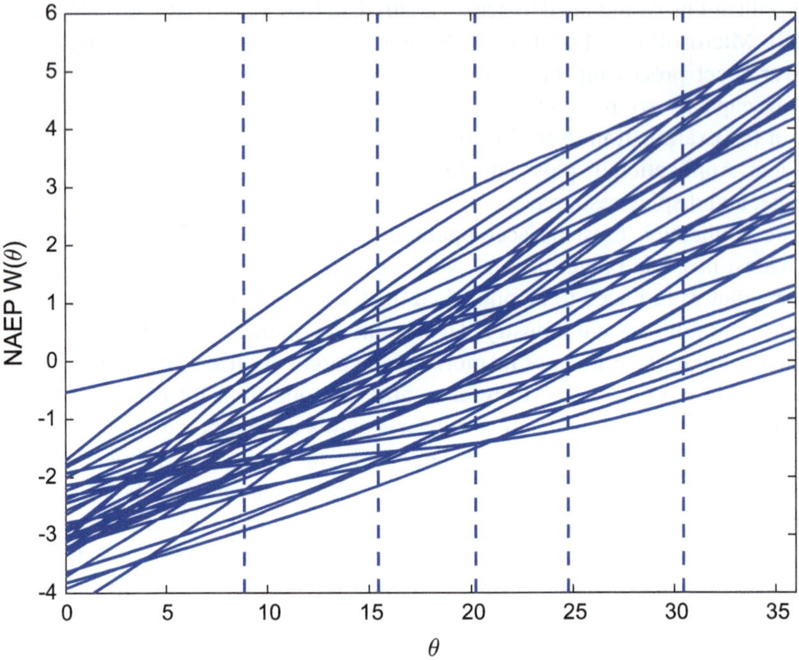

Fig. 1 The item performance curves, $W_i = \ln P_i/(1 - P_i)$ for the NAEP history test. The *vertical dashed lines* are the 5%, 25%, 50%, 75%, and 95% quantiles of the empirical distribution of the sum scores

The item performance curve W_i is rather more convenient than the profile curve P_i for both theory and computation. The 36 performance curves for the NAEP history test are shown in Fig. 1. We notice that they increase with only mild curvature and that they are unbounded. In fact, curves that were strictly increasing with common slopes would correspond to sum scoring, in which linear curves with unrestricted slopes would correspond to the two-parameter logistic model.

The need for a representation of the profile and performance functions, which is both flexible and justified by the data and is also easy to work with from a computational perspective and smooth and defined over a closed interval like $[0, 100]$, strongly suggests a basis function expansion of performance W in terms of splines. Ramsay and Wiberg (2017) discuss these issues in depth, and there are a spectrum of strategies for estimating the W_i's ranging from the quick and dirty to ones that can meet the more exacting demands of large sample statistical theory.

The maximum likelihood estimate $\hat{\theta}$ for binary-scored items with answer U_{ij} for test taker j to item i and conditional on knowing each item's performance curve has a simple expression in terms of the item performance function values $W_{ji} = W_i(\theta_j)$. The negative log likelihood for the estimation of examinee j's ability is

$$-\log L = -\sum_i [U_{ji}W_{ji} - \log(1 + e^{W_{ji}})$$

and taking its partial derivative with respect to θ leads the smart scoring equation

$$\sum_i^n [U_{ij} - P_i(\theta)] \frac{dW_i}{d\theta} = 0. \tag{2}$$

Notice that this is also the stationary equation for a continuum of nonlinear weighted least squares problems indexed by θ, where the optimal value makes the residuals $U_{ij} - P_i(\theta)$ orthogonal to the predictor values $dW_i/d\theta$. We might refer to this θ-derivative as the "item impact function." Some impact functions estimated in Ramsay and Wiberg (2017) can be viewed in Fig. 2 for a NAEP history test. We see that the test has a number of easy items whose impacts are high on the left and

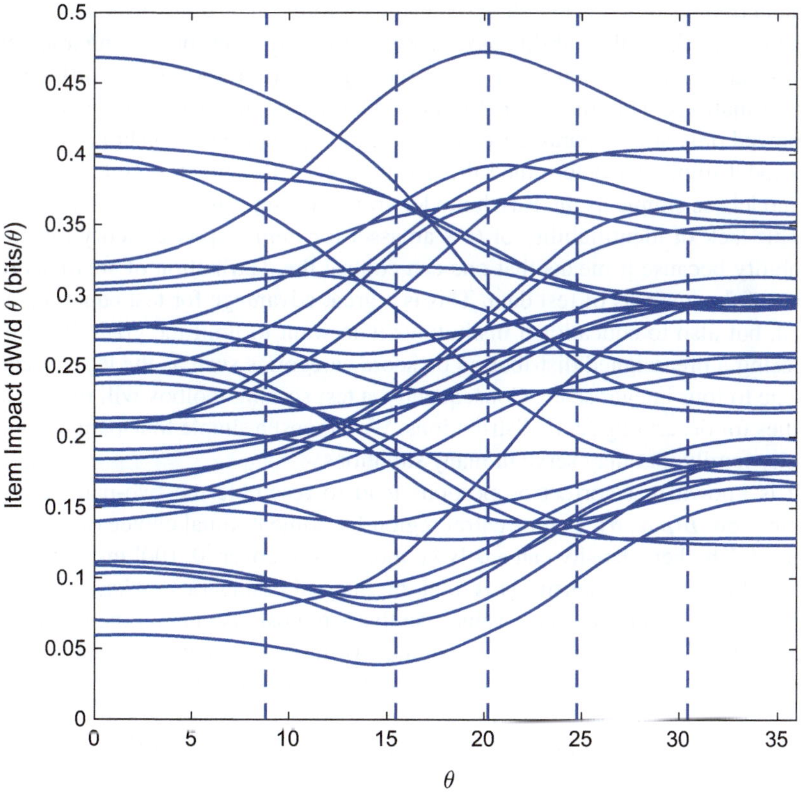

Fig. 2 The item impact curves, $dW_i/d\theta$, that provide the optimal weighting of item scores for the NAEP history test. The *vertical dashed lines* are the 5%, 25%, 50%, 75%, and 95% quantiles of the empirical distribution of the sum scores

therefore informative for the weaker of the examinees. Items for which the impact peaks near the median score of 20 inform estimation for middle-rank examinees, and difficult items high on the right provide response up-weighting for the high-end examinees.

A promising new estimation strategy is parameter cascading which combined with flexible function representation allows us to estimate the curves in Eq. (1) as accurately as the data support. The idea in parameter cascading is to represent nuisance parameters θ_j as smooth function $\theta_j(\psi)$ of the structural parameters ψ that define the log-odds functions. The largest advantage of using parameter cascading is that it speeds up the computation and thus allows test data to be analyzed in a few seconds.

4 The Smart Score Distribution

A latent trait such as θ is not observed and therefore can be transformed in ways that make sense. We think of ability as ordered, so that any order-preserving transformation is permissible, although the smart scoring equation also require differentiable transformations. This fact tended to be hidden from view when psychometricians employed only simple parametric curves for the P_i's since the algebraic structure of the model implicitly determined the distribution of θ-estimates. But, in fact, even for models as simple as the Rasch model, transforming θ is always an option.

This lack of identifiability of θ is an asset for item response theory rather than a liability because it means that one can control the distribution of smart scores as a part of the analysis of test data. This is a great advantage for test equating, as we noted, but also to educational institutions who want to limit the grade inflation by imposing a preassigned distribution on score categories such as the letter grades or the one to four integer scale. Consequently, a test scoring toolbox will require some utilities for describing score distributions. We suggest a simple four-parameter score density family that may serve in many situations.

It is known that correct proportions tend to resemble the two-parameter beta distribution, $B(p|\alpha, \beta)$, in the central region for some optimal choice of parameters α and β. The beta density can easily be modified to cover $[0, 100]$ instead of $[0, 1]$. But the fact that beta density is zero at the interval boundaries makes it unrealistic as these data can take extreme values, so that we must also model the height of the density at 0 and 100, respectively. Adding two more parameters, h_0 and h_n, solves this problem and defines the *tilted scale beta distribution* (TSβ):

$$p(S) = \frac{h_0(1 - S/n) + h_1(S/n) + (S/n)^{\alpha-1}(1 - S/n)^{\beta-1}}{n(h_0 + h_n)/2 + B(\alpha, \beta)}, \tag{3}$$

where S is the sum scores. Maximum likelihood estimation of the parameters α, β, h_0 and h_n can be obtained smoothly and reliably, given either score information from another test administration or a histogram of desired score percentages.

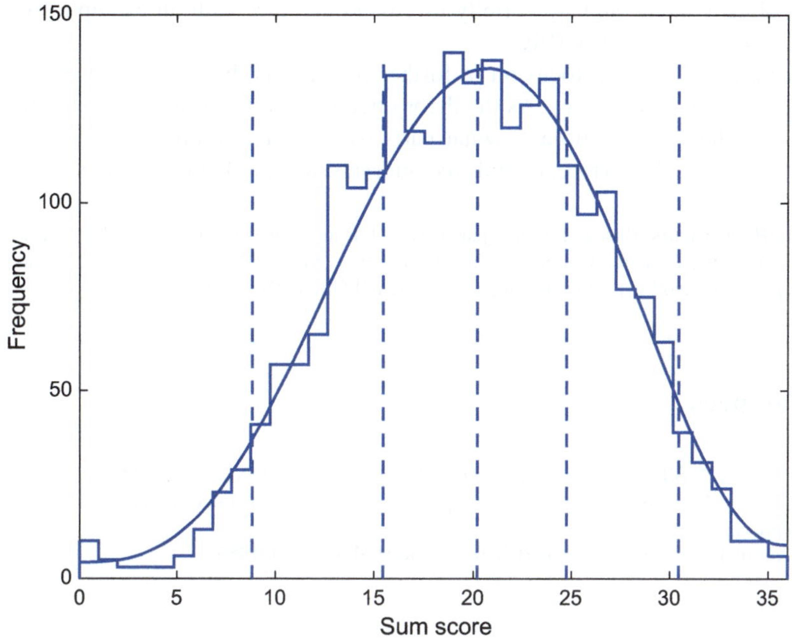

Fig. 3 The distribution of the sum scores for the 36-item NAEP, where the smooth curve is the density of the TSβ distribution

In Fig. 3 the fits of TSβ distribution to the sum scores of NAEP are given. The prior density function can be used by class teachers and test constructors to define a target for their test score distribution. But if this density does not provide an adequate picture, for example, of evident multimodality, there are many options for more flexible nonparametric representations, such as those in Silverman (1985) and Ramsay and Silverman (2005).

5 Concluding Remarks

In this paper, we have argued for replacing sum scoring with smart scoring. A major reason for using smart scoring is that weighted scores are more accurate than sum scores especially for high achievers in terms of RMSE. As many educational tests are used with the aim of selecting high achievers to colleges or to order examinees into specific grades, it is a matter of fairness that we use estimation tools for ability that are as precise as possible.

It is well known that the ability distribution is arbitrary; thus, it is possible to use different transformations of it. In line with this, we have argued that we can use this more intelligently than before. The needed theory in terms of the statistics tools already exists and was briefly discussed together with an example from the large-scale assessment NAEP.

What is essential to focus on in further research is the building of easy-to-use and accessible applications. As the theory behind smart scoring is computationally effective, the focus is more on what parts would make it attracting for the users, such as graphical interface, confidence intervals, and prediction features.

Acknowledgements Research in this paper by J.O. Ramsay was funded by the Natural Sciences and Engineering Research Council of Canada Discovery Grant 320-2012-RGPIN, and that by M. Wiberg was funded by the Swedish Research Council Grant 2014-578.

References

J.O. Ramsay, B. Silverman, *Functional Data Analysis* (Springer, New York, 2005)

J.O. Ramsay, M. Wiberg, A strategy for replacing sum scoring. J. Educ. Behav. Stat. 1–26 (2016, in press)

B. Silverman, *Density Estimation* (Chapman and Hall, London, 1985)

Overestimation of Reliability by Guttman's λ_4, λ_5, and λ_6 and the Greatest Lower Bound

Pieter R. Oosterwijk, L. Andries van der Ark, and Klaas Sijtsma

Abstract For methods using statistical optimization to estimate lower bounds to test-score reliability, we investigated the degree to which they overestimate true reliability. Optimization methods do not only exploit real relationships between items but also tend to capitalize on sampling error and do this more strongly as sample size is smaller and tests are longer. The optimization methods were Guttman's λ_4, λ_5, and λ_6 and the greatest lower bound to the reliability (GLB). Method λ_2 was used as benchmark. We used a simulation study to investigate the relation of the methods' discrepancy, bias, and sampling error with the proportion of simulated data sets in which each method overestimated true test-score reliability. Method λ_4 and the GLB often overestimated test-score reliability. When sample size exceeded 250 observations, methods λ_2, λ_5, and λ_6 provided reasonable to good statistical results, in particular when data were two-dimensional. Benchmark method λ_2 produced the best results.

Keywords Chance capitalization of reliability optimization methods • Greatest lower bound to the reliability • Guttman's λ_2 • Guttman's λ_4 • Guttman's λ_5 • Guttman's λ_6 • Reliability overestimation

P.R. Oosterwijk
Court of Audit, Lange Voorhout 8, The Hague, The Netherlands
e-mail: P.R.Oosterwijk@gmail.com

L.A. van der Ark (✉)
Research Institute of Child Development and Education, University of Amsterdam, P.O. Box 15776, 1001 NG, Amsterdam, The Netherlands
e-mail: L.A.vanderArk@uva.nl

K. Sijtsma
Tilburg School of Social and Behavioral Sciences, Tilburg University, P.O. Box 90153, 5000 LE, Tilburg, The Netherlands
e-mail: K.Sijtsma@TilburgUniversity.edu

L.A. van der Ark et al. (eds.), *Quantitative Psychology*, Springer Proceedings in Mathematics & Statistics 196, DOI 10.1007/978-3-319-56294-0_15

1 Introduction

Reliability quantifies the degree to which test scores can be repeated under identical test administration conditions, in which neither the examinee (with respect to the measured attribute) nor the test (with respect to content) has changed. Perfect repeatability is hampered by random influences beyond the test administrator's control affecting test scores, causing test scores obtained in different administrations to be different. Reliability is the correlation between two test scores, denoted X and X', obtained independently in a group of examinees (Lord and Novick 1968, p. 46), and is denoted $\rho_{XX'}$. Researchers usually collect item scores based on one test administration and use methods to approximate $\rho_{XX'}$ based on this single data set. These methods usually produce lower bounds to $\rho_{XX'}$.

Some of the approximation methods optimize a criterion based on the data in an effort to approximate $\rho_{XX'}$ as close as possible. However, because they capitalize on sample characteristics, smaller samples cause methods to overestimate $\rho_{XX'}$ more often and to a greater extent. Increasing the number of items also invites more chance capitalization. Overestimation is undesirable, because test users must be able to rely on the reported reliability estimate within the limits of statistical uncertainty, hence not providing values that are systematically too high.

Reliability overestimation has received little attention thus far. We study the degree to which four reliability methods using optimization overestimate $\rho_{XX'}$. The methods are λ_4, λ_5, and λ_6 (Guttman 1945) and the greatest lower bound to the reliability (GLB; Bentler and Woodward 1980). We recommend which method to use for reliability estimation.

2 Test-Score Reliability

Of the methods using one data set to estimate reliability, coefficient α (Cronbach 1951) is the most popular but not the best (Sijtsma 2009). Usually, these methods determine reliability based on the variance-covariance matrix of the items constituting the test. Many methods alternative to coefficient α have been proposed (e.g., Bentler and Woodward 1980; Guttman 1945; Jackson and Agunwamba 1977; Kuder and Richardson 1937; Ten Berge and Zegers 1978). Sijtsma and Van der Ark (2015) also discuss methods based on factor analysis and generalizability theory. Guttman's λ_4, λ_5, and λ_6 and the GLB employ a series of consecutive steps to optimize a method-dependent formal criterion defined on the sample data, resulting in an optimal value when the criterion is satisfied. Optimization methods are known to capitalize on chance, the more so when samples are smaller and tests are longer, hence producing more and greater overestimation effects. Such effects are unknown for Guttman's λ_4, λ_5, and λ_6 and the GLB.

2.1 Classical Reliability Definition

Test score X is the sum of J item scores, denoted X_j, with $j = 1, \ldots, J$, such that $X = \sum_{j=1}^{J} X_j$. CTT assumes that test score X can be decomposed in an unobservable, true-score part, denoted T, and an unobservable, random measurement error, denoted E, such that

$$X = T + E. \tag{1}$$

The score decomposition can be applied to any measurement value, for example, an individual item score; then, $X_j = T_j + E_j$, and Eq. (1) equals

$$\sum_{j=1}^{J} X_j = \sum_{j=1}^{J} T_j + \sum_{j=1}^{J} E_j. \tag{2}$$

Because measurement error E is random, error correlates 0 with other variables, and it follows that (a) the group variance of the test score equals

$$\sigma_X^2 = \sigma_T^2 + \sigma_E^2, \tag{3}$$

and (b) error variance for the test score equals the sum of the item error variances; that is,

$$\sigma_E^2 = \sum_{j=1}^{J} \sigma_{E_j}^2. \tag{4}$$

Measurements X and X' are parallel if (1) for each examinee, $T_i = T_i'$; hence, at the group level, $\sigma_T^2 = \sigma_{T'}^2$, and (2) at the group level, $\sigma_X^2 = \sigma_{X'}^2$. Lord and Novick (1968, p. 61) showed that $\rho_{XX'}$ can be written as

$$\rho_{XX'} = \frac{\sigma_T^2}{\sigma_X^2} = 1 - \frac{\sigma_E^2}{\sigma_X^2}. \tag{5}$$

Using Eq. (4), we can write the right-hand side of Eq. (5) as

$$\rho_{XX'} = 1 - \frac{\sum_{j=1}^{J} \sigma_{E_j}^2}{\sigma_X^2}. \tag{6}$$

Because, in practice, parallel measures usually are unavailable and because Eqs. (5) and (6) contain too many unknowns, $\rho_{XX'}$ is estimated using the data from one test administration. Methods λ_4, λ_5, and λ_6 and the GLB each seek a unique upper bound for the numerator in Eq. (6), $\sum_{j=1}^{J} \sigma_{E_j}^2$, and thus find a lower bound for $\rho_{XX'}$.

We use the following notation. Let σ_{jk} be the inter-item covariance. One can derive that $\sigma_{T_jT_k} = \sigma_{jk}$, and by definition $\sigma_{E_jE_k} = 0, j \neq k$. If Eq. (3) is rewritten for individual items, we have

$$\sigma^2_{X_j} = \sigma^2_{T_j} + \sigma^2_{E_j}. \tag{7}$$

Covariance matrices $\boldsymbol{\Sigma}_X$ and $\boldsymbol{\Sigma}_T$ are symmetrical and have order $J \times J$, and $\boldsymbol{\Sigma}_E$ is diagonal and has order $J \times J$, so that

$$\boldsymbol{\Sigma}_X = \boldsymbol{\Sigma}_T + \boldsymbol{\Sigma}_E. \tag{8}$$

Matrix $\boldsymbol{\Sigma}_X$ is positive definite (pd); that is, for any vector $\mathbf{u}$ of size J, we have $\mathbf{u}'\boldsymbol{\Sigma}_X\mathbf{u} > 0$ (i.e., $\boldsymbol{\Sigma}_X$ has a positive determinant); $\boldsymbol{\Sigma}_T$ and $\boldsymbol{\Sigma}_E$ are positive semi-definite (psd), so that $\mathbf{u}'\boldsymbol{\Sigma}_X\mathbf{u} \geq 0$ (i.e., determinants are nonnegative).

The derivation of the methods λ_4, λ_5, and λ_6 and also benchmark λ_2 can be found with Guttman (1945) and Jackson and Agunwamba (1977), and for the GLB we refer the reader to Bentler and Woodward (1980). Because derivations are known, we only provide results.

2.2 *Methods λ_4, λ_5, and λ_6, GLB, and Benchmark λ_2*

2.2.1 Method λ_4

Method λ_4 is based on splitting the J-item test in two parts, not necessarily of equal length, and finds the split that minimizes an appropriate upper bound for $\sum_{j=1}^{J} \sigma^2_{E_j}$ in Eq. (6) and consequently a lower bound for $\rho_{XX'}$. Here, we define the upper bound for $\sum_{j=1}^{J} \sigma^2_{E_j}$ typical of method λ_4. Let $\mathbf{u}$ only have elements equal to either $+1$ or -1 so that $\mathbf{u}$ selects items in either of the two test parts of a particular test split. It can be shown that

$$\mathbf{u}'\boldsymbol{\Sigma}_X\mathbf{u} = \mathbf{u}'\boldsymbol{\Sigma}_T\mathbf{u} + \mathbf{u}'\boldsymbol{\Sigma}_E\mathbf{u}, \tag{9}$$

from which it follows that

$$\mathbf{u}'\boldsymbol{\Sigma}_E\mathbf{u} = \sum_{j=1}^{J} \sigma^2_{E_j} \leq \mathbf{u}'\boldsymbol{\Sigma}_X\mathbf{u}. \tag{10}$$

The right-hand side of Eq. (10) provides an upper bound for $\sum_{j=1}^{J} \sigma^2_{E_j}$, and method λ_4 finds the vector $\mathbf{u}$ that minimizes $\mathbf{u}'\boldsymbol{\Sigma}_X\mathbf{u}$, so that

$$\lambda_4 = \max_{\mathbf{u}} \left(1 - \frac{\mathbf{u}'\boldsymbol{\Sigma}_X\mathbf{u}}{\sigma^2_X}\right). \tag{11}$$

Because $\mathbf{u}$ and $-\mathbf{u}$ provide the same value for $\mathbf{u}'\mathbf{\Sigma}_X\mathbf{u}$, and because vectors $\mathbf{u}$ containing only $+1$s or only -1s do not refer to a test split, $2^{J-1} - 2$ vectors and corresponding products $\mathbf{u}'\mathbf{\Sigma}_X\mathbf{u}$ remain to find λ_4. For test length $J < 20$, one can try all vectors $\mathbf{u}$ within reasonable computing time, and for test length $J \geq 20$, we refer to a procedure proposed by Benton (2015).

2.2.2 Method λ_5

From $\mathbf{\Sigma}_T$ being psd, it follows that every principal submatrix of $\mathbf{\Sigma}_T$ also has a nonnegative determinant, so that, for example, $\sigma^2_{T_j}\sigma^2_{T_k} \geq \sigma^2_{jk}$, all $j \neq k$. One can use this result to derive for a fixed column k of $\mathbf{\Sigma}_T$ containing $J-1$ covariances σ_{jk} $(k \neq j)$ and ignoring $\sigma^2_{X_k}$ that (noticing that $\sum_{k\neq j}$ produces summation across index k, $k \neq j$)

$$\sum_{j=1}^{J} \sigma^2_{T_j} \geq 2\left(\sum_{k\neq j} \sigma^2_{jk}\right)^{\frac{1}{2}}. \tag{12}$$

From Eq. (7) it follows that

$$\sum_{j=1}^{J} \sigma^2_{X_j} = \sum_{j=1}^{J} \sigma^2_{T_j} + \sum_{j=1}^{J} \sigma^2_{E_j}, \tag{13}$$

which implies

$$\sum_{j=1}^{J} \sigma^2_{E_j} \leq \sum_{j=1}^{J} \sigma^2_{X_j} - 2\left(\sum_{k\neq j} \sigma^2_{jk}\right)^{\frac{1}{2}}. \tag{14}$$

The right-hand side of Eq. (14) provides another upper bound for $\sum_{j=1}^{J} \sigma^2_{E_j}$, and one is free to choose the column k that minimizes this upper bound, hence finds a lower bound for $\rho_{XX'}$. To find λ_5, let k vary across each of the J columns of $\mathbf{\Sigma}_T$ and define

$$\lambda_5 = 1 - \frac{\sum_{j=1}^{J} \sigma^2_{X_j} - \max_k \left[2\left(\sum_{k\neq j} \sigma^2_{jk}\right)^{\frac{1}{2}}\right]}{\sigma^2_X}. \tag{15}$$

2.2.3 Method λ_6

Method λ_6 is based on the multiple regression of each of the J item scores X_j on the other $J-1$ item scores. By minimizing the residual variance of the model, multiple regression finds the regression weights for each of the $J - 1$ items. The residual variance of item j, $\sigma^2_{\epsilon_j}$, is an upper bound to the measurement error variance for item j, $\sigma^2_{E_j}$; that is, $\sigma^2_{E_j} \leq \sigma^2_{\epsilon_j}$, and adding across the J items, we obtain

$$\sum_{j=1}^{J} \sigma^2_{E_j} \leq \sum_{j=1}^{J} \sigma^2_{\epsilon_j} \tag{16}$$

(Jackson and Agunwamba 1977). Thus, the right-hand side of Eq. (16) provides yet another upper bound for the numerator in Eq. (6), which is $\sum_{j=1}^{J} \sigma^2_{E_j}$. Replacing the numerator in Eq. (6) by the right-hand side of Eq. (16), produces a lower bound to the reliability,

$$\lambda_6 = 1 - \frac{\sum_{j=1}^{J} \sigma^2_{\epsilon_j}}{\sigma^2_X}. \tag{17}$$

For estimation of λ_6 using covariance matrix $\boldsymbol{\Sigma}_X$, see Jackson and Agunwamba (1977) and Oosterwijk et al. (2016).

2.2.4 Greatest Lower Bound

Numerous pairs of different matrices $\boldsymbol{\Sigma}_T$ and $\boldsymbol{\Sigma}_E$ produce the same $\boldsymbol{\Sigma}_X$; see Eq. (8). Let $tr(\boldsymbol{\Sigma}_E) = \sum_{j=1}^{J} \sigma^2_{E_j}$, and let $\widetilde{\boldsymbol{\Sigma}}_E$ be the matrix of error variances for which the trace is maximized, provided that $\widetilde{\boldsymbol{\Sigma}}_E$ is psd, and $\widetilde{\boldsymbol{\Sigma}}_T$ is the corresponding covariance matrix of the item true scores, such that $\boldsymbol{\Sigma}_X = \widetilde{\boldsymbol{\Sigma}}_T + \widetilde{\boldsymbol{\Sigma}}_E$. Reliability [Eq. (5)] can be written as

$$\rho_{XX'} = 1 - \frac{tr(\boldsymbol{\Sigma}_E)}{\sigma^2_X}, \tag{18}$$

and the GLB is obtained by replacing $tr(\boldsymbol{\Sigma}_E)$ with $tr(\widetilde{\boldsymbol{\Sigma}}_E)$, so that

$$GLB = 1 - \frac{tr(\widetilde{\boldsymbol{\Sigma}}_E)}{\sigma^2_X}. \tag{19}$$

The GLB algorithm used in this chapter is due to Bentler and Woodward (1980). If the J items or the test parts in which the test is divided are essential tau-equivalent (Lord and Novick 1968, p. 90), then $GLB = \rho_{XX'}$. When essential tau-equivalence does not hold, the GLB provides the lowest possible reliability given the data, and $GLB < \rho_{XX'}$, but other methods provide lower values, hence, smaller lower bounds, and the GLB thus is the greatest lower bound (Jackson and Agunwamba 1977).

2.2.5 Benchmark Method λ_2

Guttman (1945) derived three methods, denoted λ_1, λ_2, and λ_3 (equal to coefficient α), which all use the inter-item covariances but do not optimize a statistical criterion and thus are not expected to capitalize on chance. Hence, in principle, each could serve as a benchmark for the methods λ_4, λ_5, and λ_6 and GLB. The relationship between the three methods and the reliability is

$$\lambda_1 < \lambda_3 (= \alpha) \leq \lambda_2 \leq \rho_{XX'}. \tag{20}$$

In general, λ_1 is considered practically useless, and Sijtsma (2009) and Oosterwijk et al. (2016) have recommended using λ_2 rather than λ_3. Hence, we used λ_2 as benchmark for methods λ_4, λ_5, and λ_6 and GLB. Method λ_2 equals

$$\lambda_2 = 1 - \frac{\sum_{j=1}^{J} \sigma_{X_j}^2 - \left(\frac{J}{J-1} \sum \sum_{j \neq k} \sigma_{jk}^2 \right)^{\frac{1}{2}}}{\sigma_X^2}. \tag{21}$$

2.3 *Knowledge About Reliability Methods in Samples*

For $N < 1000$ and $J > 10$, the GLB is positively biased relative to $\rho_{XX'}$ (Shapiro and Ten Berge 2000; Ten Berge and Sočan 2004). Under particular conditions, method λ_4 has values similar to the GLB (Jackson and Agunwamba 1977; Ten Berge and Sočan 2004). Hence, method λ_4 has almost the same bias relative to $\rho_{XX'}$ as the GLB; Benton (2015) found that method λ_4 is biased when $N < 3000$. Samples this size are common in the social and the behavioral sciences, and results with respect to chance capitalization are needed for smaller samples, not only for method λ_4 and the GLB but also for λ_5 and λ_6. We used a simulation study to assess this problem.

3 Method

3.1 *Population Model*

Data were simulated using the two-dimensional graded response model (De Ayala 2009, pp. 275–305). The two-dimensional graded response model expresses the probability of scoring at least x on item j as a function of latent variable θ, item location parameters β_{jx}, and item discrimination parameters α_j, such that

$$P(X_j \geq x | \theta_1, \theta_2) = \frac{\exp[\alpha_{j1}(\theta_1 - \beta_{jx}) + \alpha_{j2}(\theta_2 - \beta_{jx})]}{1 + \exp[\alpha_{j1}(\theta_1 - \beta_{jx}) + \alpha_{j2}(\theta_2 - \beta_{jx})]}. \tag{22}$$

Latent variables θ_1 and θ_2 had 101 equidistant values $(-5, -4.9, -4.8, \ldots, 5)$ and were approximately bivariate normally distributed with mean 0, variance 1, and correlation $\rho_{\theta_1\theta_2}$. Joint probability is denoted $P(\theta_1, \theta_2)$. We assumed that the test consisted of J items with five ordered item scores, $x = 0, \ldots, 4$.

Item location parameters β_{jx} consisted of an item-specific part τ_j and a category-specific part κ_x, such that $\beta_{jx} = \tau_j + \kappa_x$. We chose $\tau_j = (j-1)/(J-1) - 0.5$ and $\kappa_x = -0.75 + 0.5x$ and computed the item location parameters β_{jx} for $j = 1, \ldots, 5; x = 1, \ldots, 4$. For five items, this resulted in $\tau = (-0.5, -0.25, 0, 0.25, 0.5)$, $\kappa = (-0.75, -0.25, 0.25, 0.75)$, and 20 β values, which are readily computed.

Discrimination parameters α_j differed across latent variables. For θ_1, $\alpha_{j1} = 1.6$ for the odd-numbered items and $\alpha_{j1} = 0$ for the even-numbered items. For θ_2, $\alpha_{j2} = 1.6$ for the even-numbered items and $\alpha_{j2} = 0$ for the odd-numbered items. For 10 and 15 items, each next 5-tuple of items had the same discrimination parameters as the first 5-tuple; that is, using math operation modulo, for $k = j \mod 5$, for $j > 5$, one finds $\alpha_{j1} = \alpha_{k1}$ and $\alpha_{j2} = \alpha_{k2}$.

Equation (22) was used to compute covariance matrix $\mathbf{\Sigma}_T$, so as to obtain the numerator of $\rho_{XX'}$ [Eq. (5)]. Because $\sigma_{T_jT_k} = \sigma_{jk}$, we only require the item true-score variances, $\sigma^2_{T_j}$. It also may be noted that $\mathcal{E}(T_j) = \mathcal{E}(X_j)$. First, for fixed values of θ_1 and θ_2, the true score of item j equals

$$T_j|\theta_1, \theta_2 = \sum_x P(X_j \geq x|\theta_1, \theta_2). \tag{23}$$

Second, the true-score variance of item j then equals

$$\sigma^2_{T_j} = \sum_{\theta_1}\sum_{\theta_2} P(\theta_1, \theta_2)\left[T_j|\theta_1, \theta_2 - \mathcal{E}(T_j)\right]^2. \tag{24}$$

The conditional probability of obtaining item score x on item j equals

$$P(X_j = x|\theta_1, \theta_2) = P(X_j \geq x|\theta_1, \theta_2) - P(X_j \geq x + 1|\theta_1, \theta_2). \tag{25}$$

Equation (25) was used to compute covariance matrix $\mathbf{\Sigma}_X$, so as to obtain the denominator of $\rho_{XX'}$ [Eq. (5)], methods λ_4, λ_5, and λ_6, the *GLB*, and benchmark method λ_2. First, given discrete values for the latent variables, manifest marginal probabilities $P(X_j = x)$ and joint probabilities $P(X_j = x, X_k = y)$ were computed using

$$P(X_j = x) = \sum_{\theta_1}\sum_{\theta_2} P(\theta_1, \theta_2)P(X_j = x|\theta_1, \theta_2) \tag{26}$$

and

$$P(X_j = x, X_k = y) = \sum_{\theta_1}\sum_{\theta_2} P(\theta_1, \theta_2)P(X_j = x|\theta_1, \theta_2)P(X_k = y|\theta_1, \theta_2), \tag{27}$$

respectively. Second, the following expected values were computed using Eqs. (26) and (27): $\mathcal{E}(X_j) = \sum_x xP(X_j = x)$, $\mathcal{E}(X_j^2) = \sum_x \sum_y xyP(X_j = x, X_j = y)$, and $\mathcal{E}(X_jX_k) = \sum_x \sum_y xyP(X_j = x, X_k = y)$. Finally, $\sigma_{X_j}^2 = \mathcal{E}(X_j^2) - [\epsilon(X)]^2$, and $\sigma_{jk} = \mathcal{E}(X_jX_k) - \mathcal{E}(X_j)\mathcal{E}(X_k)$.

3.2 Data Generation

Samples of N pairs of latent variable values θ_1 and θ_2 were drawn from a bivariate normal distribution. The score for person i on item j was computed as follows. First, using Eq. (22), $P(X_j \geq x|\theta_{1i}, \theta_{2i})$ was computed for $x = 1, \ldots, 4$. Second, let I be an indicator function, and let w_{ji} be a random number between 0 and 1; then $X_{ji} = \sum_x I[(P(X_j \geq x|\theta_{1i}, \theta_{2i}) > w)]$. The resulting item scores are discrete and follow a multinomial distribution.

3.3 Design

The between-subject factors were (1) correlation $\rho_{\theta_1\theta_2}$ (values 0.30, 0.65, and 1), (2) number of items J (values 5, 10, and 15), and (3) sample size N (values 50, 250, 500, 750, and 1000). The full factorial design had $3 \times 3 \times 5 = 45$ cells. Each cell was replicated 5000 times. Note that the item scores are unidimensional if $\rho_{\theta_1\theta_2} = 1$ and two-dimensional if $\rho_{\theta_1\theta_2} < 1$. For each sample, the λs were estimated, and the GLB was estimated using function glb.algebraic from the *psych* r-package (Revelle 2015).

The dependent variables were (1) discrepancy (the difference between the population value of the reliability method and the population reliability (e.g., $\lambda_4 - \rho_{XX'}$)), (2) bias (the difference between the mean of the sample estimates (e.g., sample estimate denoted $\hat{\lambda}_4$, mean denoted $\bar{\hat{\lambda}}_4$) and the population value (e.g., λ_4) (bias equals $\bar{\hat{\lambda}}_4 - \lambda_4$)), (3) standard deviation of the coefficients (e.g., $SD(\hat{\lambda}_4)$), and (4) reliability overestimation (in each design cell, the proportion out of 5000 replicated sample values exceeding $\rho_{XX'}$ [e.g., $P(\hat{\lambda}_4 > \rho_{XX'})$]).

4 Results

4.1 Discrepancy

Table 1 provides the positive discrepancies for the lower bounds. The GLB had negligible discrepancy. Second best methods were λ_2 and λ_4. Methods λ_5 and λ_6 had the largest discrepancy. Except for the GLB, as test length increased, dis-

Table 1 Discrepancy of λ_4,λ_5, λ_6, GLB, and λ_2, as a function of correlation between latent variables and test length

$\rho_{\theta_1\theta_2}$	J	$\rho_{XX'}$	λ_4	λ_5	λ_6	GLB	λ_2
1	5	0.761	−26	−29	−45		−2
	10	0.865		−34	−14		−1
	15	0.906	−2	−30	−7		−1
0.65	5	0.725	−30	−58	−79	−2	−38
	10	0.841		−49	−26		−18
	15	0.889	−4	−41	−14		−13
0.30	5	0.678	−36	−87	−106	−4	−72
	10	0.809		−66	−35		−36
	15	0.864	−4	−53	−17		−24

Entries for λ_4,λ_5, λ_6, GLB, and λ_2 in thousandths; for example, read −26 as −0.026. Read a blank as 0

crepancy decreased, and as correlation between the two latent variables increased, discrepancy increased. Based on discrepancy alone, methods λ_5 and λ_6 probably will overestimate reliability not as often as the other methods.

4.2 *Bias and Standard Deviation*

Bias is interesting when it is positive, discrepancy is small, and SD is large. This combination of quantities produces large proportions of reliability overestimates. Tables 2, 3, and 4 show that sample size, more than test length and dimensionality, affects bias and SD; both decrease as N increases. The five lower bounds differ little with respect to SD, so that we will concentrate on bias.

Method λ_5 and benchmark method λ_2 had small negative bias, ranging across the design from 0.000 to −0.016 (λ_5) and from −0.007 to −0.018 (λ_2). Method λ_6 had bias ranging from positive when $N = 50$ (0.001 to 0.033) to negative when $N \geq 250$ (−0.002 to −0.020). Given that method λ_6 had large discrepancy, we expect the proportion of reliability overestimates to be large for $N = 50$ and small for larger N. Bias for methods λ_4 and GLB was largest and almost always positive, in particular when $J = 10, 15$. In combination with discrepancy that was almost always near 0 or equal to 0, for λ_4 and the GLB, one may expect large proportions of reliability overestimates.

4.3 *Reliability Overestimation*

For method λ_5, except when $J = 5$ and $N = 50$, overestimation was negligible (Table 5). For method λ_6, overestimation was always problematic for $N = 50$ but not for larger N. For benchmark λ_2, for unidimensionality and $N = 50$, proportions were approximately 0.4 but decreased as N increased and also decreased to 0 as

Table 2 Bias and SD of λ_4, λ_5, λ_6, GLB, and λ_2, for correlation $\rho_{\theta_1\theta_2} = 1$, as a function of test length and sample size

		Bias					SD				
J	Method	50	250	500	750	1000	50	250	500	750	1000
5	λ_4	51	13	4		−2	49	25	18	15	13
	λ_5		−9	−11	−12	−13	54	25	17	14	12
	λ_6	1	−15	−16	−16	−17	62	29	20	16	14
	GLB	40	7		−3	−5	49	25	18	15	13
	λ_2	−16	−17	−16	−16	−16	57	26	18	14	13
10	λ_4	51	19	11	7	5	20	12	9	8	7
	λ_5		−5	−7	−7	−8	29	13	9	7	6
	λ_6	17	−5	−8	−9	−9	29	14	10	8	7
	GLB	58	22	13	8	6	19	11	9	7	6
	λ_2	−9	−10	−10	−10	−10	30	13	9	8	7
15	λ_4	50	21	13	9	7	11	7	6	5	4
	λ_5	−2	−4	−5	−5	−6	20	9	6	5	4
	λ_6	23	−2	−5	−6	−6	17	9	7	5	5
	GLB	55	23	15	11	8	10	7	6	5	4
	λ_2	−7	−7	−7	−7	−7	20	9	6	5	4

Entries in thousandths; for example, read −9 as −0.009. Read a blank as 0

Table 3 Bias and SD of λ_4, λ_5, λ_6, GLB, and λ_2, for correlation $\rho_{\theta_1\theta_2} = 0.65$, as a function of test length and sample size

		Bias					SD				
J	Method	50	250	500	750	1000	50	250	500	750	1000
5	λ_4	47	9		−4	−5	60	31	22	18	16
	λ_5	−2	−13	−14	−15	−15	68	31	21	18	16
	λ_6	1	−16	−17	−18	−18	76	35	24	19	17
	GLB	33		−6	−9	−10	60	30	21	17	15
	λ_2	−16	−18	−17	−17	−17	71	33	22	18	16
10	λ_4	54	19	10	6	3	25	15	11	9	8
	λ_5	−2	−7	−9	−9	−10	38	17	12	10	8
	λ_6	21	−6	−10	−11	−11	36	18	13	10	9
	GLB	63	22	12	7	5	23	14	10	9	8
	λ_2	−9	−11	−11	−11	−12	39	17	12	10	8
15	λ_4	57	23	13	9	7	13	09	7	6	5
	λ_5	−3	−6	−7	−7	−7	26	11	8	7	6
	λ_6	28	−2	−6	−7	−8	21	11	8	7	6
	GLB	63	26	16	11	9	12	9	7	6	5
	λ_2	−8	−8	−9	−9	−9	26	12	8	7	6

Entries in thousandths; for example, read −7 as −0.007. Read a blank as 0

Table 4 Bias and SD of λ_4, λ_5, λ_6, GLB, and λ_2, for correlation $\rho_{\theta_1\theta_2} = 0.30$, as a function of test length and sample size

		Bias					SD				
J	Method	50	250	500	750	1000	50	250	500	750	1000
5	λ_4	52	11	2	−3	−5	73	36	25	21	19
	λ_5	−4	−14	−15	−16	−16	82	38	26	22	19
	λ_6	2	−17	−18	−19	−20	91	41	28	23	20
	GLB	36	1	−5	−8	−10	72	36	25	21	18
	λ_2	−13	−18	−17	−17	−18	84	39	26	22	19
10	λ_4	63	22	12	7	4	31	18	13	11	9
	λ_5	−3	−9	−10	−11	−11	48	21	15	12	11
	λ_6	25	−7	−11	−13	−13	45	21	15	13	11
	GLB	75	26	14	9	6	29	17	13	11	9
	λ_2	−9	−12	−13	−13	−13	48	21	15	12	10
15	λ_4	68	28	17	12	8	17	11	8	7	6
	λ_5	−5	−7	−8	−8	−9	34	15	10	9	7
	λ_6	33	−2	−7	−8	−9	26	14	10	8	7
	GLB	76	31	19	14	10	15	10	8	7	6
	λ_2	−8	−9	−10	−10	−10	33	15	10	8	7

Entries in thousandths; for example, read −16 as −0.016. Read a blank as 0

Table 5 Proportions of estimates of λ_4, λ_5, λ_6, GLB, and λ_2 overestimating $\rho_{XX'}$, for correlation $\rho_{\theta_1,\theta_2} = 1$, as function of test length and sample size

		$\rho_{\theta_1,\theta_2} = 1$					$\rho_{\theta_1,\theta_2} = 0.65$					$\rho_{\theta_1,\theta_2} = 0.30$				
J	Method	50	250	500	750	1000	50	250	500	750	1000	50	250	500	750	1000
5	λ_4	74	32	12	4	2	66	25	7	2	1	63	26	9	3	1
	λ_5	31	5				18	1				12				
	λ_6	24	1				14					10				
	GLB	82	64	52	44	37	74	49	37	28	22	71	49	39	28	23
	λ_2	41	25	16	11	8	23	3				14				
10	λ_4	98	94	89	83	78	97	89	83	75	68	96	89	83	74	68
	λ_5	9					6					5				
	λ_6	59	8	100			49	2				45	1			
	GLB	99	97	93	87	84	99	94	88	81	75	98	93	87	80	74
	λ_2	42	22	12	8	5	25	3	1			16				
15	λ_4	100	99	96	91	84	100	98	92	84	71	100	98	93	86	76
	λ_5	2					2					2				
	λ_6	85	19	4	1		77	7				76	7			
	GLB	100	100	99	99	98	100	100	99	97	95	100	100	99	98	95
	λ_2	39	20	10	6	3	22	2				15				

Entries in hundredths; for example, read 31 as 0.31. Read a blank as 0
Note. The reliability was (left to right, top to bottom) 0.761, 0.725, 0.678, 0.865, 0.841, 0.809, 0.906, 0.889, and 0.864

$\rho_{\theta_1\theta_2}$ decreased. For method λ_4 and the GLB, irrespective of N, when $J = 10, 15$, proportions varied between 0.78 and 1.00. Dimensionality only had little effect on proportions; they were invariably high.

5 Discussion

Discrepancy, bias, and standard deviation together determine proportion of overestimation, but exact numerical results had to be computed using a simulation study. Table 6 provides qualifications of the results based on Tables 1, 2, 3, 4, and 5 and enables us to summarize each of the methods' results and draw conclusions with respect to their practical usefulness.

For method λ_4, discrepancy is small, but bias is substantial to large, and, moreover, it is positive, thus driving estimates of λ_4 to overestimate $\rho_{XX'}$; this happens often, and proportion of overestimation is large. The GLB is closer to $\rho_{XX'}$ and has the same statistical properties as λ_4, hence producing many gross overestimates of $\rho_{XX'}$. Method λ_4 and GLB both suffer greatly from their tendency to capitalize on chance. This renders their application to moderate sample sizes questionable.

Methods λ_5 and λ_6 have a large discrepancy, whereas the former method has small bias irrespective of sample size N, and the latter method has substantial bias (small N) to small bias (moderate N). Combined with a standard deviation that is small enough to have little effect on overestimation if $N > 250$, this combination of properties causes methods λ_5 and λ_6 to rarely overestimate $\rho_{XX'}$. Their large discrepancy speaks to their disadvantage as practical estimates of $\rho_{XX'}$.

For benchmark method λ_2, Table 6 suggests that discrepancy is small, bias is small irrespective of sample size, and variance did not differ notably between λ_2 and other reliability estimation methods. The magnitude of overestimation usually is small, but for unidimensional data, overestimation may be larger due to λ_2 having small discrepancy.

Compared to method λ_2, methods λ_4, λ_5, and λ_6 and the GLB all seem to underperform. We only studied small samples; hence, a study of the large-sample performance of the methods may be useful. Chance capitalization caused by realistic numbers of items seems to be a principled problem and not easily fixed. Given that

Table 6 Summary of discrepancy, bias, variance, and reliability overestimation

	Discrepancy	Bias $N = 50$	Bias $N > 250$	Variance	Overestimation
λ_2	Small	Small	Small	No effect	Variable
λ_4	Small	Large	Substantial	No effect	Large
λ_5	Large	Small	Small	No effect	Small
λ_6	Large	Substantial	Small	No effect	Small
GLB	Negligible	Large	Substantial	No effect	Large

one does not want to report a reliability boosted by chance, reporting a lower bound that does not capitalize on chance, such as λ_2, may be recommendable. Even method λ_3 (coefficient α), not studied here, may be considered. Such recommendations require further study. In addition, future studies are required to investigate the degree to which the results in this study generalize to continuous item scores. It can be expected that the degree of discrepancy, bias, and precision is slightly different for continuous item scores and for the discrete item scores considered in this study. Methods from the factor analysis approach (Bollen 1989; McDonald 1999) probably also suffer from chance capitalization and require the same kind of evaluation as the methods studied in this article.

References

P.M. Bentler, J.A. Woodward, Inequalities among lower bounds to reliability: with applications to test construction and factor analysis. Psychometrika **45**, 249–267 (1980). doi:10.1007/BF02294079

T. Benton, An empirical assessment of Guttman's lambda 4 reliability coefficient, in *Quantitative Psychology Research: The 78th Annual Meeting of the Psychometric Society*, ed. by R.E. Millsap, D.M. Bolt, L.A. van der Ark, W.-C. Wang (Springer, New York, 2015), pp. 301–310

K.A. Bollen, *Structural Equations with Latent Variables* (Wiley, New York, 1989)

L.J. Cronbach, Coefficient alpha and the internal structure of tests. Psychometrika **16**, 297–334 (1951). doi:10.1007/BF02310555

R.J. De Ayala, *The Theory and Practice of Item Response Theory* (Guilford Press, New York, 2009)

L. Guttman, A basis for analyzing test-retest reliability. Psychometrika **10**, 255–282 (1945). doi:10.1007/BF02288892

P.H. Jackson, C.C. Agunwamba, Lower bounds for the reliability of the total score on a test composed of non-homogeneous items: I. Algebraic lower bounds. Psychometrika **42**, 567–578 (1977). doi:10.1007/BF02295979

G.F. Kuder, M.W. Richardson, The theory of estimation of test reliability. Psychometrika **2**, 151–160 (1937). doi:10.1007/BF02288391

F.M. Lord, M.R. Novick, *Statistical Theories of Mental Test Scores* (Addison-Wesley, Reading, 1968)

R.P. McDonald, *Test Theory: A Unified Treatment* (Erlbaum, Mahwah, 1999)

P.R. Oosterwijk, L.A. Van der Ark, K. Sijtsma, Numerical differences between Guttman's reliability coefficients and the GLB, in *Quantitative Psychology Research: The 80th Annual Meeting of the Psychometric Society*, Beijing, 2015, ed. by L.A. van der Ark, D.M. Bolt, W.-C. Wang, J.A. Douglas, M. Wiberg (Springer, New York, 2016), pp. 155–172

W. Revelle, psych: Procedures for personality and psychological research Version 1.5.8 [computer software] (2015). Evanston, IL. Retrieved from https://cran.r-project.org/web/packages/psych/index.html

A. Shapiro, J.M.F. Ten Berge, The asymptotic bias of minimum trace factor analysis, with applications to the greatest lower bound to reliability. Psychometrika **65**, 413–425. doi:10.1007/BF02296154 2000

K. Sijtsma, On the use, the misuse, and the very limited usefulness of Cronbach's alpha. Psychometrika **74**, 107–120 (2009). doi:10.1007/s11336-008-9101-0

K. Sijtsma, L.A. Van der Ark, Conceptions of reliability revisited and practical recommendations. Nurs. Res. **64**, 128–136 (2015). doi:10.1097/NNR.0000000000000077

J.M.F. Ten Berge, G. Sočan, The greatest lower bound to the reliability of a test and the hypothesis of unidimensionality. Psychometrika **69**, 613–625 (2004). doi:10.1007/BF02289858

J.M.F. Ten Berge, F.E. Zegers, A series of lower bounds to the reliability of a test. Psychometrika **43**, 575–579 (1978). doi:10.1007/BF02293815

The Performance of Five Reliability Estimates in Multidimensional Test Situations

Shuying Sha and Terry Ackerman

Abstract This paper investigates the estimation biases of five reliability indices, Cronbach's α, Guttman's λ_2 and λ_4, glb, and McDonald's ω. The factors included are test dimensionality, population ability distribution, sample size, test length, and test discrimination. It was found that estimation biases of Alpha, λ_4, and glb were correlated with the test's true reliability, whereas λ_2 and ω were not. Estimation biases were larger in two-dimensional tests than in unidimensional tests. Alpha overall had the largest estimation bias, but glb displayed similar bias in unidimensional tests. In light of the findings in the simulation study, we recommend McDonald's ω because it had the smallest estimation bias in most test condition.

Keywords Reliability • Lower bounds to reliability • ω • λ_2, Cronbach's α • λ_4

1 Introduction

Measurement precision, referred to as reliability, is a major issue in educational assessment and psychological research. Spearman (1904) defined reliability to be the correlation between observed total scores of two parallel tests or between test scores on two repeated administrations. However, it is rare to have either parallel tests or repeated testing. Instead, reliability is usually based on single administration and is defined as the proportion of observed score variance explained by true score variance. Therefore, based on the contribution of Spearman (1904) to classical test theory, the reliability of a test is also defined as the ratio of true score variance and observed score variance.

Paper presented at 80th annual meeting of Psychometric Society, July 2015.

S. Sha (✉)
Center for Educational Measurement, Excelsior College, 7 Columbia Circle, Albany, NY 12203, USA
e-mail: ssha@excelsior.edu

T. Ackerman
ACT, 500 ACT Drive, Iowa City, IA 52243-0168, USA

L.A. van der Ark et al. (eds.), *Quantitative Psychology*, Springer Proceedings in Mathematics & Statistics 196, DOI 10.1007/978-3-319-56294-0_16

Since Spearman's (1904) definition of reliability, many reliability indices have been developed by researchers, such as Guttman's (1945) six λs $\lambda_1-\lambda_6$ that were developed by bounding the estimation of true score variance or error variance of the test score to derive the estimates of reliability, McDonald's ω (McDonald 1970, 1999), and the greatest lower bound (glb; Jackson and Agunwamba 1977; Bentler 1972; Woodhouse and Jackson 1977; Bentler and Woodward 1980). Guttman's (1945) λ_3 is actually Cronbach's α.

Among all the reliability estimates, Cronbach's α is the mostly widely used since its publication (Cronbach 1951; Raykov 2001; Sijtsma 2009). Cronbach (1951) proved that alpha is the mean of all possible Flanagan-Rulon split-half reliabilities under a strong assumption that all items are equally loaded and unidimensional. In reality, the assumption of equal discrimination power for all test components is rarely met. Plus, many tests measure more than one dimension. Sijtsma (2009) proved that alpha does not change while dimensionality increases, whereas glb decreases dramatically.

The purpose of this study is to investigate the performance of Cronbach's α, in comparison to four other alternative reliability estimates: Guttman's λ_2, λ_4, McDonald's ω, and glb in estimating the reliability of a multidimensional test.

2 Background

2.1 Guttman's λ_2, λ_3, and λ_4

Guttman's λ_2 is defined as

$$\lambda_2 = \lambda_1 + \frac{\sqrt{\frac{J}{J-1}\sum_{i\neq}^{n}\sum_{j}^{n}\sigma_{ij}^2}}{\sum_{i=1}^{n}\sum_{j=1}^{n}\sigma_{ij}} = \frac{\sum_{i=1}^{k}\sigma_i^2 - \sqrt{\frac{J}{J-1}\sum_{i\neq}^{k}\sum_{j}^{k}\sigma_{ij}^2}}{\sum_{i=1}^{k}\sum_{j=1}^{k}\sigma_{ij}}, \tag{1}$$

where i and j are the ith and jth item; σ_i^2 and σ_j^2 are the variances of item score x_{j} and item score x_{j}, respectively; and σ_{ij} is the covariance between item scores x_i and x_j.

Guttman's λ_3, commonly known as Cronbach's α, is defined as

$$\begin{aligned} \lambda_3 &= \tfrac{J}{J-1}\lambda_1 \\ &= \tfrac{J}{J-1}\left[1 - \frac{\sum_{j=1}^{J}\sigma_j^2}{\sigma_t^2}\right] \\ &= \tfrac{J}{J-1}\frac{\sum\sum_{j\neq k}\sigma_{jk}}{\sigma_t^2} \end{aligned} \tag{2}$$

where σ_t^2 is the total score variance.

Guttman (1945) proved that λ_2 is always larger or equal to λ_3. Therefore, λ_2 should be a better estimate than λ_3.

Another lower bound λ_4 was originally proposed as any split-half reliability (Guttman 1945). It is often taken as the split-half that maximizes the reliability coefficient,

$$\lambda_4 = \frac{4r_{12}}{\sigma_1^2 + \sigma_2^2 + 2r_{12}\sigma_1\sigma_2} \quad (3)$$

where σ_1^2 is the variance of the first part of the test, σ_2^2 is the variance of second part, and r_{12} is the correlation of the two. The statistic λ_4 does not assume unidimensionality or tau-equivalence (Guttman 1945) and varies according to how the test is split.

Guttman proposed λ_4 but didn't develop an algorithm to compute its value. Several algorithms have been developed by other researchers. In the "psych" package (Revelle 2014) of the R programming software, the calculation of λ_4 is approached by combining the output from three different approaches, and λ_4 is reported as the max of these three algorithms.

Ten Berge and Socan (2004) suggested that λ_4 is a better lower bound than alpha, with an exception when the number of items is odd. Ten Berge and Socan (2004) also found that when the sample size was small and the item number was large, λ_4 was severely positively biased. However, Benton (2015) found that when sample size is 1000 and reliability over 0.85, positive bias of λ_4 is not an issue.

2.2 Greatest Lower Bound

The greatest lower bound (glb) was originally proposed by Bentler (1972), in which the covariance matrix C_x is decomposed to be $C_x = C_T + C_E$. Here C_x is the inter-item covariance matrix, C_T is the true score covariance matrix, and C_E is the error inter-item covariance matrix. It was shown that glb can be found by maximizing the trace of inter-item error covariance matrix (C_E) while keeping both true score variance and error variance to be positive semi-definite (Woodhouse and Jackson 1977; Bentler and Woodward 1980),

$$\text{glb} = 1 - \frac{tr\,(C_E)}{S_x^2} \quad (4)$$

There are three ways in the "psych" package (Revelle 2014) to calculate glb: "glb," "glb.algebraic," and "glb.fa." This study used "glb.fa" to find the glb. "glb.fa" estimates the communalities of the variables from a factor model in which the number of factors is the number of positive eigenvalues. Then reliability is found by the equation

$$\text{glb} = 1 - \frac{\sum_{i=1}^{k} e^2}{\sigma_t^2} = 1 - \frac{\sum_{i=1}^{k} 1 - h^2}{\sigma_t^2} \quad (5)$$

where e represents error variance, and h represents the communality. Revelle (2014) indicated that "glb.fa" has larger positive bias when the sample size is large (n >1000).

2.3 McDonald's ω

McDonald (1970, 1999) proposed a factor analytical approach to estimate reliability. He called his formulation of reliability omega (ω),

$$\omega = 1 - \frac{\sum_{j=1}^{J}\left(1 - h_j^2\right)}{\sigma_t^2} = 1 - \frac{\sum_{j=1}^{J}\mu_j^2}{\sigma_t^2} \quad (6)$$

where h is the communality and μ is the unique variance of the item, which is considered to be the error variance of the item.

A few studies have compared the performance of different lower bounds of reliability. Sijtsma (2009) and Bendermacher (2010) considered glb the best reliability estimate because it is the largest among all the lower bounds. However, Revelle and Zinbarg (2009) showed that glb is systematically smaller than ω and thus recommended using *omega* rather than glb in reporting test reliability. Tang and Cui (2012) evaluated three lower bounds including λ_2, Cronbach's α, and the glb under different simulation conditions where sample size, test length, and dimensionality were manipulated. Among them, λ_2 showed the least bias in most of the conditions. However, most of the tests in the abovementioned studies have extremely low true reliabilities (<0.70). The problem is that when true reliability is below 0.80, the difference between reliability estimators probably does not matter. Because the test has severely low measurement precision, large modifications to the test should be made in order to improve its quality.

Shorter tests, depending on the examinees, often yield lower reliability. In most studies (Sijtsma 2009; Tang and Cui 2012), test length was very small with maximum number of items on a test equal to 12. However, in educational measurement, tests with this length are not common. In addition, though not discussed and systematically investigated, the results in some studies (Benton 2015; Tang and Cui 2012) suggested that estimation bias may vary according to the test's true reliability. For the results to be more generalizable, studies need to investigate relative performance of lower bounds in tests with more items and higher reliability. It is the purpose of this study to investigate how the five reliability estimates perform in unidimensional tests and two-dimensional tests with the test's true reliability being manipulated.

3 Methodology

This study investigated the estimation of reliability in unidimensional tests and two-dimensional (2D) tests by manipulating three factors: (1) the correlation between different dimensions of ability in two-dimensional tests ($r = 0, 0.5, 0.8$); (2) the dispersion of ability that has three levels, N (0, 0.8), N (0, 1), and N (0, 1.5); and (3) sample size (N = 500, 1000). Hence, there are 3 (correlation) × 3 (variance) × 2 (sample size) = 18 conditions for two-dimensional tests and 3 (variance) × 2 (sample size) = 6 conditions for unidimensional tests.

$$P\left(x_{ij} = 1 | a_1, a_{2,} d_{j,} \theta_1, \theta_2\right) = \frac{\exp\left[1.7\left(a_1\theta_1 + a_2\theta_2 + d_j\right)\right]}{1 + \exp\left[1.7\left(a_1\theta_1 + a_2\theta_2 + d_j\right)\right]} \tag{7}$$

The 2PL multidimensional IRT model (Eq. (7)) and unidimensional 2PL IRT model were used to generate the two-dimensional item responses and unidimensional items, respectively. A total of 30 items were generated in all conditions, and for each condition, there were 30 replications.

Both bias and root mean square error of estimation (RMSE) were calculated for the five reliability estimates. Bias was calculated by subtracting true reliability from reliability estimates, and RMSE is defined as

$$RMSE = \sqrt{\frac{\sum_1^{30}\left(\widehat{R}_i - \rho_i\right)^2}{30}} \tag{8}$$

where $\widehat{R}_i$ is the estimated reliability and ρ_i is the true reliability.

4 Results

Tables 1 and 2 provide the five reliability estimates and their RMSEs with respect to the correlation of abilities, population variance, and sample size. As can be observed from Table 1, unidimensional tests overall had higher reliability than the two-dimensional tests; Cronbach's α had the smallest value, followed by λ_2, ω, and λ_4; glb had the largest value in almost all conditions. This is different from what's found in Revelle and Zinbarg (2009) which showed that glb was systematically smaller than ω. By comparing estimated reliabilities with true reliabilities, it can be seen that Cronbach's α, λ_2, and ω tended to underestimate the true reliability, whereas glb and λ_4 overestimated the true reliability in most two-dimensional test conditions.

RMSEs (Table 2) tell the relative bias size of the five reliability estimates. As can be observed, in unidimensional tests, ω showed the smallest bias, while Cronbach's

Table 1 Average true reliability and reliability estimates with respect to correlation, variance, and sample size (replication = 30)

Correlation	Variance	N	True	Omega	λ_2	Alpha	λ_4	glb
Two dimensional								
0	0.64	1000	0.905	0.900	0.894	0.891	0.913	0.918
0.5	0.64	1000	0.923	0.925	0.920	0.917	0.934	0.937
0.8	0.64	1000	0.942	0.934	0.929	0.926	0.942	0.944
0	1	1000	0.928	0.924	0.918	0.915	0.934	0.935
0.5	1	1000	0.950	0.943	0.938	0.936	0.950	0.951
0.8	1	1000	0.952	0.950	0.945	0.943	0.957	0.957
0	1.25	1000	0.949	0.942	0.936	0.934	0.950	0.950
0.5	1.25	1000	0.963	0.958	0.954	0.952	0.964	0.964
0.8	1.25	1000	0.968	0.963	0.959	0.958	0.968	0.968
0	0.64	500	0.898	0.905	0.905	0.900	0.911	0.925
0.5	0.64	500	0.931	0.922	0.922	0.917	0.927	0.939
0.8	0.64	500	0.941	0.939	0.938	0.934	0.943	0.951
0	1	500	0.934	0.920	0.920	0.915	0.926	0.937
0.5	1	500	0.954	0.945	0.943	0.940	0.948	0.956
0.8	1	500	0.954	0.950	0.949	0.946	0.953	0.960
0	1.25	500	0.942	0.944	0.943	0.939	0.948	0.955
0.5	1.25	500	0.959	0.959	0.957	0.954	0.962	0.966
0.8	1.25	500	0.966	0.964	0.962	0.960	0.966	0.971
Unidimensional								
	0.64	1000	0.928	0.903	0.904	0.899	0.907	0.921
	1	1000	0.940	0.932	0.932	0.928	0.937	0.944
	1.25	1000	0.959	0.945	0.944	0.940	0.947	0.953
	0.64	500	0.923	0.903	0.898	0.895	0.916	0.920
	1	500	0.943	0.932	0.927	0.925	0.941	0.944
	1.25	500	0.960	0.945	0.940	0.938	0.952	0.953

Note: N = sample size

α and glb have similar bias. In two-dimensional tests, one of ω, glb, and λ_4 has smallest bias depending on the conditions. Cronbach's α had the largest RMSE, followed by λ_2.

The correlations between the true reliability and the five estimation biases were calculated to examine if there is a relationship between reliability estimation bias and the true reliability. As we can observe in Table 3, the correlation between Cronbach's α and true reliability was positive and significant, suggesting that higher reliability comes with larger estimation bias of Cronbach's α. There were significant negative correlations between the true reliabilities and biases of λ_4 and glb, suggesting the higher the true reliability, the smaller the bias. No significant correlation was found between the true reliability and the bias of ω and λ_2. This is consistent with the findings in Tables 1 and 2. For the test with the lowest true reliability, Cronbach's *a* has the smallest bias and glb has the largest bias.

Table 2 RMSEs of five reliability estimators with respect to correlation, variance, and sample size (replication = 30)

r	σ^2	N	True	Omega	λ_2	Alpha	λ_4	glb
Two dimensional								
0	0.64	1000	0.905	0.017	0.020	0.022	0.018	0.020
0.5	0.64	1000	0.923	0.015	0.015	0.016	0.019	0.021
0.8	0.64	1000	0.942	0.015	0.018	0.020	0.012	0.012
0	1	1000	0.928	0.018	0.020	0.022	0.018	0.019
0.5	1	1000	0.95	0.013	0.016	0.018	0.011	0.011
0.8	1	1000	0.952	0.010	0.012	0.013	0.010	0.010
0	1.25	1000	0.949	0.013	0.017	0.019	0.011	0.011
0.5	1.25	1000	0.963	0.011	0.014	0.015	0.010	0.010
0.8	1.25	1000	0.968	0.009	0.012	0.013	0.008	0.008
0	0.64	500	0.898	0.028	0.028	0.027	0.031	0.039
0.5	0.64	500	0.931	0.019	0.019	0.022	0.019	0.019
0.8	0.64	500	0.941	0.017	0.017	0.018	0.018	0.020
0	1	500	0.934	0.024	0.024	0.027	0.021	0.019
0.5	1	500	0.954	0.017	0.018	0.020	0.017	0.015
0.8	1	500	0.954	0.012	0.013	0.014	0.012	0.013
0	1.25	500	0.942	0.016	0.016	0.017	0.019	0.020
0.5	1.25	500	0.959	0.012	0.012	0.012	0.013	0.013
0.8	1.25	500	0.966	0.014	0.015	0.016	0.014	0.014
Unidimensional								
	0.64	1000	0.928	0.017	0.017	0.018	0.017	0.023
	1	1000	0.94	0.015	0.015	0.016	0.016	0.017
	1.25	1000	0.959	0.010	0.011	0.012	0.010	0.012
	0.64	500	0.923	0.019	0.021	0.023	0.021	0.022
	1	500	0.943	0.012	0.013	0.015	0.014	0.015
	1.25	500	0.96	0.012	0.013	0.014	0.014	0.014

Table 3 Correlation between estimators and true reliability

	True	Omega	λ_2	Alpha	λ_4
True	1.000				
Omega	−0.061				
λ_2	0.049	0.953***			
Alpha	0.164***	0.882***	0.965***		
λ_4	−0.380***	0.770***	0.607***	0.464***	
glb	−0.626***	0.523***	0.360***	0.166***	0.829***

Note: ***$p < 0.001$

5 Discussion and Conclusion

For test reliability reporting, some researchers (Green and Yang 2009; Sijtsma 2009; Ten Berge and Socan 2004) recommended the use of glb along with coefficient α, and some (Tang and Cui 2012) recommended λ_2. We recommend the use of ω as well as λ_2 because we found the performance of ω and λ_2 is similar in most of the situations examined in this study, and their biases are the smallest.

Although Sijtsma (2009) showed that Cronbach's α does not reflect multidimensionality of the tests, in the two-dimensional tests of this study, Cronbach's α showed the largest bias, whereas in unidimensional test, the estimation bias of α was quite close to that of glb and λ_4. This suggests that we should consider reporting other reliability estimates for two-dimensional tests.

Another major finding of this study is that when reliability increased, estimation bias became smaller for all five estimates. However, in most of the simulated tests, reliabilities are above 0.90 and biases are very small. Therefore, one should be careful with generalizing the results to tests with lower reliabilities.

References

N. Bendermacher, Beyond alpha: lower bounds for the reliability of tests. J. Mod. Appl. Stat. Methods **9**(1), ii–vi (2010). Article 11. Available at: http://digitalcommons.wayne.edu/jmasm/vol9/iss1/11

P.M. Bentler, A lower-bound method for the dimension-free measurement of internal consistency. Soc. Sci. Res. **1**, 343–357 (1972)

P.M. Bentler, J.A. Woodward, Inequalities among lower bounds to reliability: with applications to test construction and factor analysis. Psychometrika **45**, 249–267(1980)

T. Benton, An empirical assessment of Guttman's Lambda 4 reliability coefficient, in *Quantitative Psychology Research: The 78th Annual Meeting of the Psychometric Society*, ed. by R.E. Millsap, D.M. Bolt, L.A. van der Ark, W.-C. Wang (Springer, New York, 2015), pp. 301–310

K. Sijtsma, On the use, the misuse, and the very limited usefulness of Cronbach's Alpha. Psychometrika **74**, 107–120 (2009)

L. Cronbach, Coefficient alpha and the internal structure of tests. Psychometrika **16**, 297–334 (1951)

S.A. Green, Y. Yang, Commentary on coefficient alpha: a cautionary tale. Psychometrika **74**, 121–135 (2009)

L. Guttman, A basis for analyzing test-retest reliability. Psychometrika **10**, 255–282 (1945)

P. Jackson, C. Agunwamba, Lower bounds for the reliability of the total score on a test composed of non-homogeneous items: I: algebraic lower bounds. Psychometrika **42**, 567–578 (1977)

R.P. McDonald, The theoretical foundations of principal factor analysis, canonical factor analysis and alpha factor analysis. Br. J. Math. Psychol. **23**, 1–21 (1970)

R.P. McDonald, *Test Theory: A Unified Treatment* (Erlbaum, Mahwah, NJ, 1999)

T. Raykov, Bias of coefficient α for fixed congeneric measures with correlated errors. Appl. Psychol. Meas. **25**, 69–76 (2001)

W. Revelle, R.E. Zinbarg, Coefficients alpha, beta, omega and the glb: comments on Sijtsma. Psychometrika **74**, 145–154 (2009)

W. Revelle, psych: Procedures for personality and psychological research. R package version 1.3.2 (2014). http://personality-project.org/r, http://personality-project.org/r/psych-manual.pdf

C. Spearman, The proof and measurement of association between two things. Am. J. Psychol. **15**, 72–101 (1904)

W. Tang, Y. Cui, A simulation study for comparing three lower bounds to reliability. Paper presented at the AERA, Vancouver, British Columbia, 2012

J. Ten Berge, G. Socan, The greatest lower bound to the reliability of a test and the hypothesis of unidimensionality. Psychometrika **69**, 613–625 (2004)

B. Woodhouse, P. Jackson, Lower bounds for the reliability of the total score on a test composed of non-homogeneous items: II: a search procedure to locate the greatest lower bound. Psychometrika **42**, 579–591 (1977)

Weighted Guttman Errors: Handling Ties and Two-Level Data

Letty Koopman, Bonne J. H. Zijlstra, and L. Andries van der Ark

Abstract We provide an introduction to weighted Guttman errors and discuss two problems in computing weighted Guttman errors that are currently not handled correctly by all software: Handling ties—that is, computing weighted Guttman errors when two items have the same estimated popularity—and computing weighted Guttman errors when the data have a two-level structure. Handling ties can be incorporated easily in existing software. For computing weighted Guttman errors for two-level data, we provide an R function.

Keywords Guttman errors • Item ordering • Mokken scale analysis • Multilevel test data • Nonparametric item response theory

1 Introduction

For a pair of dichotomous items in descending order of popularity, a Guttman error (Guttman 1950) occurs if a respondent answers negatively to the first (more popular or easier) item and positively to the second (less popular or more difficult) item. Hence, if item 1 is more popular than item 2, the item-score vector $(0, 1)$ constitutes a Guttman error, whereas $(0, 0)$, $(1, 0)$, and $(1, 1)$ are admissible item-score vectors. Guttman errors are violations of the deterministic Guttman (1950) scale. Guttman errors are used for detecting outliers (e.g., Zijlstra et al. 2007) and aberrant response patterns (e.g., Meijer 1994 Karabatsos 2003) and for computing Mokken's (1971) scalability coefficients in Mokken scale analysis (Sijtsma and Molenaar 2002; also see Sijtsma and Van der Ark 2017; Snijders 2001a). For a pair of polytomous items, multiple item-score vectors can constitute a Guttman error, making both the calculation and the interpretation of Guttman errors more complicated. Molenaar (1991) proposed to weight the Guttman errors to acknowledge that the degree in which item-score vectors are aberrant may differ. For example, consider two polytomous items, each having ordered answer categories $0, 1, 2, 3, 4$. Suppose item

L. Koopman (✉) • B.J.H. Zijlstra • L.A. van der Ark
Research Institute of Child Development and Education, University of Amsterdam, P.O. Box 15776, 1001 NG, Amsterdam, The Netherlands
e-mail: V.E.C.Koopman@uva.nl

L.A. van der Ark et al. (eds.), *Quantitative Psychology*, Springer Proceedings in Mathematics & Statistics 196, DOI 10.1007/978-3-319-56294-0_17

1 is more popular than item 2, then item-score vector $(0,4)$ is more aberrant than item-score vector $(0,1)$.

In recent work on deriving standard errors for two-level scalability coefficients (Koopman 2016), we encountered two problems in estimating the weights of Guttman errors: Estimated weights depend on the value of a random seed when two or more estimated *item popularities* are equal, and estimated weights may be biased for two-level data. In this chapter, we first introduce weighted Guttman errors, then we discuss the two problems and offer a solution for each problem, and finally we discuss some additional features of (two-level) weight computations.

2 Weighted Guttman Errors

2.1 Theory

Let a test consist of J items with $m+1$ ordered response categories indexed by x $(x=0,1,\ldots,m)$. Let X_j denote the item score of item j. Each item score consists of m item steps (Molenaar 1983), binary variables denoted Z_{jx} $(j=1,\ldots,J; x=1,\ldots,m)$. $Z_{jx}=1$ if $X_j \geq x$ (the item step was passed) and $Z_{jx}=0$ if $X_j < x$ (the item step was failed). It follows that $Z_{j,x-1} \geq Z_{jx}$ and $X_j = \sum_x Z_{jx}$. For example, if $X_j=1$ and $m=3$, then $Z_{j1}=1$, $Z_{j2}=0$, and $Z_{j3}=0$. Let the popularity of item step Z_{jx} be the probability of having a score of at least x on item j: $P(Z_{jx}) \equiv P(X_j \geq x)$. Note that by definition, $P(X_j \geq 0)=1$. Let z_{njx} denote the realization of Z_{jx} for person n, then, in a sample of N respondents, $P(X_j \geq x)$ is estimated by

$$\widehat{P}\left(X_j \geq x\right) = \frac{1}{N}\sum_{n=1}^{N} z_{njx}. \tag{1}$$

Item pair (i,j) has $2m$ item steps: $Z_{i1},\ldots,Z_{im},Z_{j1},\ldots,Z_{jm}$. For the purpose of determining weighted Guttman errors, the $2m$ item steps are put in descending order of their popularity. For example, Table 1 shows $J=2$ items with $m+1=3$ ordered response categories, for which $P(Z_{i1}) > P(Z_{j1}) > P(Z_{i2}) > P(Z_{j2})$. Hence, the order of the item steps is

$$Z_{i1}, Z_{j1}, Z_{i2}, Z_{j2}. \tag{2}$$

Table 1 Probabilities of item scores X_i and X_j, with $m+1=3$ ordered answer categories

	X_j				
X_i	0	1	2	$P(X_i=x)$	$P(X_j \geq x)$
0	0.08	0.16	0.00	0.24	1.00
1	0.04	0.04	0.24	0.32	0.76
2	0.36	0.08	0.00	0.44	0.44
$P(X_i=x)$	0.48	0.028	0.24		
$P(X_j \geq x)$	1.00	0.52	0.24		

For notational convenience the subscripts jx in the item steps may be replaced by subscripts (1), (2), ..., (2m) indicating the order of the item steps in an item pair. In this notation, Eq. (2) equals $Z_{(1)}, Z_{(2)}, Z_{(3)}, Z_{(4)}$. For each item pair, item-score pattern (x, y) corresponds a specific realization of the ordered item steps. For example, for Eq. (2), item-score pattern $(0, 2)$ corresponds to $Z_{i1} = 0, Z_{j1} = 1, Z_{i2} = 0, Z_{j2} = 1$. In a Guttman scale, the ordered item steps are strictly nonincreasing: Once a more popular item step is failed, a less popular item step cannot be passed. For example, in a Guttman scale, admissible values for Eq. (2) are 0,0,0,0; 1,0,0,0; 1,1,0,0; 1,1,1,0; and 1,1,1,1, which correspond to item-score patterns $(0, 0)$, $(1, 0)$, $(1, 1)$, $(2, 1)$, and $(2, 2)$, respectively. A Guttman error occurs if a less popular item step is passed while a more popular item step is failed. For Eq. (2), realizations 0,1,0,0; 0,1,0,1; 1,1,0,1; and 1,0,1,0 which correspond to item-score patterns $(0, 1)$, $(0, 2)$, $(1, 2)$, and $(2, 0)$, respectively—are Guttman errors.

The weight of a Guttman error, denoted w_{ij}^{xy}, indicates the degree of deviation from the perfect Guttman scale (Molenaar 1991). Let $z_{(h)}^{xy}$ denote the realization of the hth $(1 \leq h \leq 2m)$ item step corresponding to the item-score pattern (x, y). The weight is computed as

$$w_{ij}^{xy} = \sum_{h=2}^{2m} \left\{ z_{(h)}^{xy} \left[\sum_{g=1}^{h-1} \left(1 - z_{(g)}^{xy} \right) \right] \right\} \tag{3}$$

(see, e.g., Kuijpers et al. 2013). Note that Eq. (3) counts the number of times a more difficult item step was passed, while an easier item step was failed. For admissible item-score patterns, the corresponding weights are zero, whereas for Guttman errors, the weights are positive. For example, assuming the order of the item steps in Eq. (2) is correct, for item-score pattern $(0, 2)$, $z_{(1)}^{02} = 0$, $z_{(2)}^{02} = 1$, $z_{(3)}^{02} = 0$, and $z_{(4)}^{02} = 1$. Hence, following Eq. (3), $w_{ij}^{02} = 1 \times [1] + 0 \times [1 + 0] + 1 \times [1 + 0 + 1] = 3$. Also note that for dichotomous items, the only item-score pattern that constitutes a Guttman error (i.e., either $(0, 1)$ or $(1, 0)$) receives a weight 1 by definition. Hence, for dichotomous items weighting the Guttman errors has no effect.

In samples, weights w_{ij}^{xy} are estimated from the order of the item steps in the sample with Eq. (1) and denoted $\widehat{w}_{ij}^{xy}$. Typically, w_{ij}^{xy} and $\widehat{w}_{ij}^{xy}$ are the same, but when the sample is small or when the popularities of two item steps are close, w_{ij}^{xy} and $\widehat{w}_{ij}^{xy}$ may differ (for more information on this topic, we refer to Kuijpers et al. 2016).

2.2 *Applications*

Weighted Guttman errors are used to compute scalability coefficients in Mokken scale analysis. Mokken (1971) discussed scalability coefficients for dichotomous items, Molenaar (1983, 1991, 1997) generalized the scalability coefficients to polytomous items, and Snijders (2001a, also, see Crisan et al. 2016) generalized the scalability coefficients to two-level data. The scalability coefficients are implemented in several software packages, including the stand-alone package MSP (Molenaar and Sijtsma 2000) and the R package `mokken` (Van der Ark 2012). Mokken's (1971) item-pair scalability coefficient H_{ij} can be written as a function of the Guttman weights and the univariate and bivariate item probabilities:

$$H_{ij} = 1 - \frac{\sum_x \sum_y w_{ij}^{xy} P\left(X_i = x, X_j = y\right)}{\sum_x \sum_y w_{ij}^{xy} P\left(X_i = x\right) P\left(X_j = y\right)}. \tag{4}$$

Note that if unweighted Guttman errors were used, weights w_{ij}^{xy} only take on the values 0 and 1. By using weighted Guttman errors, H_{ij} equals the ratio of the inter-item correlation and the maximum inter-item correlation given the marginal distributions of the two items (Molenaar 1991).

In a sample of size N, the item-pair scalability coefficient is estimated by replacing the weights in Eq. (4) by the estimated weights and replacing the probabilities by the sample proportions, that is,

$$\widehat{H}_{ij} = 1 - \frac{\sum_x \sum_y \widehat{w}_{ij}^{xy} \widehat{P}\left(X_i = x, X_j = y\right)}{\sum_x \sum_y \widehat{w}_{ij}^{xy} \widehat{P}\left(X_i = x\right) \widehat{P}\left(X_j = y\right)} = 1 - \frac{F_{ij}}{E_{ij}}. \tag{5}$$

$F_{ij} = N \sum_x \sum_y \widehat{w}_{ij}^{xy} \widehat{P}\left(X_i = x, X_j = y\right)$ expresses the weighted sum of observed Guttman errors, and $E_{ij} = N \sum_x \sum_y \widehat{w}_{ij}^{xy} \widehat{P}\left(X_i = x\right) \widehat{P}\left(X_j = y\right)$ the weighted sum of Guttman errors expected when the two items are marginally independent.

Weighted Guttman errors are also used as an index to detect outliers and as a person-fit statistic. In these applications, the total of estimated Guttman weights within a response pattern is used. Let x_{ni} denote the observed score of person n on item i, and let y_{nj} denote the observed score of person n on item j. Using the notation of Zijlstra et al. (2007), index G_+ for respondent n equals

$$G_{n+} = \sum \sum_{i<j} \widehat{w}_{ij}^{x_{ni} y_{nj}}. \tag{6}$$

The function `check.errors()` in the R package `mokken` provides weighted Guttman errors for each observation.

3 Computational Problems

3.1 Problem 1: Ties

Estimating Guttman weights can be problematic if two estimated item steps have the same popularity. If the estimated item steps pertain to the same item, $\widehat{P}\left(X_j \geq x\right) = \widehat{P}\left(X_j \geq x + 1\right)$, it means that no one in the sample had score x on item j. The ordering of the estimated item steps is not affected because item steps have a fixed order within an item, and estimating Guttman errors is not problematic. However, if

Table 2 Cross-classification of item scores X_i and X_j, with $m+1=3$ ordered answer categories, for $N=15$ respondents

	X_j				
X_i	0	1	2	Total	$\widehat{P}(X_i \geq x)$
0	2	4	0	6	1.00
1	1	1	0	2	0.60
2	3	2	2	7	0.47
Total	6	7	2	15	
$\widehat{P}(X_j \geq y)$	1.00	0.60	0.13		

Table 3 Observed and expected frequencies, Guttman weights under two possible item-step orderings, and their mean, for each response pattern in Table 2

	Item-score vector								
	00	01	02	10	11	12	20	21	22
$N \times \widehat{P}(X_i = x, X_j = y)$	2	4	0	1	1	0	3	2	2
$N \times \widehat{P}(X_i = x)\widehat{P}(X_j = y)$	2.40	2.80	0.80	0.80	0.93	0.27	2.80	3.27	0.93
w_{ij}^{xy} 1	0	1	3	0	0	1	1	0	0
w_{ij}^{xy} 2	0	0	2	1	0	1	2	0	0
$\bar{w}_{ij}^{xy}$	0	0.5	2.5	0.5	0	1	1.5	0	0

For details, see text

the equally popular estimated item steps pertain to two different items, $\widehat{P}(X_i \geq x) = \widehat{P}(X_j \geq y)$, the item-step ordering cannot be determined. As an example, Table 2 shows the frequencies of the response patterns of $N=15$ respondents, for two polytomous items with three response categories. For these data, $\widehat{P}(X_i \geq 1) = \widehat{P}(X_j \geq 1) = 0.6$, so the order of the item steps is either $Z_{i1}, Z_{j1}, Z_{i2}, Z_{j2}$ or $Z_{j1}, Z_{i1}, Z_{i2}, Z_{j2}$.

Currently, the software program `mokken` (Van der Ark 2012) adds a small random value to the estimated popularities to avoid equal item steps. There are two downsides to this approach. First, one item step is randomly assigned to be more popular than the other item step without theoretical justification. Second, analyzing the same data twice may result in different weights and, thus, different scalability coefficients.

Molenaar (1991) suggested computing the weights for all combinations of equivalent item-step orderings. For each item-score vector in Table 2, Table 3 shows the observed frequencies ($N \times \widehat{P}(X_i = x, X_j = y)$), the expected frequencies under marginal independence ($N \times \widehat{P}(X_i = x)\widehat{P}(X_j = y)$), the resulting weights given item-step ordering $Z_{i1}, Z_{j1}, Z_{i2}, Z_{j2}$ ($\widehat{w}_{ij}^{xy}$ 1), the resulting weights given item-step ordering $Z_{j1}, Z_{i1}, Z_{i2}, Z_{j2}$ ($\widehat{w}_{ij}^{xy}$ 2), and the average of the two weights. For both item-step orderings, the weighted sum of Guttman errors results in $F_{ij} = 7$ and $E_{ij} = 8.27$ (yielding $\widehat{H}_{ij} \approx 0.15$). Therefore, for scalability coefficients, the item-step order does not affect the outcome (Molenaar 1991). However, for individual-level statistics, such as the person-fit index G_+ (Eq. (6)), the item-step order matters. For example, a person with item-score vector (0,2) has value $G_+ = 3$ for the first item-step ordering and $G_+ = 2$ for the second item-step ordering. Because both item-step orderings are equally likely in the population, the average weight (Table 3, last row) is considered more appropriate as opposed to randomly favouring one ordering over the other, and results in a value of $G_+ = 2.5$.

3.2 Problem 2: Estimating the Item Ordering for Two-Level Test Data

In Mokken scale analysis for two-level data, X_{srj} denotes the response of subject s ($s = 1, \ldots, S$) to item j ($j = 1, \ldots J$) scored by rater ($r = 1, \ldots, R_s$). As with one-level data, item step $Z_{jx} = 1$ if $X \geq x$, and $Z_{jx} = 0$, otherwise. The problem is that the order of the item steps, and hence the value of the Guttman weights, depends on the estimation method for $P(X_j \geq x)$. $P(X_j \geq x)$ can be estimated in two ways (Snijders 2001a), possibly yielding different estimates. Let Z_{srjx}, with realization z_{srjx}, be a binary variable that takes on the value one if $X_{srj} \geq x$, and zero otherwise. First, $P(X_j \geq x)$ can be estimated by averaging the relative frequencies for all subjects, that is,

$$\widehat{P}\left(X_j \geq x\right) = \frac{1}{S}\sum_{s=1}^{S}\frac{1}{R_s}\sum_{r=1}^{R_s} z_{srjx}, \tag{7}$$

and, second, $P(X_j \geq 1)$ can be estimated by averaging the absolute frequencies for all subjects, that is,

$$\widehat{P}\left(X_j \geq x\right) = \frac{1}{\sum_{s=1}^{S} R_s}\sum_{s=1}^{S}\sum_{r=1}^{R_s} z_{srjx}. \tag{8}$$

The example in Table 4 (last two rows) shows that the estimation methods do not only result in different estimates but also in different ordering of item steps. When averaging the relative frequencies of all subjects in Eq. (7), the ordering of the item steps is $Z_{i1}, Z_{j1}, Z_{i2}, Z_{j2}$, and when averaging the absolute frequencies of

Table 4 Values of $\sum_{r=1}^{R_s} z_{srjx}$ for $J = 2$ items, each having three ordered response categories, $S = 3$ subjects who are rated by $R_s = 10, 3, 10$ raters, respectively, and the values of $\widehat{P}\left(X_j \geq x\right)$ using Eqs. (7) and (8), respectively

	X_i			X_j			
s	$x \geq 0$	$x \geq 1$	$x \geq 2$	$x \geq 0$	$x \geq 1$	$x \geq 2$	R_s
1	10	4	2	10	3	3	10
2	3	2	2	3	3	2	3
3	10	4	2	10	3	3	10
Equation (7)	1.00	0.49	0.36	1.00	0.43	0.26	
Equation (8)	1.00	0.53	0.42	1.00	0.39	0.35	

all subjects in Eq. (8), the ordering of the item steps is $Z_{i1}, Z_{i2}, Z_{j1}, Z_{j2}$. Snijders (2001a) argued that averaging the relative frequencies in Eq. (7) is the preferred method, as averaging the absolute frequencies is biased under certain conditions.

4 Discussion

Two problems with the weighted Guttman errors have been addressed and described in this chapter. The solution to the problem of ties can be incorporated in the software easily. The software program MSP prints a warning when ties are present. As far as we know, the DOS program TWOMOK (Snijders 2001b) is the only software for two-level scalability coefficients. Because it pertains to dichotomous items only, weighted Guttman errors are not an issue. A new R function to compute weighted Guttman errors for dichotomous and polytomous two-level item scores is called `MLweight()`. The function is described in Koopman (2016). The main goal of `MLweight()` is to allow the computation of Mokken's scalability coefficients for two-level data in the function `MLcoefH()`. Both functions have been implemented in the R package `mokken`.

References

D.R. Crisan, J.E. Van de Pol, L.A. Van der Ark, Scalability coefficients for two-level polytomous item scores: an introduction and an application, in *Quantitative Psychology Research: The 80th Annual Meeting of the Psychometric Society, Beijing, China, 2015*, ed. by L.A. van der Ark, D.M. Bolt, W.-C. Wang, J.A. Douglas, M. Wiberg (Springer, New York, 2016), pp. 139–153. doi: 10.1007/978-3-319-38759-8_11

L. Guttman, The basis for scalogram analysis, in *Measurement and Prediction*, ed. by S.A. Stouffer, L. Guttman, E.A. Suchman, P.F. Lazarsfeld, S.A. Star, J.A. Clausen (Princeton University Press, Princeton, 1950), pp. 60–90

G. Karabatsos, Comparing the aberrant response detection performance of thirty-six person-fit statistics. Appl. Meas. Educ. **16**, 277–298 (2003). doi:10.1207/S15324818AME1604_2

L. Koopman, Standard errors of scalability coefficients in two-level Mokken scale analysis. Unpublished master's thesis, Research Institute of Child Development and Education, University of Amsterdam, The Netherlands, 2016

R.E. Kuijpers, L.A. Van der Ark, M.A. Croon, Standard errors and confidence intervals for scalability coefficients in Mokken scale analysis using marginal models. Sociol. Methodol. **43**, 42–69 (2013). doi:10.1177/0081175013481958

R.E. Kuijpers, L.A. Van der Ark, M.A. Croon, K. Sijtsma, Bias in estimates and standard errors of Mokken's scalability coefficients. Appl. Psychol. Meas. **40**, 331–345 (2016). doi:10.1177/0146621616638500

R.R. Meijer, The number of Guttman errors as a simple and powerful person-fit statistic. Appl. Psychol. Meas. **18**, 311–314 (1994). doi:10.1177/014662169401800402

R.J. Mokken, *A Theory and Procedure of Scale Analysis: With Applications in Political Research* (De Gruyter, Berlin, 1971)

I.W. Molenaar, *Item Steps. Heymans Bulletin HB-83-630-EX* (University of Groningen, Groningen, 1983)

I.W. Molenaar, A weighted Loevinger H-coefficient extending Mokken scaling to multicategory items. Kwantitatieve Methoden **12**(37), 97–117 (1991)

I.W. Molenaar, Nonparametric models for polytomous responses, in *Handbook of Modern Item Response Theory*, ed. by W.J. van der Linden, R.K. Hambleton (Springer, New York, 1997), pp. 369–380. doi: 10.1007/978-1-4757-2691-6_21

I.W. Molenaar, K. Sijtsma, *User's Manual MSP5 for Windows* (iec ProGAMMA, Groningen, 2000)

K. Sijtsma, I.W. Molenaar, *Introduction to Nonparametric Item Response Theory* (Sage, Thousand Oaks, CA, 2002)

K. Sijtsma, L.A. Van der Ark, A tutorial on how to do a Mokken scale analysis on your test and questionnaire data. Br. J. Math. Stat. Psychol. **70**(1), 137–158 (2017). doi:10.1111/bmsp.12078

T.A.B. Snijders, Two-level nonparametric scaling for dichotomous data, in *Essays on Item Response Theory*, ed. by A. Boomsma, M.A.J. van Duijn, T.A.B. Snijders (Springer, New York, 2001a), (pp. 319–338). doi: 10.1007/978-1-4613-0169-1_17

T.A.B. Snijders, TWOMOK [computer software]. Retrieved from https://www.stats.ox.ac.uk/~snijders/ (2001b)

L.A. Van der Ark, New developments in Mokken scale analysis in R. J. Stat. Softw. **48**(5), 1–27 (2012). doi:10.18637/jss.v048.i05

W.P. Zijlstra, L.A. Van der Ark, K. Sijtsma, Outlier detection in test and questionnaire data. Multivar. Behav. Res. **42**, 531–555 (2007). doi:10.1080/00273170701384340

Measuring Cognitive Processing Capabilities in Solving Mathematical Problems

Susan Embretson

Abstract Understanding the sources of processing complexity in mathematical problem solving items is an important aspect of test validity. The sources of cognitive complexity may be either construct relevant or construct irrelevant. Studies have shown that the levels and sources of cognitive complexity predict item difficulty (e.g., S.E. Embretson, R.C. Daniel, Understanding and quantifying cognitive complexity level in mathematical problem solving items. Psychol. Sci. **50**, 328–344 (2008)) and, further, that items can be selected or designed for difficulty in different sources of cognitive complexity. Although these results are relevant to the *response processes* aspect of construct validity, potential impact on the other aspects of validity was not addressed. That is, the modeling procedures did not include multidimensional measurement of individual differences in processing capabilities. In the current study, the multicomponent latent trait model for diagnosis (MLTM-D; S.E. Embretson, X. Yang, A multicomponent latent trait model for diagnosis. Psychometrika **78**, 14–36 (2013)) was applied to measure cognitive processing capabilities in processing mathematical items. Individual differences in patterns of processing capabilities were significantly related to examinee background variables, thus indicating potential impact on the *consequential* aspect of validity. Implications of the findings for item design and test development are discussed.

Keywords Cognitive processes • Item difficulty modeling • Multicomponent latent trait model

1 Measuring Cognitive Processing Capabilities in Solving Mathematical Problems

Cognitive complexity in mathematical problem solving items is recognized as an important aspect of test design. For achievement tests, for example, multiple levels of cognitive complexity are often explicit in the test blueprints. For eighth

S. Embretson (✉)
Georgia Institute of Technology, Atlanta, GA, 30332, USA
e-mail: susan.embretson@psych.gatech.edu

L.A. van der Ark et al. (eds.), *Quantitative Psychology*, Springer Proceedings in Mathematics & Statistics 196, DOI 10.1007/978-3-319-56294-0_18

grade mathematical achievement, the specifications for the *National Assessment of Educational Progress* (NAEP) include five content areas and three levels of cognitive complexity (see National Assessment Governing Board (NAGB) 2015). The cognitive complexity levels, low complexity, moderate complexity, and high complexity, are defined globally. For example, "high complexity items make heavy demands on students, because they are expected to use reasoning, planning, analysis, judgment and creative thought" (NAGB 2015, p. 46).

Cognitive complexity is also included under one of the five aspects of construct validity, namely, the *response processes aspect* (Standards for Educational and Psychological Testing 2014). Studying the impact of varying sources of cognitive complexity on item difficulty and response times is a major method for understanding the cognitive processes applied by examinees. If evidence on response processes is available, construct-relevancy can be evaluated. Further, the response processes involved in responding to items can impact other aspects of test validity, such as *internal structural*, *relationship to external variables*, and *test consequences* for different groups of examinees (Embretson 2007, 2016).

Yet, despite the importance of cognitive complexity in items, Leighton and Gierl (2007) note that it is rarely studied empirically. That is, according to Leighton and Gierl (2007), typically it is assumed that examinees apply the intended processes. But, without empirical support, the intended cognitive processes may not represent the actual thinking processes applied by examinees.

Studying cognitive processes requires a theoretical perspective. For mathematical problem solving items, prior studies have supported a multistage theory of cognitive processing (Mayer 2003). In the processing theory, originally developed by Mayer et al. (1984), two major stages of processing are postulated, with two substages each. Mayer (2003) notes that empirical support has been obtained for each substage. Mayer's theory has also been applied to mathematical items on a high-stakes aptitude test (e.g., Embretson and Daniel 2008; Daniel and Embretson 2010). The sources of cognitive complexity associated with the stages significantly predicted item difficulty. Further, items could be selected or designed for involving different sources of cognitive complexity. These results are relevant to the *response processes* aspect of construct validity for the aptitude test. However, potential impact on the other aspects of validity was not addressed in these studies. That is, the modeling procedure did not include measurement of individual differences in processing capabilities.

In the current study, the *response processes* aspect of validity is examined for high-stakes mathematical achievement tests using the Mayer (2003) theory. Then, by applying an item response theory (IRT) model appropriate for measuring processing, the impact of *response processes* on other aspects of validity—*internal structure*, *relationship to external variables*, and the *consequential* aspects—is examined. Prior to presenting the study, some background on cognitive processing in mathematical problem solving items and IRT models for measuring processing is presented.

2 Background

2.1 Multistage Theory of Mathematical Problem Solving

Mayer's et al. (1984) theory of processing includes two global stages, Problem Representation and Problem Execution. Each global stage is further divided into substages. For Problem Representation, the substages are Translation, converting item content into a meaningful form in short-term memory, and Integration, producing the equation(s) to be solved. For Problem Execution, the substages are Solution Planning, developing a procedure to solve the equation(s), and Solution Execution, computing a solution to the problem. According to the theory, the stages are processed sequentially, as follows:

Translation → Integration → Solution Planning → Solution Execution.

As noted by Mayer (2003), several studies in the context of teaching and learning strategies have supported the plausibility of each stage.

The plausibility of the Mayer (2003) processing model for mathematical problem solving on test items can be examined by item difficulty modeling. That is, variables that represent content features that are postulated to impact processing difficulty are scored on the items. For the Solution Execution stage in the Mayer (2003) theory, for example, items can be scored for the number and the knowledge level of the computations required for solution.

Embretson and Daniel (2008) modeled the difficulty of *Graduate Record Examination* (GRE) quantitative items from a set of variables associated with the substages. To apply the Mayer (2003) theory to solving mathematical test items, Embretson and Daniel (2008) added a fifth stage, Decision. That is, some multiple choice test items that cannot be solved in a top-down fashion from the information in the stem. That is, the response options must be compared and evaluated. Using the linear logistic test model (LLTM; Fischer 1973), as described below, a moderately strong relationship was found. That is, a likelihood ratio fit statistic (Δ) was equal to 0.724, which is similar in magnitude to a multiple correlation coefficient. Thus, the Mayer (2003) theory was supported for mathematical test items.

In a second study, Daniel and Embretson (2010) examined the impact of designing items for cognitive complexity source on item difficulty. A subset of GRE items was selected to be redesigned by varying either Solution Execution difficulty (i.e., the number of subgoals) or Integration difficultly (i.e., the equation source). Strong effects on item difficulty and item response time were observed for both design changes. The overall predictability of item difficulty was strong, as the likelihood ratio fit (Δ) was 0.941.

2.2 Psychometric Models for Cognitive Processing

Some item response theory (IRT) models can be used to estimate parameters to represent item differences in cognitive processing complexity or individual differences in processing capabilities. To apply these models, items are scored on variables that represent the existence or complexity of the postulated cognitive processes in the items. In some models, parameters are estimated for the impact of the cognitive complexity variables on item psychometric properties, such as difficulty or discrimination. Other IRT models can estimate individual differences in processing capabilities.

A unidimensional model that can be applied to model item difficulty is the linear logistic test model (LLTM; Fischer 1973). LLTM can be applied to binary items for which a plausible theory of cognitive processes exists and stimulus features (i.e., q_{im}) can be scored to represent processing difficulty. LLTM is a generalization of the Rasch model and can be written in two parts to show the Rasch model and the prediction model as follows:

$$P\left(X_{is}=1\right)=\frac{e^{\theta_s-\beta_i'}}{1+e^{\theta_s-\beta_i'}} \tag{1a}$$

and

$$\beta_i'=\sum_m \eta_m q_{im}+\eta_0 \tag{1b}$$

where θ_s is the trait level for person *s*, β_i' is predicted item difficulty from scored item features, q_{im} is the score on predictor *m* for item *i*, η_m is the weight of feature *m*, and η_0 is a normalization constant. It should be noted that β_i' in Eq. (1a) is obtained from the weights and scores in Eq. (1b) and is not estimated directly. Parameters may be estimated for LLTM by conditional maximum likelihood (CML) or marginal maximum likelihood (MML).

Fit of the model to the data supports the plausibility of the cognitive model. Item fit can be evaluated in multiple methods, including chi-square tests, standardized residuals, and plots. Overall fit indices that are based on the likelihood of the data also can be used to assess the plausibility of the cognitive model compared to other models. Nested models, in which the predictors in one model are a subset of the other model, can be compared statistically with likelihood ratio chi-square tests. Non-nested models can be compared with other fit statistics, such as the Akaike Information Criterion. An overall fit statistic for IRT models that is similar to the fit indices in structural equation modeling also can be used. The Δ index (Embretson 1997), which ranges from 0 to 1, is based on the ratio of the likelihood of alternative models of item difficulty: (1) a null model, in which items have equal difficulties; (2) a saturated model, in which each item has uniquely estimated parameters; and (3) the target cognitive model. The statistic is written as follows:

$$\Delta = \sqrt{\frac{(-2\ln L_{null}) - (-2\ln L_{t\arg et})}{(-2\ln L_{null}) - (-2\ln L_{saturated})}}. \tag{2}$$

The Δ statistic in Eq. (2) is analogous to a multiple correlation coefficient and often has the same magnitude.

If examinees' response processes involve two or more processing stages, each of which involves a separate trait, a multidimensional model is appropriate. If correct responses from each stage are needed to solve items, as in the cognitive model for mathematical problem solving, a conjunctive model is appropriate. That is, the overall probability of item solution is the product of the probabilities of the various stages.

Conjunctive models can be applied in two different situations. First, the subtask response modeling is appropriate if responses to the various substages are available. Although this type of model is most direct, in practice test items typically are not administered as subtasks. Second, the total item response method is appropriate if items are heterogeneous with respect to which processing outcomes are needed for solution. In this method, items must be scored for process involvement.

Table 1 presents three items that vary in the Problem Representation and Problem Solution global stages of the cognitive model for mathematics. Solving Item 3 involves both Problem Representation and Problem Solution. However, solving Item 1 and Item 2 involves only one global process, Problem Representation and Problem Solution, respectively.

The multicomponent latent trait model for diagnosis (MLTM-D; Embretson and Yang 2013) is an example of a conjunctive multidimensional model that does not require subtasks responses. Two types of scores are required for the full MLTM-D, c_{ik}, the involvement of component *k* in item *i*, and q_{imk}, a score for stimulus feature *m* in component *k* for item *i*. Thus, matrix C_{ixk} is scored to represent the involvement of the components in each item and within components; matrices $Q_{ixm(k)}$ are scored to represent stimulus features that impact difficulty in the various components. MLTM-D may be written as follows, given that X_{isT} is the response to the total item and X_{isk} is the (possibly unobserved) response to component *k*:

Table 1 Three items with different involvement of global processing components

Item 1 Kyle will travel 100 miles in 2 unequal segments. The second segment is 40 miles shorter than the first segment. Which equation could be used to find the length of the first segment (m)? (A) m + m + 40 = 100 (B) m + m = 100 (C) m − 40 = 100 (D) X m + (m − 40) = 100
Item 2 What is the value of m in the equation m + (m − 40) = 100? (A) X 30 (B) 35 (C) 40 (D) 60
Item 3 Kyle will travel 100 miles in 2 unequal segments. The second segment is 40 miles shorter than the first segment. What is the length of the first segment? (A) X 30 (B) 35 (C) 40 (D) 60

$$P\left(X_{isT} = 1\right) = \Pi_k P(X_{isk} = 1)^{cik} \tag{3a}$$

and

$$P\left(X_{isk} = 1\right) = \frac{e^{\theta_{sk} - \sum_{m(k)} \eta_{mk} q_{imk} + \eta_{0k}}}{1 + e^{\theta_{sk} - \sum_{m(k)} \eta_{mk} q_{imk} + \eta_{0k}}} \tag{3b}$$

where θ_{sk} is the competency of examinee *s* on component *k*, q_{imk} is a score for stimulus feature *m* in component *k*, η_{mk} is the weight of feature *m* in item difficulty on component *k*, and η_{0k} is the normalization constant for component *k*. In Eq. (3a), it can be seen that c_{ik} defines whether or not a particular component is involved in an item. Equation (3b) is an LLTM at the component level. If the q_{imk} are dummy variables, scored uniquely for each item, then Eq. (3b) is a Rasch model at the component level.

Standard MML estimation is possible for the MLTM-D item parameters, using a standard normal scale (i.e., $\boldsymbol{\theta}$ $MVN(0, \sum)$). Either uncorrelated or correlated traits may be specified. Trait levels for the components can be estimated with a multidimensional *expected a posteriori* (EAP) algorithm. MLTM-D was developed in the context of diagnosis, which can be obtained by developing cutlines, γ_k, on the trait level estimates, θ_{sk}. A criterion for the diagnosis must be specified as a probability of item solving, *y*. Then the cutline on trait level, γ_k, is established such that if $\theta_{sk} \geq \gamma_k$, then $P(X_{.sk}) \geq y$.

3 Study: Cognitive Processing on Mathematical Achievement Items

This study had several goals: (1) to examine the cognitive complexity of mathematical achievement items, (2) to estimate individual differences in cognitive processing on mathematical achievement test items, (3) to examine the internal structure of these estimates, and (4) to examine relationships to external variables, particularly group membership to assess the potential consequential aspect of validity.

3.1 Method

3.1.1 Test

The test was a year-end mathematical achievement test for eighth grade students. The test was a standards-based test used for accountability purposes by the cooperating state. The test consisted of 86 operational items was administered in three parts, in separate sessions. As typical for year-end achievement tests, the items on the test have survived multiple levels of review, including a mathematics editor, a panel of educators, and empirical tryout and review by the state department of education.

3.1.2 Examinees

The examinees were students taking the operational test at the end of the school year. A sample of 4000 Grade 8 students, randomly selected from the participating state, was available for this study.

3.1.3 Cognitive Scores

The cognitive variables used in the study were originated by Embretson and Daniel (2008) for use on the GRE. For the current study, it became apparent that some adaptation of the system was required, as some variables were not relevant or were defined at a level too high for items on an eighth grade mathematics achievement items. Thus, a redefined set of cognitive variables was developed and scored on each item for use with the current test. Definitions of the variables are presented on Table 2. Items also were scored for the involvement of components at the global level, Problem Representation and Problem Execution.

The cognitive variables were scored for each item by a team of three raters with expertise in cognitive psychology. For the binary cognitive variables, Cohen's (1960) kappa statistic was computed between each pair of raters to identify patterns of consistency. The Fleiss index was computed for overall rater consistency. For the continuous cognitive variables, correlations were computed between each pair of raters and then averaged across raters. The mean Fleiss index was 0.787 for the nine binary variables. The mean rater correlation, across forms, for the six continuous variables, was 0.825. The final scores used in the analysis represent the consensus scoring of each of the items on the cognitive variables.

3.2 Results

3.2.1 Descriptive Statistics

Table 3 presents descriptive statistics on the scored variables in the 86 items. It can be seen that variability of the cognitive complexity variables was found in each substage. Also shown on Table 3 are descriptive statistics on component involvement in items. It can be seen that Problem Representation was involved in a larger percentage of the items than Problem Execution. A cross-tabulation of process involvement showed 41.8% of the items involved both components.

Table 2 Definitions of cognitive attributes

Attribute	Definition
Translation	
Mathematical	The total number of mathematical terms in the stem and all answer options. This includes numerals, variables (e.g., x, y, m, etc.), axis labels, comparators (e.g., <, >, =), and implicit and explicit operators
Context	The total number of words, excluding variables, in the stem and all answer options
Encode diagram	Indicator of presence of a diagram, graph, or other figure, excluding tables, in the stem or answer options
Integration	
Equation given in symbols	Indicator of whether a mathematical equation is provided in the stem of an item
Equation given in words	Indicator of whether the examinee needs to interpret an equation given in word (context) form
Generate equations or plausible values	Indicator of whether equations or plausible values for variables must be generated to answer the item
Recall equations/knowledge	Indicator of whether the examinee must recall known equations (e.g., formula for slope of a line, the Pythagorean theorem, etc.) or definitions
Translate diagram	Indicator for whether presented diagram or figure is necessary for problem solution
Visualization	Indicator for whether a diagram or figure must be visualized or drawn to understand or answer the item
Solution planning	
Number of subgoals	The total number of sub-steps necessary for answering an item (e.g., finding a slope for the equation of a line)
Relative definition of variables	Indicator of whether one variable is defined only in terms of another
Solution execution	
Procedural knowledge	The maximum procedural knowledge necessary in solving the item (1) integers, (2) fractions, (3) proportions, (4) decimals, (5) negative numbers, and (6) squares/square roots, as outlined below
Number of procedures	The total number of unique procedures necessary for solving the item
Number of computations	The total number of computations necessary for solving the item, including computations necessary in evaluating distractors and stem
Decision	
Decision processing	Indicator for whether information found in distractors is necessary to eliminate options or answer item

3.2.2 Item Cognitive Complexity

LLTM was applied to the item response data for examinees. The scored variables on Table 2 were used to represent cognitive complexity. The cognitive model was estimated with an intercept, weights for each of the variables on Table 3 and a constant item discrimination parameter. Two comparison models were estimated for the likelihood ratio fit statistic Δ. The saturated model was a 1PL model with 86 item difficulties and a constant discrimination, and the null model was a 1PL model with a constant item discrimination parameter. Parameters for three models were estimated by MML, using a one parameter logistic (1PL) variant of LLTM in the normal metric, with a constant item discrimination value, α, and the trait distribution specified as N(0,1). The cognitive model (−2lnL = 341,543, AIC = 341,577, #parameters = 17) had moderately strong prediction as indicated by the likelihood ratio statistic (Δ = 0.606).

Table 3 presents the weights for the cognitive complexity variables from the LLTM cognitive model. It can be seen that the most strongly significant variable (t-statistic) was procedural knowledge, with a positive weight. However, significant weights were found for all cognitive complexity variables, thus supporting the

Table 3 Descriptive statistics and parameter estimates for cognitive complexity and component model

	Descriptives		Item parameter estimates		
Variables	Mean	Standard deviation	Estimate	Standard error	t_{obs}
Cognitive complexity model (intercept)			−2.104	0.026	−82.26*
Translation					
Mathematical encoding	20.17	15.351	0.012	0.001	20.25*
Contextual encoding	35.05	20.275	0.010	0.001	31.92*
Encode diagram	0.29	0.457	0.679	0.020	33.17*
Integration					
Given equation: words	0.20	0.457	−1.287	0.036	−36.23*
Given equation: symbols	0.23	0.401	−0.441	0.024	−18.69*
Generate equation	0.35	0.479	0.151	0.023	6.52*
Recall/access equations	0.30	0.462	0.396	0.021	18.48*
Translate diagram	0.27	0.445	−0.304	0.019	−15.43*
Visualization	0.07	0.256	0.818	0.026	32.03*
Solution planning					
Number of subgoals	0.24	0.588	0.618	0.011	53.77*
Relative definition	0.01	0.108	1.413	0.049	28.70*
Solution execution					
Procedural knowledge	1.85	1.979	0.429	0.001	65.98*
Number of procedures	1.09	0.953	−0.869	0.015	−56.83*
Computations	1.98	2.666	−0.066	0.003	−21.92*
Decision processing	0.30	0.462	−0.682	0.027	−25.26*
α			0.762	0.006	129.48*
Cognitive component model					
Problem representation	0.87	0.336	−1.8328	0.024	−76.86*
Problem execution	0.55	0.497	−2.1325	0.033	−63.93*

*p<.01

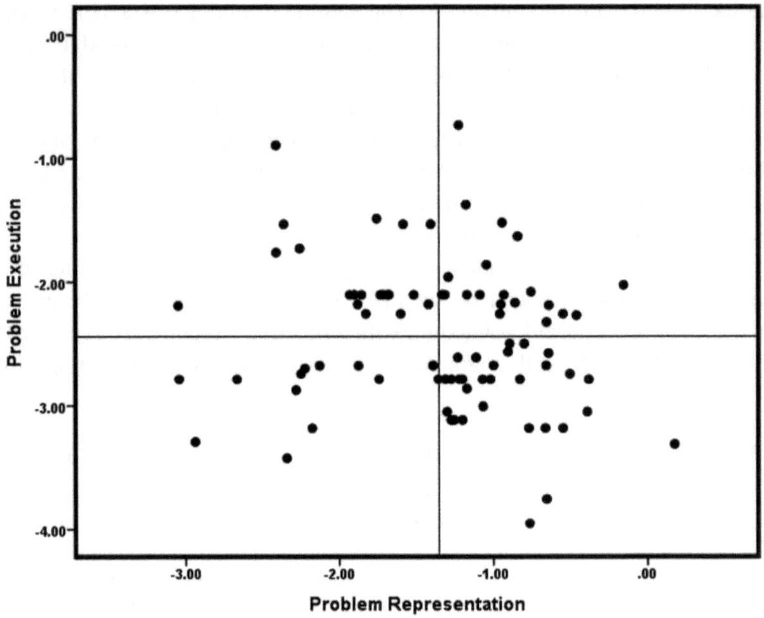

Fig. 1 Scatterplot of processing component difficulty in items

importance of each processing substages. Some weights are negative. For example, given equation: words and given equation: symbols were associated with less difficult items, as expected.

Figure 1 presents a scatterplot of the predicted item difficulty from the cognitive complexity model for Problem Execution and Problem Representation. That is, the weights on Table 3 were applied to the scored variables for items to predict the cognitive complexity within each global. On Fig. 1, the mean predicted value for each global stage is shown by the lines within the plot. It can be seen that item difficulties are widely scattered and that items with different primary sources of difficulty are shown. That is, items with difficulty primarily dependent on Problem Execution or Problem Representation can be identified.

3.2.3 Individual Differences in Cognitive Processing

Similar to the LLTM analysis, parameters for three variants of MLTM-D were estimated by MML. All models contain two trait dimensions (Problem Representation and Problem Execution); hence, the trait distribution specified as $N(\mathbf{0},\Sigma)$ where Σ is a covariance matrix of theta with 1's on the diagonal. All MLTM-D models were 1PL variants with constant item discriminations and the normal metric.

As in the LLTM analysis, the three models include varying parameters for item difficulty, including a null model, a cognitive complexity model, and a

Table 4 Descriptive statistics for processing component competencies

Component	Mean	Standard deviation	Mean standard error	Empirical reliability
Problem representation	0.0464	1.107	0.36	0.901
Problem execution	0.5660	1.003	0.50	0.800

saturated model. That is, the full MLTM-D was estimated the cognitive model using the scored cognitive complexity variables as predictors of item difficulty within components. The cognitive model (−2lnL = 341,543, AIC = 341,577, #parameters = 17) yielded moderate prediction of item difficulty within components as indicated by the likelihood ratio statistic ($\Delta = 0.487$), which was lower than LLTM.

The constant item difficulty estimates for the null model are shown on Table 3. It can be seen that the item difficulty for the Problem Representation component is somewhat higher than for the Problem Execution component.

The theta estimates for MLTM-D were estimated by expected a posteriori (EAP) method using the item parameters from the saturated model. The Table 4 presents descriptive statistics on the theta estimates for processing component competencies. Problem Execution had a statistically significant ($t = 41.84$, df = 3999, $p < 0.001$) and substantially higher mean than Problem Representation. Empirical reliabilities were relatively strong (>0.80) for both components but Problem Representation somewhat higher. The correlation between the two thetas was 0.701, which is typical for cognitive abilities.

Multivariate analyses of variance using Wilks' lambda were conducted on the relationship of processing component competencies to other variables. Overall proficiency on the test, with examinees placed in one of five categories, was related to processing component individual differences. Significant effects were observed for component type ($F_{1,3995} = 1683.19$, $p < 0.001$, $\eta^2 = 0.296$) and for the interaction of component type with proficiency category ($F_{4,3995} = 125.96$, $p < 0.001$, $\eta^2 = 0.112$). Figure 2 shows that Problem Execution is generally higher than Problem Representation but that the differences decrease with increasing proficiency.

Special learning status (emotionally disabled, learning disabled, special learning disability, other disability, gifted, and none) was also available for all examinees. The categories differed significantly overall ($F_{5,3994} = 92.90$, $p < 0.001$), with highest scores for the gifted category and lower scores for all disabilities, as expected. The interaction of component type with special learning status was also statistically significant ($F_{5,3994} = 31.27$, $p < 0.001$). For gifted students, somewhat higher scores for Problem Representation than for Problem Execution were observed. For all other categories, Problem Execution was higher.

Background variables were available for 3851 examinees. Main effects for English language status, racial-ethnic background, and gender were as expected. That is, statistically significant overall effects were observed for English language status ($F_{1,3849} = 81.54$, $p < 0.001$), with higher overall scores for native English speakers than for English language learners. Similarly, a statistically significant

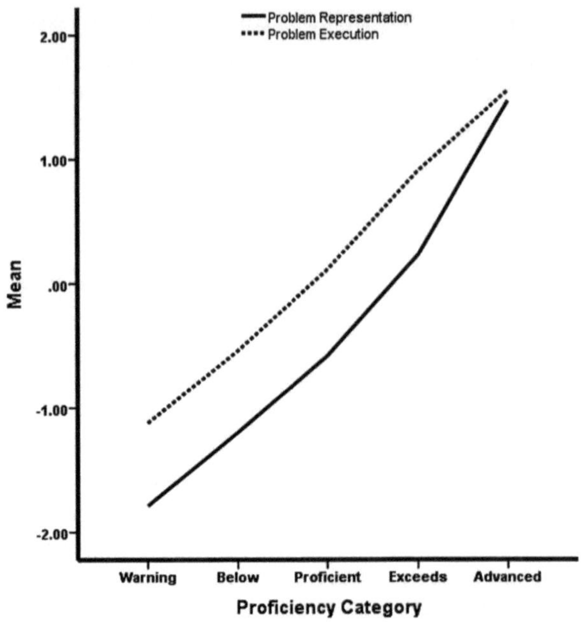

Fig. 2 Cognitive component competencies by overall proficiency categories

effect was also observed for racial ethnic background ($F_{4,3846} = 64.22$, $p < 0.001$), with Asian and Caucasian students scoring higher than African-American and Hispanic students, as expected. For gender, the overall scores did not differ significantly ($F_{1,3849} = 0.003$, $p = 0.954$).

More pertinent to the goals of the study were the interactions of component type with the background variables. Several statistically significant interactions of component type with background variables were observed. For English language status, the interaction was statistically significant ($F_{1,3849} = 11.42$, $p < 0.001$), with Problem Execution higher than Problem Representation for both groups. However, the difference was greater for English language learners. For racial ethnic background, the interaction was also statistically significant ($F_{4,3846} = 10.50$, $p < 0.001$), with Problem Execution higher than Problem Representation for all groups, but the difference was greater for African-American students. Finally, for gender, a statistically significant crossover interaction was observed ($F_{1,3849} = 29.65$, $p < 0.001$). Figure 3 shows that h female students score relatively higher on Problem Execution than male students, whereas male students are relatively higher on Problem Representation.

3.3 Discussion

Several aspects of validity for a test used to assess mathematical achievement in middle school were examined in this study. The *response processes* aspect was examined by modeling item difficulty from a multistage theory of cognitive

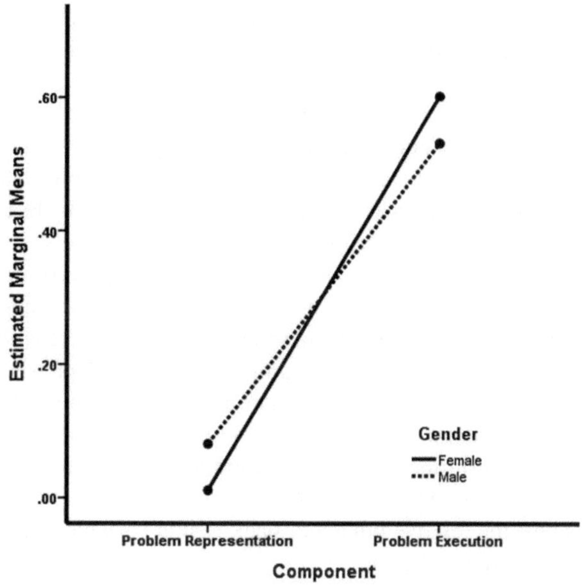

Fig. 3 Gender differences in the processing components

processing in solving mathematical items. The stimulus features of items that were postulated to impact processing in the substages in the Mayer (2003) model were statistically significant in predicting item difficulty. That is, an LLTM analysis of item responses found moderately strong overall prediction and statistically significant weights for variables from each of the substages. Thus, the multistage theory of cognitive processing in solving mathematical items was supported.

The relative impact of the global stages of processing, Problem Representation and Problem Execution, was also examined. Many items were primarily difficult in one stage. These findings imply that items can be selected or designed to emphasize primarily one aspect of cognitive processing.

A primary goal of the study was to examining the impact of individual differences in processing on performance. Estimates from the multivariate conjunctive MLTM-D (Embretson and Yang 2013) were obtained for the two global stages of processing, Problem Representation and Problem Execution. Adequate empirical reliability was found for the estimates of trait levels on each dimension. The results also indicated that the mean was substantially higher on Problem Execution than on Problem Representation. Interestingly, the mean difference between the trait levels for the processing stages was related to overall proficiency levels. That is, Problem Execution competency was higher than Problem Representation for all proficiency levels except at the highest level of proficiency, where the differences were minimal. Future research is needed to determine if the differences in the highest category are a ceiling effect or true differences in trait levels.

Finally, the relationship of processing competencies in Problem Representation and Problem Execution to student background variables was examined. These variables included special education status, English language status, race/ethnicity, and gender. The multivariate analyses found expected overall effects for the background variables on the two processing component competencies, with significant differences for all background variables except gender. However, interactions with component processing type were also found, which have potential impact on the potential consequential aspect of validity. For example, the relatively higher trait levels on Problem Execution than on Problem Representation was greater for English language learners and African-American students. Because Problem Representation involves the Translation substage, reduced levels of cognitive complexity due to language may boost performance of these groups. Also, the crossover interaction of gender is small but interesting. That is, female students are relatively higher on Problem Execution than male students, whereas the reverse effect is found for Problem Representation. Again, the relative emphasis of the skills involved in these two stages in a particular test could lead to gender differences.

In summary, individual differences in cognitive processing capabilities have potential implications for performance levels and test consequences. Items can be designed to emphasize either of the two processes that were examined in this study. Thus, the relative impact of background variables on item responses will depend on item and test design.

Acknowledgment The research in this report was partially supported by a Goal 5 (measurement grant from the *Institute of Educational Science* to Susan Embretson (Georgia Institute of Technology)).

References

American Education Research Association, National Council on Measurement in Education, American Psychological Association, *Standards for Educational and Psychological Tests* (American Educational Research Association, Washington, DC, 1999, 2014)

R.C. Daniel, S.E. Embretson, Designing cognitive complexity in mathematical problem solving items. Appl. Psychol. Meas. **34**(5), 348–364 (2010)

S.E. Embretson, Construct validity: a universal validity system or just another test evaluation procedure? Educ. Res. **36**(8), 449–455 (2007)

S.E. Embretson, in Multicomponent latent trait models, ed. by W. van der Linden, R. Hambleton. Handbook of Modern Item Response Theory, (Springer-Verlag, New York, NY, 1997), pp. 305–322

S.E. Embretson, in The impact of cognitively based item and test development on validity and reliability, ed. by A. Rupp, J. Leighton. The Handbook of Cognition and Assessment, (Wiley-Blackwell, New York, NY, 2016)

S.E. Embretson, X. Yang, A multicomponent latent trait model for diagnosis. Psychometrika **78**, 14–36 (2013)

S.E. Embretson, R.C. Daniel, Understanding and quantifying cognitive complexity level in mathematical problem solving items. Psychol. Sci. **50**, 328–344 (2008)

G.H. Fischer, The linear logistic test model as an instrument in educational research. Acta Psychol. **37**, 359–374 (1973)

C. Jacob, A coefficient of agreement for nominal scales. Educ. Psychol. Meas. **20**(1), 37–46 (1960)

J.P. Leighton, M.J. Gierl, *Cognitive Diagnostic Assessment for Education: Theory and Applications* (Cambridge University Press, New York, NY, 2007)

R. Mayer, Mathematical problem solving, in *Mathematical Cognition: A Volume in Current Perspectives on Cognition, Learning, and Instruction* (Information Age Publishing, Inc., Charlotte, NC, 2003), pp. 69–92.

R.E. Mayer, J. Larkin, J.B. Kadane, in *Advances in the Psychology of Human Intelligence*, ed. by R. Sternberg. A cognitive analysis of mathematical problem solving ability, vol V2 (Lawrence Erlbaum Associates, Hillsdale, NJ, 1984), pp. 741–749

National Assessment Governing Board (NAGB), *Mathematics Framework for the 2015 National Assessment of Educational Progress* (U.S. Department of Education, Washington, DC, 2015)

Parameter Constraints of the Logit Form of the Reduced RUM

Hans-Friedrich Köhn

Abstract The Reduced Reparameterized Unified Model (Reduced RUM) has received considerable attention among educational researchers. Markov chain Monte Carlo (MCMC) or Expectation Maximization (EM) is typically used for estimating the Reduced RUM. Implementations of the EM algorithm are available in the latent class analysis (LCA) routines of commercial software packages (e.g., `Latent GOLD`, `Mplus`). Using a commercial LCA routine as a vehicle for fitting the Reduced RUM with the EM algorithm requires that it be reparameterized as a logit model, with complex constraints imposed on the parameters. This article summarizes the general parameterization of the Reduced RUM as a logit model and the associated parameter constraints.

Keywords Cognitive diagnosis • Reduced RUM • EM algorithm

1 Introduction

In the past decade, cognitive diagnosis (CD) has emerged as a new paradigm of educational measurement that seeks to combine rigorous psychometric standards with the goals of formative assessment (DiBello et al. 2007; Haberman and von Davier 2007; Leighton and Gierl 2007; Rupp et al. 2010). Cognitively diagnostic tests target mastery of the instructional content and provide immediate feedback on students' strengths and weaknesses in a knowledge domain in terms of skills learned and skills needing study. The Reduced Reparameterized Unified Model (Reduced RUM Hartz 2002; Hartz and Roussos 2008) is one of the CD models—or "Diagnostic Classification Models" (DCMs), as they are called here—that has received considerable attention among educational researchers (e.g., Feng et al. 2014; Henson and Douglas 2005; Henson and Templin 2007; Henson et al. 2008, 2007; Kim 2011; Liu et al. 2009; Templin et al. 2008). Compared with a simple conjunctive DCM like the DINA model (Junker and Sijtsma 2001; Macready and Dayton 1977), the Reduced RUM offers greater flexibility in modeling the

H.-F. Köhn (✉)
University of Illinois at Urbana-Champaign, Champaign, IL 61820, USA
e-mail: hkoehn@illinois.edu

L.A. van der Ark et al. (eds.), *Quantitative Psychology*, Springer Proceedings in Mathematics & Statistics 196, DOI 10.1007/978-3-319-56294-0_19

probability of correct item responses for different skill profiles. Concretely, the DINA model cannot distinguish between examinees who master none and those who master a subset of the skills required for an item. Only if all required skills are mastered can an examinee realize a high probability of answering the item correctly. This restriction has been relaxed in case of the Reduced RUM, as it allows for incremental probabilities of a correct response along with an increasing number of required skills mastered.

However, this flexibility of the Reduced RUM comes at the cost of a significant increase in the complexity of the model estimation process. In fact, with the exception of Feng et al. (2014), the studies referenced above all use Markov chain Monte Carlo (MCMC) techniques for fitting the model. But MCMC requires advanced technical skills so that its usefulness is likely restricted to researchers with a solid background in statistics. Alternatively, marginal maximum likelihood estimation relying on the EM algorithm (MMLE-EM) can be used for fitting the Reduced RUM. Commercial packages like `Latent GOLD` [Vermunt and Magidson 2005 and `Mplus` (Muthén and Muthén 1998–2015)] provide implementations of the EM algorithm for fitting (constrained) latent class models. Using a latent class analysis routine as a vehicle for MMLE-EM, however, requires that the Reduced RUM be reexpressed as a logit model with rather complex constraints imposed on the parameters of the logistic function (Chiu and Köhn 2016; Henson et al. 2009; Rupp et al. 2010). This article provides a summary of the constraints on the model parameters if the Reduced RUM is expressed as a logit model.

2 The Reduced RUM as a Logit Model

Within the CD framework, ability is perceived as a composite of "attributes," a collective term for knowledge, aptitude, and specific skills—any cognitive characteristic required to perform tasks—that an examinee may or may not possess. Attributes are denoted by α_k, $k = 1, 2, \ldots, K$; distinct profiles of attributes, $\boldsymbol{\alpha} = (\alpha_1, \alpha_2, \ldots, \alpha_K)'$, define classes of proficiency to which examinees are to be assigned based on their test performance.

The item response function (IRF) of the Reduced RUM is

$$P(Y_{ij} = 1 \mid \boldsymbol{\alpha}_i) = \pi_j^* \prod_{k=1}^{K} r_{jk}^{*\ q_{jk}(1-\alpha_{ik})}$$

with Y_{ij} denoting the response of examinee i, $i = 1, 2, \ldots, n$, to item j, $j = 1, 2, \ldots, J$, $\alpha_{ik} = 1$ if examinee i possesses attribute k, 0 otherwise; $q_{jk} = 1$ if attribute k is required for item j, 0 otherwise. In this "traditional" parameterization of the Reduced RUM, $0 < r_{jk}^* < 1$ is a penalty parameter for lacking attribute k required for item j, and $0 < \pi_j^* < 1$ is the probability of a correct response if an examinee has mastered all the attributes required for item j because then $\prod_{k=1}^{K} r_{jk}^{*\ q_{jk}(1-\alpha_{ik})} = 1$.

For two attributes, α_1 and α_2, the IRF of the Reduced RUM as a logit model can be found in Henson et al. (2009) (in omitting the examinee index i for succinctness):

$$P(Y_j = 1 \mid \boldsymbol{\alpha}) = \frac{e^{\beta_{j0}+\beta_{j1}q_{j1}\alpha_1+\beta_{j2}q_{j2}\alpha_2+\beta_{j12}q_{j1}q_{j2}\alpha_1\alpha_2}}{1+e^{\beta_{j0}+\beta_{j1}q_{j1}\alpha_1+\beta_{j2}q_{j2}\alpha_2+\beta_{j12}q_{j1}q_{j2}\alpha_1\alpha_2}} \tag{1}$$

where $q_{jk} = 0, 1$ indicates whether attribute α_k is required for item j. Verbally stated, the probability of a correct response to item j is modeled as a linear combination of the attribute main effects and their interaction $\alpha_1\alpha_2$. (Including the interaction term allows for modeling a possibly nonadditive effect of the two attributes on the probability of a correct item response.)

2.1 Parameter Constraints

The IRF given in Eq. (1) is subject to these constraints:

$$\beta_{jk} > 0 \quad k = 1, 2 \qquad \text{(to ensure monotonicity)}$$

$$\beta_{j12} = \ln\left(\frac{1+e^{\beta_{j0}}}{1+e^{\beta_{j0}+\beta_{j1}}+e^{\beta_{j0}+\beta_{j2}}-e^{\beta_{j0}+\beta_{j1}+\beta_{j2}}}\right)$$

The first constraint on the β_{jk} is mathematically not required, because $0 < \pi_j^*, r_{jk}^* < 1$ but $\beta_{jk} > 0$ is necessary to guarantee monotonicity. (Monotonicity means that the probability of a correct response for an examinee who masters certain attributes must be equal to or higher than the probability of a correct response if these attributes are not mastered.) Second, the coefficient of the interaction term, β_{j12}, is constrained to be a function of the main-effect coefficients; hence, the traditional and the logit parameterization of the Reduced RUM have the same number of parameters. Third, as $1 + e^{\beta_{j0}} > 0$, the denominator of

$$\beta_{j12} = \ln\left(\frac{1+e^{\beta_{j0}}}{1+e^{\beta_{j0}+\beta_{j1}}+e^{\beta_{j0}+\beta_{j2}}-e^{\beta_{j0}+\beta_{j1}+\beta_{j2}}}\right)$$

must be strictly positive:

$$1+e^{\beta_{j0}+\beta_{j1}}+e^{\beta_{j0}+\beta_{j2}}-e^{\beta_{j0}+\beta_{j1}+\beta_{j2}} > 0$$

Otherwise, $\ln(\cdot)$ is not defined. [This constraint was not explicitly listed in Henson et al. (2009).]

The (strict) inequality format of the constraint

$$1+e^{\beta_{j0}+\beta_{j1}}+e^{\beta_{j0}+\beta_{j2}}-e^{\beta_{j0}+\beta_{j1}+\beta_{j2}} > 0$$

as a function of one or several model parameters is not supported, for example, by Mplus. But rephrasing the constraint as an upper bound (UB) on one of the model parameters is supported:

$$\beta_{j2} < \ln\left(1 + e^{\beta_{j0}+\beta_{j1}}\right) - \ln\left(e^{\beta_{j1}} - 1\right) - \beta_{j0}$$

The convention adopted here is to rephrase the inequality constraint as a UB on the last coefficient with the largest index $k = K$.

3 The Reduced RUM as a Logit Model: The General Case

The reparameterization of the Reduced RUM as a logit model for $K > 2$ attributes was derived in Chiu and Köhn (2016).

3.1 *The Case of K = 3 Attributes*

Consider an item having the vector of required attributes $\boldsymbol{q} = (111)$. Then, as was proven in Chiu and Köhn (2016), the logit form of the IRF of the Reduced RUM must contain the three main effects, the three two-way interactions, and the three-way interaction:

$$P(Y_j = 1 \mid \boldsymbol{\alpha}) = \frac{e^{\beta_{j0}+\sum_{k=1}^{3}\beta_{jk}\alpha_k+\sum_{k'=k+1}^{3}\sum_{k=1}^{2}\beta_{jkk'}\alpha_k\alpha_{k'}+\beta_{j123}\alpha_1\alpha_2\alpha_3}}{1 + e^{\beta_{j0}+\sum_{k=1}^{3}\beta_{jk}\alpha_k+\sum_{k'=k+1}^{3}\sum_{k=1}^{2}\beta_{jkk'}\alpha_k\alpha_{k'}+\beta_{j123}\alpha_1\alpha_2\alpha_3}}$$

where the coefficients of the main effects must be strictly positive, $\beta_{jk} > 0$. Like in the case of $K = 2$ attributes, the coefficients of the interaction terms are constrained to be functions of the related main-effect coefficients. Hence, for all k and k', the coefficients of the two-way interactions, $\beta_{jkk'}$, must be functions of β_{jk} and $\beta_{jk'}$:

$$\beta_{jkk'} = \ln\left(\frac{(1 + e^{\beta_{j0}})^{2-1}}{(1 + e^{\beta_{j0}+\beta_{jk}})(1 + e^{\beta_{j0}+\beta_{jk'}}) - (1 + e^{\beta_{j0}})^{2-1}e^{\beta_{j0}+\beta_{jk}+\beta_{jk'}}}\right)$$

In the same manner, the coefficient of the three-way interaction, β_{j123}, must equal

$$\beta_{j123} = \ln\left(\frac{(1 + e^{\beta_{j0}})^{3-1}}{(1 + e^{\beta_{j0}+\beta_{j1}})(1 + e^{\beta_{j0}+\beta_{j2}})(1 + e^{\beta_{j0}+\beta_{j3}}) - (1 + e^{\beta_{j0}})^{3-1}e^{\beta_{j0}+\beta_{j1}+\beta_{j2}+\beta_{j3}}}\right)$$

$$-\beta_{j12} - \beta_{j13} - \beta_{j23}$$

Thus, besides the K constraints $\beta_{jk} > 0$, there are $\binom{3}{2} + \binom{3}{3} = 4$ additional constraints if $K = 3$.

Like for the model with $K = 2$, the interaction coefficients are mathematically legitimate only if the argument of the log function is strictly positive. Hence, the denominators must be strictly positive. Therefore, for all three two-way interactions involving k and k',

$$(1 + e^{\beta_{j0}+\beta_{jk}})(1 + e^{\beta_{j0}+\beta_{jk'}}) - (1 + e^{\beta_{j0}})e^{\beta_{j0}+\beta_{jk}+\beta_{jk'}} > 0$$

must hold. However, recall that this specific format of (strict) inequality constraints is not supported, for example, by `Mplus`. Thus, the inequality constraint on the denominator is rephrased as a UB on the coefficient with index $k' > k$:

$$\beta_{jk'} < \ln\left(1 + e^{\beta_{j0}+\beta_{jk}}\right) - \ln\left(e^{\beta_{jk}} - 1\right) - \beta_{j0}$$

The constraint on the denominator of the log expression of β_{j123} is

$$(1 + e^{\beta_{j0}+\beta_{j1}})(1 + e^{\beta_{j0}+\beta_{j2}})(1 + e^{\beta_{j0}+\beta_{j3}}) - (1 + e^{\beta_{j0}})^2 e^{\beta_{j0}+\beta_{j1}+\beta_{j2}+\beta_{j3}} > 0$$

which must also be rephrased as a UB on one of the main-effect coefficients. In following the earlier convention, this (strict) inequality constraint is rephrased as a UB on the last main-effect coefficient—that is, with index $k = K$: β_{j3}:

$$\begin{aligned}\beta_{j3} < \ln\left((1 + e^{\beta_{j0}+\beta_{j1}})(1 + e^{\beta_{j0}+\beta_{j2}})\right) - \ln\big(e^{\beta_{j1}+\beta_{j2}}(1 + e^{\beta_{j0}})^2 \\ -(1 + e^{\beta_{j0}+\beta_{j1}})(1 + e^{\beta_{j0}+\beta_{j2}})\big) - \beta_{j0}\end{aligned} \tag{2}$$

However, there are two additional UBs on β_{j3} that can be derived from β_{j13} and β_{j23}:

$$\beta_{j3} < \ln(1 + e^{\beta_{j0}+\beta_{j1}}) - \ln(e^{\beta_{j1}} - 1) - \beta_{j0} \qquad \text{from } \beta_{j13}$$
$$\beta_{j3} < \ln(1 + e^{\beta_{j0}+\beta_{j2}}) - \ln(e^{\beta_{j2}} - 1) - \beta_{j0} \qquad \text{from } \beta_{j23}$$

Are all three UBs on β_{j3} needed? And if not, then which one(s) should be used?

Recall that the UB on β_{j3} was derived from

$$(1 + e^{\beta_{j0}+\beta_{j1}})(1 + e^{\beta_{j0}+\beta_{j2}})(1 + e^{\beta_{j0}+\beta_{j3}}) - (1 + e^{\beta_{j0}})^2 e^{\beta_{j0}+\beta_{j1}+\beta_{j2}+\beta_{j3}} > 0$$

Because this expression can also be manipulated into a UB for β_{j1} or β_{j2}, the form of a UB on β_{j3} implies UBs on β_{j1} and β_{j2}. Chiu and Köhn (2016) showed that the UB on β_{j3} in Eq. (2) that was derived from the highest-order interaction term is the least UB and, therefore, is the only one required and to be used.

3.2 The Case of $K > 3$ Attributes

The complexity of the constraint structure of the logit form of the Reduced RUM increases with the number of attributes K. The following guidelines can be formulated:

1. All main-effect coefficients β_{jk}, $k = 1, 2, \ldots, K$, must be strictly positive.
2. The coefficients of the interaction terms are constrained to be functions of the main-effect coefficients:

$$\beta_{j1\ldots K'} = \ln\left(\frac{\left(1+e^{\beta_{j0}}\right)^{K'-1}}{\prod_{k=1}^{K'}\left(1+e^{\beta_{j0}+\beta_{jk}}\right)-\left(1+e^{\beta_{j0}}\right)^{K'-1}e^{\beta_{j0}+\sum_{k=1}^{K}\beta_{jk}}}\right)-\sum_{k'=k+1}^{K'}\sum_{k=1}^{K'-1}\beta_{jkk'}-\cdots$$

$$-\sum_{k_{K'-1}=k_{K'-2}+1}^{K'}\cdots\sum_{k_2=k_1+1}^{3}\sum_{k_1=1}^{2}\beta_{jk_1\ldots k_{K'-1}} \qquad 1 < K' \leq K$$

3. Some software packages require that the inequality constraints on the denominators of the log expressions are defined as a UB on one of the model parameters. By convention, the coefficient of the last main effect, the one with index $k = K$, is chosen.
4. The UB should be derived from the highest-order interaction coefficient because it is the least UB:

$$\beta_{jK} < \sum_{k=1}^{K-1}\ln\left(1+e^{\beta_{j0}+\beta_{jk}}\right)-\ln\left(\prod_{k=1}^{K-1}\left(e^{\beta_{jk}}+e^{\beta_{j0}+\beta_{jk}}\right)-\prod_{k=1}^{K-1}\left(1+e^{\beta_{j0}+\beta_{jk}}\right)\right)-\beta_{j0}$$

4 Conclusion

A summary of the general parameterization of the Reduced RUM as a logit model and the associated parameter constraints was presented. This allows a potential user to fit the Reduced RUM with MMLE-EM relying on the implementation of the EM algorithm in the LCA routines of commercial software packages like `Latent GOLD` and `Mplus`. Chiu et al. (2016) provide a tutorial on how to use `Mplus` for fitting the Reduced RUM as a logit model to educational data. This tutorial also includes a detailed description of how to retrieve the estimates of the traditional parameters of the Reduced RUM from the logit model-based estimates. The traditional parameterization of the Reduced RUM has two immediate practical advantages. First, the parameters, π_j^* and r_{jk}^*, are bounded by 0 and 1. They are well defined and have a direct and meaningful interpretation as probabilities (as opposed to the parameters of the Reduced RUM as a logit model that are on a logit scale). Second, the clear and intuitive meaning of the traditional parameters of the Reduced RUM allows for the immediate detection of low-quality items.

References

C.-Y. Chiu, H.-F. Köhn, The reduced RUM as a logit model: parameterization and constraints. Psychometrika **81**, 350–370 (2016)

C.-Y. Chiu, H.-F. Köhn, H.M. Wu, Fitting the reduced RUM using Mplus: a tutorial. Int. J. Test. **16**, 331–351 (2016)

L.V. DiBello, L.A. Roussos, W.F. Stout, Review of cognitively diagnostic assessment and a summary of psychometric models, in *Handbook of Statistics*, ed. by C.R. Rao, S. Sinharay. Psychometrics, vol. 26. (Elsevier, Amsterdam, 2007), pp. 979–1030

Y. Feng, B.T. Habing, A. Huebner, Parameter estimation of the reduced RUM using the EM algorithm. Appl. Psychol. Meas. **38**, 137–150 (2014)

S.J. Haberman, M. von Davier, Some notes on models for cognitively based skill diagnosis, in *Handbook of Statistics*, ed. by C.R. Rao, S. Sinharay. Psychometrics, vol. 26 (Elsevier, Amsterdam, 2007), pp. 1031–1038

S.M. Hartz, A Bayesian framework for the Unified Model for assessing cognitive abilities: Blending theory with practicality. Doctoral dissertation. Available from ProQuest Dissertations and Theses database. (UMI No. 3044108), 2002

S.M. Hartz, L.A. Roussos, The Fusion Model for skill diagnosis: Blending theory with practicality. (Research report No. RR-08-71), Educational Testing Service, Princeton, NJ, October 2008

R. Henson, J. Douglas, Test construction for cognitive diagnosis. Appl. Psychol. Meas. **29**, 262–277 (2005)

R.A. Henson, J. Templin, Large-scale language assessment using cognitive diagnosis models. Paper presented at the Annual Meeting of the National Council on Measurement in Education, Chicago, IL, April 2007

R. Henson, J.L. Templin, J. Douglas, Using efficient model based sum-scores for conducting skills diagnoses. J. Educ. Meas. **44**, 361–376 (2007)

R.A. Henson, J.L. Templin, J.T. Willse, Defining a family of cognitive diagnosis models using log-linear models with latent variables. Psychometrika **74**, 191–210 (2009)

R. Henson, L.A. Roussos, J. Douglas, X. He, Cognitive diagnostic skill-level discrimination indices. Appl. Psychol. Meas. **32**, 275–288 (2008)

B.W. Junker, K. Sijtsma, Cognitive assessment models with few assumptions, and connections with nonparametric item response theory. Appl. Psychol. Meas. **25**, 258–272 (2001)

Y.-H. Kim, Diagnosing EAP writing ability using the Reduced Reparameterized Unified Model. Lang. Test. **28**, 509–541 (2011)

J. Leighton, M. Gierl, *Cognitive Diagnostic Assessment for Education: Theory and Applications*. (Cambridge University Press, Cambridge, 2007)

Y. Liu, J.A. Douglas, R.A. Henson, Testing person fit in cognitive diagnosis. Appl. Psychol. Meas. **33**, 579–598 (2009)

G.B. Macready, C.M. Dayton, The use of probabilistic models in the assessment of mastery. J. Educ. Stat. **33**, 379–416 (1977)

L.K. Muthén, B.O. Muthén, `Mplus` *User's Guide*, 7th edn. (Muthén & Muthén, Los Angeles, 1998–2015)

A.A. Rupp, J.L. Templin, R.A. Henson, *Diagnostic Measurement. Theory, Methods, and Applications* (Guilford, New York, 2010)

J.L. Templin, R.A. Henson, S.E. Templin, L.A. Roussos, Robustness of hierarchical modeling of skill association in cognitive diagnosis models. Appl. Psychol. Meas. **32**, 559–574 (2008)

J.K. Vermunt, J. Magidson, `Latent GOLD 4.0` *Users's Guide* (Statistical Innovations Inc., Belmont, 2005)

Hypothesis Testing for Item Consistency Index in Cognitive Diagnosis

Lihong Song and Wenyi Wang

Abstract Conjunctive and disjunctive condensation rules are cognitive assumptions about how attributes interact with each other on specific items. The existing item consistency index (ICI) in cognitive diagnostic assessment is developed under the conjunctive condensation rule. This study introduced two item consistency indices combining hypothesis testing (SICI and MICI) for the conjunctive and disjunctive rules to help identify the underlying condensation rules and to assist in screening items with attribute misspecification. A simulation study was conducted to assess the performance of the SICI, MICI, and ICI. Results showed that (a) given a Q-matrix, the ICI and SICI precisely identified the correct condensation rule and (b) the item consistency indices, especially the SICI, were successfully applied in evaluating the misspecification of a Q-matrix as well as identifying items with attribute misspecification. The promising results indicate that the item consistency indices can also help provide more information for practitioners in model determination.

Keywords Hypothesis testing • Condensation rules • Item consistency index • Cognitive diagnosis

1 Introduction

Cognitive diagnostic assessment (CDA) is a new paradigm for educational and psychological testing aiming to provide more specific and individualized feedback on attribute mastery for later on instruction and learning. The latent attributes in the domain of interest were usually specified by a cognitive model of task

L. Song
Elementary Education College, Jiangxi Normal University, 99 Ziyang Road, Nanchang, Jiangxi, People's Republic of China
e-mail: viviansong1981@163.com

W. Wang (✉)
College of Computer Information Engineering, Jiangxi Normal University, 99 Ziyang Road, Nanchang, Jiangxi, People's Republic of China
e-mail: wenyiwang2009@gmail.com

L.A. van der Ark et al. (eds.), *Quantitative Psychology*, Springer Proceedings in Mathematics & Statistics 196, DOI 10.1007/978-3-319-56294-0_20

performance (Leighton et al. 2004). For clearly describing the relationship between the latent attributes and response variables, two interesting condensation rules, the conjunctive and disjunctive condensation rules, were formally defined for dichotomous item response variables (Maris 1995, 1999). In a popular diagnostic model, the deterministic inputs, noisy "AND" gate (DINA) model (Haertel 1989; Junker and Sijtsma 2001), latent response variables are defined as conjunctive. Another diagnostic model, the deterministic inputs, noisy "OR" gate (DINO) model (Templin and Henson 2006), follows the disjunctive rule.

The choice of condensation rules clearly depends on the diagnostic setting, including the purpose of the assessment, and how the skills or attributes are defined. The conjunctive rule holds that a respondent has to master all the attributes required by an item to give a correct answer, which means that lacking one attribute cannot be compensated by a preponderance of another attribute. This rule frequently appears in educational testing (Chen et al. 2013; Maris 1999), such as the fraction subtraction test (Tatsuoka 1990). However, the disjunctive model assumes that having satisfied either of the required attributes can lead to a positive response. This hypothesis is well suited to applications of diagnosis in psychological disorders, such as diagnosis of pathological gambling. Individuals are diagnosed as being pathological if a set of dichotomous criteria or latent attributes has been satisfied (Templin and Henson 2006).

Cognitive diagnostic models should be carefully chosen for different testing situations by examining its appropriateness both theoretically and empirically. Understanding of the type of condensation rules within tasks is important because it will help determine which psychometric model is most appropriate and interpretable for the intended diagnostic assessment. From an empirical perspective, indices that meet the need of evaluating a model-data fit can aid the selection of a model. The hierarchy consistency index (HCI) (Cui 2007; Cui and Leighton 2009) has already been successfully used to evaluate fit. However, the HCI is an index designed only for the conjunctive rule (Crawford 2014). Therefore, indices that can be used for the disjunctive rule need further exploration (Cui and Roberts 2013).

The HCI, which is designed for the conjunctive rule, cannot directly generalize to the disjunctive rule. It is because, on the one side, essential differences exist between the underlying assumptions of the two condensation rules. On the other side, the HCI itself has deficiencies. When student i correctly answered an item, say item j, that is, $X_{ij} = 1$, the HCI only assumes that student i has mastered all the attributes required by item j, and the incorrect responses he/she made to items that measure the subset of attributes of item j are misfits. However, when student i answered item j incorrectly, that is, $X_{ij} = 0$, there also exist misfits when student i have correctly answered items including the set of attributes measured by item j. Hence, some researchers proposed the new HCI (NHCI) (Ding et al. 2012; Mao 2011) and item consistency index (ICI) (Lai et al. 2012) by considering both aspects. In fact, because the behavior of item response is sometimes stochastic with slip, guessing, and even cheating, it is still hard to infer that he/she really mastered the required attributes when a correct answer on a specific item was given by an examinee. Instead, it will be more reliable to make use of information from other associated items based on statistical inferences.

The purposes of this study were (a) to introduce two modified item consistency indices combining hypothesis testing for the conjunctive and disjunctive rules, respectively, and (b) to conduct a simulation study to assess the performance of the item consistency indices by comparison. The paper is organized as follows: the modified ICI combining hypothesis testing is introduced for the conjunctive and disjunctive rules, respectively, in Sect. 2 after a brief review to the HCI and NHCI; a simulation study designed to assess the performance of the item consistency indices is described, and its results are shown in Sect. 3; and the final section gives a summary on this study.

2 Methods

2.1 A Review to the HCI and NHCI

2.1.1 The Hierarchy Consistency Index

The HCI for the conjunctive model is designed to detect the number of misfits between students' item response vectors and the expected response associated with a Q-matrix (Cui 2007; Cui and Leighton 2009). Cui and Leighton (2009) defined the rate of misfit for student i as follows:

$$r_i = \sum_{j \in C_i} \sum_{g \in S_j} X_{ij} \left(1 - X_{ig}\right) / N_i, \tag{1}$$

where N_i is the total number of comparisons for all the items that are correctly answered by student i; C_i is an index set that includes items correctly answered by student i; S_j is an index set that includes items requiring the subset of attributes measured by item j; X_{ij} is student i's score (1 or 0) to item j, where item j belongs to C_i; and X_{ig} is student i's score (1 or 0) to item g, where item g belongs to S_j.

The numerator of Formula (1) represents the number of misfit between student i's item responses and expected responses, and the denominator is the member of all possible comparisons of the items correctly answered by student i. Then, Cui and Leighton (2009) gave the following HCI for student i by converting misfits to fits and restricted the value of the HCI to the range of [−1, 1]:

$$HCI_i = 1 - 2r_i. \tag{2}$$

Here, the assumption behind the HCI is that when student i gives correct responses to item j, it is possible to infer that he/she has mastered all the attributes measured by item j and he/she will answer correctly on the items requiring the subset of attributes measured by item j. A test-level HCI can be calculated by averaging the HCI of all students. A model-data fit is regarded as good when the value of the HCI is larger than 0.6 and as excellent when it reaches 0.8 (Cui 2007). The HCI has been widely applied to model determination, evaluating and verifying hierarchical attributes structures (Gierl et al. 2008; Wang and Gierl 2011). Researches on exploring approaches that synthesize verbal reports and the HCI to validate student score inferences also have been reported (Cui and Roberts 2013).

2.1.2 The NHCI

Some researchers stated that the HCI has weaknesses (Ding et al. 2012). Ding et al. (2012) pointed out that misfits also exist when students give incorrect answers on item j but correct answers on other items that include the set of attributes measured by item j. Therefore, they present the NHCI based on the HCI as follows (Mao 2011):

$$NHCI_i = 1 - \left[\sum_{j \in C_i} \sum_{g \in S_j} x_{ij} \left(1 - x_{ig}\right) / N_i + \sum_{j \in C_j^*} \sum_{h \in S_j^*} x_{ih} \left(1 - x_{ij}\right) / N_i^* \right], \quad (3)$$

where C_i^* is an index set that includes items answered incorrectly by student i, including item j; S_j^* is an index set that includes items requiring the set of attributes measured by item j, including item j; N_i^* is the total number of comparisons for all the items that are wrongly answered by student i; and other notations are the same as above. The NHCI was applied to analyze a Chinese vocabulary test of colors taken by non-Chinese speaking oversea students (Liu and Bian 2014).

2.1.3 The Item Consistency Index (ICI)

The development of ICI was inspired by the idea that an item can have an item-fit index just like a test taker can have a person-fit statistic. Enlightened by the HCI, researches introduced the following item-fit index to evaluate item-model fit for cognitive diagnostic assessment (Lai et al. 2012):

$$ICI_j = 1\text{-}2 \sum_i \left[\sum_{g \in S_j} x_{ij} \left(1 - x_{ig}\right) + \sum_{h \in S_j^*} x_{ih} \left(1 - x_{ij}\right) \right] / M_j, \quad (4)$$

where M_j is the number of comparisons for item across all students and other notations have the same meaning in the formulas mentioned above.

The HCI, NHCI, and ICI are used to detect aberrant examinees or items. Both the NHCI and ICI include misfits derived from inconsistent responses on item j and S_j as well as misfits resulted from inconsistent response on item j and S_j^*. A small difference exists between the two that the former computes a total number of misfits, while the latter calculates the ratio of misfits to all possible comparisons of the items.

2.2 *ICI Combining Hypothesis Testing*

Because randomness exists in the process of item response, it is not sound to make decisions solely based on the limited information provided by one item that a student had really mastered the attributes required by the item when he/she gives a correct response to the item. Actually, there is a certain probability for a test taker to show a correct answer when he/she has not really mastered the attributes required, and vice versa. Because statistical inferences are generally more precise than everyday inferences, it will be more reasonable to make statistical inferences firstly on whether the examinee answered each question by his/her own knowledge or just by chance. Then we can continue to make more reliable classification on the student's attribute mastery from his/her response behaviors. We, therefore, introduce a revision of item consistency index (SICI and MICI) combining hypothesis testing to evaluate item-data fit for the two different condensation rules.

2.2.1 ICI for the Conjunctive Rule

This study introduced a hypothesis testing of proportion p_{ij}, that is, the probability of examinee *i* giving a correct response to item *j*. According to the results of related research (Cui and Leighton 2009; Templin and Henson 2006), the lower bound of the probability p_1 of correct item response for mastery students in this study was set to 0.75, and the upper bound of the probability p_0 of correct item response for nonmastery students was set to 0.25. Inferences then can be made based on the rationale below: If examinee *i* really masters all the attributes required by item *j* and gives a correct response to item *j* ($X_{ij} = 1$), then he will give correct responses to the majority of items that measure the subset of attributes of item *j*. Similarly, if examinee *i* in fact does not master all the attributes required by item *j* and gives a wrong answer on item *j* ($X_{ij} = 0$), then he will give incorrect responses to the majority of items that include the set of attributes measured by item *j*. Based on the two assumptions, hypothesis testing on item *j* can be carried out under these two situations, respectively: a right answer is observed ($X_{ij} = 1$), and a wrong answer is observed ($X_{ij} = 0$).

(1) When $X_{ij} = 1$, one-sided null and alternative hypotheses are of the form:

$$H_0 : p_{ij} \geq p_1 \;\; vs \;\; H_1 : p_{ij} < p_1.$$

If the null hypothesis H_0 cannot be rejected, we can infer that examinee *i* has mastered all the attributes required by item *j*. Let examinee *i*'s score on an arbitrary item *j* be a variable; let ξ denote the total score examinee *i* achieved on the n_j items, S_j, which measure the subset of attributes of item *j*, and ξ is also a random variable. According to local independency assumption, item responses on S_j for examinee *i* can be regarded as n_j independent experiments with same probability p_1. Here, variable ξ is from a binomial distribution with parameters (n_j, p_1). Then a *p*-value can be calculated based on the observed total score ($\xi = t_{ij}$) on the items S_j on the basis of the null hypothesis:

$$p = P\left(\xi \le t_{ij}\right) = \sum_{k=0}^{t_{ij}} C_{n_j}^{k} {p_1}^{k} (1 - p_1)^{n_j - k}. \tag{5}$$

We can make inferences then by comparing the p-value with a chosen significant level α, such as $\alpha = 0.1$. The null hypothesis would not be rejected when $p > \alpha$, indicating there is an item-data misfit, and $\sum_{g \in S_j} x_{ij}\left(1 - x_{ig}\right)$ takes part in the computation of the ICI. Otherwise, the null hypothesis should be rejected and $\sum_{g \in S_j} x_{ij}\left(1 - x_{ig}\right)$ doesn't take part in the computation of the ICI.

(2) When $X_{ij} = 0$, the following hypotheses are to be tested:

$$H_0 : p_{ij} \le p_0 \;\; vs \;\; H_1 : p_{ij} > p_0.$$

A p-value can be computed for the specific test and the observed score ($\xi^* = t_{ij}^*$) on the n_j^* items S_j^*:

$$p = P\left(\xi^* \ge t_{ij}^*\right) = 1 - \sum_{k=0}^{t_{ij}} C_{n_j^*}^{k} {p_0}^{k} (1 - p_0)^{n_j - k}. \tag{6}$$

The null hypothesis would not be rejected when $p > \alpha$, indicating that there is an item-data misfit, and $\sum_{h \in S_j^*} x_{ih}\left(1 - x_{ij}\right)$ takes part in the computation of the ICI. Otherwise, the null hypothesis should be rejected and $\sum_{h \in S_j^*} x_{ih}\left(1 - x_{ij}\right)$ doesn't take part in the computation of the ICI.

2.2.2 ICI for the Disjunctive Rule

The disjunctive condensation rule assumes that examinees that have not mastered any of the required attributes will have high probabilities to give incorrect responses, while examinees that have mastered any one or more of the required attributes will have high probabilities to give correct responses (DiBello et al. 2007). According to this assumption, we can make inferences based on the rationale below: If examinee i really masters the attributes required by item j and gives a correct response to item j ($X_{ij} = 1$), he will give correct responses to the majority of the items that include the set of attributes measured by item j. Similarly, if examinee i actually does not master any of the attributes required by item j and gives an incorrect answer on item j ($X_{ij} = 0$), then he will give incorrect responses to the majority of items that measure the subset of attributes of item j. We here adapt the HCI and ICI to the disjunctive rule, and for reading convenience, we do not change the original abbreviate terms and notations:

$$HCI_i = 1\text{-}2 \sum_{j \in C_i^*} \sum_{g \in S_j} x_{ig}\left(1 - x_{ij}\right) / N_i, \tag{7}$$

$$NHCI_i = 1\text{-}\left[\sum_{j\in C_i^*}\sum_{g\in S_j} x_{ig}\left(1-x_{ij}\right)/N_i + \sum_{j\in C_i}\sum_{h\in S_j^*} x_{ij}\left(1-x_{ih}\right)/N_i^*\right], \quad (8)$$

$$ICI_j = 1\text{-}2\sum_i\left[\sum_{g\in S_j} x_{ig}\left(1-x_{ij}\right) + \sum_{h\in S_j^*} x_{ij}\left(1-x_{ih}\right)\right]/M_j. \quad (9)$$

According to the rationale described above, hypothesis testing can be carried out for $X_{ij} = 0$ and $X_{ij} = 1$ as follows:

(1) When $X_{ij} = 0$, and given that the total score of all the n_j items that measure the subset of attributes of item *j* is t_{ij}, one-sided hypothesis is tested:

$$H_0 : p_{ij} \leq p_0 \;\; vs \;\; H_1 : p_{ij} > p_0.$$

A misfit exists when $\sum_{k=0}^{t_{ij}} C_{n_j}^k p_0{}^k(1-p_0)^{n_j-k} \leq 1-\alpha$, and $\sum_{g\in S_j} x_{ig}\left(1-x_{ij}\right)$ takes part in the computation of the ICI. Otherwise, $\sum_{g\in S_j} x_{ig}\left(1-x_{ij}\right)$ doesn't take part in the computation.

(2) When $X_{ij} = 1$, and suppose that the total score of all the $n_j{}^*$ items that include the set of attributes of item *j* is $t_{ij}{}^*$, the hypotheses are:

$$H_0 : p_{ij} \geq p_1 \;\; vs \;\; H_1 : p_{ij} < p_1.$$

$\sum_{h\in S_j^*} x_{ij}\left(1-x_{ih}\right)$ takes part in the computation of the ICI when $\sum_{k=0}^{t_{ij}^*} C_{n_j^*}^k p_1{}^k$ $(1-p_1)^{n_j-k} \geq \alpha$; otherwise, it doesn't take part in the computation.

If examinee *i* responds to item *j* incorrectly and gives incorrect answers on the majority of the items that measure the subset of attributes of item *j*, we can infer that he/she has not mastered any of the attributes required by item *j*, and the correctly answered items are misfit items. If examinee *i* responds to item *j* correctly and gives correct answers on the majority of items that include the set of attributes measured by item *j*, we can infer that he/she has mastered a part of or all the attributes required by item *j*, and the wrongly answered items are misfit items.

3 Simulation Study

3.1 Study Design

The purposes of this study were (a) to verify whether the item consistency indices could help identify condensation rules, especially when the Q-matrix of the test is misspecified, and (b) to check whether the item consistency indices could be used to evaluate the misspecification of a Q-matrix and to distinguish wrongly specified items from correctly specified items.

3.1.1 Four Factors

There are four factors manipulated in this study: sample size (N = 500, 1000), cognitive diagnostic model (the DINA model and the DINO model), quality of item parameters (high and low quality), and percentage of misspecified q-entries (0.1, 0.2, 0.3, and 0.4).

3.1.2 Data Generation

The number of attributes was set to five in this study. The correct Q-matrix with 31 rows includes all possible combinations of five independent attributes, and the universal set of attribute patterns (Tatsuoka 2009) contains 32 rows. A total sample size of 500 and 1000 were generated, respectively, from a discrete uniform distribution of attribute patterns. The DINA model was selected for the conjunctive rule, and the DINO model was adopted for the disjunctive rule. The quality of item parameters was represented by different values of parameters. For both models, the distributions of item parameters of high and low quality were U(0.05, 0.25) and U(0.05, 0.40), respectively. Using these two diagnostic models, the study generated a total number of 800 response matrices across 8 ($2 \times 2 \times 2$) conditions (100 replications per condition). In order to study the effect of the misspecification of Q-matrix, Q-matrices with four percentages of misspecified q-entries (0.1, 0.2, 0.3, and 0.4) are simulated.

3.1.3 Item Consistency Indices

For both the conjunctive and disjunctive models, the following indices are compared in this study: (a) ICI calculated by Formula (4) or (9) without hypothesis testing, (b) SICI calculated by Formula (4) or (9) combining hypothesis testing only for $X_{ij} = 1$, and (c) MICI calculated by Formula (4) or (9) combining hypothesis testing for $X_{ij} = 1$ and $X_{ij} = 0$. Note that $p_1 = .75$ and $p_0 = .25$ were used in the hypothesis testing when calculating the SICI and MICI. The notations in the formulas did not distinguish between the conjunction and disjunction rules, and the condensation rules were represented by the psychometric models used in data analyzing (the DINA model for conjunctive and the DINO model for disjunctive).

3.1.4 Evaluation Criteria

The average overall accuracy, type I and type II errors of 100 replications were reported to assess the performance of the three indices.

3.2 Results

Due to similar results, the following only presents results with sample size of 500. Table 1 presents the thresholds (i.e., the lower 10 percentile) of the item consistency indices obtained under a correct Q-matrix and correct condensation rules. A higher threshold indicates a better item-fit when the quality of item parameters is high, and a lower threshold indicates a worse item-fit when the quality of item parameters is low. As can be seen in Table 1, an average difference of about 0.15 existed between the thresholds of items with high and low quality under the same model. When observing the thresholds under different conditions, it was found that the ICI was remarkably lower than the SICI and MICI. It is because without doing a hypothesis testing, examinees' responses on all items took part in the computation of the ICI and resulted in a larger misfit and a smaller ICI. When examining the items with high quality and low quality, respectively, the data showed that the thresholds of these three indices under the DINA model and the DINO model were quite similar, and the SICI and MICI were especially in the case.

Tables 2 and 3 show the average values of the item consistency indices with 100 replications for the DINA model and the DINO model, respectively. An apparent difference emerged when the indices were compared between the correct and incorrect condensation rules. The values of indices obtained from a correct condensation rule were remarkably larger than the values obtained from an incorrect condensation rule, especially for the ICI and SICI. For example, when the Q-matrix was correctly

Table 1 The thresholds of ICI, SICI, and MICI with sample size of 500

Model	Parameter	ICI	SICI	MICI
DINA	U(0.05, 0.25)	0.09	0.62	0.68
	U(0.05, 0.40)	−0.09	0.43	0.53
DINO	U(0.05, 0.25)	0.00	0.62	0.64
	U(0.05, 0.40)	−0.06	0.46	0.50

Table 2 Averages of the item consistency indices with 100 repetitions under the DINA model

		Conjunctive			Disjunctive		
Parameter	Q	ICI	SICI	MICI	ICI	SICI	MICI
U(0.05, 0.25)	0.0	0.50	0.68	0.73	−0.01	0.08	0.58
	0.1	0.41	0.60	0.68	0.05	0.15	0.57
	0.2	0.35	0.55	0.64	0.08	0.17	0.57
	0.3	0.28	0.50	0.62	0.12	0.20	0.59
	0.4	0.23	0.47	0.60	0.15	0.23	0.59
U(0.05, 0.40)	0.0	0.36	0.54	0.62	−0.01	0.09	0.52
	0.1	0.30	0.49	0.59	0.03	0.13	0.51
	0.2	0.26	0.45	0.58	0.05	0.15	0.53
	0.3	0.20	0.41	0.55	0.09	0.18	0.53
	0.4	0.17	0.39	0.55	0.10	0.19	0.53

Table 3 Averages of the item consistency indices with 100 repetitions under the DINO model

		Conjunctive			Disjunctive		
Parameter	Q	ICI	SICI	MICI	ICI	SICI	MICI
U(0.05, 0.25)	0.0	0.03	0.09	0.47	0.46	0.67	0.69
	0.1	0.08	0.13	0.49	0.39	0.62	0.65
	0.2	0.11	0.16	0.51	0.33	0.58	0.61
	0.3	0.14	0.19	0.54	0.27	0.54	0.59
	0.4	0.15	0.20	0.54	0.24	0.51	0.57
U(0.05, 0.40)	0.0	−0.04	0.04	0.41	0.34	0.54	0.57
	0.1	0.00	0.07	0.41	0.28	0.50	0.54
	0.2	0.03	0.09	0.43	0.23	0.47	0.52
	0.3	0.06	0.12	0.44	0.18	0.43	0.49
	0.4	0.08	0.13	0.46	0.15	0.40	0.48

Table 4 The accuracy of the ICI, SICI, and MICI with 100 repetitions under the DINA model

		Ac			I			II		
Parameter	Q	ICI	SICI	MICI	ICI	SICI	MICI	ICI	SICI	MICI
U(0.05, 0.25)	0.0	0.94	0.94	0.94	0.06	0.06	0.06	0.00	0.00	0.00
	0.1	0.57	0.54	0.53	0.10	0.57	0.50	0.88	0.30	0.42
	0.2	0.40	0.63	0.58	0.10	0.86	0.72	0.84	0.13	0.27
	0.3	0.33	0.77	0.70	0.11	0.92	0.78	0.78	0.09	0.20
	0.4	0.32	0.87	0.77	0.17	0.95	0.82	0.73	0.06	0.18
U(0.05, 0.40)	0.0	0.94	0.94	0.94	0.06	0.06	0.06	0.00	0.00	0.00
	0.1	0.55	0.57	0.55	0.09	0.24	0.23	0.94	0.68	0.73
	0.2	0.35	0.50	0.46	0.09	0.36	0.30	0.92	0.57	0.66
	0.3	0.26	0.59	0.47	0.10	0.52	0.38	0.86	0.38	0.56
	0.4	0.20	0.66	0.48	0.13	0.54	0.35	0.85	0.33	0.53

specified, the ICI and SICI under the DINA model were 0.5 and 0.68 for the conjunctive rule that were much larger than their values of −0.01 and 0.08 for the disjunctive rule. The data in Table 2 also showed that the misspecification rate of the Q-matrix exerted a positive impact on the values of the indices. An obvious decrease was observed when the misspecification rate of the Q-matrix increased. Interestingly, the ICI and SICI obtained from the correct condensation rule even under the highest Q-matrix misspecification rate were still larger than the values obtained from a wrong condensation rule. This result indicates that the ICI and SICI can identify the correct condensation rule even though the Q-matrix is partially misspecified.

Tables 4 and 5 present the averaged overall accuracy and type I and type II errors of the ICI, SICI, and MICI with 100 repetitions. As can be seen in these two tables, the three indices showed an obvious decrease tendency when the misspecification rate of the Q-matrix increased. Among them, the SICI had the highest average classification accuracy and smallest type II error, while the ICI had the smallest type I error. This finding suggests that combining these indices will be helpful to evaluate the misspecification of a Q-matrix.

Table 5 The accuracy of the ICI, SICI, and MICI with 100 repetitions under the DINO model

		Ac			I			II		
Parameter	Q	ICI	SICI	MICI	ICI	SICI	MICI	ICI	SICI	MICI
U(0.05, 0.25)	0.0	0.94	0.94	0.94	0.06	0.06	0.06	0.00	0.00	0.00
	0.1	0.55	0.55	0.54	0.10	0.49	0.44	0.92	0.40	0.48
	0.2	0.37	0.62	0.61	0.09	0.72	0.66	0.88	0.22	0.27
	0.3	0.29	0.74	0.72	0.12	0.85	0.82	0.83	0.14	0.17
	0.4	0.25	0.82	0.79	0.14	0.86	0.83	0.8	0.11	0.14
U(0.05, 0.40)	0.0	0.94	0.94	0.94	0.06	0.06	0.06	0.00	0.00	0.00
	0.1	0.55	0.57	0.56	0.10	0.26	0.21	0.92	0.65	0.75
	0.2	0.36	0.51	0.47	0.09	0.44	0.34	0.89	0.51	0.61
	0.3	0.27	0.61	0.54	0.12	0.63	0.52	0.84	0.35	0.45
	0.4	0.25	0.70	0.63	0.12	0.69	0.60	0.81	0.26	0.35

4 Conclusion

In this study, we introduced the modified item consistency indices combining hypothesis testing, SICI and MICI, for both the conjunctive and disjunctive models in cognitive diagnostic assessment. This paper also conducted a simulation study to evaluate the performance of the indices by comparing with the existing index ICI under different situations.

Results of the study showed that given a Q-matrix, the ICI and SICI precisely identified the underlying correct condensation rule. As we know, the idea response patterns are dependent on condensation rules, and correctly identifying the condensation rule is crucial to model determination and thus to the classification accuracy. This result helps to find a way to make correct decision on the underlying condensation rule and to choose a proper psychometric model or a nonparametric classification method (Chiu and Douglas 2013). Our results also showed that the item consistency indices were successfully used in evaluating the misspecification of a Q-matrix, and they were also applied to identify items with attribute misspecification. The SICI was especially in the case with a higher overall accuracy and a lower type II error. This result suggests that the item consistency indices can be used to assess the amount of random error in response data and can be used to evaluate systematic error existing in the specification of a Q-matrix.

There are some limitations to this research that are worth noting. Firstly, the hypothesis testing cannot be achieved under some circumstances. For the disjunctive rule, hypothesis testing cannot be conducted for $X_{ij} = 1$ when there is no other item including the set of attributes required by item *j*. It also cannot be conducted for $X_{ij} = 0$ when there is no other item measuring the subset of attributes of item *j*. On the contrary, for the conjunctive rules, hypothesis testing cannot be carried out for $X_{ij} = 1$ when there is no other item measuring the subset of attributes of item *j* and for $X_{ij} = 0$ when there is no other item including the set of attributes measured by item *j*. Secondly, the revised item consistency indices introduced in this paper were successfully used to screen the problematic attribute specification in a Q-matrix. However, it still needs more studies to explore how to modify the Q-matrix based on the ICI. In addition, the ICI can only provide an empirical evidence for the quality of Q-matrix, and other approaches, especially theoretical methods, such as the theoretical construct validity (Ding et al. 2012) should be combined to help design and modify a Q-matrix. Factor analysis may be used to determine the number of constructs (or factors) that a set of test items really measures (Cui 2016), developing or selecting an appropriate method of determining the number of latent attributes is deserved further study.

Because the real probability for examinees correctly answering an item is unknown, this study assumed that the probability of correct responding to item j was the same with that to the items measuring the subset of attributes of item j. The correct responding probability was restricted to a lower bound of 0.75, and the incorrect responding probability was restricted to an upper bound of 0.25. With these assumptions, a Bernoulli test based on a binomial distribution could be easily carried out. Tatsuoka (1987) also adopted a binomial distribution in building the distribution of errors in the Rule Space Method. In testing practice, the ICI introduced in this paper can be obtained by the hypothesis testing of proportion based on a compound binomial distribution (Tatsuoka 1987) provided examinees' correct responding probabilities. Other methods (Cui and Li 2015) can be used to accomplish this purpose as well.

Acknowledgments This research is supported by the key project of the National Education Science "Twelfth Five-Year Plan" of the Ministry of Education of China (Grant No. DHA150285), the Jiangxi Education Science Foundation (Grant No. 13YB032), the China Scholarship Council (CSC No. 201509470001), the National Natural Science Foundation of China (Grant Nos. 31500909, 31360237, and 31160203), the Humanities and Social Sciences Research Foundation of the Ministry of Education of China (Grant Nos. 13YJC880060 and 12YJA740057), the National Natural Science Foundation of Jiangxi Province (Grant No. 20161BAB212044), the Science and Technology Research Foundation of Education Department of Jiangxi Province (Grant No. GJJ13207), and the Youth Growth Fund and the Doctoral Starting Up Foundation of Jiangxi Normal University. The authors thank Prof. Hua-Hua Chang for his kind support. Finally, we would like to thank Prof. Steven A. Culpepper for his helpful suggestions.

References

J.-S. Chen, J. de la Torre, Z. Zhang, Relative and absolute fit evaluation in cognitive diagnosis modeling. J. Educ. Meas. **50**(2), 123–140 (2013)

C.-Y. Chiu, J.A. Douglas, A nonparametric approach to cognitive diagnosis by proximity to ideal response patterns. J. Classif. **30**, 225–250 (2013)

A. Crawford, Posterior predictive model checking in Bayesian networks. Unpublished doctoral dissertation, Arizona State University, Tempe, AZ, 2014

Y. Cui, The hierarchy consistency index: a person fit statistic for the attribute hierarchy method. Unpublished Doctoral Dissertation, University of Alberta, Edmonton, AB, 2007

Y. Cui, A simulation approach to selecting the appropriate method of determining the number of factors to retain in factor analysis. J. Jiangxi Norm. Univ. (Nat. Sci.) **40**(5), 456–464 (2016)

Y. Cui, J.P. Leighton, The hierarchy consistency index: evaluating person fit for cognitive diagnostic assessment. J. Educ. Meas. **46**, 429–449 (2009)

Y. Cui, J. Li, Evaluating person fit for cognitive diagnostic assessment. Appl. Psychol. Meas. **39**(3), 223–238 (2015)

Y. Cui, M.R. Roberts, Validating student score inferences with person-fit statistic and verbal reports: a person-fit study for cognitive diagnostic assessment. J. Educ. Meas. **32**(1), 34–42 (2013)

L.V. DiBello, L.A. Roussos, W. Stout, in *Handbook of Statistics*, ed. by C. R. Rao, S. Sinharay. Review of cognitively diagnostic assessment and a summary of psychometric models, vol 26 (Elsevier, Amsterdam, 2007), pp. 979–1030

S.L. Ding, M.M. Mao, W.Y. Wang, F. Luo, Y. Cui, Evaluating the consistency of test items relative to the cognitive model for educational cognitive diagnosis. Acta Psychol. Sin. **44**(11), 1535–1546 (2012)

M.J. Gierl, C. Wang, J. Zhou, Using the attribute hierarchy method to make diagnostic inferences about examinees' cognitive skills in algebra on the SAT. J. Technol. Learn. Assess. **6**(6) (2008.) Retrieved June 1, 2011, from http://www.jtla.org

E.H. Haertel, Using restricted latent class models to map the skill structure of achievement items. J. Educ. Meas. **26**(4), 301–321 (1989)

B.W. Junker, K. Sijtsma, Cognitive assessment models with few assumptions, and connections with nonparametric item response theory. Appl. Psychol. Meas. **25**(3), 258–272 (2001)

H. Lai, M.J. Gierl, Y. Cui, Item consistency index: an item-fit index for cognitive diagnostic assessment. Paper presented at the annual meeting of the National Council on Measurement in Education, Vancouver, BC, Canada, 2012

J.P. Leighton, M.J. Gierl, S.M. Hunka, The attribute hierarchy method for cognitive assessment: a variation on Tatsuoka's rule-space approach. J. Educ. Meas. **41**(3), 205–237 (2004)

H. Liu, Y. Bian, A research on diagnosis of international students' attribute-mastery patterns of basic Chinese color terms. Psychol. Explor. **34**(1), 29–35 (2014)

M. Mao, The introduction of granular computing and formal concept analysis for research on cognitive diagnosis. Unpublished Doctoral Dissertation, Jiangxi Normal University, Nanchang, Jiangxi, China, 2011

E. Maris, Psychometric latent response models. Psychometrika **60**(4), 523–547 (1995)

E. Maris, Estimating multiple classification latent class models. Psychometrika **64**, 187–212 (1999)

K.K. Tatsuoka, Bug distribution and statistical pattern classification. Psychometrika **52**(2), 193–206 (1987)

K.K. Tatsuoka, in *Diagnostic Monitoring of Skill and Knowledge Acquisition*, ed. by N. Frederiksen, R. L. Glaser, A. M. Lesgold, M. G. Safto. Toward an integration of item-response theory and cognitive error diagnosis (Erlbaum, Hillsdale, NJ, 1990), pp. 453–488

K.K. Tatsuoka, *Cognitive Assessment: An Introduction to the Rule Space Method* (Taylor & Francis Group, New York, NY, 2009)

J.L. Templin, R.A. Henson, Measurement of psychological disorders using cognitive diagnosis models. Psychol. Methods **11**, 287–305 (2006)

C. Wang, M.J. Gierl, Using the attribute hierarchy method to make diagnostic inferences about examinees' cognitive skills in critical reading. J. Educ. Meas. **48**(2), 165–187 (2011)

Irreplaceability of a Reachability Matrix

Shuliang Ding, Wenyi Wang, Fen Luo, Jianhua Xiong, and Yaru Meng

Abstract Q-matrix is a critical concept for cognitive diagnosis. Reachability matrix R is special in its two important properties: (1) any column in the Q-matrix can be expressed by a linear combination of R's columns; (2) there is a one-to-one mapping from the set of knowledge states to the set of ideal response patterns if R is a submatrix of the test Q-matrix. It is proved that these properties are irreplaceable, i.e., no other upper triangular Q-matrices have these properties.

Keywords Cognitive diagnosis • Reachability matrix (R) • Irreplaceability • Sufficient Q-matrix • Necessary Q-matrix

1 Introduction

Diagnostic test design is critical for cognitive diagnosis. There are some related studies addressing this (e.g., Chiu et al. 2009; Henson and Douglas 2005; Madison and Bradshaw 2015), and two shared features characterize these researches: (1) The test design is based on the independent attribute hierarchical structure; (2) identity matrix plays an important role in test design. However, there are also some other attribute hierarchical structures (Leighton et al. 2004) whose optimal test design may not be the same as that of the independent structure. The proposed solution by Chiu et al. (2009) may not suffice here, but the reachability matrix R developed

S. Ding (✉) • F. Luo • J. Xiong
School of Computer and Information Engineering, Jiangxi Normal University, 99 Ziyang Ave., Nanchang, Jiangxi 330022, China
e-mail: ding06026@163.com; luofen312@163.com; 270281168@qq.com

W. Wang
College of Computer Information Engineering, Jiangxi Normal University, 99 Ziyang Road, Nanchang, Jiangxi, People's Republic of China
e-mail: wenyiwang2009@gmail.com

Y. Meng
School of Foreign Studies, Xi'an Jiaotong University, Xian, Shanxi 710049, China
e-mail: yarum@163.com

L.A. van der Ark et al. (eds.), *Quantitative Psychology*, Springer Proceedings in Mathematics & Statistics 196, DOI 10.1007/978-3-319-56294-0_21

by Ding et al. (2010) can make a big difference under a certain condition. It is a generalized version of Chiu et al. (2009). This paper discusses the properties of the reachability matrix R in detail.

Suppose there are only K attributes in the domain of interest and the attribute hierarchy among these K attributes is given. Let R, Q_p, Q_s, and Q_t be the reachability matrix, the potential Q-matrix (coinciding with the reduced Q-matrix defined by Tatsuoka (2009)), the student Q-matrix, and the test Q-matrix, respectively. Each column of Q_p is a latent item attribute vector, and each column of Q_s is a knowledge state.

The Augment algorithm (Ding et al. 2008) establishes the relationship between matrix R and Q_p. In the Augment algorithm, the Boolean addition for some 0–1 vectors is element-wise Boolean addition, i.e., $0 + 0 = 0; 0 + 1 = 1 + 0 = 1 + 1 = 1$.

The Augment algorithm (Ding et al. 2008) to derive the potential Q-matrix from R is described in detail as follows:

Step 1. Partition R according to its columns.

Step 2. Let Q = R.

Step 3. Let $j = 1$.

Step 4. Add (Boolean addition) r_j to every column from $(j + 1)$th column to the last column of Q, and if a new column is produced, put the new column to the far-right side of Q, i.e., augment Q.

Step 5. $j := j + 1$, if $j \leq K$, then go to step 4; stop otherwise.

It has been proved from the Augment algorithm that an arbitrary column in Q, say x, can be represented by a linear combination of the columns in R with combinational coefficients being 0 or 1 (Yang and Ding 2011). Any column of R, if it is the linear combination corresponding to x, is called a combination component of x. It is found that for a column of Q, the combination of the columns of R is often not unique.

2 Main Results

Given that x and y are two K-dimension vectors, x is called greater than y, denoting as $y \leq x$, if all elements of x–y difference is nonnegative.

If x is an item attribute vector of item *i* and y is a knowledge state of examinee j, $y \circ x$ denotes the ideal response of j on item *i* without guessing or slipping.

2.1 Properties of Combination Components

Proposition 1 *If* r *is a combination component of* x, *then* x *is greater than* r. *Suppose it's a* 0–1 *scoring rubric and its attributes are non-compensatory, then* $x \circ r = 1$, *and for any arbitrary column vector of* R, *say* r_0, *not being in the combination component of* x, *then* $x \circ r_0 = 0$.

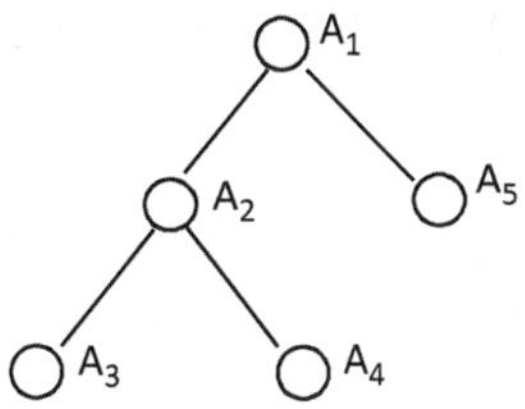

Fig. 1 Divergent

Suppose that x *is a knowledge state, let* $S_x = \{r | (r \text{ is a column of } R) \text{ and } (r \leq x)\}$ *be a set of combination components of* x.

Definition of a redundant expression of x. The Boolean union of all r in the set S_x is called a redundant expression of x.

Definition of a concise expression of x. Let S'_x be a subset of S_x and if any two different elements are not comparable in S'_x, then the Boolean union of the elements in S'_x is a concise expression of x.

Example (see Fig. 1).

$$R = \begin{bmatrix} 1\,1\,1\,1\,1 \\ 0\,1\,1\,1\,0 \\ 0\,0\,1\,0\,0 \\ 0\,0\,0\,1\,0 \\ 0\,0\,0\,0\,1 \end{bmatrix} = [r_1 r_2 r_3 r_4 r_5.]$$

$$Q_p = \begin{bmatrix} 1\,1\,1\,1\,1\,1\,1\,1\,1\,1 \\ 0\,1\,1\,1\,0\,1\,1\,1\,1\,1 \\ 0\,0\,1\,0\,0\,0\,1\,1\,0\,1 \\ 0\,0\,0\,1\,0\,0\,1\,0\,1\,1 \\ 0\,0\,0\,0\,1\,1\,0\,1\,1\,1 \end{bmatrix} = [r_1 r_2 r_3 r_4 r_5 q_6 q_7 q_8 q_9 q_{10}]$$

$q_6 = r_2 \bigvee r_4$ (concise expression of q_6) $= r_1 \bigvee r_2 \bigvee r_4$ (redundant expression of q_6).

In fact, there must be a both redundant and concise expression of a knowledge state x, because the sets S_x and S'_x are not empty. It can be proved that if the redundant expression of x equals to the concise expression of x, then there is only one combination component in the redundant expression, and x is a column of the identity matrix.

A possible use of the redundant expression and concise expression of a knowledge state, say x, is that they exhibit some different paths from some status of knowledge state to x in remedy course.

Proposition 2 *In the redundant expression of knowledge state, say* x, *the length of* x *is longer than that of any of its combination component.*

2.2 Irreplaceability of Reachability Matrix

Suppose matrix R is an upper triangular 0–1 matrix without the loss of generality because R is a partial relation matrix. The Augment algorithm gives the fact that the columns of R are representative of attribute vectors in a cognitive diagnostic test. Here the first question emerges: is there another K-order upper triangular matrix in the potential Q-matrix and can it represent any other column in the Q-matrix besides the matrix R? The answer is negative.

Theorem 1 *Suppose* Q_0 *is a* K-order *upper triangular sub-matrix of* Q_p, *then any column of* Q_s *can be represented linearly by the columns of* Q_0 *if and only if* $Q_0 = R$.

Proof Because any column of Q_s can be represented by the columns of Q_0, so does any column of R.

"$\Leftarrow$" It is validity from the Augment algorithm (Ding et al. 2008).

"$\Rightarrow$" Its contrapositive is: If Q_0 is not equal to R, then there is at least one column of Q_s, say q, not being able to be represented linearly by the columns of Q_0.

Firstly, any column of R can't be represented by other columns of R. If Q_0 is not equal to R, then at least one column of R, say r_j, is not in Q_0. Note that $r_{jj}=1$ and for all t > j, $r_{tj}=0$. If for Q_0, there is no column with j-th element being 1 and for all t, t > j, then t-th element being 0, and r_j can't be represented by the columns of Q_0.

Take one column, say q, out from all columns of R with the j-th element being 1, and for all t, t > j, $q_t = 0$, because q is a column augmented by R, the length of q is longer than that of r_j, and r_j can't be represented by the columns of Q_0.

Matrix R's important role is demonstrated in cognitive diagnosis that follows.

Theorem 2 *(Ding et al.* 2010*) Suppose a* 0–1 *scoring rubric is adopted and the attributes are non-compensatory. Given a test matrix Q, let* $\alpha \circ Q$ *be the expected response vector of knowledge state (attribute mastery pattern)* α *on the matrix* Q. *If* R *is the test* Q-matrix, *then for any knowledge state* α, $\alpha \circ R = \alpha^T$. *Otherwise, if* R *is a sub-matrix of the test matrix Q, and* α, β *are different knowledge states, then* $\alpha \circ Q \neq \beta \circ Q$.

Proof

(a) If α contains h positive elements, then the redundant expression of α contains h and only h columns of R (Yang and Ding 2011). From Proposition 1, in the ideal response pattern $\alpha \circ R$, the ideal response of α on each of these h columns of R is 1, otherwise it is 0. So $\alpha \circ R = \alpha^T$.
(b) If $\alpha \neq \beta$, then $\alpha \circ R = \alpha T \neq \beta T = \beta \circ R$, because R is a sub-matrix of Q, based on the definition of equality of two vectors, then $\alpha \circ Q \neq \beta \circ Q$.

Remarks The part (a) of the Theorem 2 has been proved by Ding et al. (2010) through several lemmas. The present new form of proof is comparatively clear and concise.

The second problem emerges: is there another square sub-matrix Q_1 of Q_p, $Q_1 \neq R$, and Q_1 that satisfies the property listed in Theorem 2? The answer is negative, too.

Theorem 3 *Under the conditions that the* 0–1 *scoring rubric is adopted and there is no compensation among the attributes, suppose* Q_1 *is a* K-order *upper triangular sub-matrix of* Q_p *and* α *is an arbitrary column vector of* Q_s, *i.e.,* α *is a knowledge state, then* $\alpha \circ Q_1 = \alpha^T \Longleftrightarrow Q_1 = R$.

Proof Let $\alpha = (\alpha_1, \ldots, \alpha_K)^T$, $\beta = (\beta_1, \ldots, \beta_K)^T$ be K-dimension 0–1 vectors.

"$\Leftarrow$" Suppose that $\alpha \circ Q_1 = \beta^T$.

It must be proved that for all i = 1, 2, …, K, $\alpha_i = 1 \Longleftrightarrow \beta_i = 1$, and $\alpha_i = 0 \Longleftrightarrow \beta_i = 0$.

Because that α and β are 0–1 vectors, then the contrapositive of $\alpha_i = 1 \Longleftrightarrow \beta_i = 1$ is $\alpha_i = 0 \Longleftrightarrow \beta_i = 0$, so it is enough to prove that $\alpha_i = 1 \Longleftrightarrow \beta_i = 1$.

Let $Q_1 = (q_1, \ldots, q_K) = (r_1, \ldots r_K)$, $\beta_i = 1 \Longleftrightarrow \alpha \circ r_i = \alpha \circ q_i = 1 \Longleftrightarrow r_i = q_i \le \alpha \Longleftrightarrow r_i$ is a combination component of α, and because $r_{ii} = 1 \Rightarrow \alpha_i = 1$.

Because $\alpha_i = 1$ implies that r_i is a combination component of $\alpha \Rightarrow r_i \le \alpha$, so $\beta_i = 1$.

So for all $\alpha \epsilon Qs$, $\alpha \circ R = \alpha^T$.

"$\Rightarrow$" If there is a K-order matrix Q_1, and Q_1 satisfies that for arbitrary $\alpha \epsilon Q_s$, $\alpha \circ Q_1 = \alpha$, it must be proved that $Q_1 = R$. Let $\alpha \circ Q_1 = \beta^T$.

Suppose that $Q_1 \neq R$,

1. Q_1 is an upper triangular matrix and if any of its diagonal elements is 1, $Q_1 = (q_1, \ldots, q_k)$, then there is a column of Q_1, say q_j, $q_j \neq r_j$, the fact that $q_{jj} = r_{jj} = 1$, and $q_j \neq r_j$ implies that r_j is a combination component of q_j, then based on Proposition 2, $r_j^T r_j < q_j^T q_j$, so $r_j \circ q_j = 0$. It is said that the j-th element of $r_j \circ Q_1$, $\beta_j = (r_j \circ Q_1)_j = 0 \neq r_{jj} = 1$. So when $Q_1 \neq R$, Q_1 is an upper triangular matrix, and all of its diagonal elements are 1, it is impossible that for arbitrary column α of Qs satisfies $\alpha \circ Q_1 = \alpha^T$, i.e., the j-th element of $r_j \circ Q_1$, $\beta_j = (r_j \circ Q_1)_j = 0 \neq r_{jj} = 1$.
2. Q_1 can't be rearranged as an upper triangular matrix with all diagonal elements being 1, which implies that there is a column, say (j + 1)-th column of Q_1, say q_{j+1}, $q_{j,j+1} = 1$ and for all t, when t > j, $q_{t,j+1} = 0$. Under this condition, the problem could fall into two parts: the first part is that if there is a column of Q_1 equal to q_{j+1}, it can be supposed that the j-th column of Q_1 is equal to q_{j+1} without the loss of generality because exchanging columns of Q_1 does not change the result; the second part is that if no column of Q_1 is equal to q_{j+1}.

If $q_j = q_{j+1}$, the ideal response pattern of q_j on the test Q-matrix Q_1 is considered. Because

$\beta_{j+1} \widehat{=} (q_j \circ Q_1)_{j+1} = q_j \circ q_{j+1} = 1 > q_{j+1,j+1} = 0$, let $\alpha = q_{j+1}$, so there is a column $\alpha \epsilon Q_s$, and $\alpha \circ Q_1 \neq \alpha^T$.

(b) If $q_j \neq q_{j+1}$, because that $q_{j,j+1} = 1$, and for any t and t > j, then $q_{t,j+1} = 0$, so or $(q_{j+1} = r_j)$ or $(r_j \le q_{j+1})$ and $(r_j \neq q_{j+1})$.

When $q_{j+1} = r_j$, consider the ideal response pattern of r_j on the test matrix Q_1. It can be obtained that $r_j \circ q_{j+1} = 1 = \beta_{j+1} > r_{j+1} = 0$, so $r_j \circ Q_1 \neq r_j^T$. When $(r_j < = q_{j+1})$ and $(r_j \neq q_{j+1})$, consider the ideal response pattern of q_{j+1} on the test matrix Q_1. It is obtained that $\beta_{j+1} = (q_{j+1} \circ Q_1)_{j+1} = 1 > q_{j+1,j+1} = 0$, let$\alpha = q_{j+1}$, then there is a column α, $\alpha \epsilon Q_s$, and $\alpha \circ Q_1 \neq \alpha^T$. The proof is complete.

2.3 *Sufficient Q-Matrix and Necessary Q-Matrix*

Definition of a necessary Q-matrix. If a reachability matrix R is a sub-matrix of the Q-matrix (Ding et al. 2016), it is called a necessary Q-matrix.

Definition of R-equivalent class (REC). The reachability matrix R and all of the permutations of the columns of R are called R-equivalent class (REC).

Any matrix in the R-equivalent class could be correct provided the modification of the above results is offered to the irreplaceability of a reachability matrix R.

Definition of a sufficient Q-matrix. (Tatsuoka 2009). Suppose K attributes have a prerequisite relation with a reachability matrix R, and their involvement with n items are expressed by a Q-matrix. If the pairwise comparison of attribute vectors in the Q-matrix with respect to the inclusion relation yields the reachability matrix R, then the Q-matrix is said to be sufficient for representing the cognitive model of the domain of interest.

Tatsuoka (2009) pointed out "it is important to note that a sufficient Q matrix is the core of a knowledge structure." A necessary Q-matrix is a sufficient Q-matrix; however, a sufficient Q-matrix may not be a necessary Q-matrix. If a sufficient Q-matrix is augmented, some knowledge states may be lost. At this time, the knowledge structure is not integral. Sometimes, a test Q-matrix is a sufficient Q-matrix, but it does not work well for improving the classification accuracy. To illustrate these and to probe the differences between the necessary and the sufficient Q-matrix, Monte Carlo simulation is conducted to find a good cognitive diagnostic test design and to investigate the loss of the accuracy of the estimated knowledge states if the reachability matrix R is not used.

3 Simulation Study

3.1 *Purpose*

A simulation study consisting of various conditions was conducted to investigate the impact of three kinds of test Q-matrices on correct classification rate of attribute patterns (CCRAPs). The three kinds of test Q-matrices are a sufficient Q-matrix (Tatsuoka 2009), a necessary Q-matrix, and a Q-matrix that is insufficient (neither the sufficient nor the necessary).

3.2 *Experiment Design*

Five factors were considered:

a. The number of attributes (K): K = 4, 5, 6, and 7.
b. The length of test (L): L = 1K, 3K, and 5K.

Table 1 Eighteen conditions of the simulation study

Conditions																		
	1	2	3	4	5	6	7	8	9	10	11	12	13	14	15	16	17	18
I	0.05						0.15						0.25					
N	500			1000			500			1000			500			1000		
L	1K	3K	5K	1K	3K	5K	1K	3K	5K	1K	3K	5K	1K	3K	5K	1K	3K	5K

Note: I = the values of slip and guessing parameters; L = test length; N = sample size

Table 2 Three types of test Q-matrices

	Q_t (sufficient)				Q_t (insufficient)				Q_t (necessary)			
	I1	I2	I3	I4	I1	I2	I3	I4	I1	I2	I3	I4
A1	1	0	0	1	1	0	0	0	1	0	0	0
A2	1	1	0	0	0	1	0	0	0	1	0	0
A3	0	1	1	0	0	0	1	1	0	0	1	0
A4	0	0	1	1	0	0	0	1	0	0	0	1

Table 3 Numbers of equivalent classes resulted from three types of test Q-matrices

	K = 4	K = 5	K = 6	K = 7
Q_t (sufficient)	9 + 1	16 + 1	28 + 1	50 + 1
Q_t (insufficient)	11 + 1	23 + 1	47 + 1	95 + 1
Q_t (necessary)	16	32	64	128

c. The quality of items or the item parameters in the deterministic inputs, noisy "AND" gate (DINA; Junker and Sijtsma 2001) model (I): I = 0.05, 0.15, and 0.25.
d. The sample size (N): N = 500 and 1000.
e. The types of test Q-matrix: sufficient, insufficient, and necessary Q-matrix.

Eighteen conditions of simulation study arising from these factors were showed in Table 1.

Suppose that the attribute hierarchical structure is independent. When K = 4, the representatives of the three types of Q-matrices are listed in Table 2. Using the Augment algorithm (Ding et al. 2008), all of the knowledge states (i.e., the student Q-matrix) are obtained; based on the student Q-matrix, the test Q-matrix, and the relationship among the attributes (compensatory or not), the equivalent classes of knowledge state (Tatsuoka 2009) are obtained and listed in Table 3. Table 4 gives equivalent classes of attribute patterns for three Q-matrices shown in Table 2.

3.3 *Results*

Table 5 shows the CCRAPs for the three types of test Q-matrices when K = 4, and the CCRAPs from the sufficient Q-matrix is significantly lower than the results from both the insufficient Q-matrix and the necessary Q-matrix. The results for K = 5,

Table 4 Equivalent classes of attribute patterns

Q_t (sufficient)		Q_t (insufficient)		Q_t (necessary)	
AP	EC/IRP	AP	EC/IRP	AP	EC/IRP
0000	0000	0000	0000	0000	0000
0001		0001		0001	0001
0010		0100	0100	0010	0010
0100		0101		0101	0100
1000		1000	1000	1000	1000
1010		1001		1001	1010
0101		1100	1100	0101	0101
1100	1000	1101		1100	1100
0110	0100	0010	0010	0110	0110
0011	0010	1010	1010	0011	0011
1001	0001	0110	0110	1001	1001
1110	1100	0011	0011	1110	1110
0111	0110	1110	1110	0111	0111
1101	1001	0111	0111	1101	1101
1011	0011	1011	1011	1011	1011
1111	1111	1111	1111	1111	1111

Note: *AP* attribute pattern, *EC* equivalent classes, *IRP* ideal response pattern. The number of attributes equals to 4

6, and 7 are not given here due to limited space. The results from these three types of test Q-matrices had the similar trends as is showed in Table 5. Interestingly, the differences of performance between any two types of test Q-matrices became more significant as the number of attributes increased from 4 to 7.

4 Conclusion and Discussion

Reachability matrix R plays an important role in test design for cognitive diagnostic purposes. Each item attribute vector can be expressed as a linear combination of the columns of R, and when the test Q-matrix includes R as a sub-matrix, R exhibits some optimality, and its role is irreplaceable.

For Tables 3 and 4, let N_1 be the number of equivalent classes, and N_2 be the number of the knowledge states. Consider the ratio N_1/N_2, and this ratio is called the theoretical construct validity (TCV) by Ding et al. (2012). The TCV is dependent on the relationship among the attributes, i.e., whether compensatory or not, and N_1 and N_2 are defined as above. For $K = 4$, in the sufficient, the insufficient, and the necessary Q-matrix, the TCVs are 0.625, 0.75, and 1.00, respectively. Note that the highest mean (M) in Table 4, the CCRAPs for the three matrices, are 0.63, 0.75, 1.00, respectively. It implies that the highest CCRAP is no larger than the corresponding TCV. This is also true for $K = 5$, 6, and 7. Under some conditions,

Table 5 CCRAPs for the three types of test Q-matrices (K = 4)

			Q_t (sufficient)				Q_t (insufficient)				Q_t (necessary)			
Parameters	N	Length	*M*	*MIN*	*MAX*	*SD*	*M*	*MIN*	*MAX*	*SD*	*M*	*MIN*	*MAX*	*SD*
0.05	500	1K	0.53	0.48	0.57	0.02	0.62	0.58	0.68	0.02	0.82	0.78	0.86	0.02
		3K	0.61	0.56	0.66	0.02	0.73	0.69	0.78	0.02	0.97	0.96	0.99	0.01
		5K	0.63	0.57	0.67	0.02	0.75	0.70	0.78	0.02	10.00	0.99	10.00	0.00
	1000	1K	0.53	0.48	0.57	0.02	0.63	0.60	0.65	0.01	0.82	0.78	0.83	0.01
		3K	0.62	0.59	0.64	0.01	0.73	0.70	0.76	0.01	0.97	0.96	0.98	0.01
		5K	0.62	0.61	0.65	0.01	0.75	0.73	0.77	0.01	0.99	0.99	10.00	0.00
0.15	500	1K	0.36	0.33	0.42	0.02	0.42	0.38	0.45	0.02	0.51	0.47	0.56	0.02
		3K	0.52	0.48	0.58	0.02	0.60	0.57	0.65	0.02	0.78	0.75	0.82	0.02
		5K	0.58	0.54	0.63	0.02	0.68	0.64	0.74	0.02	0.89	0.86	0.94	0.02
	1000	1K	0.36	0.32	0.40	0.02	0.41	0.37	0.45	0.02	0.52	0.49	0.56	0.02
		3K	0.52	0.49	0.57	0.02	0.60	0.58	0.63	0.01	0.78	0.76	0.80	0.01
		5K	0.58	0.56	0.61	0.01	0.69	0.65	0.71	0.02	0.90	0.88	0.92	0.01
0.25	500	1K	0.24	0.20	0.27	0.02	0.26	0.23	0.31	0.02	0.31	0.26	0.36	0.03
		3K	0.37	0.34	0.43	0.02	0.41	0.37	0.44	0.02	0.48	0.38	0.54	0.05
		5K	0.45	0.40	0.49	0.02	0.52	0.47	0.58	0.03	0.64	0.61	0.67	0.02
	1000	1K	0.24	0.22	0.27	0.01	0.26	0.24	0.29	0.01	0.32	0.28	0.36	0.02
		3K	0.37	0.35	0.41	0.02	0.41	0.33	0.45	0.02	0.49	0.41	0.54	0.03
		5K	0.45	0.44	0.49	0.01	0.51	0.48	0.55	0.02	0.64	0.61	0.67	0.01

the optimality in terms of CCRAPs in the test Q-matrix design can be predicted using the index of TCV.

The conclusion is that each column of the student Q-matrix can be represented by a combination of matrix R columns, whether the attributes are compensatory or not. Even for a polytomous Q-matrix, there are analogical results (Ding et al. 2016).

For a 0–1 rubric scoring and non-compensatory cognitive models, the optimal design of cognitive diagnostic test has also been explored. However, how to construct an optimal design of cognitive diagnostic test for compensatory cognitive models, polytomous rubric scoring, testlet situation, and even a polytomous Q-matrix (e.g., Ding et al. 2016) is still a demanding but exciting challenge for us.

Acknowledgments This research is supported by the National Natural Science Foundation of China (Grant No. 31360237,31500909, and 31160203), the National Social Science Fund of China (Grant No. 16BYY096,13BYY087), the Humanities and Social Sciences Research Foundation of Ministry of Education of China (Grant No. 13YJC880060 and 12YJA740057), the National Natural Science Foundation of Jiangxi Province (Grant No. 20161BAB212044), Jiangxi Education Science Foundation (Grant No. 13YB032), the Science and Technology Research Foundation of Education Department of Jiangxi Province (Grant No. GJJ13207, GJJ150356), the China Scholarship Council (CSC No. 201509470001), and the Youth Growth Fund and the Doctoral Starting up Foundation of Jiangxi Normal University.

References

C. Chiu, J.A. Douglas, X. Li, Cluster analysis for cognitive diagnosis: theory and applications. Psychometrika **74**(4), 633–665 (2009)

S.L. Ding, F. Luo, Y. Cai, H.J. Lin, X.B. Wang, Complement to Tatsuoka's Q matrix theory, in *New Trends in Psychometrics*, ed. by ed. by K. Shigemasu, A. Okada, T. Imaizumi, T. Hoshino, (Universal Academy Press, Tokyo, 2008), pp. 417–423

S.L. Ding, S.Q. Yang, W.Y. Wang, The importance of reachability matrix in constructing cognitively diagnostic testing. J. Jiangxi Normal Univ. **34**, 490–495 (2010)

S.L. Ding, M.M. Mao, W.Y. Wang, F. Luo, Y. Cui, Evaluating the consistency of test items relative to the cognitive model for educational cognitive diagnosis. Acta Psychol. Sin. **44**(11), 1535–1553 (2012)

S.L. Ding, F. Luo, W.Y. Wang, J.H. Xiong, Dichotomous and polytomous Q matrix theory, in *Quantitative Psychology Research. The 80th Annual Meeting of the Psychometric Society, Beijing, 2015*, ed. by ed. by L. A. van der Ark, D. M. Bolt, W.-C. Wang, J. A. Douglas, M. Wiberg, (Springer International Publishing, Cham, 2016), pp. 277–290

R. Henson, J. Douglas, Test construction for cognitive diagnosis. Appl. Psychol. Meas. **24**(9), 262–277 (2005)

B.W. Junker, K. Sijtsma, Cognitive assessment models with few assumptions, and connections with nonparametric item response theory. Appl. Psychol. Meas. **25**(3), 258–272 (2001)

J.P. Leighton, M.J. Gierl, S.M. Hunka, The attribute hierarchy method for cognitive assessment: a variation on Tatsuoka's rule space approach. J. Educ. Meas. **41**(3), 205–237 (2004)

M.J. Madison, L.P. Bradshaw, The effects of Q-matrix design on classification accuracy in the log-linear cognitive diagnosis model. Educ. Psychol. Meas. **75**(3), 491–511 (2015)

K.K. Tatsuoka, *Cognitive Assessment: An Introduction to the Rule Space Method* (Taylor & Francis Group, New York, NY, 2009)

S.Q. Yang, S.L. Ding, Theory and method for predicating of valid objects. J. Jiangxi Norm. Univ. **35**(1), 1–4 (2011)

Ensuring Test Quality over Time by Monitoring the Equating Transformations

Marie Wiberg

Abstract One important part of ensuring test quality over consecutive test administrations is to make sure that the equating procedure works as intended, especially when the composition of the test taker groups might change over the administrations. The aim of this study was to examine the equating transformations obtained using one or two previous administrations of a college admissions test that is given twice a year. The test has an external anchor, and thus a nonequivalent group with anchor test design is typically used to equate the test, although other data collection designs are possible. This study examined the use of different equating methods with different data collection designs and different braiding plans. The methods included traditional equating methods and (item response theory) kernel equating methods. We found that different equating methods and different braiding strategies gave somewhat different results, and some reflections on how to proceed in the future are given.

Keywords Kernel equating • IRT kernel equating • NEAT design • NEC design • Braiding plan

Abbreviations

CE	Chained equipercentile equating
FEG	Frequency estimation with equivalent groups design
FEN	Frequency estimation with nonequivalent groups with anchor test design
KEC	Chained kernel equating
KECIRT	Chained IRT kernel equating
KEG	kernel equating with equivalent groups design
KEP	Poststratification kernel equating
KEPIRT	Poststratification IRT kernel equating
NEC	Nonequivalent groups with raw covariates

M. Wiberg (✉)
Department of Statistics, USBE, Umeå University, Umeå, Sweden
e-mail: marie.wiberg@umu.se

L.A. van der Ark et al. (eds.), *Quantitative Psychology*, Springer Proceedings in Mathematics & Statistics 196, DOI 10.1007/978-3-319-56294-0_22

1 Introduction

When a standardized achievement test is administered consecutively over several years, it is important to examine the quality of the test over time in terms of validity and reliability (Wiberg and von Davier 2017). Several aspects of reliability and validity can be examined, and in this paper the focus is on the equatings performed over time. In order to examine this, the test used in this study included an anchor test that was administered over several administrations, and because the anchor test was identical over time, the nonequivalent groups with anchor test (NEAT) design could be used when performing equating. In this paper, traditional equating methods were compared with kernel equating methods under two different braiding plans. The examined test is a college admissions test; thus, the labor market tends to affect how many people take the test. If unemployment is low, fewer persons take the test; thus, the groups of test takers are not necessarily homogenous in terms of their background over the years. To determine if this has an impact on the equating, a kernel equating with the nonequivalent groups with covariates (NEC) design (Wiberg and Bränberg 2015) was also included in the comparison. In the past, this test has used an equivalent groups (EG) design, so this was included even though it has been shown that the EG assumption is questionable (Lyrén and Hambleton 2011). The aim of this study was to examine the equating transformations of a college admissions test with an external anchor using different equating methods with different data collection designs and two different braiding plans.

In previous research, Livingston et al. (1990) examined which combinations of sampling and equating methods work best by comparing the Tucker, Levine equally reliable, chained equipercentile, frequency estimation, and item response theory (IRT) equating methods using the three-parameter logistic IRT model with either representative samples or matched samples. In their study, the IRT and Levine methods showed agreement with each other, and the chained equipercentile method had low bias in the representative samples. Mao et al. (2006) found only trivial differences when comparing the results of traditional equipercentile equating with those of kernel equating in the EG design and with poststratification equating results using a NEAT design with real data. Liu and Low (2008) compared the use of traditional and kernel equating methods in two scenarios—equating to a very different population and equating to a similar population. The overall conclusions were that traditional and kernel equating methods were comparable and that they gave similar results when the populations were similar on the anchor score distribution even though they rest on different assumptions. If the test group changed, the equating methods gave different results.

This study is different from these previous studies in several important aspects. First, it used the whole group who took the anchor test to perform the equating. Second, it compares not only common kernel equating methods with traditional methods but also with equating with covariates under a NEC design. Third, it includes IRT observed-score kernel equating instead of the more studied IRT observed-score equating. Fourth, because the same anchor test was used for a large number of administrations, comparisons could be made over several administrations using different braiding plans.

The rest of the paper is structured as follows. In Sect. 2, the college admissions test is described. Section 3 contains brief descriptions of the examined equating methods. In Sect. 4, the empirical study is described followed by the results in Sect. 5. Section 6 contains some concluding remarks.

2 The College Admissions Test

The Swedish Scholastic Assessment Test (SweSAT) is a college admissions test that is given twice a year (A = spring and B = Fall). The SweSAT is a paper-and-pencil multiple-choice test with binary responses with 160 items divided into two equally sized subsections—quantitative and verbal. The two sections are equated separately. In this paper we used the quantitative section, which consists of items covering data sufficiency, mathematics, quantitative comparisons, diagrams, tables, and maps.

Each time the SweSAT is administered, a subsample of the test groups is given a 40-item quantitative anchor test the rest of the test takers get tryout items. All test takers fill out a background questionnaire, which includes questions about their age, gender, and educational background. The test scores are assumed to be independent between test administrations. Test takers are allowed to repeat the test as many times as they like, and only the highest score is used for admissions to university. A test score can be used for 5 years. Although test takers can repeat the test, it is unlikely that there are any repeaters in the sample who took the anchor test because the anchor test is administered in different cities at each administration. Table 1 gives some descriptive statistics for the examined anchor test and total test for the quantitative section of the SweSAT. It is evident that the anchor test score distributions are not identical over time, and this might affect the equating. In the later subsequent empirical study, the focus was on administrations 1 (11B), 7 (14B), and 9 (15B), which are marked in bold in Table 1.

3 Equating Methods

The equating methods used in this study are all equipercentile equating methods and are described briefly below for the NEAT design, including frequency estimation, chained equipercentile equating, chained kernel equating, poststratification kernel equating, IRT observed-score poststratification kernel equating, and IRT observed-score chained kernel equating. In addition, a NEC design with raw covariates is described, and the EG design was used for both traditional and kernel equating. Details for conducting the described methods in practice are described in González and Wiberg (2017).

Table 1 Means and standard deviations of the anchor scores and the total scores of the quantitative section of the SweSAT

		Anchor quant			Total quant		
Adm	Test season	M	SD	*N*	M	SD	*N*
1	**11 B Fall**	18.40	6.55	5263	37.91	13.43	40,431
2	12 A Spring	18.46	5.98	6465	37.05	12.10	56,358
3	12 B Fall	19.25	6.68	1175	38.07	12.35	43,957
4	13 A Spring	17.97	6.76	6664	37.66	12.28	59,475
5	13 B Fall	16.88	6.28	1997	36.83	12.29	54,033
6	14 A Spring	16.37	6.32	2016	38.28	12.72	76,094
7	**14 B Fall**	16.64	6.62	2783	42.52	13.31	58,840
8	15 A Spring	16.71	6.44	2826	43.00	12.35	74,437
9	**15 B Fall**	17.37	6.11	1052	42.90	12.54	60,008

Adm = Administration, *N* = Number of test takers, *M* = Mean, *SD* = Standard deviation, *Quant* = Quantitative section of the SweSAT

3.1 Frequency Estimation

In frequency estimation (Angoff 1971), one estimates the score distributions of the test forms X and Y in populations P and Q, respectively, for a target population T when a common anchor test A is administered. The necessary assumption is that for both test forms X and Y, the conditional distribution of the total score given each anchor score is the same in both populations. Let x be the scores on X, let y be the scores on Y, and let a be the scores on A. The cumulative distribution functions (CDFs) can then be defined as $F_{XT}(x) = \int F_P(x|\,a)dF_{AT}(a)$ and $F_{YT}(y) = \int F_Q(y|\,a)dF_{AT}(a)$. Percentile ranks are obtained from the CDFs, and the equipercentile equating is defined as

$$\varphi_Y(x) = F_{YT}^{-1}\left(F_{XT}(x)\right). \tag{1}$$

Frequency estimation should only be conducted if the two populations are reasonably similar because it tends to give biased results when group differences are large (Powers and Kolen 2014; Wang et al. 2008). If the populations differ considerably, it is better to use other methods (Kolen and Brennan 2014, p. 146). However, an advantage of frequency estimation under a NEAT design is that the standard errors of equating are somewhat lower than for chained equipercentile equating (Wang et al. 2008).

3.2 *Chained Equipercentile Equating*

The chained equipercentile equating (Dorans 1990; Livingston et al. 1990) connects the CDF of test form X, F_P, to the CDF of test form Y, F_Q, through the CDFs of the anchor test forms H_P and H_Q in populations P and Q, respectively, as follows:

$$\varphi_Y(x) = F_Q^{-1}\left(H_Q\left(H_P^{-1}\left(F_P(x)\right)\right)\right). \tag{2}$$

Chained equipercentile equating does not require a joint distribution of total scores and anchor item scores, and it is less computationally intensive than frequency estimation. A drawback is that it equates a long (total) test with a short (anchor) test, and in general one typically avoids equating tests with large differences because the obtained scores are not necessarily interchangeable. However, von Davier et al. (2004b) have shown that the results obtained from chained equipercentile equating and equipercentile frequency estimation methods are similar if the two populations are equivalent and if the scores on the anchor test and the total test are perfectly correlated. An advantage with chained equipercentile equating is that it tends to give more accurate results in terms of smaller bias than frequency estimation if the two groups of test takers differ substantially (Wang et al. 2008).

3.3 *Poststratification Kernel Equating and Chained Kernel Equating*

Kernel equating (von Davier et al. 2004a) consists of the following five steps:

1. *Presmoothing*. The observed score distributions are fitted to a feasible model. Log-linear models have typically been used, although it is also possible to fit IRT models instead (Andersson and Wiberg 2016).
2. *Estimation of score probabilities*. The estimates of the score probabilities are obtained from the models in step one.
3. *Continuization*. The discrete distributions obtained in step two are made continuous. In traditional equating, linear interpolation is used, while in kernel equating, Gaussian kernel continuization is typically used to estimate a continuous CDF of X, i.e., $F_{h_X}(x)$, and likewise for Y, $F_{h_Y}(y)$, where h_X and h_Y are bandwidths. Gaussian kernel continuization involves selecting bandwidths to control the smoothness of the curves, and a penalty function has traditionally been used to select the bandwidths (von Davier et al. 2004a), although other alternatives exist that give similar results (Häggström and Wiberg 2014).
4. *Equating*. The actual equating is performed. The poststratification kernel equating (KEP) can be defined as

$$\varphi_Y(x) = F_{h_Y}^{-1}\left(F_{h_X}(x)\right), \tag{3}$$

and the chained kernel equating (KEC) can be defined as

$$\varphi_Y(x) = F_{h_Y}^{-1}\left(H_{h_Y}\left(H_{h_X}^{-1}\left(F_{h_X}(x)\right)\right)\right), \tag{4}$$

where H_{h_Y} and H_{h_X} represent the continuized score CDFs for the anchor test taken by either the group who took test form X or the group who took test form Y.

5. *Calculating standard errors of equating (SEE).* Accuracy measures are obtained, including SEE, percent relative error, and mean squared errors. For more details on how to assess an equating transformation, refer to Wiberg and González (2016). An advantage of using either KEP or KEC over the traditional equating method is that one uses a comprehensive framework with easy access to the SEE and other accuracy measures such as the percent relative error.

3.4 IRT Observed-Score Kernel Equating

IRT observed-score kernel equating (Andersson and Wiberg 2016) uses an IRT model in the presmoothing step instead of a log-linear model in the kernel equating framework. If unidimensional IRT models are used, it is important that the assumptions of unidimensionality and local independence are fulfilled when modeling the items. In this study we used the two-parameter logistic IRT model

$$p_{ji} = \frac{\exp\left(a_i\left[\theta_j - b_i\right]\right)}{1 + \exp\left(a_i\left[\theta_j - b_i\right]\right)}, \tag{5}$$

where θ_j is the ability of test taker j, a_i is the item discrimination, and b_i is the item difficulty for item i. In the empirical study, both IRT observed-score kernel equating with poststratification (KEPIRT) and with chained equating (KECIRT) were used. A clear advantage with IRT kernel equating as opposed to kernel equating with log-linear models in the presmoothing step is the possibility to presmooth the data with the same IRT model as is used in the item analysis of the test. A potential disadvantage could be if some of the items do not fit reasonably well the chosen IRT model.

3.5 Kernel Equating with the NEC Design

In kernel equating with raw covariates with a NEC design, one categorizes the test takers into different groups depending on the values on the available covariates. Here, the method proposed by Wiberg and Bränberg (2015) was used. In their method, a poststratification kernel equating with covariates is used, which means that the categorized covariates are used instead of an anchor test in Eq. (3). The assumption is that we can adjust the nonequivalent groups through the test takers' covariates and thus obtain equivalent test takers. A clear advantage with this method is the direct use of covariates to adjust for differences between test taker groups in situations where we know that the groups are nonequivalent and we do not have access to an anchor test. A disadvantage is that the number of categories tends to grow quickly if there are many covariates of interest. A solution to this problem, although not used here, is to use propensity scores in the NEC design instead of the covariates directly, as proposed by Wallin and Wiberg (2017a, b). From the SweSAT administration, we had access to the covariates of gender, age, and educational background, which have previously been shown to have an impact on the SweSAT results (Bränberg et al. 1990). The test takers' gender was coded as 0 for women and 1 for men. Age was divided into five categories (<21, 21–24, 25–29, 30–39, and $\geq$40 years). Educational background was categorized into six categories where the lower levels represent high school programs categorized in the order of the amount of theory in the core subjects of mathematics and language and the higher levels represent education after high school.

4 Empirical Study

The examined equating designs were EG, NEAT, and NEC designs, and the equated methods were frequency estimation with the EG design (FEG), frequency estimation with the NEAT design (FEN), chained equipercentile equating (CE), kernel equating with an EG design (KEG), KEP, KEC, KEPIRT, KECIRT, and NEC with raw covariates (NEC).

Two different braiding plans were examined, including equating from test form 15B directly to test form 11B and equating from test form 15B to test form 11B via test form 14B, indicated by a "v" in the names in Fig. 2 and Table 3. When examining the two different braiding plans, the following equating methods were examined: FEG, KEG, KEP, KEC, and NEC. Because we had access to nine test administrations, a large number of possible braiding plans could have been used, but in this study we chose to focus more closely on these two possible braiding plans. When comparing the equated values, we used the difference that matters (DTM) criterion, which means that score differences larger than $|0.5|$ are of concern. The R package *equate* (Albano 2016) was used to perform the traditional equating methods, and the R package *kequate* (Andersson et al. 2013) was used to perform the kernel equating methods.

5 Results

The CDFs of the three considered tests are given in the upper left corner of Fig. 1, from which it is noticeable that test forms 15B and 14B are similar, while test form 11B is different. In the rest of Fig. 1, the equating transformations from using the nine examined equating methods are shown for the case of equating test form 15B to 14B, equating test form 14B to 11B, and equating test form 15B to 11B. Details of every tenth equated value are given in Table 2 for equating test form 15B directly to 11B. The excluded values follow the same pattern and can be obtained upon request. It is evident that there is a difference in equated scores between different methods and depending on which equating design is used. The equated values for FEG were furthest away from the other methods, although they were somewhat similar to when a NEC design was used. The two IRT kernel equating methods (KEPIRT and KECIRT) gave similar results with almost no DTM between them. The KEP and KEC showed several DTM when compared with their traditional

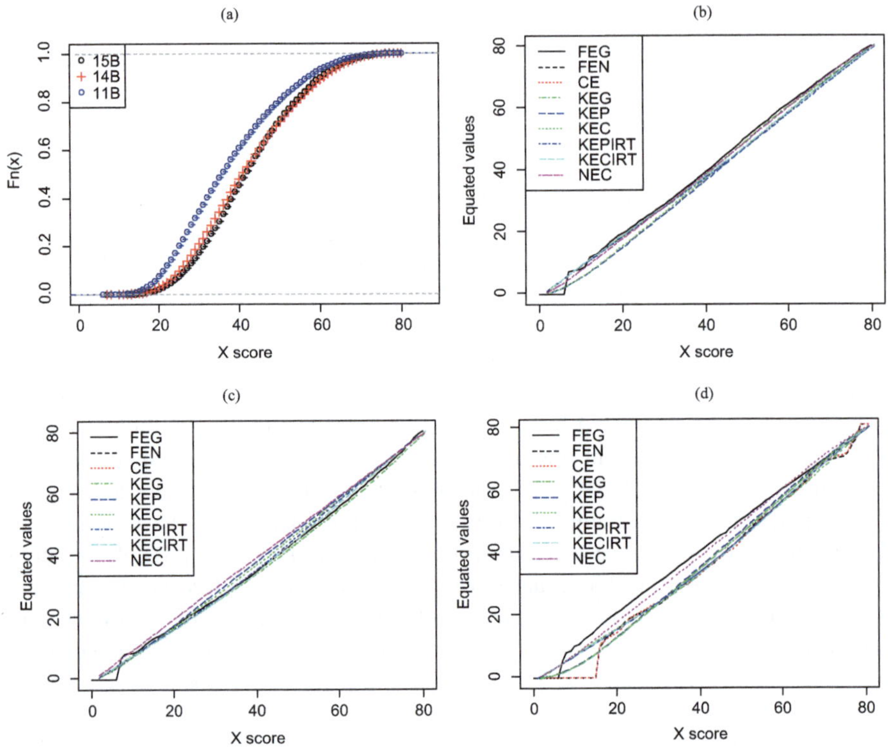

Fig. 1 CDFs of the three administrations 15B, 14B, and 11B. (**a**) Equating transformations for 15B to 14B (**b**), 14B to 11B (**c**), and 15B to 11B (**d**) for FEG, FEN, CE, KEG, KEP, KEC, KEPIRT, KECIRT, and NEC

Table 2 Equated values for the different examined methods showing every tenth value for equating from administration 15B to administration 11B

X	FEG	FEN	CE	KEG	KEP	KEC	KEPIRT	KECIRT	NEC
0	−0.50	−0.50	−0.50	−1.24	−1.02	−1.02	−0.73	−0.65	−0.48
10	9.74	−0.50	−0.50	7.79	3.78	3.72	7.40	7.54	8.09
20	20.40	16.24	15.15	16.11	13.63	13.45	15.79	15.82	18.40
30	30.21	23.96	24.09	24.15	24.81	24.46	24.62	24.40	28.84
40	40.08	34.08	33.83	33.85	36.20	35.58	34.40	33.91	39.41
50	50.24	45.86	45.00	45.48	47.65	46.81	45.38	44.88	50.13
60	59.97	57.07	56.95	56.71	59.07	58.34	57.27	57.08	60.99
70	69.23	68.97	69.52	68.06	70.31	69.86	69.12	69.23	71.53
80	79.15	80.50	80.50	79.98	80.18	80.13	79.90	80.00	80.22

FEG = Frequency estimation in EG design, *FEN* = Frequency estimation in NEAT design, *CE* = Chained equipercentile equating, *KEG* = kernel equating in EG design, *KEP* = Poststratification kernel equating, *KEC* = Chained kernel equating, *KEPIRT* = Poststratification IRT kernel equating, *KECIRT* = Chained IRT kernel equating, *NEC* = kernel equating with raw covariates

equating counterparts FEN and CE, although not for all equated values. Note that the equated values for X scores below seven are all zero because this is a multiple-choice test and no test taker had such low test scores.

In Fig. 2 and Table 3, some of the previously described equating methods are shown when using two different braiding plans. The overall observation is that we obtained different results depending on the braiding plan that was used, although all methods gave similar results for the highest score values. When comparing the direct equating with the equating via test form 14B, all kernel equating methods gave similar results for the higher score values—especially for score values of 60 and above. This is good news because this is an admission test and to have similar results in the upper score scale is more important than to have similar values in the lower score scale. The frequency estimation with an EG design gave very different results if a direct equating was performed (FEG) as compared to an equating via test form 14B (FEGv). Interestingly, this was not seen for the kernel equating with an EG design (KEG and KEGv) that had similar values as the FEGv—which was the braiding plan that equated test form 15B to 11B via test form 14B. NEC and NECv had DTM for many values as well as in comparison to the NEAT design methods.

6 Concluding Remarks

One aspect of the quality over time of a college admission test is the consistency of the equatings over time. This was examined here by comparing the results from different equating methods using two different braiding plans. Because the same common anchor test was administered on each administration, a NEAT design is

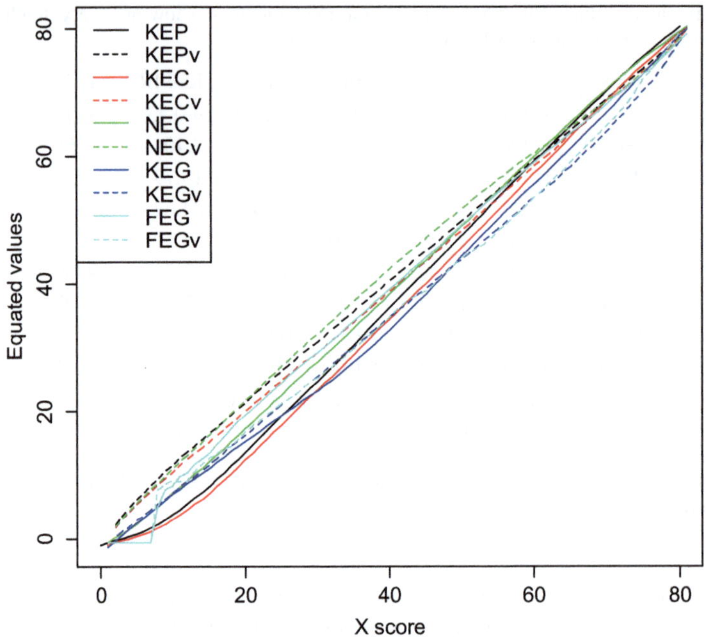

Fig. 2 Different braiding plans when equating directly from test form 15B to 11B or equating from 15B to 11B via 14B (designated by "v") using NEAT, NEC, and EG designs with different equating methods

Table 3 Every tenth equated value when two different braiding plans were used for equating directly from test form 15B to 11B or from 15B to 11B via 14B

X	FEG	FEGv	KEG	KEGv	KEP	KEPv	KEC	KECv	NEC	NECv
0	−0.5	−0.5	−0.32	−0.31	−1.02	0.56	−1.02	0.39	−0.48	0.42
10	9.74	9.03	7.61	8.09	3.78	12.58	3.72	11.44	8.09	12.04
20	20.40	17.67	16.79	17.24	13.63	22.38	13.45	20.96	18.40	22.73
30	30.21	26.21	26.21	26.42	24.81	31.91	24.46	30.28	28.84	33.18
40	40.08	35.08	35.69	35.60	36.20	41.38	35.58	39.69	39.41	43.25
50	50.24	45.03	45.20	44.83	47.65	50.82	46.81	49.38	50.13	52.71
60	59.97	54.52	54.79	54.36	59.07	60.28	58.34	59.28	60.99	61.34
70	69.34	65.98	64.76	65.10	70.31	69.82	69.86	69.34	71.53	69.49
80	79.15	80.24	78.69	79.55	80.18	79.92	80.13	79.80	80.22	79.74

FEG=Frequency estimation with EG design, *KEG*=kernel equating with NEAT design, *KEP*=Poststratification kernel equating, *KEC*=Chained kernel equating, *NEC*=kernel equating with raw covariates, *v* via test form 14B

currently used when performing the equating. Also, because SweSAT in the past has been equated with an EG design, such a comparison was also included here even though the EG assumption has been shown to be questionable (Lyrén and Hambleton 2011). Finally, a NEC design was included because the composition of the test taker groups is known to change over administrations. Both traditional equating and kernel equating were examined, and we found that different methods gave different results. An EG design gave equated values furthest away from the NEAT design results—although somewhat close to the NEC design results. This could possibly be because the EG assumption is violated, something Lyrén and Hambleton (2011) found when they examined other SweSAT test forms. Interestingly, the IRT kernel equating methods had almost no DTM between each other, and this should be explored in the future because the item analysis of this test includes IRT analyses.

The use of different braiding plans gave different results for all methods, and this is something that should be taken into consideration when equating test scores. A way to handle this is to use more than one link (Kolen and Brennan 2014). However, one should probably not use circular equating because Wang et al. (2000) showed that such an approach cannot handle systematic errors very well. Interestingly there were large differences in the kernel equating methods both when comparing equating methods and when using different braiding plans. This is contrary to the study of Mao et al. (2006) in which they only found trivial differences in a real data example when comparing the equating results of traditional equipercentile equating with those of kernel equating in the EG design and to poststratification equating with a NEAT design. It is, however, possible that the examined groups are different, and thus the result is in line with Liu and Low (2008) who compared the use of traditional and kernel equating methods. They found that if the groups taking two test forms are quite different from each other, then different equating methods tend to give different results. Because neither of those two studies examined different braiding plans, and because neither of them examined KEC, IRT observed-score kernel equating, or a NEC design, it is important to examine this more closely in the future—especially with a simulation study where one can keep different factors under control and where one knows the true equated values. In addition, longer equating chains and different braiding plans should be examined. One could also examine the possibility of using more than one equating transformation and the possibility of averaging them, in line with the proposal of Holland and Strawderman (2011). Finally, one could examine other equating methods and other covariates in the NEC design because that might be a good alternative if we do not have access to an anchor test and the test groups cannot be assumed to be equivalent.

Acknowledgments This research was funded by the Swedish Research Council (Grant No. 2014-578).

References

A.D. Albano, Equate: an R package for observed-score linking and equating. J. Stat. Softw. **74**(8), 1–36 (2016)

B. Andersson, M. Wiberg, Item response theory observed-score kernel equating. Psychometrika **82**(1), 48–66 (2016). doi:10.1007/s11336-016-9528-7

B. Andersson, K. Bränberg, M. Wiberg, Performing the kernel method of test equating with the package kequate. J. Stat. Softw. **55**(6), 1–25 (2013)

W.H. Angoff, in *Educational Measurement*, 2nd edn., ed. by R. L. Thorndike. Scales, norms and equivalent scores (American Council on Education, Washington, DC, 1971), pp. 508–600

K. Bränberg, W. Henriksson, H. Nyquist, I. Wedman, The influence of sex, education and age on the scores on the Swedish scholastic aptitude test. Scand. J. Educ. Res. **34**, 189–203 (1990)

N.J. Dorans, Equating methods and sampling designs. Appl. Meas. Educ. **3**(1), 3–17 (1990)

J. González, M. Wiberg, *Applying Test Equating Methods Using R* (Springer, Cham, 2017). doi:10.1007/978-3-319-51824-4

J. Häggström, M. Wiberg, Optimal bandwidth in observed-score kernel equating. J. Educ. Meas. **51**(2), 201–211 (2014)

P.W. Holland, W.E. Strawderman, in *Statistical Models for Test Equating, Scaling, and Linking*, Chapter 6, ed. by A. A. von Davier. How to average equating functions, if you must (Springer, New York, NY, 2011), pp. 109–122

M. Kolen, R. Brennan, *Test Equating, Scaling, and Linking: Methods and Practices*, 3rd edn. (Springer-Verlag, New York, NY, 2014)

J. Liu, A.C. Low, A comparison of the Kernel equating method with traditional equating methods using SAT® data. J. Educ. Meas. **45**(4), 309–323 (2008)

S.A. Livingston, N.J. Dorans, N.K. Wright, What combination of sampling and equating methods works best? Appl. Meas. Educ. **3**(1), 73–95 (1990)

P.-E. Lyrén, R.K. Hambleton, Consequences of violated the equating assumptions under the equivalent group design. Int. J. Test. **36**(5), 308–323 (2011)

X. Mao, A.A. von Davier, S.L. Rupp, Comparisons of the kernel equating method with the traditional equating methods on Praxis data. ETS Research Report, RR-06-30, 2006

S. Powers, M.J. Kolen, Evaluating equating accuracy and assumptions for groups that differ in performance. J. Educ. Meas. **51**, 39–56 (2014)

A.A. von Davier, P. Holland, D. Thayer, *The Kernel Method of Test Equating* (Springer-Verlag, New York, NY, 2004a)

A.A. von Davier, P. Holland, D. Thayer, The chain and post-stratification methods for observed-score equating: their relationship to population invariance. J. Educ. Meas. **41**, 15–32 (2004b)

G. Wallin, M. Wiberg, in *Quantitative Psychology–The 81st Annual Meeting of the Psychometric Society, Asheville, North Carolina, 2016*, ed. by L. A. van der Ark, M. Wiberg, S. A. Culpeppar, J. A. Douglas, W.-C. Wang. Non-equivalent groups with covariates design using propensity scores for kernel equating (Springer, New York, NY, 2017a)

G. Wallin, M. Wiberg, Propensity scores in kernel equating under the non-equivalent groups with covariates design. Manuscript submitted for publication, 2017b

T. Wang, M.J. Hanson, D.J. Harris, The effectiveness of circular equating as a criterion for evaluating equating. Appl. Psychol. Meas. **24**, 195–210 (2000)

T. Wang, W.-C. Lee, R.L. Brennan, M. Kolen, A comparison of the frequency estimation and chained equipercentile methods under the common-item nonequivalent groups design. Appl. Psychol. Meas. **32**(8), 632–651 (2008)

M. Wiberg, K. Bränberg, kernel equating under the non-equivalent groups with covariates design. Appl. Psychol. Meas. **39**(5), 349–361 (2015)

M. Wiberg, J. González, Statistical assessment of estimated transformations in observed-score equating. J. Educ. Meas. **53**(1), 106–125 (2016)

M. Wiberg, A.A. von Davier, Examining the impact of covariates on anchor tests to ascertain quality over time in a college admissions test. Int. J. Test., 1–22 (2017). doi:10.1080/15305058.2016.1277357

An Illustration of the Epanechnikov and Adaptive Continuization Methods in Kernel Equating

Jorge González and Alina A. von Davier

Abstract Gaussian kernel continuization of the score distributions has been the standard choice in kernel equating. In this paper we illustrate the use of both the Epanechnikov and adaptive kernels in the actual equating step using the R package **SNSequate** (González, J Stat Softw 59(7):1–30, 2014). The two new kernel equating methods are compared with each other and with the Gaussian, logistic, and uniform kernels.

Keywords Kernel equating • Epanechnikov kernel • Adaptive kernel • Continuization

1 Introduction

Equating methods are commonly used to ensure the comparability of test scores from different test forms (Kolen and Brennan 2014; von Davier 2011; González and Wiberg 2017). Such comparability is obtained using a so-called equating function, which maps the scores of one test form into the scale of the other. If X and Y are the random variables representing the scores in test forms X and Y, then the *equipercentile* equating function is defined as $\varphi(x) = F_Y^{-1}(F_X(x))$, where F_X and F_Y are the cumulative distribution functions (CDFs) of X and Y, respectively. Typically, X and Y are *number-correct scores* whose possible values are consecutive integers, yielding to discrete score distributions. In common practice, continuous approximations of F_X and F_Y are used to obtain the equating function φ by *continuizing* the discrete score random variables X and Y. Different continuization methods

J. González (✉)
Faculty of Mathematics, Pontificia Universidad Católica de Chile, Av. Vicuña Mackenna 4860, Macul, Santiago, Chile
e-mail: jorge.gonzalez@mat.uc.cl

A.A. von Davier
ACTNext by ACT, Inc., 500 ACT Dr. (18), Iowa City, IA 52243-0168, USA
e-mail: Alina.vonDavier@act.org

L.A. van der Ark et al. (eds.), *Quantitative Psychology*, Springer Proceedings in Mathematics & Statistics 196, DOI 10.1007/978-3-319-56294-0_23

define different equating methods, each producing parametric, nonparametric, or semiparametric statistical inference about φ (González and von Davier 2013).

A popular semiparametric approach is kernel equating (von Davier et al. 2004), in which kernel smoothing techniques are used to obtain continuous approximations of the test score distributions. Gaussian kernel continuization of the score distributions has been the standard selection in kernel equating, although alternative kernels such as the logistic and uniform have recently been proposed for continuization (Lee and von Davier 2011). Cid and von Davier (2015) explored the use of the Epanechnikov and adaptive kernels for the estimation of score densities, as potential alternative approaches to reducing boundary bias.

Rather than exploring kernel density estimation of score distributions, in this paper we propose the use of both the Epanechnikov (Epanechnikov 1969) and adaptive kernels (Silverman 1986) as two new continuization methods for kernel equating. In comparison with the Gaussian, the Epanechnikov kernel has bounded support, and it is optimal in the sense that it minimizes the asymptotic mean integrated squared error (e.g., Silverman 1986). On the other hand, adaptive kernels are a more flexible alternative to fixed kernel density estimators as they allow the smoothing parameter to vary across the data points in the distribution. It is of interest to evaluate the performance of the Epanechnikov and adaptive continuization in the context of kernel equating due to the fact that test scores are usually bounded above and below and that sometimes the score distributions might show atypical scores, such as gaps and spikes.

The rest of this paper is organized as follows. The first section gives a brief overview of the kernel equating framework, including the details on the continuization step. Then, the Epanechnikov and adaptive kernels are introduced as two alternatives for continuization in kernel equating. The two new continuization methods are illustrated and compared in a subsequent section. The paper concludes with final comments and discussion.

2 Kernel Equating and Continuization

In this section we briefly describe the steps involved in the kernel equating framework, paying special attention to the continuization step.

2.1 A Quick Overview of Kernel Equating

In the kernel equating approach (Holland and Thayer 1989; von Davier et al. 2004), continuous approximations of the discrete score distributions F_X and F_Y are obtained using kernel smoothing techniques (Silverman 1986). Such approximation is achieved by defining a *continuized* random variable which is a function of: (1) the originally discrete score random variable, (2) a continuous random variable characterizing the kernel, and (3) a parameter controlling the degree of smoothness for the continuization. The conversion of the scores is based on the estimated equating transformation:

$$\hat{\varphi}(x; \mathbf{r}, \mathbf{s}) = F_{h_Y}^{-1}(F_{h_X}(x; \hat{\mathbf{r}}); \hat{\mathbf{s}}) = \hat{F}_{h_Y}^{-1}(\hat{F}_{h_X}(x)), \quad (1)$$

where h_X and h_Y are parameters which control the degree of smoothness in the continuization and $\hat{\mathbf{r}}$ and $\hat{\mathbf{s}}$ are vectors of estimated score probabilities with coordinates defined as $r_j = \Pr(X = x_j)$ $(j = 1, \ldots, J)$ and $s_k = \Pr(Y = y_k)$ $(k = 1, \ldots, K)$, respectively, with x_j and y_k being possible values of X and Y, respectively. Both $\hat{\mathbf{r}}$ and $\hat{\mathbf{s}}$ are obtained using the so-called *design functions* (DF), which take into account the chosen data collection design in the estimation. This process is made after presmoothing the discrete (univariate and/or bivariate) observed score frequency distributions by typically using log-linear models. The accuracy of the estimated $\hat{\varphi}(x)$ is assessed with different measures, particularly the standard error of equating.

The main stages in the previous description of the kernel equating method have been summarized in the following five steps (see e.g., von Davier et al. 2004): (1) presmoothing, (2) estimation of score probabilities, (3) continuization, (4) computing the equating transformation, and (5) computation of accuracy measures. In the following section, we give more details on the continuization step.

2.2 Continuization

The continuization step involves the use of a continuous random variable which characterizes the kernel that will be used for equating. More precisely, a *continuized* score $X(h_X)$ is defined as a function of the originally discrete score X, a continuous random variable V, and a parameter h_X in the form $X(h_X) = X + h_X V$. However, in order to preserve the first two moments of X, the following definition is used in practice:

$$X(h_X) = a_X(X + h_X V) + (1 - a_X)\mu_X , \quad (2)$$

where $a_X^2 = \frac{\sigma_X^2}{\sigma_X^2 + \sigma_V^2 h_X^2}$, $\mu_X = \sum_j x_j r_j$, and $\sigma_X^2 = \sum_j (x_j - \mu_X)^2 r_j$ $(j = 1, \ldots, J)$. If $R_{jX}(x) = \frac{x - a_X x_j - (1 - a_x)\mu_X}{a_X h_X}$, then Theorem 4.2 in von Davier et al. (2004) establishes that the kernel smoothing of F_X, defined by

$$F_{h_X}(x) = \sum_j r_j K(R_{jX}(x)) , \quad (3)$$

is exactly the CDF of the continuized variable $X(h_X)$. Here, $K(\cdot)$ is the kernel associated to the random variable V. Similar definitions are used to obtain $F_{h_Y}(y)$, the continuous approximation of F_Y.

It should be noted that neither the presmoothing (step 1) nor the estimation of score probabilities (step 2) depends on the kernel used in the continuization (step 3). The type of kernel used will only influence the computation of the equating transformation and the calculation of the standard error of equating. The flexibility and modularity of the kernel equating approach allow for the ability to easily extend existing methods and implement new ones. In the following sections, two alternative continuization methods are explored to be used in the kernel equating framework.

3 Epanechnikov and Adaptive Continuization

3.1 Epanechnikov Kernel

The Epanechnikov kernel is defined for a random variable V with density function

$$f(v) = \frac{3}{4}(1 - v^2)1_{|v|\leq 1}$$

and corresponding CDF

$$F(v) = \begin{cases} 0 & v < -1 \\ \dfrac{3v - v^3 + 2}{4} & -1 \leq v \leq 1 \\ 1 & v > 1 \end{cases}$$

for which it is easily verified that $E(V) = 0$ and $Var(V) = 1/5$. The Epanechnikov continuization thus becomes

$$F_{h_X}(x) = \sum_{\mathcal{J}} \frac{r_j\left(3R_{jX}(x) - R_{jX}^3(x) + 2\right)}{4} + \sum_{\mathcal{K}} r_j \tag{4}$$

where $\mathcal{J}$ is the set of all j such that $-1 \leq R_{jX} \leq 1$ and $\mathcal{K}$ is the set of j such that $R_{jX} > 1$. Similar steps are followed for Y scores to obtain $F_{h_Y}(y)$ so that one obtains $\varphi(x) = F_{h_Y}^{-1}(F_{h_X}(x))$. Note that the derivative of $F_{h_X}(x)$ in (4) can easily be obtained so that the penalty function method for the selection of the bandwidth parameter applies straightforwardly.

3.2 Adaptive Kernel

Adaptive kernels (e.g., Silverman 1986) allow the bandwidth parameter h_X to vary across the data points in the score distribution. The kernel continuization has the form

$$F_{h_{jX}}(x) = \sum_j r_j K\left(\frac{x - a_{jX}x_j - (1 - a_{jX})\mu_X}{a_{jX}h_{jX}}\right),$$

where $a_{jX} = \frac{\sigma_X^2}{\sigma_X^2 + h_{jX}^2}$, $h_{jX} = \lambda_j h_X$, $(j = 1, \ldots, J)$, and λ_j are local bandwidth factors. For illustrations, we will consider a Gaussian adaptive kernel so that $K(\cdot) = \Phi(\cdot)$.

Silverman (1986) suggested the following steps to obtain λ_j: first, find a pilot estimate of the density, $\widetilde{f}(t)$, such that $\widetilde{f}(X_j) > 0\ \forall j$; second, define a local bandwidth factor λ_j as

$$\lambda_j = \left(\frac{\widetilde{f}(X_j)}{g}\right)^{-\alpha},$$

where g is the geometric mean of $\widetilde{f}(X_j)$ and α is a sensibility parameter satisfying $0 \leq \alpha \leq 1$. Silverman's recommendation is to use $\alpha = 0.5$. To obtain λ_j, we propose as a pilot estimate

$$\widetilde{f}(x) = \sum_j r_j \phi\left(\frac{x - a_X x_j - (1 - a_X)\mu_X}{a_X h_X}\right)\frac{1}{a_X h_X}, \tag{5}$$

where h_X can be obtained using any bandwidth selection method. In the illustrations we use the penalty method to select the bandwidth parameters.

Following the strategy described above, we can also obtain $F_{h_{jY}}$, so that the adaptive kernel equating transformation becomes $\varphi(x) = F_{h_{jY}}^{-1}(F_{h_{jX}}(x))$.

4 Illustrations

4.1 Data Generation

To illustrate the use of the Epanechnikov and adaptive kernel equating functions, data were simulated using the beta-binomial model (Keats and Lord 1962). The simulated data came from five different score distributions that covered different types of shapes in the score distributions including one symmetric, one positively skewed, one negatively skewed, and two slightly negatively skewed distributions.

Table 1 Information used to simulate the score distributions

Type of distribution	Test form	Number of items	*n*	Mean	SD
Symmetric	X	50	2000	25.82	7.28
	Y	50	1800	25.82	7.28
Positively skewed	X	30	1000	6.76	5.12
	Y	30	1200	6.75	5.11
Negatively skewed	X	30	1000	23.75	5.59
	Y	30	1200	23.78	5.62
Slightly negatively skewed	X	40	2354	27.06	8.19
	Y	40	2000	27.06	8.19
Slightly negatively skewed	X	50	6103	32.93	8.04
	Y	50	6103	32.93	8.04

Table 1 shows the number of items, sample size (n), mean, and standard deviation (SD) used to generate each of the five score distributions. Figure 1 shows the shape of these score distributions.

4.2 *Evaluation Criteria*

The two proposed kernel equating methods were compared to each other and with the Gaussian, logistic, and uniform kernels. Differences in equated values were evaluated using equipercentile equating as criterion, and compared against the *difference that matters* (DTM, Dorans and Feigenbaum 1994), originally defined as the difference between equated scores and scale scores that are larger than half of a reported score unit. The standard error of equating (SEE) was also evaluated for each of the five methods and is defined as:

$$\mathrm{SEE}_Y(x) = \hat{\sigma}_Y(x) = \sqrt{\mathrm{Var}(\hat{\varphi}(x))}, \tag{6}$$

where $\hat{\varphi}(x)$ is defined in Eq. (1); the delta method was used to calculate the variance. The **R** package **SNSequate** (González 2014) was used in all calculations.

4.3 *Results*

The results are graphically summarized in Figs. 2, 3, 4, 5, 6 for the five score distributions. In all cases, differences in equated values are practically identical for all the kernel equating functions evaluated and for each of the score distributions, except for some ranges of the score scale in the symmetric, positively and negatively skewed case where the adaptive kernel shows to be slightly different from the others.

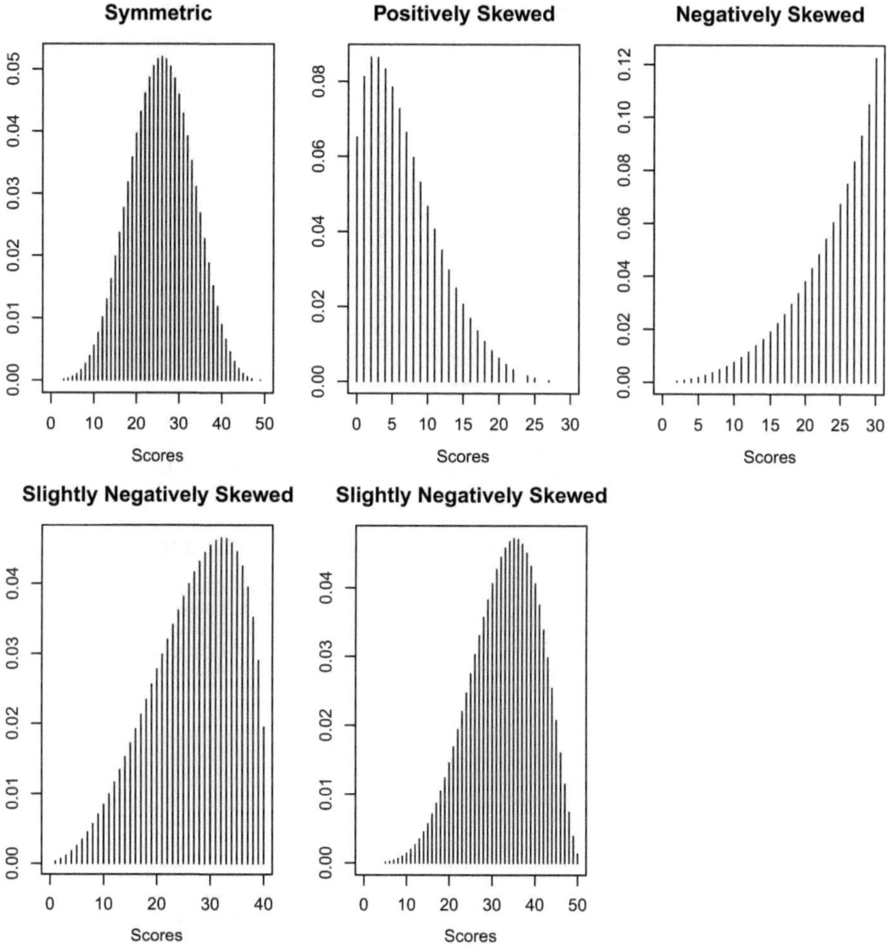

Fig. 1 Distributions used in the simulations

The differences in equated values are larger at the extremes of the score scales, at the upper part for the positively skewed distributions (Fig. 3) and at the lower part for the negatively and slightly negatively skewed distributions (Figs. 4, 5, and 6). In all cases these differences are larger than one score point, which could be problematic according to DTM.

Regarding the SEE, it can be seen that in all cases for most of the score scale, the adaptive kernel produced more accurate results than the other kernel equating methods. The other four kernels yield to very similar results in terms of SEE, except in the case of symmetric and positively skewed distribution when the Epanechnikov kernel performs slightly better than the Gaussian, logistic, and uniform kernels at the upper part of the score scale.

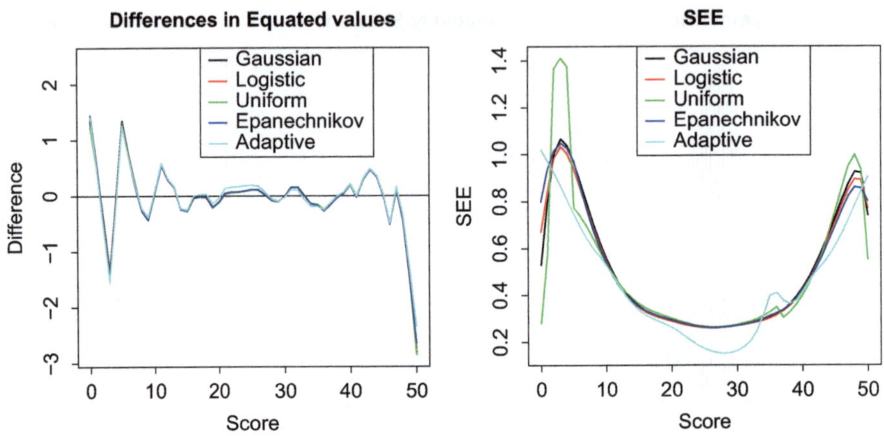

Fig. 2 Differences in equated values (*left*) and SEE (*right*) for score data coming from a symmetric distribution

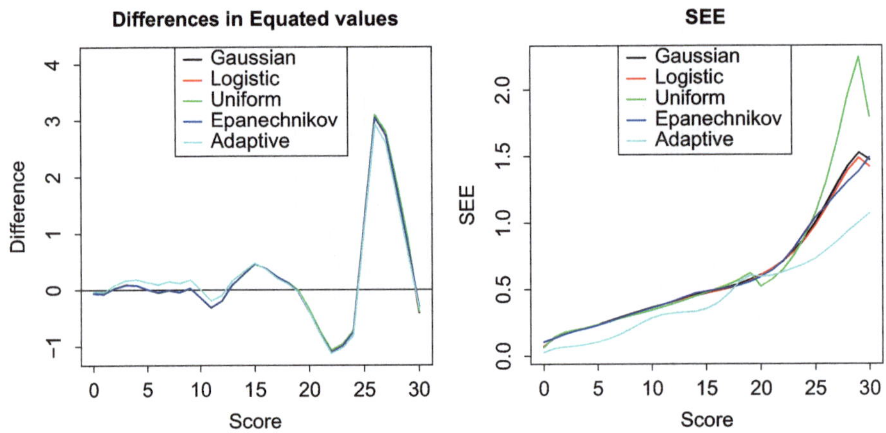

Fig. 3 Differences in equated values (*left*) and SEE (*right*) for score data coming from a positively skewed distribution

5 Concluding Remarks

Cid and von Davier (2015) examined the used of the Epanechnikov and the adaptive kernels for density estimation of score distributions. In this paper we have evaluated the performance of five kernel equating functions in the actual equating step. In all cases, all the continuization kernels lead to similar equated values, and when comparing the Gaussian kernel with equipercentile equating, the results agree with those in Cid and von Davier (2015). The adaptive kernel provided more accurate equated values, especially when both positively and negatively skewed score distributions are concerned. The Epanechnikov kernel performed slightly

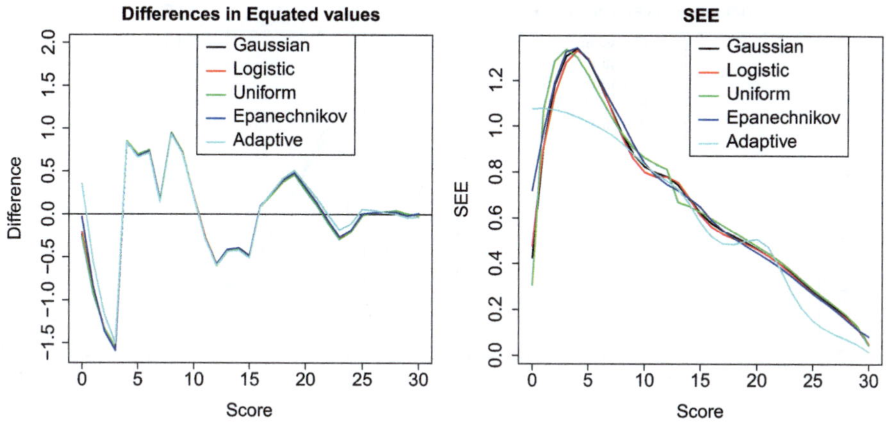

Fig. 4 Differences in equated values (*left*) and SEE (*right*) for score data coming from a negatively skewed distribution

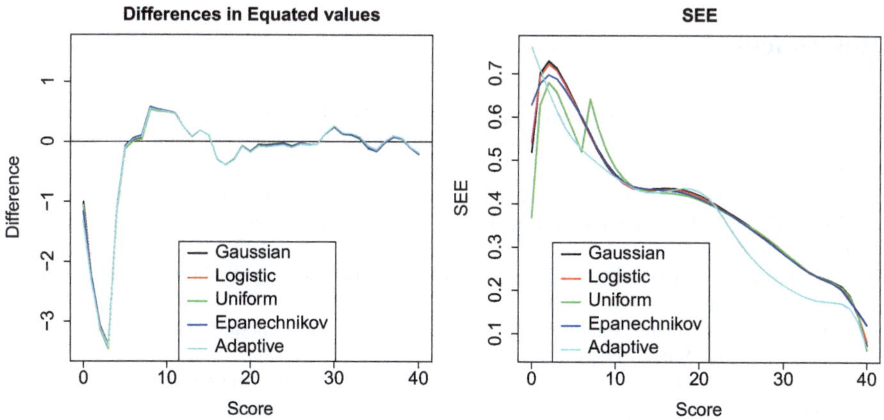

Fig. 5 Differences in equated values (*left*) and SEE (*right*) for score data coming from a slightly negatively skewed distribution

better than the Gaussian, logistic, and uniform kernels at the upper part of the score scale for the case of symmetric and positively skewed distribution. The Epanechnikov and adaptive continuization methods proposed seem to be a valid and competitive alternative to current continuization methods.

Acknowledgements The first author acknowledges partial support of grant Fondecyt 1150233. The authors thank Ms. Laura Frisby, ACT, for editorial help.

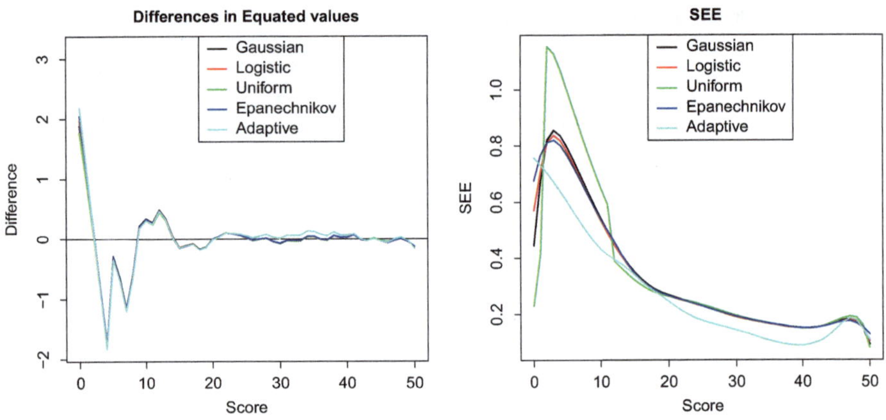

Fig. 6 Differences in equated values (*left*) and SEE (*right*) for score data coming from a slightly negatively skewed distribution

References

J.A. Cid, A.A. von Davier, Examining potential boundary bias effects in kernel smoothing on equating: an introduction for the adaptive and Epanechnikov kernels. Appl. Psychol. Meas. **39**(3), 208–222 (2015)

N.J. Dorans, M.D. Feigenbaum, Equating issues engendered by changes to the SAT and PSAT/NMSQT. Technical issues related to the introduction of the new SAT and PSAT/NMSQT (1994), pp. 91–122

V.A. Epanechnikov, Non-parametric estimation of a multivariate probability density. Theory Probab. Appl. **14**, 153–158 (1969)

J. González, SNSequate: standard and nonstandard statistical models and methods for test equating. J. Stat. Softw. **59**(7), 1–30 (2014)

J. González, M. von Davier, Statistical models and inference for the true equating transformation in the context of local equating. J. Educ. Meas. **50**(3), 315–320 (2013)

J. González, M. Wiberg, *Applying Test Equating Methods Using R* (Springer, New York, 2017)

P.W. Holland, D.T. Thayer, The kernel method of equating score distributions. Technical report, Educational Testing Service, Princeton, NJ, 1989

J.A. Keats, F.M. Lord, A theoretical distribution for mental test scores. Psychometrika **27**(1), 59–72 (1962)

M.J. Kolen, R.L. Brennan, *Test Equating, Scaling, and Linking: Methods and Practices*, 3rd edn. (Springer, New York, 2014)

Y.H. Lee, A.A. von Davier, Equating through alternative kernels, in *Statistical Models for Test Equating, Scaling, and Linking*, vol. 1, ed. by A.A. von Davier (Springer, New York, 2011), pp. 159–173

B.W. Silverman, *Density Estimation for Statistics and Data Analysis* (Chapman and Hall, London, 1986)

A.A. von Davier, *Statistical Models for Test Equating, Scaling, and Linking* (Springer, New York, 2011)

A.A. von Davier, P.W. Holland, D.T. Thayer, *The Kernel Method of Test Equating* (Springer, New York, 2004)

(The Potential for) Accumulated Linking Error in Trend Measurement in Large-Scale Assessments

Lauren Harrell

Abstract Trend measurement is one of the key priorities in large-scale survey assessments such as the National Assessment of Educational Progress (NAEP); however, minimal research has been conducted into the stability of the trend comparisons over a long chain of links as new items are phased into the item pool and older items are discontinued. The potential for linking error in trend hypothesis tests in large-scale assessments is evaluated through a simulation study across ten assessment cycles in which items are discontinued and replaced. The data are generated based on the observed means of Grade 8 Reading and Grade 8 Mathematics over ten cycles. The estimated difference in means between two years is compared to the true data-generating difference for 500 replications under each condition. The mean squared error of trend comparisons tends to modestly increase with the number of linkages. The bias of trend comparisons across a chain of links appeared to be small yet proportional to the magnitude of the difference rather than the length of the chain.

Keywords NAEP • Accumulated linking error • Trend measurement • IRT linking

1 Introduction

One of the primary purposes of the National Assessment of Educational Progress (NAEP) is to compare group and subgroup mean scale scores over time. The text of the NAEP report card focuses on comparisons with the immediately preceding year as well as the comparisons to the first year of the trend line. For example, the headline of the results overview for the Grades 4 and 8 2015 Mathematics Report Card stated, "Both fourth- and eighth- grade students score lower in mathematics than in 2013; scores higher than in 1990" (U.S. Department of Education, National Center for Education Statistics 2015c). In the graphics of the release, additional

L. Harrell (✉)
National Center for Education Statistics, 550 12th St. SW, Washington, DC 20202, USA
e-mail: lauren.harrell@ed.gov

L.A. van der Ark et al. (eds.), *Quantitative Psychology*, Springer Proceedings in Mathematics & Statistics 196, DOI 10.1007/978-3-319-56294-0_24

comparisons are made between the current assessment cycle and all preceding assessment years for which there is data. Comparisons across all combinations of years can be found in the NAEP Data Explorer (U.S. Department of Education, National Center for Education Statistics 2015a).

This manuscript is intended to highlight the potential for increased uncertainty in trend comparisons that the existing comparison procedures in NAEP do not adjust for. Several research studies are underway within the NAEP program to evaluate and develop methods for estimating linking error as NAEP transitions to digitally based assessments in the future, and this manuscript is focused on highlighting some of the challenges and potential issues for cross-year comparisons across a chain of links. Specifically, a simulation study was conducted to evaluate whether there is potential for accumulated linking error in NAEP trend comparisons. The guiding question behind this line of research is: should longer-term comparisons have the same degree of accuracy as comparisons between two consecutive years, or should additional uncertainty over an increased number of links be expected?

1.1 NAEP Plausible Values

NAEP is designed to provide inferences on populations and subgroups, not individual students. To limit the response burden on the test takers, NAEP utilizes a balanced-incomplete-block (BIB) design in which each student receives only two 25-minute blocks, instead of up to ten operational blocks for a given grade level and subject administered within a year. Information on the individual students is brevity of the assessment, and individual score estimates, if produced, would be subject to a high degree of measurement error. Plausible values (PVs) were introduced to provide consistent estimates of population characteristics and associated standard errors, which account for both sampling and measurement variability (Mislevy et al. 1992). Instead of an individual point estimate of a scale score, 20 PVs are drawn from a posterior distribution of the latent trait θ, which has the form

$$f(\theta_i|x_i, y_i, \beta, \Gamma, \Sigma) \propto \phi(\theta_i; \Gamma x_i, \Sigma) \prod_{j=1}^{n_i} f(y_{ij}|\theta_i; \beta_j), \quad (1)$$

where:

- y_i are the responses of student i to the n_i test items presented,
- x_i are the responses of student i to the survey questionnaire,
- β_j are the item response theory (IRT) parameters for item j,
- Γ is the matrix of the latent regression parameters,
- Σ is the variance-covariance matrix (for multidimensional models, otherwise Σ represents the scalar variance), and
- $\phi(\theta_i; \Gamma x_i, \Sigma)$ is a normal distribution with mean $\Sigma' x_i$ and variance Σ.

The posterior distribution and PVs incorporate information from the survey questionnaire and latent regression in addition to the likelihood function based on the observed test item responses and estimated IRT parameters.

1.2 Comparing Means Across Years in NAEP

For any pair of years i and j, we compare the group means using an independent sample t-test,

$$T = \frac{A_i - A_j}{\sqrt{S(A_i)^2 + S(A_j)^2}}, \tag{2}$$

where A_i and A_j are the group mean (or quantity) for years i and j and $S(A_i)^2$ and $S(A_j)^2$ are the associated standard errors for the mean within each year (U.S. Department of Education, National Center for Education Statistics 2015b). The standard errors of any cross-sectional quantity in NAEP have two components due to the PV imputation:

$$S(A_i)^2 = U_i^* + \left(1 + \frac{1}{M}\right) B_i, \tag{3}$$

where $S(A_i)^2$ is the total variance of the quantity A_i, U_i^* is the average sampling variance (generally estimated through jackknife procedures), M is the number of PV imputations, and B_i is the between-imputation variance of the quantity estimate. More details about the computation of standard errors in NAEP can be found in Mislevy et al. (Mislevy et al. 1992), based on inferences from multiple imputation (Little and Rubin 2002).

The standard error of the t-test assumes that there is no linking error in the comparison across years. This may be a tenable assumption if measurement error is small and the number of overlapping items between comparison years is large. However, it is uncertain if there is additional bias (systematic error) or uncertainty from random error across a chain of links inherent in a comparison. For reference, the overall standard deviation (SD) within a year and grade level of the NAEP Mathematics and Reading Assessments is approximately 37. For national-level overall mean comparisons, differences of less than 1 scale score point (or less than 3% of a standard deviation) may be reported as a significant change.

1.3 Current NAEP Linking Procedure

The current procedure employed within NAEP for linking data from a new assessment cycle involves concurrent IRT calibration with data from the previous

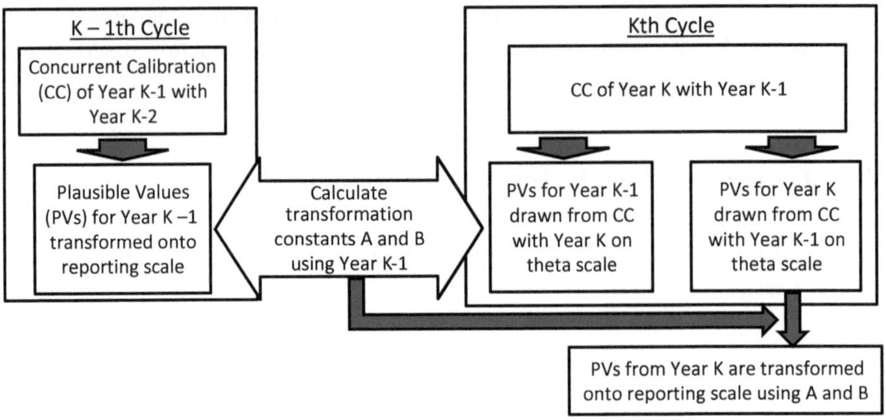

Fig. 1 Diagram of the NAEP linking and transformation procedure

assessment cycle. NAEP Grades 4 and 8 Mathematics and Reading assessments have been administered every two years since 2002, with trends going back to 1990 and 1992, respectively, so the consecutive pair of year-to-year comparisons are two years apart. For most operational assessments, there are approximately ten blocks of 10–15 items, and individual students are only presented two blocks. Additionally, two blocks are typically new to that assessment cycle, replacing two blocks that were retired and released to the public after the previous assessment year's results were released.

A concurrent calibration is performed with data from both year k and the previous assessment year $k - 1$. The resulting item parameters are used to draw PVs, separately, for both years on the calibration, or *provisional*, metric. To place the PVs for year k on the *reporting* metric (i.e., the NAEP scale from either 0–500 or 0–300), transformation constants are estimated using the two sets of PVs from year $k - 1$. A diagram of the NAEP linking procedure is displayed in Fig. 1, and the procedure is further described in the NAEP Technical Documentation (U.S. Department of Education, National Center for Education Statistics 2015b).

1.4 *Evaluating Linking Error NAEP: Challenges and Opportunities*

The purpose of linking and equating in large-scale survey assessments such as NAEP is not same as traditional applications of linking and equating. In NAEP, test forms or booklets are linked across assessment cycles (years) in NAEP in order to compare the performances of populations and subgroups over time, not to ensure consistent placement of individual scores over multiple forms. NAEP trend linking procedures must ensure distributions of scale scores can be compared over time and provide stable, consistent, and unbiased contrasts of group performances over specified time periods.

To that extent, traditional approaches for evaluating and addressing linking error may not be appropriate for NAEP (Kolen and Brennan 2014). First, NAEP does not produce estimates of scores for individual students. Up to 20 PVs per student are drawn given their responses to both the assessment items and the survey questionnaire items. Second, the same response patterns to the same assessment items would not produce the same posterior distribution (and distribution of PVs) for an individual student unless all survey responses are the same as well, which is highly unlikely due to the large number of variables. Referring to the posterior distribution from which PVs are drawn in Eq. (1), even though the IRT-based likelihood functions would remain the same for the same set of assessment item responses, the latent regression prior distribution would change in response to different values of x_i for different students. To perfectly equate the distributions of PVs would be to ignore and underestimate meaningful differences in subgroup performance distributions. Thus the traditional approaches to linking and estimating the standard error of equating are not desirable for the purposes of large-scale survey assessments.

Another concern is the issue of scale drift. In NAEP Mathematics and Reading assessments, approximately two blocks (of ten) are released to the public and discontinued in each operation cycle. Even with trend items, the performance of students and thus the item measurement properties are expected to change over time. Estimates of scale drift may be inherently confounded with changes in the population characteristics and performance over time.

Past research of linking error in trend measurements in NAEP assessments have focused on either (or both) the link between two consecutive assessment cycles or the linking procedure employed in NAEP long-term trend, which is a separate assessment dating back to the 1970s. NAEP long-term trend differs in a number of important ways from the NAEP Mathematics and Reading assessments administered every two years in Grades 4 and 8, including the stability of item pool and trend estimation procedures; thus research published using NAEP long-term trend may not be applicable to any other NAEP assessment. Hedges and Vevea (Hedges and Vevea 1997) examined the equating error between two consecutive cycles through a simulation study. While the results showed that the procedure employed currently in NAEP minimizes the bias of trend comparisons, biases may still be present of up to several scale score points, particularly at the tails of the distribution. Donoghue and Mazzeo (Donoghue and Mazzeo 1992) conducted a similar investigation in which the pairwise linking error was evaluated through jackknife procedures. More recently, Xu and von Davier (Xu and Davier 2010) examined the impact of common item sampling in pairwise linking through double jackknife procedures. However, none of these studies examined the potential for increased bias or random error across a chain of links for trend comparisons.

Accumulated linking error of an assessment is defined as the compounding of linking error over a chain of links. Guo (2010) derived an approximation for the asymptotic accumulated standard error of equating from a series of links in a nonequivalent groups with anchor test (NEAT) design and showed that the accumulation of error may be non-negligible even when the standard error of equating is small in a pairwise link. While the standard error of equating may not be estimated using traditional approaches in NAEP, we can use this work as the theoretical inspiration for the potential impact of accumulated linking error on trend inferences.

2 Simulation Study Methods

2.1 Data-Generating Model

The simulation study was designed to mimic ten assessment cycles of an operational NAEP assessment. Starting with 150 items (10 blocks of 15 items) presented in the first cycle, 30 items (2 blocks) are discontinued and replaced with 30 new items in the next cycle. If this procedure is repeated for each assessment cycle, year 5 would have only 30 items in common with year 1, and year 6 would have no common items with year one. The resulting "item pool" over 10 assessment cycles has 420 unique items. Two patterns for the overall means and standard deviations were simulated. The first was based on ten cycles of the NAEP Grade 8 Mathematics national average composite scale scores. In this pattern, the means are numerically increasing each year, with minor fluctuations in the standard deviations. The second pattern for the means were based loosely on the Grade 8 Reading national average composite scale scores, in which the national average both decreases and increases over a period of ten cycles. The ten cycles means and standard deviations were transformed to start with a mean of 0 and standard deviation of 1 in the first cycle, and the values for each cycle are listed in Table 1.

The simulation study is conducted under an ideal condition in which the data-generating parameters for the 420 are constant across all years in which the item is presented. The items are all generated under unidimensional two-parameter logistic (2PL) IRT models. It should be noted that in practice the estimated item parameters vary from year to year to detect differences in subgroup performance, and some items are "split" (treated as separate items) due to differential performance across years. The data-generating model for the individual item responses does

Table 1 Data-generating means and standard deviations

Year	1	2	3	4	5	6	7	8	9	10
Mathematics mean	0.000	0.162	0.220	0.293	0.417	0.452	0.522	0.565	0.591	0.612
Mathematics SD	1.000	1.007	1.041	1.057	1.006	1.008	1.001	1.010	1.005	1.013
Reading mean	0.000	−0.011	0.100	0.081	0.120	0.091	0.059	0.077	0.111	0.144
Reading SD	1.000	1.024	0.965	0.979	0.943	0.986	0.980	0.971	0.957	0.956

not include a population structure model, which is employed to generate PVs in NAEP. Because we are interested in evaluating the overall mean comparisons, this additional complexity was not introduced (yet).

The item responses were generated for $n = 9000$ students per year (given the 2PL IRT parameters), consistent with the minimum sample size for national-level-only NAEP studies. It should be noted that for main NAEP Grades 4 and 8, sample sizes may be as large as 250,000, which may decrease the magnitude and impact of linking error. A balanced-incomplete-block design was imposed on the full set of item responses to emulate the sparse matrix of item responses in practice, where each student has observed responses to two out of ten blocks presented in a given year for a total of thirty items per "student."

2.2 *Calibration, Transformation, and Scoring*

Items were calibrated onto a unidimensional IRT scale for each simulation iteration using the software flexMIRT 3.0 (Houts and Cai 2015). Concurrent calibration was conducted for each pair of consecutive years, and scores were placed on the same scale as the first year using the same operational transformation equation detailed in Sect. 1.3.

The means for each year were estimated using expected a posteriori (EAP) scoring for individuals rather than PVs. Since there was no population structure model, there was no need to conduct latent conditioning (regressing the latent trait onto the set of survey covariates). The data-generating model within each year is a normal distribution, and the average of the PVs from the same posterior would converge to the EAP; thus overall means are expected to converge to the same results regardless of using EAP or PV scoring. In planned future work, population structure models will be incorporated into the simulation, at which point PV scoring will be conducted.

2.3 *Evaluating Results*

Let $k = 1, \ldots, K$ index the simulation number, where in this study $K = 500$ replications were conducted. For a pair of years i and j, the true difference in means is represented by $\mu_i - \mu_j$ for all pairs $i = 1, \ldots, 10$ and $j = 1, \ldots, 10$. Let $\hat{\mu}_i, k$ and $\hat{\mu}_j, k$ be the estimated means of years i and j, respectively, from replication k, calculated from the EAP scores after linking. The magnitude of error in trend comparisons was evaluated using the average bias, percent relative bias, mean squared error (MSE), and root mean squared error (RMSE), given as follows:

$$\text{Bias} = \frac{1}{500}\sum_{k=1}^{500}\left[\left(\hat{\mu}_{i,k} - \hat{\mu}_{j,k}\right) - \left(\mu_i - \mu_j\right)\right], \tag{4}$$

$$\text{PercentRelativeBias} = \frac{1}{500}\sum_{k=1}^{500}\frac{\left[\left(\hat{\mu}_{i,k} - \hat{\mu}_{j,k}\right) - \left(\mu_i - \mu_j\right)\right]}{\left(\mu_i - \mu_j\right)}, \tag{5}$$

$$\text{MSE} = \frac{1}{500}\sum_{k=1}^{500}\left[\left(\hat{\mu}_{i,k} - \hat{\mu}_{j,k}\right) - \left(\mu_i - \mu_j\right)\right]^2, \tag{6}$$

$$\text{RMSE} = \sqrt{\frac{1}{500}\sum_{k=1}^{500}\left[\left(\hat{\mu}_{i,k} - \hat{\mu}_{j,k}\right) - \left(\mu_i - \mu_j\right)\right]^2}. \tag{7}$$

3 Results

For both sets of data-generating means, all of the results converged to maximum likelihood solutions during the IRT calibration. The estimated means for each "year," after linking and transformation, are plotted in Fig. 2 along with the data-generating means and average means across all simulations for both Reading- and Mathematics-based trend lines. The bias for the each iteration and the average bias are plotted in Fig. 2 as well. For the Mathematics-based trend, almost all simulations showed an overestimation of the means across years, and the magnitude of this bias increased over the number of links. However, for the Reading-based simulations, the estimates of the means tended to be highly variable over a number of links, but relatively unbiased on average.

The estimation of all combinations of pairwise differences was also evaluated. Figure 3 displays the matrices of the RMSE, bias, data-generating differences, and percent relative bias (bias divided by the true difference) for each pairwise comparison for both Mathematics- and Reading-based trends. The scales for bias, RMSE, and data-generating mean differences are printed on a scale emulating the NAEP reporting metric, which has a standard deviation of 37 rather than 1, due to the constrained space. For example, biases and RMSEs of 3.7 in the matrix would represent 10% of a standard deviation.

In the case of the Grade 8 Mathematics-based trend, both the RMSE and bias of the comparisons tend to increase when the number of links between the two years increases. However, this effect may be confounded with the magnitude and direction of the trend. When examining the relative bias of the mean differences, the bias tends to be approximately 5.6–9% of the true difference, and the relative bias appears to be fairly constant across any number of links. The highest RMSE occurs when comparing year 10 to year 1, with an estimated error of approximately 5.5% of a standard deviation. Because the true difference between years 1 and 10 is 61.2% of a standard deviation, this amount of error is unlikely to impact inferences.

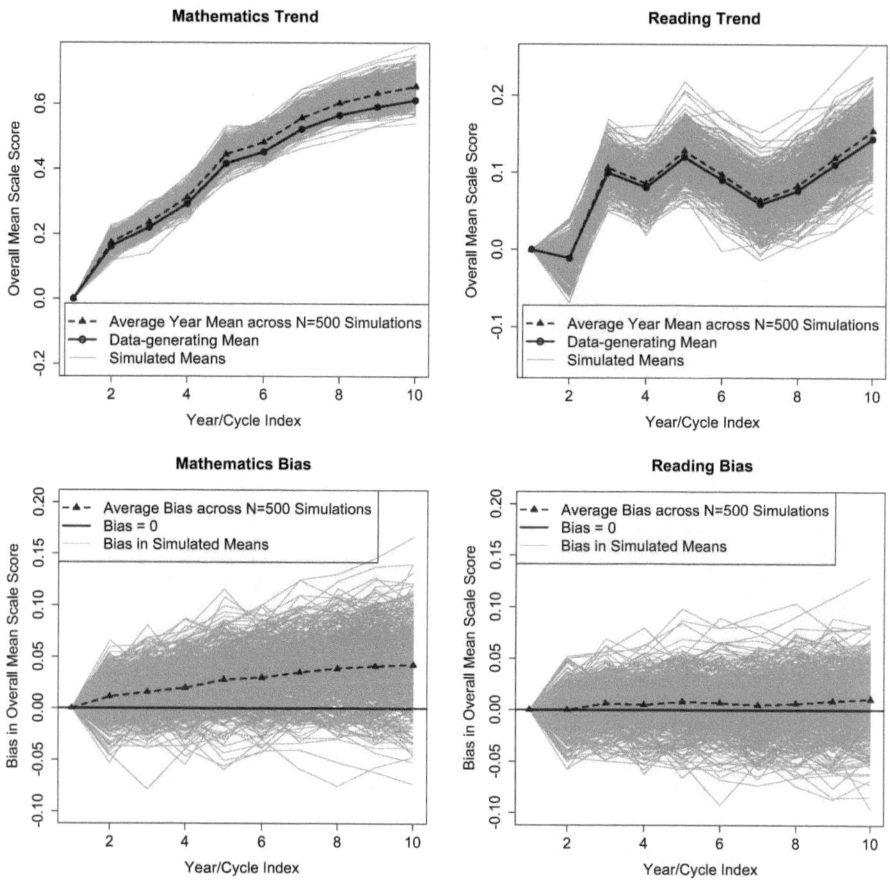

Fig. 2 Average scale scores and bias for 10 years/cycles across N = 500 simulations

The Reading-based pairwise comparisons (Fig. 3) typically showed increasing RMSEs with the number of links, where the maximum magnitude of the errors was approximately 3.2% of a standard deviation when comparing year 10 to year 1. The bias of Reading pairwise comparisons was relatively small, and while that bias appeared to increase with the number of links, the relative bias showed no pattern with some outliers. The estimated difference between years 4 and 8 had a relative bias of 32.3% of the true difference, which was only 0.5% of a standard deviation and thus sensitive to even small magnitudes of bias. The pattern from the Reading-based trend pairwise comparisons showed almost no bias, but the potential for accumulated linking error through random variability still persisted, as evidenced in the RMSE of the comparisons.

Not presented in the figures are the results when students receive all 150 items per year instead of only 2 blocks. When IRT parameters are calibrated and the means estimated using fully observed data for the Mathematics-based trend, the bias and

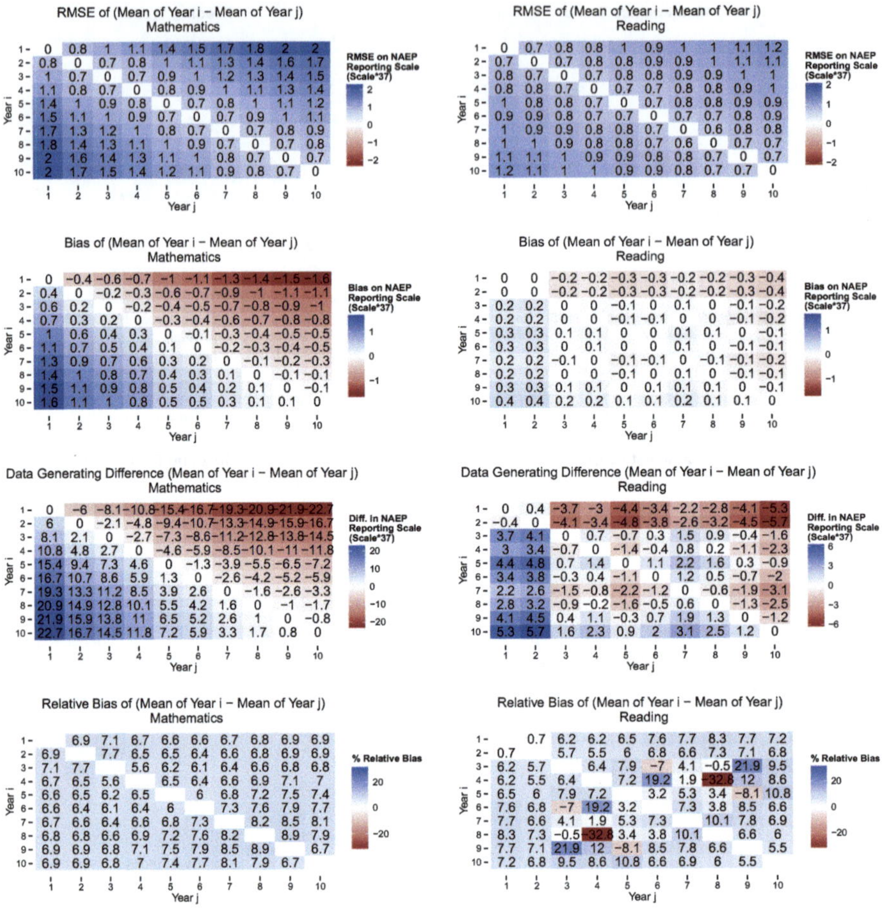

Fig. 3 Average bias, RMSE, true values, and percent relative bias of mean differences across N = 500 simulations under mathematics and reading data-generating models

MSEs are reduced to maximums of 0.5–0.8 on the NAEP reporting scale (under 1.5–2.2% of a standard deviation). Thus the linking error may also be related to the rate of missing information from unobserved item responses on the full set of items.

4 Concluding Remarks and Future Extensions

The potential for accumulated linking error in NAEP trend estimates was evaluated through simulation study under some idealized conditions. In trends similar to Grade 8 Mathematics from 1990 to 2011, where the means increase numerically each cycle, both bias and RMSE increase over time with magnitudes of up to 5% of a standard deviation across ten links. In trends similar to NAEP Grade 8 Reading, where the national average oscillates over time, RMSE also increases with the number of links, up to 3% of a standard deviation. However, the bias does not appear to consistently increase when the mean fluctuates over time. The percent relative bias is generally between 5 and 10% of the true difference between means and in general in the direction of the true difference (positive or negative). The observed difference in means from the data may be inflated relative to the true difference by 5–10%, regardless of the number of links. The magnitudes of the random linking errors found could be diminished with an increased sample size and should be interpreted with caution. Since results of the simulation study do seem to differ by the trend patterns, other patterns should be investigated in future simulation studies.

As the design and analysis of NAEP data is more complicated than traditional assessments, there are a number of additional conditions for simulation that were not discussed in this manuscript. First, NAEP items in practice are a mixture of 2PL, three-parameter logistic (3PL) IRT, and generalized partial credit items, not just 2PL as assumed in this study. Next, this study assumed all 30 items per student measured the same overall unidimensional scale. In practice, NAEP Mathematics has five distinct subscales, and the IRT models and linking are conducted separately for each subscale. Although the Mathematics subscales are highly correlated, students may receive as few as 3–4 items per subscale. Given the impact of the BIB design on the simulation results compared to fully observed item responses, the limited information on the subscales may likely increase the impact of linking error on trend comparisons. In future simulation studies, IRT parameters will not be held constant across all years in the data-generating model; rather additional random fluctuations over time will be incorporated.

As mentioned previously, there are no individual scale scores for NAEP, rather 20 PVs are drawn for each student given the responses to the survey questionnaires and the assessment items. Future investigations into linking error for NAEP will incorporate complex population distributions similar to the latent conditioning model and utilize PV imputation instead of EAP scoring. Finally, the number of replications for this simulation study was relatively limited due to computational time and speeds; future work will increase the number of replications for comparison.

The results of this study demonstrate that there may be additional uncertainty in trend comparisons due to accumulated linking error. As the NAEP trend comparisons in subjects such as Grade 8 Mathematics and Reading continue forward across an increasing number of links, the potential impact of linking error on inferences and conclusions should be further investigated.

Acknowledgements The author would like to thank her colleagues Xueli Xu, John Donoghue, Helena Jia, Dave Freund, Ed Kulick, Samantha Burg, and Emmanuel Sikali for their comments and input during various phases of this study and related research topics. Additionally, members of the NAEP Design and Analysis Committee provided valuable discussions and recommendations for the development and future of this work. The opinions expressed in this article are the author's own and do not reflect the view of the National Center for Education Statistics, the US Department of Education, or the US government.

References

J. Donoghue, J. Mazzeo, Comparing IRT-based equating procedures for trend measurement in a complex test design, in *Annual Meeting of the National Council on Measurement in Education*, San Francisco, 1992

H. Guo, Accumulative equating error after a chain of linear equatings. Psychometrika **75**(3), 438–453 (2010)

L.V. Hedges, J.L. Vevea, A Study of Equating in NAEP. Technical report, American Institutes for Research, Washington, DC, 1997

C.R. Houts, L. Cai, *flexMIRT: Flexible Multilevel Item Factor Analysis and Test Scoring Users Manual Version 3.0* (Vector Psychometric Group, LLC, Chapel Hill, 2015)

M.J. Kolen, R.L. Brennan, *Test Equating, Scaling, and Linking: Methods and Practices* (Springer Science & Business Media, New York, 2014)

R. Little, D. Rubin, *Statistical Analysis with Missing Data* (Wiley, Hoboken, 2002)

R.J. Mislevy, E.G. Johnson, E. Muraki, Scaling procedures in NAEP. J. Educ. Behav. Stat. **17**(2), 131–154 (1992)

U.S. Department of Education, National Center for Education Statistics, The NAEP Data Explorer (2015a). Retrieved from http://nces.ed.gov/nationsreportcard/naepdata/

U.S. Department of Education, National Center for Education Statistics, NAEP Technical Documentation (2015b). Retrieved from https://nces.ed.gov/nationsreportcard/tdw/

U.S. Department of Education, National Center for Education Statistics, The Nation's Report Card (2015c). Retrieved from http://www.nationsreportcard.gov/

X. Xu, M. Davier, Linking errors in trend estimation in large-scale surveys: a case study. ETS Res. Rep. Ser. **2010**(1), 1–12 (2010)

IRT Observed-Score Equating with the Nonequivalent Groups with Covariates Design

Valentina Sansivieri and Marie Wiberg

Abstract Nonequivalent groups with anchor test (NEAT) design is typically preferred in test score equating, but there are tests which do not administer an anchor test. If the groups are nonequivalent, an equivalent groups (EG) design cannot be recommended. Instead, one can use a nonequivalent groups with covariates (NEC) design. The overall aim of this work was to propose the use of item response theory (IRT) with a NEC design by incorporating the mixed-measurement IRT with covariates model within IRT observed-score equating in order to model both test scores and covariates. Both simulations and a real test example are used to examine the proposed test equating method in comparison with traditional IRT observed-score equating methods with an EG design and a NEAT design. The results show that the proposed method can be used in practice, and the simulations show that the standard errors of the equating are lower with the proposed method as compared with traditional methods.

Keywords NEC design • Item response theory • Collateral information

1 Introduction

Test score equating is used to compare different test scores from different test forms (Kolen and Brennan 2014; González and Wiberg 2017). If the test groups which have taken the different test forms can be considered similar, the equivalent groups (EG) design can be used. A problem is that test groups who get different test forms might be nonequivalent (Lyrén and Hambleton 2011) and thus a nonequivalent groups with anchor test (NEAT) design should be used if an anchor test is distributed. There are however large-scale assessments that do not distribute an

V. Sansivieri (✉)
Department of Statistics, University of Bologna, Bologna, Italy
e-mail: valentina.sansivier2@unibo.it

M. Wiberg
Department of Statistics, USBE, Umeå University, Umeå, Sweden
e-mail: marie.wiberg@umu.se

L.A. van der Ark et al. (eds.), *Quantitative Psychology*, Springer Proceedings in Mathematics & Statistics 196, DOI 10.1007/978-3-319-56294-0_25

anchor test, e.g. Armed Services Vocational Aptitude Battery (Quenette et al. 2006), the American College Testing (ACT 2007) and the previous version of the Swedish Scholastic Aptitude Test (SweSAT). Wiberg and Bränberg (2015) proposed the nonequivalent groups with covariates (NEC) design. Instead of an anchor test, they used kernel equating with covariates which correlated high with the test scores. A problem with their proposed method is that we may not always have covariates which correlate high with the test scores but rather covariates which affect the test scores—such as gender or educational background. For example, Differential item functioning (DIF; Holland and Wainer 1993) is used to examine if an item does not favour a specific group, e.g. gender. In the past, covariates have been included differently in equating, for example, through propensity score matching (Longford 2015), in linear kernel equating (Bränberg and Wiberg 2011), for matching (e.g. Wright and Dorans 1993), and used as surrogate variables (Liou et al. 2001).

Items in large-scale assessments are typically modelled with item response theory (IRT), and IRT observed-score equating or the IRT true-score equating appears as good choices. A problem with IRT true-score equating is that we do not know what the true score is; instead observed scores are typically used as estimates of the true scores. Also, Hsu et al. (2002) examined the use of collateral information to improve IRT true-score equating. None of the previous cited studies have incorporated covariates in IRT observed-score equating. Recently, Tay et al. (2016) have proposed the inclusion of covariates in IRT models. The aim of this paper is to propose IRT observed-score equating method with covariates, using a NEC design. Further, it aims to compare the proposed method with the traditional IRT observed-score equating in the EG and NEAT design using both simulations and data from a real test.

2 IRT Observed-Score Equating

To find equivalent test scores between test forms X and Y with IRT observed-score equating, the recursion formula described in Lord and Wingersky (1984) is typically used although other alternatives exist (González et al. 2016). Let θ_j the ability of examinee j, let x be a test score, and $f_r(x|\,\theta_j)$ is the distribution of number-correct scores over the first r items for examinees with ability θ_j. If we use the three-parameter logistic (3PL) IRT model, the probability for answering item i correctly is defined as

$$p_{ji} = c_i + \frac{1 - c_i}{1 + \exp\left(-a_i\left[\theta_j - b_i\right]\right)}, \tag{1}$$

where a_i is the item discrimination, b_i is the item difficulty and c_i is a pseudo-guessing parameter for item i. Setting $c_i = 0$ yields the two-parameter logistic (2PL) IRT model. The probability of earning a score of 1 on the first item can be defined as $f_1(x = 1|\,\theta_j) = p_{j1}$; likewise $f_1(x = 0|\,\theta_j) = (1 - p_{j1})$ is the probability of earning a score of 0 on the first item. For $r > 1$, the observed-score distribution for examinees of a given ability is obtained from the recursion formula as follows:

$$
\begin{aligned}
f_r\left(x\mid\theta_j\right) &= f_{r-1}\left(x\mid\theta_j\right)\left(1-p_{jr}\right), x=0\\
&= f_{r-1}\left(x\mid\theta_j\right)\left(1-p_{jr}\right)+f_{r-1}\left(x-1\mid\theta_j\right)p_{jr}, 0<x<r\\
&= f_{r-1}\left(x-1\mid\theta_j\right)p_{jr}, x=r.
\end{aligned}
\tag{2}
$$

The observed-score distribution for examinees at each ability is found, and then these are accumulated and if the ability distribution is continuous, then $f(x) = \int f(x|\theta)\psi(\theta)d\theta$, where $\psi(\theta)$ is the distribution of θ. When performing IRT observed-score equating, observed-score distributions are found for test forms X and Y, and then equipercentile equating is used to find score equivalents.

3 IRT Observed-Score Equating with Covariates

The traditional IRT models observed-score equating methods do not use information from covariates. Recently, Tay et al. (2016) proposed the IRT-C model which is an IRT model which contains information about covariates by modelling uniform and non-uniform DIF. The probability of answering an item correctly, given the ability of an examinee θ_j and its vector of covariates z_j, is defined

$$
p\left(y_{ji}|\theta_j, z_j\right) = c_i + \frac{1-c_i}{1+\exp\left(-a_i\left[\theta_j+b_i+d_iz_j+e_iz_j\theta_j\right]\right)},
\tag{3}
$$

where a_i, b_i and c_i represent the item discrimination, item location and the item pseudo-guessing, respectively. $d_iz_j$and e_iz_j represent the direct and interaction effects for modelling uniform and non-uniform DIF, respectively. Equation (3) implies that we assume the presence of DIF in some of the items which compose the test. Although it is true that one should not include DIF items in a test, it is also true that DIF occurs in regular tests; for an example when DIF items were included in a test, refer to Gnaldi and Bacci (2015).

We propose the following adjustment for the 2PL IRT model, which has not been used for the IRT-C model before:

$$
p\left(y_{ji}|\theta_j, z_j\right) = \frac{1}{1+\exp\left(-a_i\left[\theta_j+b_i+d_iz_j+e_iz_j\theta_j\right]\right)}.
\tag{4}
$$

To perform IRT observed-score equating with the IRT-C models, simply use Eq. (3) or (4) in Eq. (2), which yields the updated recursion formula for $r > 1$ as follows:

$$
\begin{aligned}
f_r\left(x|\theta_j, z_j\right) &= f_{r-1}\left(x|\theta_j, z_j\right)\left(1-p_{jr}\right), x=0\\
&= f_{r-1}\left(x|\theta_j, z_j\right)\left(1-p_{jr}\right)+f_{r-1}\left(x-1|\theta_j, z_j\right)p_{jr}, 0<x<r\\
&= f_{r-1}\left(x-1|\theta_j, z_j\right)p_{jr},\ x=r.
\end{aligned}
\tag{5}
$$

The accumulated observed-score distributions for various abilities are likewise replaced with a distribution function which also includes the different covariate values $f(x) = \sum_j f\left(x|\theta, z_j\right) \psi\left(\theta\right) d\theta$. Once these distributions are known for test forms X and Y, equipercentile methods are used to conduct IRT observed-score equating with covariates.

4 The Examined Test

The college admission test SweSAT consists of 160 multiple-choice binary-scored items divided into an 80-item quantitative and an 80-item verbal section, which are equated separately. The quantitative section was used here. SweSAT only recently added an anchor test; previously an EG design was performed in different groups with specific values on certain covariates including gender and education (see Lyrén and Hambleton 2011). The fact that an anchor test is now administered with the SweSAT to a small sample of examinees gives us a unique opportunity to compare the results from the proposed method with a NEC design with the results of both a NEAT design and an EG design. Table 1 shows the mean scores of the subpopulations and proportions used in the simulations. Gender was coded 0 for males and 1 for females. Education categories used were *EL*, for elementary education, *HS* for high school education and *UNI* for university education. Females had lower mean test scores than males.

5 Simulation Study

The aim of the simulation study was to evaluate the performance of the proposed method in comparison to traditional IRT observed-score equating using parametric bootstrap (Efron and Tibshirani 1993) standard errors. The simulated data mimics

Table 1 Mean scores and proportions within parentheses for the different subpopulations used in the simulations and the real test studies from the SweSAT test 2012 and 2013

Education	Male	Female
SweSAT 2012		
EL	32.41	26.16
HS	39.08	33.19
UNI	35.38	35.91
SweSAT 2013		
EL	35.00 (7.2)	27.34 (6.9)
HS	39.21 (81.8)	32.63 (83.2)
UNI	39.07 (11.0)	36.61 (9.9)

EL = Elementary school education, *HS* = High school education, *UNI* = University education

the real SweSAT data. Let N_X and N_Y be number of examinees taking test forms X and Y and n_X and n_Y number of items in test forms X and Y, respectively. The simulation was conducted using the following steps:

1. For test forms X and Y, probabilities p_{ji} are generated from the IRT-C model (or the traditional IRT models) with fixed values of the parameters.
2. Randomly select $N_X \cdot n_X$ and $N_Y \cdot n_Y$ probabilities from the models from step 1.
3. Conduct IRT observed-score equating without and with covariates as defined in Eq. (1) for the 3PL IRT model (and Eq. (1) with $c_i = 0$ for the 2PL IRT model) and Eq. (3) for the 3PL IRT model (and Eq. (4) for the 2PL IRT model) using the parametric bootstrap samples drawn in step 2.
4. Repeat steps 2 and 3 R times. The estimated standard error (SE) for the IRT observed-score equating is defined as

$$SE_{boot}\left[\widehat{\varphi}_Y(x_i)\right] = \sqrt{\frac{\sum_r \left[\widehat{\varphi}_{Yr}(x_i) - \widehat{\varphi}_{Y.}(x_i)\right]^2}{R-1}}, \tag{6}$$

where $\widehat{\varphi}_{Yr}(x_i)$ indicate the r bootstrap estimate of the equated value and

$$\widehat{\varphi}_{Y.}(x_i) = \sum_r \frac{\widehat{\varphi}_{Yr}(x_i)}{R}. \tag{7}$$

Both the simulation study and the real test example were carried out in R (R Core Development Team 2016), and the code can be obtained upon request.

5.1 *Data, Design and Ability Distributions Used in the Simulation Study*

In order to simulate relationships between the external covariates and the latent trait that have high fidelity to test data, we used descriptive statistics from a sample of 2014 examinees taking the 2013 SweSAT test to predict the standardized SweSAT scores (SS). These predicted scores were used as simulated ability estimates. To predict the *SS*, an additive regression model was fit to the data with predictor variables gender (GEN) and education using dummy coding:

$$SS = 0.315 - 0.496(GEN) - 0.585(EL) + 0.442(HS) + e \tag{8}$$

where e represents the error term. Only the coefficient of the *UNI* variable was not significantly different from zero.

The proposed method using a NEC design was compared with the traditional IRT observed-score equating using a NEAT and an EG design. To simulate a θ distribution that conforms to the estimated additive model, random normal distribution was simulated for each of the six cells (gender $\times$ education) shown in Table 1 using the following model

$$\theta \cdot mean = 0.315 - 0.496(GEN) - 0.585(EL) + 0.442(HS). \tag{9}$$

The number of simulated examinees in each cell was based on the proportions in Table 1.

5.2 Item Parameters Used in the Simulation Study

Following Tay et al. (2016), for a traditional 3PL IRT model, item discriminations a_i were sampled from a truncated normal (mean = 1.2, SD = 0.3) distribution with the lowest and highest possible values set to 0.5 and 1.7, respectively; item difficulty b_i was sampled from a uniform (−2, 2) distribution, and c_i was sampled from a logit-normal (−1.1; 0.5) distribution. For a 2PL IRT model, the parameters were sampled in the same manner.

5.3 Specified DIF Used Across Multiple Covariates

To simplify the simulation, we only examined the presence of uniform DIF; thus we simulated DIF on the first 15 items shown through the design matrix in Eq. (10), which specify the DIF coefficients $d_i, i = 1, \ldots, 15$. This matrix is used for illustrative purpose and does not reflect how the real test is biased for or against different groups of examinees. The same matrix was used for the two test forms, although it does not have to be the same. For moderate DIF, we specified uniform DIF of the magnitude 0.40 on items 1; 2; 3; 4; 5; 12 and 14; 15 biased against females (Tay et al. 2016). Items 6; 7; 8; 9; 10; 11 and 14; 15 are biased in favour of EL, and, finally, items 11 and 13; 14; 15 are biased against HS. In this last case, uniform DIF of magnitude 0.20 was used. For large DIF, we specified uniform DIF of the magnitude 0.60 in place of 0.40 and 0.30 in place of 0.20; for low DIF we specified uniform DIF of the magnitude 0.20 in place of 0.40 and 0.10 in place of 0.20, and finally to use no DIF, one can place all values very close to zero in the design matrix.

$$\begin{bmatrix} -.4 & -.4 & -.4 & -.4 & -.4 & 0 & 0 & 0 & 0 & 0 & 0 & -.4 & 0 & -.4 & -.4 \\ 0 & 0 & 0 & 0 & 0 & .4 & .4 & .4 & .4 & .4 & .4 & 0 & 0 & .4 & .4 \\ 0 & 0 & 0 & 0 & 0 & 0 & 0 & 0 & 0 & 0 & -.2 & 0 & -.2 & -.2 & -.2 \end{bmatrix}^T \begin{bmatrix} GEN \\ EL \\ HS \end{bmatrix} \tag{10}$$

5.4 Conditions in the Simulation Study

A simulated test length of 45 items was used where the first 15 items have DIF of the form specified in the DIF matrix but the remaining items do not have DIF. Our simulations focus on the boundary conditions with tests that have a large proportion of DIF items (50%) and a moderately large proportion (30%) of DIF items, respectively, in line with Tay et al. (2016). A simulated test length of 80 items was also included to mimic the number of items of the real test used. The sample size for the 3PL IRT model should be at least 1500 per form (Kolen and Brennan 2014, p. 304); thus we choose to use 2000 per form (as in effect we have to estimate the three parameters of the 3PL IRT model and the DIF coefficient d_i). To evaluate how the method works with small samples, a sample size of 600 per test form was also included. Finally, to examine the impact of the item parameters, the cases of less difficult items (by subtracting 0.5 to the item difficulty) and less discriminating items (by dividing the item discrimination with 2) were examined.

6 Results from the Simulation Study

In general, most NEC designs had smaller standard errors than when an EG or a NEAT design was used as seen in Figs. 1, 2 and 3. This was true, regardless of the amount of DIF (low, moderate, high), and thus only a limited number of figures are shown as the rest follows the displayed pattern and can be obtained upon request. Figure 1 shows the standard errors when equating 45 items tests modelled with the 3PL IRT models when we have moderate DIF and either 2000 or 600 examinees. Clearly all methods using different combinations of covariates with a NEC design gave smaller standard errors than when EG design or NEAT design was used. The right-hand plot of Fig. 1 shows the only exception, which was for moderate DIF and 600 examinees. In that plot, the NEC with gender and high school education had slightly higher standard errors around test score 30 although the standard errors were still in general lower than the standard errors for the EG and NEAT designs for the other scores.

In Fig. 2, standard errors for the equated values for low DIF when we have less difficult items and less discriminating items are shown. Regardless of the amount of DIF, almost identical plots were obtained for the case of less difficult and less discriminating items; thus only two plots are shown. Notice the slightly different scales on the y-axis.

The left-hand plot of Fig. 3 shows low DIF when 2PL IRT models are used instead of 3PL IRT models. The case of moderate and high DIF was similar to the low case scenario and thus excluded. Comparing this plot to Fig. 1, it is evident that they give similar standard error patterns although the magnitude is in general higher. The right-hand plot of Fig. 3 illustrates when we have an 80-item test with low DIF. The moderate and high DIF cases were similar with the exception that for high DIF the standard errors were slightly higher (0.002) for NEC with GEN and HS when there was moderate DIF. Please be aware of the scale differences in the y-axis.

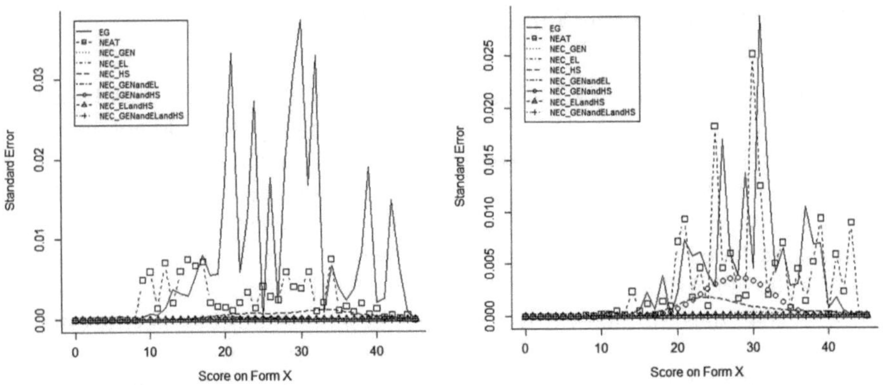

Fig. 1 Standard errors for equated values of 45 items test modelled with the 3PL IRT models when we have moderate DIF and either 2000 (*left*) or 600 examinees (*right*)

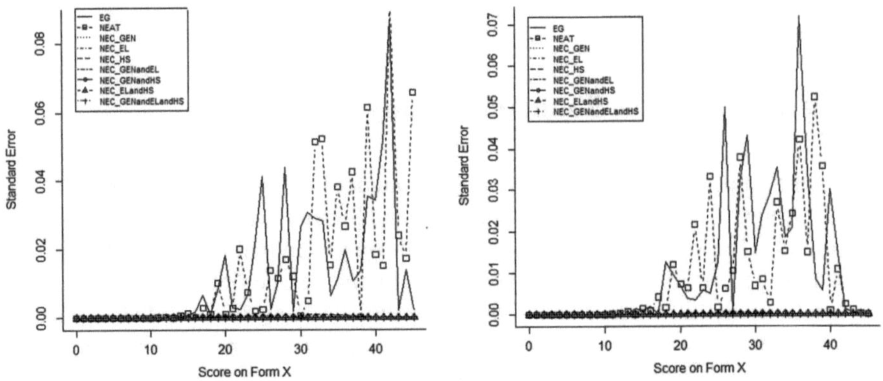

Fig. 2 Standard errors for equated values of 45 items test modelled with the 3PL IRT models with low DIF for less difficult items (*left*) and less discriminating items (*right*)

7 Real Test Example

A real test example using SweSAT was used to examine how the method could be used in practice. Two samples of size 1997 and 2014 examinees were used, with the same covariates as in the simulation study. The equated values from the proposed method were compared with traditional IRT observed-score equating using both the EG design and the NEAT design with 3PL IRT models. Standard errors were examined following the same steps as described in the simulation study with one important difference. In the first two steps, the parameters of the IRT-C models were not fixed; instead they were estimated from the real test data.

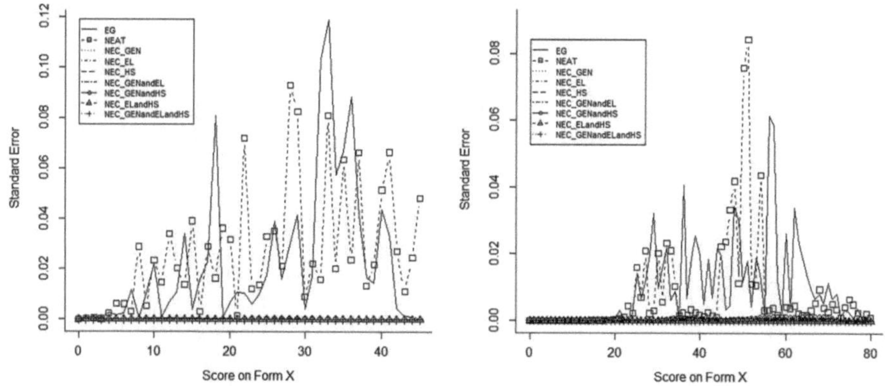

Fig. 3 Standard errors for equated values with low DIF for a 45-item test modelled with the 2PL IRT models (*left*) and an 80-item test modelled with the 3PL IRT models (*right*)

8 Results from the Real Test Example

For the sample of 1997 examinees used in the real example, the items have the following mean DIF (calculated on the absolute values), 0.31, 0.34 and 0.36, for the covariates GEN, EL and HS, respectively. For the sample of 2014 examinees, the items have the following mean DIF (calculated on the absolute values), 0.31, 0.30 and 0.29, for the covariates GEN, EL and HS, respectively. Compared to the thresholds used in the simulation study, we can say that the mean DIF across the three covariates has medium-high magnitude for the two real test samples. Figure 4 shows the standard errors for the real data with the proposed method with different combinations of covariates in the NEC design in comparison with the EG design and the NEAT design using traditional IRT observed-score equating. It is evident that if we do not have an anchor, it is much better to incorporate information from the covariates than to use an EG design. If all covariates are used, the standard errors are close to zero and thus smaller than if a NEAT design is used as well.

9 Concluding Remarks

The objective was to show that using a NEC design with IRT observed-score equating and thus using the information in the covariates could increase the accuracy of an equating. Both the simulation study and the real test study supported the proposed method as they gave in general lower standard errors than using a NEAT design or an EG design.

The results that we can increase the accuracy in the equatings with the help of covariates are in line with Wiberg and Bränberg (2015) although they did not examine IRT observed-score equating. As many large-scale assessments are

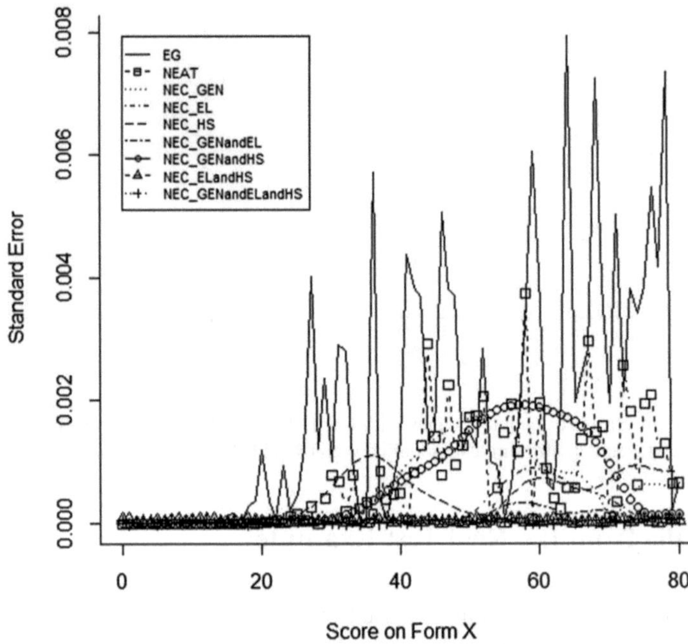

Fig. 4 SE for the real test example with the NEC design with different covariates in comparison with using an EG design and a NEAT design

modelled with IRT, the results have a clear practical implication, which was supported from the results of the real test example. Throughout this paper, we have compared the results with using an EG design and a NEAT design. As several large-scale assessments lack an anchor test, they might only have an EG design as a possibility. The fact that we can lower the standard errors by the inclusion of covariates is promising. In this study, we had the unique opportunity to compare with a NEAT design as well. Surprisingly, the standard errors were sometimes smaller when a NEC design was used instead of a NEAT design. In the future one should examine the proposed methods with respect to bias and other assessment measures as proposed by Wiberg and González (2016). Other large-scale assessment tests as well as other conditions including different DIF matrices should also be included in further studies.

Acknowledgement The research in this paper by Marie Wiberg was funded by the Swedish Research Council grant: 2014-578.

References

ACT, *ACT Technical Manual* (ACT, Iowa City, IA, 2007)
K. Bränberg, M. Wiberg, Observed score linear equating with covariates. J. Educ. Meas. **48**(4), 419–440 (2011)
B. Efron, R. Tibshirani, *An Introduction to the Bootstrap* (Chapman Hall, New York, NY, 1993)
M. Gnaldi, S. Bacci, Joint assessment of the latent trait dimensionality and observable differential item functioning of students' national tests. Qual. Quant. **50**(4), 1429–1447 (2015)
J. González, M. Wiberg, A.A. von Davier, A note on the Poisson's binomial distribution in item response theory. Appl. Psychol. Meas. **40**(4), 302–310 (2016)
P.W. Holland, H. Wainer, *Differential Item Functioning* (Lawrence Erlbaum Associates, Hillsdale, NJ, 1993)
T. Hsu, K. Wu, J. Yu, M. Lee, Exploring the feasibility of collateral information test equating. Int. J. Test. **2**(1), 1–14 (2002)
M. Kolen, R. Brennan, *Test Equating, Scaling, and Linking: Method and Practice*, 3rd edn. (Springer-Verlag, New York, NY, 2014)
M. Liou, P.E. Cheng, M. Li, Estimating comparable scores using surrogate variables. Appl. Meas. Educ. **25**(2), 197–207 (2001)
N.T. Longford, Equating without an anchor for nonequivalent groups of examinees. J. Educ. Behav. Stat. **40**(3), 227–253 (2015)
F.M. Lord, M.S. Wingersky, Comparison of IRT true-score and equipercentile observed-score "equatings". Appl. Psychol. Meas. **8**, 452–461 (1984)
P.-E. Lyrén, R.K. Hambleton, Consequences of violated the equating assumptions under the equivalent group design. Int. J. Test. **36**(5), 308–323 (2011)
M.A. Quenette, W.A. Nicewander, G.L. Thomasson, Model-based versus empirical equating of test forms. Appl. Psychol. Meas. ***30***, 167–182 (2006)
R Core Development Team, *R: A Language and Environment for Statistical Computing* (R Foundation for Statistical Computing, Vienna, 2016.) http://www.R-project.org/
L. Tay, D. Newman, J.K. Vermunt, Item response theory with covariates (IRT-C): assessing item recovery and differential item functioning for the three-parameter logistic model. Educ. Psychol. Meas. **76**(1), 22–42 (2016)
M. Wiberg, K. Bränberg, Kernel equating under the non-equivalent groups with covariates design. Appl. Psychol. Meas. **39**(5), 349–361 (2015)
M. Wiberg, J. González, Statistical assessment of estimated transformations in observed-score equating. J. Educ. Meas. **53**(1), 106–125 (2016)
J. González, M. Wiberg *Applying Test Equating Methods Using R* (Springer, Berlin, 2017). doi:10.1007/978-3-319-51824-4
N.K. Wright, N.J. Dorans, *Using the Selection Variable for Matching or Equating (ETS Research Report RR-93–04)* (Educational Testing Service, Princeton, NJ, 1993)

Causal Inference with Observational Multilevel Data: Investigating Selection and Outcome Heterogeneity

Jee-Seon Kim, Wen-Chiang Lim, and Peter M. Steiner

Abstract Causal inference with observational data is challenging, as the assignment to treatment is not random, and people may have different reasons to receive or be assigned to the treatment. The multilevel structure adds complexity to the issue, as the assignment to treatment can be determined by various sources across levels. In multilevel analysis, it is critical to account for both the nested structure of the data and potential heterogeneity in selection processes and treatment effects. This study presents methodology for classifying level-1 and level-2 units into homogeneous groups called "classes" with regard to class-specific selection and outcome models. The classification into homogeneous groups can take place at a cluster level or at the lowest level, depending on the main sources of heterogeneity in the selection and outcome mechanisms. This chapter introduces the methods, examines their properties, and provides recommendations for their proper use.

Keywords Propensity score analysis • Multilevel matching • Hierarchical linear models • Latent class analysis • Mixture models • Selection bias • Quasi-experimental design

1 Introduction

In the social sciences, randomized experiments are considered the canonical model for estimating causal effects of treatments. Due to ethical or practical reasons, however, random assignment of treatment cannot always be conducted. Instead, quasi-experiments like regression discontinuity designs, interrupted time series designs, instrumental variables, or nonequivalent control group designs are frequently used as alternative methods (Shadish et al. 2002). In particular, propensity score (PS) techniques for matching nonequivalent groups have become increasingly popular during the past three decades across various disciplines (Guo and Fraser 2015). Although an extensive literature exists for PS design and analysis, most

J.-S. Kim (✉) • W.-C. Lim • P.M. Steiner
University of Wisconsin-Madison, 1025 West Johnson Street, Madison, WI 53706, USA
e-mail: jeeseonkim@wisc.edu; ivan.lim@wisc.edu; psteiner@wisc.edu

L.A. van der Ark et al. (eds.), *Quantitative Psychology*, Springer Proceedings in Mathematics & Statistics 196, DOI 10.1007/978-3-319-56294-0_26

of the PS techniques deal with single-level data without any nested or clustered structure, and corresponding strategies for matching nonequivalent control groups in the context of multilevel data are still limited and underdeveloped (Hong and Raudenbush 2006; Kim and Seltzer 2007; Thoemmes and West 2011; Steiner et al. 2012; Keele and Zubizarreta 2014; Kim and Steiner 2015; Kim et al. 2016). This is problematic considering many large-scale surveys and assessment data are collected in clustered settings.

This study concerns PS matching strategies for nonequivalent control group designs to correct for bias due to confounding and make proper causal inferences with observational multilevel data. Kim et al. (2016) investigated heterogeneity in the treatment versus comparison condition assignment by implementing latent class multilevel logit models as selection models. The current study furthers Kim et al. (2016) by examining heterogeneity in the outcome process in addition to the selection process and also investigating at different levels in multilevel data.

1.1 Potential Outcomes in Multilevel Settings

To formalize the treatment effects of interest, we use the Rubin causal model (Rubin 1974, 1978; Rosenbaum and Rubin 1983; Holland 1986) with its potential outcome notation and its extension to multilevel settings by Hong and Raudenbush (2006). In this framework, level-1 units (e.g., students) $i = 1, \ldots, N_j$ within level-2 units (e.g., schools) $j = 1, \ldots, J$ have a set of potential treatment and control outcomes that can be denoted as $Y_{ij}(Z_{ij}, \mathbf{Z}_{-ij}, \mathbf{S})$. That is, the potential outcomes depend on three factors: (1) unit i's treatment assignment Z_{ij}, where $Z_{ij} = 0$ for the control condition and $Z_{ij} = 1$ for the treatment condition; (2) the treatment assignments of the other units within cluster j (e.g., peer assignments) to treatment Z_{-ij}; and (3) a matrix $\mathbf{S}$ that indicates the allocation of units across the clusters. The rows and columns of the $\mathbf{S}$ matrix correspond to level-1 units and level-2 clusters, respectively, representing unit i's membership to cluster j.

As unit i's potential outcomes depend both on $\mathbf{Z}_{-ij}$ and $\mathbf{S}$, the resulting set of potential outcomes to be estimated is often very large. It is therefore common to restrict the set of potential outcomes by using some form of *stable-unit-treatment-value assumption*, SUTVA (Imbens and Rubin 2015), and by restricting the generalizability of estimated treatment effects to the observed student allocation to schools as in Hong and Raudenbush (2006) and Steiner et al. (2012), for example. In the most restrictive case we get only two potential outcomes for each student: the potential control outcome $Y_{ij}(0) = Y_{ij}(Z_{ij} = 0, \mathbf{S} = \mathbf{s}^*)$ and the potential treatment outcome $Y_{ij}(1) = Y_{ij}(Z_{ij} = 1, \mathbf{S} = \mathbf{s}^*)$, where $\mathbf{s}^*$ indicates the observed allocation of unit i to cluster j. The assignment of the other units within the same cluster is no longer considered under SUTVA, implying no interference among level-1 units. This formulation is used in this paper to simplify the discussion of issues involved in multilevel matching strategies.

1.2 Causal Estimands

Given the two potential outcomes $Y_{ij}(0)$ and $Y_{ij}(1)$, we can define the average treatment effect (ATE) for the entire population as the expected difference in potential outcomes:

$$\tau = E[Y_{ij}(1) - Y_{ij}(0)]. \tag{1}$$

Frequently, of interest is not only the average across all clusters but also the average treatment effect for each cluster, that is,

$$\tau_j = E[Y_{ij}(1) - Y_{ij}(0)|J = j], \ \ j \in J. \tag{2}$$

In addition to ATE, the average treatment effects for the treated (ATT) is another causal quantity of interest. For example, ATT should be estimated as the causal quantity of interest if we are interested in the retention effect on the actually retained students as opposed to all students, both promoted and retained. The overall ATT for the population is defined as

$$\tau_T = E[Y_{ij}(1) - Y_{ij}(0)|Z_{ij} = 1] \tag{3}$$

and the cluster-specific ATT is defined as

$$\tau_{T_j} = E[Y_{ij}(1) - Y_{ij}(0)|Z_{ij} = 1, J = j], \ \ j \in J. \tag{4}$$

1.3 Strong Ignorability and Propensity Score Matching

As two potential outcomes $Y_{ij}(0)$ and $Y_{ij}(1)$ are never observed simultaneously, the treatment effects cannot be directly estimated without further assumptions. In general, we can estimate unbiased treatment effects only if the pair of potential outcomes $(Y_{ij}(0), Y_{ij}(1))$ is independent of treatment assignment Z_{ij}. Block randomized experiments or multisite randomized trials achieve this independence by the randomization of treatment assignment within blocks or sites. For observational multilevel data, we require conditional independence, also referred to as *strong ignorability* in the causal inference literature (Rosenbaum and Rubin 1983; Rubin 1978), which implies that the potential outcomes are independent of treatment assignment, given the observed vector of level-1 covariates $\boldsymbol{X}$ and level-2 covariates $\boldsymbol{W}$:

$$(Y_{ij}(0), Y_{ij}(1)) \perp Z_{ij} \,|\, \boldsymbol{X}, \boldsymbol{W}. \tag{5}$$

The formulation of strong ignorability directly suggests an exact matching of units on level-1 and level-2 covariates (Hong and Raudenbush 2006). However, an (approximately) exact matching on a large set of covariates is frequently not feasible, and we may switch to *propensity score matching* by matching units on the conditional probability of being assigned to the treatment group given the observed covariates, that is, the propensity score (PS), denoted as $e_{ij}(\boldsymbol{X}, \boldsymbol{W})$, where

$$e_{ij}(\boldsymbol{X}, \boldsymbol{W}) = \frac{P(Z_{ij} = 1, \boldsymbol{X}, \boldsymbol{W})}{P(Z_{ij} = 0, \boldsymbol{X}, \boldsymbol{W}) + P(Z_{ij} = 1, \boldsymbol{X}, \boldsymbol{W})} = P(Z_{ij} = 1 | \boldsymbol{X}, \boldsymbol{W}). \quad (6)$$

If treatment selection is strongly ignorable given an observed set of covariates $(\boldsymbol{X}, \boldsymbol{W})$, then selection is also strongly ignorable with respect to the propensity score $e_{ij}(\boldsymbol{X}, \boldsymbol{W})$:

$$(Y_{ij}(0), Y_{ij}(1)) \perp Z_{ij} \,|\, e_{ij}(\boldsymbol{X}, \boldsymbol{W}), \quad (7)$$

and

$$0 < e_{ij}(\boldsymbol{X}, \boldsymbol{W}) < 1, \quad (8)$$

and matching on the PS alone also identifies the average treatment effect (Rosenbaum and Rubin 1983; Hong and Raudenbush 2006).

The true PS is rarely known in practice and needs to be estimated from observed pretreatment covariates using a parametric binomial regression model (e.g., a logit or probit model) or more flexible semi- or nonparametric approaches like generalized additive models or statistical learning algorithms (McCaffrey et al. 2004; Berk 2008; Keller et al. 2015). It is important to note that conditioning on covariates $\boldsymbol{X}$ and $\boldsymbol{W}$ (instead of the PS) in Eq. (5) implies a within-cluster matching as long as cluster-level covariates $\boldsymbol{W}$ uniquely identify clusters (either via variations in cluster characteristics or fixed effects dummies). This is no longer the case if we condition on the PS as in Eq. (7), as level-1 units with identical PS, $e_{ij} = e_{i'j'}$ $(i \neq i', j \neq j')$, might actually come from different clusters (Steiner et al. 2012). As discussed in Thoemmes and West (2011), a pair of PS-matched treatment and control students might be very different with regard to student- and school-level covariates (despite having the same PS). This led Thoemmes and West to the conclusion that across-school matching should only be used if we can reasonably assume that the selection mechanism is identical across schools. Similarly, Kim and Seltzer (2007) argue that across-school matching makes an unbiased estimation of school-specific treatment effects difficult or even impossible.

More recently, however, Steiner et al. (2012) showed that across-cluster matching produces consistent estimates of both overall and cluster-specific treatment effects, given a correctly specified joint PS model and sufficient overlap of treatment and comparison cases within each cluster. When the overlap is lacking within clusters, Kim and Steiner (2015) and Kim et al. (2016) used latent class approaches to identify finite groups of clusters with similar selection processes and pooled the cases across clusters but within the homogeneous groups of clusters, referred to as "classes." Kim and Steiner (2015); Kim et al. (2016) also demonstrated that these *across-cluster within-class multilevel matching techniques* can be effective in examining the heterogeneity of treatment effects in observational multilevel data. The current study furthers these recent advances by investigating heterogeneity in both selection and outcome processes and classification of units at the cluster as well as the individual levels.

Although the current study considers multilevel matching only in a two-level structure where individuals are nested within clusters to simplify matters, the principles of our matching strategies can be applied to any multilevel setting. These settings include three- or higher-level data, longitudinal data where the lowest level corresponds to repeated measures over time, and dyadic data, for example.

2 Multilevel Matching Strategies

2.1 The Levels of Treatment Implementation

In comparison to single-level data, treatment implementation can be at different levels in multilevel data. In two-level data such as students nested within schools or patients nested within clinics, treatments or interventions can be implemented at the cluster level or the individual level. Implementation at the cluster level implies that the treatment status only varies across clusters and that all individuals within a cluster are assigned to either the treatment or control condition. By contrast, if a treatment is implemented at the individual level, individuals are assigned or self-select into the treatment or control conditions within each cluster and therefore both treatment and control individuals are observed within clusters. Depending on the level of treatment implementation and selection, the general matching strategy differs (Steiner et al. 2012). If treatment is implemented at the cluster level, one should match comparable treatment and control clusters, as selection takes place at the cluster level (Stuart 2007). A cluster-level matching strategy mimics a cluster randomized controlled trial where clusters are randomly assigned to treatment. Mahalanobis-distance matching on observed cluster-level covariates or standard PS techniques might be directly used since only clusters need to be matched.

However, if treatment is administered at the individual level, then individuals should be matched within clusters since selection into treatment occurs at the individual level. Matching students within schools mimics a randomized block design or multisite randomized trial where individuals are randomly assigned to the treatment condition within clusters (blocks/sites). Once individuals are matched within clusters and cluster-specific treatment effects τ_j, $j = 1, \ldots, J$, are estimated, we can compute the ATE across clusters by pooling cluster-specific estimates using meta-analytic approaches (Cooper et al. 2009) through multilevel modeling (Raudenbush and Bryk 2002). As before, standard matching methods like Mahalanobis-distance matching or PS techniques might be used.

Although the within-cluster matching strategy is simple and theoretically sound, it is not always applicable in practice for two reasons: First, if extreme selection processes take place (e.g., retention of poorly performing students), we might lack comparable treatment and control individuals within schools. Second, with small sample sizes, we might not be able to find satisfactory matches for most individuals within a cluster (Kelcey 2009; Kim and Seltzer 2007; Thoemmes and West 2011). Considering that within-cluster matching strategies might fail in practice, across-cluster matching strategies that also allow for borrowing individuals from other clusters or pooling individuals across clusters might offer a practical solution. Given that only standard matching techniques for single-level data are required whenever treatment is implemented at the cluster level, we will exclusively focus in this study on matching strategies where level-1 units select or are assigned to treatment within clusters.

2.2 *Within-Cluster Matching*

Within-cluster matching approaches match individuals in the same cluster, and thus the treatment and control conditions share the environment that might affect the selection and outcome mechanisms. A separate PS model is fit for each cluster using only level-1 covariates and is not affected by differences among clusters. This leads to several important advantages in PS matching: First, within-cluster matching requires a weaker identification assumption than across-cluster matching for the estimation of treatment effects; that is, strong ignorability is more likely to be met. Second, selection and outcome models for each cluster are single-level regression models that do not include level-2 covariates, cross-level interactions, or random slopes, and the risk of model misspecification will be lower than with multilevel models. Third, we can investigate treatment effect heterogeneity by estimating cluster-specific treatment effects. The ATE can be estimated by pooling or averaging across cluster-specific treatment effects. Simulation studies in Kim and Seltzer (2007) demonstrated the effectiveness of within-cluster matching when the selection processes were different across clusters.

However, within-cluster matching is often infeasible in practice due to a lack of overlap between treated and control units when cluster sizes are small or selection processes are strong and only a small percentage of individuals receive treatments. Even in large-scale data, some or many clusters may have small sample sizes or lack sufficient overlap. Small sample sizes may result in unreliable estimates and finite sample bias due to imperfect matches. Lack of overlap does not allow us to match all treatment and control cases and, thus, requires us to delete nonoverlapping cases which results in a lack of generalizability of results. If nonoverlapping cases are not deleted, bias will result (essentially due to the violation of the positivity assumption, $0 < P(Z_{ij}|\boldsymbol{X}) < 1$). Moreover, some clusters may only have treatment or control units, making it impossible to estimate treatment effects within these clusters.

2.3 Across-Cluster Matching

If sample sizes are small and/or overlap is lacking within clusters, across-cluster matching can be conducted in observational multilevel data by borrowing units from other clusters. Across-cluster matching pools units across clusters and matches units within and across clusters, such that one common multilevel model is fit as a selection model for the entire data, as is the case in most multilevel analyses. Across-cluster matching has contrasting advantages compared to within-cluster matching such that it improves overlap between treatment and control conditions and can be used when sample sizes are small for some or many clusters. As a result of improved overlap, across-cluster matching can reduce bias and provide a reliable estimate of ATE. Steiner et al. (2012) showed that across-cluster matching can provide a consistent treatment effect even when the distributions of the covariates and outcome are different across clusters, as long as the selection processes are monotonic and the PS ranks of units are not reversed across the clusters.

Despite these advantages, across-cluster matching violates the idea of block randomized experiments and multisite randomized trials and requires a stronger identification assumption for estimating the ATE and ATT than within-cluster matching. As one common multilevel model is fit to all units in the data, both level-1 and level-2 covariates $(\boldsymbol{X}, \boldsymbol{W})$ are needed to establish strong ignorability in the selection model as in Eq. (7), but the risk of model misspecification is higher than for cluster-specific single-level models. When there exists heterogeneity in the data and the selection processes differ substantially across clusters, we might fail to correctly specify the multilevel selection model and across cluster matching may result in large bias in the estimation of treatment effects.

2.4 Multilevel Matching Continuum

This chapter introduces a general framework for a multilevel matching continuum that covers various multilevel matching approaches, consisting of within-cluster matching and across-cluster matching as two opposite ends of the continuum. This continuum includes the combination of within-cluster and across-cluster matching as in Arpino and Cannas (2016), where units are matched within clusters first and then across clusters additionally if needed. It also comprises the manifest and latent class matching approaches by Kim et al. (2016) and Lim (2016) where homogeneous groups of clusters (called "classes") are identified first with respect to the selection process, outcome process, or both, and then units are pooled across clusters but within the homogeneous classes of clusters.

Although level-1 units and clusters belong to classes, classes do not constitute a third level as the number of classes is generally small and classes do not follow a continuous (e.g., normal) distribution like clusters in multilevel models, except when each cluster corresponds to its own class as in within-cluster matching on the one end of the continuum. On the other end of the continuum, the number of classes is one, and across-cluster within-class matching is equivalent to across-cluster matching. Except for these extreme cases, differences among classes are accounted for as multiple-group fixed effects in multilevel models. Specifically, the across-cluster within-class selection model for two-level data is defined as follows:

$$logit(\pi_{ijs}) = \alpha_s + \boldsymbol{X}'_{ijs}\boldsymbol{\beta}_{js} + \boldsymbol{W}'_{js}\boldsymbol{\gamma}_s + \boldsymbol{XW}'_{ijs}\boldsymbol{\delta}_{js} + T_{js}, \tag{9}$$

and the outcome model is defined as:

$$Y_{ijs} = \zeta_s + \tau_s Z_{ijs} + \boldsymbol{X}'_{ijs}\boldsymbol{\kappa}_{js} + \boldsymbol{W}'_{js}\boldsymbol{\phi}_s + \boldsymbol{XW}'_{ijs}\boldsymbol{\lambda}_{js} + U_{js} + \epsilon_{ijs}, \tag{10}$$

where

- subscripts $i = 1, \ldots, n_{js}, j = 1, \ldots, M_s, s = 1, \ldots, K$ denote level-1 unit, cluster, and class, respectively
- π_{ijs} is the probability of being assigned to the treatment condition for a level-1 unit i in cluster j in class s
- Y_{ijs} is the outcome for a level-1 unit i in cluster j in class s
- Z_{ijs} is the treatment assignment variable for a level-1 unit i in cluster j in class s, $Z_{ijs} \in 0, 1$ (0 = untreated; 1 = treated) $\sim$ Bernoulli(π_{ijs}).
- τ_s is the class-specific treatment effect
- $\boldsymbol{X}_{ijs}$, $\boldsymbol{W}_{js}$, and $\boldsymbol{XW}_{ijs}$ are level-1 covariates, level-2 covariates, and their cross-level interactions, respectively
- T_{js} and U_{js} are random effects for cluster j,
- α_s and ζ_s are class-specific intercepts
- $\boldsymbol{\beta}_{js}$, $\boldsymbol{\delta}_{js}$, $\boldsymbol{\kappa}_{js}$, and $\boldsymbol{\lambda}_{js}$ are level-1 regression coefficients and may vary across clusters (i.e., random slopes).
- $\boldsymbol{\gamma}_s$ and $\boldsymbol{\phi}_s$ are level-2 regression coefficients for class s
- ϵ_{ijs} is the error term

Note that both the selection and outcome models are presented as two-level univariate models for simplicity, but either can be generalized to higher-level multivariate models. When the class membership is known (e.g., regions or districts, participation in different policies, etc.), class memberships can be added directly to the model, for example, using $K - 1$ fixed effect dummies. When the class memberships are unknown, latent class memberships can be estimated by multilevel latent class models.

As explained above, the selection and outcome models in Eqs. (9) (10) for across-cluster within-class matching consist of within-cluster and across-cluster matching as two special extreme cases. On the one hand, when sample sizes are not small and overlap is sufficient for all clusters, each cluster can be considered as its own class, $j = s$ and $M_s = K$, resulting in within-cluster matching. On the other hand, when clusters are homogeneous with regard to selection and outcome mechanisms, the number of classes can be one ($K = 1$), and both selection and outcome models are one-class models, and across-cluster within-class matching is identical to across-cluster matching. In practice, a finite number of classes may be most appropriate by encompassing the advantage of within-cluster matching and across-cluster matching. Therefore, the across-cluster and within-class matching continuum provides a theoretical framework to examine the heterogeneity of selection and outcome mechanisms as well as practical tools for pooling compatible units from other clusters if needed.

We conducted simulation studies to examine the properties of across-cluster and within-class matching strategies in various settings, and this chapter presents two results of the investigation: (1) the consequences of classification of units at the lowest level versus cluster level (i.e., level-1 vs. level-2 classifications) and (2) findings related to the sequential classification of selection and outcome processes (i.e., the estimations of class-specific selection models and then class-specific outcome models).

3 Classification at the Cluster Level vs. Individual Level

In multilevel settings, selection and outcome processes may (1) simultaneously take place at different levels, (2) differ from cluster to cluster, and (3) introduce biases of different directions at different levels. An example would be a school retention policy where student retention is the "treatment," with student academic performance as the outcome. While some schools may adhere to state recommended policies on retention, other schools may place more emphasis on teachers' evaluations to determine retention. In addition, students and their parents may also influence retention decisions, if they are allowed to seek extra credit to gain promotion, for example. Likewise, even in schools that practice similar retention policies, outcomes may differ, as the implementation of academic support for retained students may vary widely, depending on school resources and funding. Students' motivation and academic self-concept could also play a part in their own academic outcomes.

Therefore, we investigated selection and outcome heterogeneity at both the cluster level and individual level in two-level data. The decision to determine the suitability of classification at the cluster-level or individual-level would depend on the theoretical benefits as to the level at which the effect is more likely to take place as well as the hypothesis being researched. For example, if the main interest is in investigating school-level effectiveness of school-determined retention policies, a cluster-level classification would seem most appropriate.

3.1 Evaluation of Current Techniques and Programs

As multilevel latent class regression requires computationally intensive numerical methods to find an optimal solution, we rely on software programs to estimate our models. Existing software we can use for multilevel latent class models include Mplus, Latent GOLD, and R packages such as FlexMix (Grün and Leisch 2008).

Table 1 Comparison of software

Software	Level-2 selection classification		Level-1 selection classification	
	MGLM[a] fixed effects[b] only	MGLM with random effects	MGLM fixed effects only	MGLM with random effects
FlexMix	✓			
Latent GOLD	✓	✓	✓	✓
Mplus		✓		✓
Software	**Level-2 outcome classification**		**Level-1 outcome classification**	
	MLM[c] fixed effects only	MLM with random effects	MLM fixed effects only	MLM with random effects
FlexMix	✓	✓		
Latent GOLD	✓	✓	✓	✓
Mplus		✓		✓

Note:
[a]multilevel generalized linear mixture
[b]restricting $\text{Var}(Uj) = 0$
[c]multilevel linear mixture

Table 1 shows a brief comparison of the three programs and their capacities to fit various multilevel generalized linear mixture (MGLM) models and multilevel linear mixture (MLM) models as selection and outcome models, respectively, at the cluster and individual levels.

FlexMix performs classification at the cluster level but not at the individual level. Random effects in multilevel models are limited to a continuous dependent variable in FlexMix 2.3–13. Although Mplus and Latent GOLD allow the estimation of nonlinear multilevel latent class models at both levels, we could not obtain stable estimation for classification at the individual level and encountered difficulties using either program for the investigation of level-1 classification.

3.2 *Difficulties and Limitations of Classification at Individual Level*

3.2.1 Data Simulation

Two scenarios were simulated and analyzed to evaluate the correct proportion of classification and parameter recovery of multilevel latent class models, varying in the strength of selection. To generate the data with the selection mechanism at the lowest level (i.e., level-1), a sample of 200 clusters (level-2) with varying cluster sizes [the average of 300 and the SD of 50, $N_j \sim N(300, 50)$] was randomly drawn from a population. Each unit in the first 120 clusters had a probability of between 0.6 and 0.8 for being assigned to the first selection mechanism and a probability of between 0.2 and 0.4 to the second selection mechanism. In the remaining 80 clusters, each unit had a probability of between 0.2 and 0.4 for being assigned to the first selection mechanism and a probability of between 0.6 and 0.8 to the second selection mechanism. The likelihood of being assigned to treatment would therefore depend on the unit's selection mechanism and its level-1 and level-2 covariates.

To investigate the characteristics of level-1 classification under different conditions, two selection mechanisms were generated using random intercept models with regression coefficients in opposite directions, reflecting two selection classes. The first data-generating model reflects a strong selection process, which implies little overlap between treated and control cases, and the second data-generating model reflects a weaker selection process and provides a large overlap between treated and control cases.

3.2.2 Classification and Parameter Recovery Results

We modified the settings of Latent GOLD to be able to use the problem for conducting across-cluster within-class matching with level-1 classification that may not be entirely intuitive. Although the data and model of interest both have two levels, we had to use a three-level environment where "Case ID" and "Group ID" correspond to level-2 and level-3 identification values, respectively (Vermunt and Magidson 2010). To estimate multilevel latent class models with level-1 classification, "Case ID" and "Group ID" were instead used to indicate the level-1 and level-2 identification values but with only level-1 unit in each level-2 unit, forcing the software to identify the level-1 unit as level-2.[1] Using Mplus and Latent GOLD, we estimated the multilevel latent class logistic selection models with strong and weak selection processes. The results of the individual level classification are summarized in Table 2.

Even in a condition where two selection processes are strong and in opposite directions, implying that the two selection processes are very different, the misclassification of the individual level-1 units was not very small (Latent GOLD 12.3%, Mplus 21.3%). When the selection mechanisms are weak yet in opposite directions, i.e., the two selection processes are more similar than before, the misclassification rises to as high as 45.3% (Latent GOLD) and 48.7% (Mplus). Table 2 also shows that Mplus and Latent GOLD provide different classifications of the level-1 units, where only 75.1% of the level-1 units were classified the same way for the strong selection mechanisms scenario. This proportion drops to 66.2% for weak

[1]This caused some problems with the multilevel logistic model specification in the software as random effects had to be modeled as a level-3 "GCFactor" and kept class-independent, so as to keep the number of parameters estimated comparable to the model fitted in Mplus. This in turn allowed the random effects of the intercept to vary between the classes, because Latent GOLD automatically adds an estimation of random effects ("CFactor") at the level of "Case ID." This effect should not have been estimated in a logistic regression. However, this was the only way to obtain a comparable level-1 classification in Latent GOLD that was comparable to Mplus in terms of the parameters estimated.

Table 2 Model statistics for simulations of level-1 classification

		Strong selection	Weak selection
True proportion of units in class 1		49.8%	53.8%
Mplus	Estimated proportion of units in class 1	59.2%	68.1%
	Proportion classified correctly (overall)	78.7%	51.3%
	Log likelihood	−38,634.6	−38,992.6
	AIC	77,295.3	78,011.2
	BIC	77,412.2	78,128.1
Latent GOLD	Estimated proportion of units in class 1	54.5%	73.2%
	Proportion classified correctly (overall)	87.7%	55.7%
	Log likelihood	−38,123.3	−38,943.2
	AIC	76,272.9	77,912.5
	BIC	76,389.8	78,029.4
MPlus vs	Proportion of units classified in agreement for both software	75.1%	66.2%
Latent GOLD	Proportion of units classified correctly	70.1%	36.1%
Common support	Class 1	Class: 19.6%	Class: 99.7%
		Cluster: 5.1%	Cluster: 90.9%
	Class 2	Class: 19.7%	Class: 99.9%
		Cluster: 4.2%	Cluster: 92.6%
	Overall	Overall: 20.4%	Overall: 99.8%
		Cluster: 5.7%	Cluster: 99.8%

selection mechanisms, suggesting the classification at the individual level may not be sufficiently robust for analysis.

We also examined the estimation of regression coefficients. The recovery of parameters was not good for either program. Table 3 shows the regression coefficients recovered using multilevel latent class logistic models with Mplus and Latent GOLD for the strong selection mechanism simulated data. The results showed that neither recovered the true parameters closely and that the two programs provided different estimates.

Therefore, although there is much value in investigating selection and outcome classification at the individual level theoretically, the estimation of such models faces a severe challenge in practice. Neither Latent GOLD or Mplus recovered the model parameters nor classified units correctly even when the classes were clearly separated. Kim and Steiner (2015); Kim et al. (2016) found that FlexMix performed well in multilevel latent class outcome models, but the R package allows for classification only at the cluster level, not for the individual level. It thus cannot be used for level-1 classification.

Table 3 Regression coefficients recovered by multilevel latent class logistic regression

Class 1	True	Mplus	Latent GOLD
Intercept	4.0	−7.755 (2.265)	−0.715 (0.139)
X1	1.5	1.052 (0.128)	0.189 (0.001)
X2	−1.5	−1.024 (0.122)	−0.187 (0.001)
W1	0.3	2.090 (0.460)	0.357 (0.002)
W2	−0.3	−1.096 (0.268)	−0.248 (0.002)
Class 2	True	Mplus	Latent GOLD
Intercept	−4.0	0.122 (0.106)	0.988 (0.198)
X1	−1.5	−0.031 (0.002)	−0.126 (0.001)
X2	1.5	0.027 (0.002)	0.209 (0.001)
W1	−0.3	−0.043 (0.012)	−0.391 (0.002)
W2	0.3	0.031 (0.009)	0.252 (0.002)

X1, X2: Level-1 covariates; W1, W2: Level-2 covariates
Standard errors are reported in parentheses

4 Heterogeneity in Selection and Outcome Mechanisms

We conducted a series of simulation studies to investigate the effectiveness of multilevel latent class regression modeling for identifying heterogeneity in the outcome processes. Because of the difficulties in estimating individual-level classification models as explained in the previous section, we examined selection and outcome process classification only at the cluster level in this section. As the selection process would occur prior to the outcome process theoretically, our simulation reflects this natural order and the selection and outcome models were considered sequentially rather than simultaneously.

One important aspect of this sequential analysis is that the misclassification of selection classes will carry over to the outcome process analysis when selection classes are unknown and estimated. As the identification of unknown selection classes requires the implementation of nonlinear multilevel latent class models with binary outcomes, the separation of classes might not be perfect, and some units can be misclassified to the wrong selection classes especially when the selection mechanisms are not very distinctive across classes, some selection classes are small, or the sample sizes are not sufficiently large for reliable estimation. Therefore, when the selection classes can be determined, we recommend to use manifest selection class models. It is important to note that the manifest selection class approach does not imply known selection mechanisms but only known class memberships. The specific selection processes can be estimated separately for different selection classes (Kim and Steiner 2015; Kim et al. 2016).

We examined the heterogeneity of outcome processes and estimated class-specific treatment effects by multilevel latent class models in various conditions. Due to space limits, this section focuses on results of the simulation study where we investigated properties of latent class outcome models in combination with manifest class and latent class selection models. For details of the simulation design and data-generating processes, we refer to Lim (2016).

4.1 Detecting Outcome Heterogeneity with Multilevel Latent Class Models

Initially, we generated two selection classes. For those two selection classes, we generated two outcome classes for the first class and one outcome class for the second class. This results in two selection classes and three-outcome classes in total. These can also be viewed as three sets of selection and outcome classes (i.e., selection1outcome1, selection1outcome2, selection2outcome3). We examined the properties of the latent class modeling approach in identifying the selection and outcome classes and classifying units into the corresponding classes as well as the consequences of misclassification in estimating treatment effects.

When selection classes were known and the correct number of latent outcome classes was specified in the models, only four out of the total 300 simulations (100 each for the three selection strength conditions; strong, medium, and weak) had outcome heterogeneity misclassification, suggesting that estimating heterogeneity in outcome is fairly reliable for this study's selected sample size.

Using the comparative model fit statistics of the Akaike information criterion (AIC) and the Bayesian information criterion (BIC) for the first selection class which contained two-outcome classes, one would favor the correct number of outcome classes about three-fifths of the time, regardless of the selection strength condition in Selection Class 1 (see Table 4), with the remaining favoring a three-outcome-class solution. However, checks of the last Mplus output generated in the simulation runs reveal that these three-class solutions favored were effectively two-class solutions with the third class of size zero, implying two-class solutions were recovered in most of the replications. For the second selection class which contained one-outcome class, a one-class solution was favored 100% of the time. Therefore, when the selection class memberships are known, the outcome classes can be estimated quite reliably by multilevel latent class models.

This was not the case when selection classes had to be estimated. With unknown and thus estimated latent selection classes, there is a need to account for misclassification occurring at the stage of multilevel latent class logistic regression to detect selection heterogeneity. With a stronger selection mechanism, selection class misclassification was expected to be low, whereas a weaker selection mechanism

Table 4 Proportion of identified correct number of outcome classes by AIC and BIC when selection classes are known

Selection	AIC (%)	BIC (%)
Strong	66	68
Medium	60	63
Weak	65	67

Table 5 Proportions of misclassification when selection classes are unknown

Selection	Proportion of misclassification (%)	Average proportion of units misclassified (%)
Strong	8	1.05
Medium	32	1.21
Weak	88	5.04

Table 6 Proportion of correct identification of outcome classes and model fit statistics in estimated Selection Class 1

Selection	Classification	AIC (%)	BIC (%)
Strong	Correctly classified (92%)	65	67
	Misclassified (8%)	75	75
Medium	Correctly classified (68%)	60	65
	Misclassified (32%)	56	56
Weak	Correctly classified (12%)	58	58
	Misclassified (88%)	66	68

was expected to provide a higher selection class misclassification. The simulation results for the selection class misclassification and the average proportion of units that are misclassified are shown below in Table 5, and the spread of the misclassifications are shown in Fig. 1.

As selection misclassification resulted in the estimated selection classes possibly having more outcome classes than they were supposed to have, it was difficult to estimate misclassification for outcome classes. A request for a two-outcome-class solution in Mplus appeared to require the solution to collapse two of the three-outcome classes into one class, leading to potentially very low or very high proportions of misclassification, depending on how the collapsing of the outcome classes occurred. However, using the comparative model fit statistics of AIC and BIC, one would still favor a two-outcome-class solution for the first estimated selection class in the majority of the simulations, due to the small proportion of misclassifications for each simulation (see Table 6). In contrast, for the second estimated selection class, one would favor a two-outcome-class solution 100% of the time if there was misclassification, and a one-outcome-class solution if there was no misclassification. This suggests that it was much easier to detect heterogeneity from one-outcome class to two-outcome classes, as compared to differentiating two-outcome classes from three-outcome classes.

4.2 *Estimating Treatment Effects with Multilevel Latent Class Models*

The unadjusted mean difference, referred to as the *prima facie* causal effect in the causal inference literature, and across-cluster matching treatment effects were estimated during the simulation. Both approaches performed very poorly in the presence of selection processes and heterogeneity, replicating the results in Kim

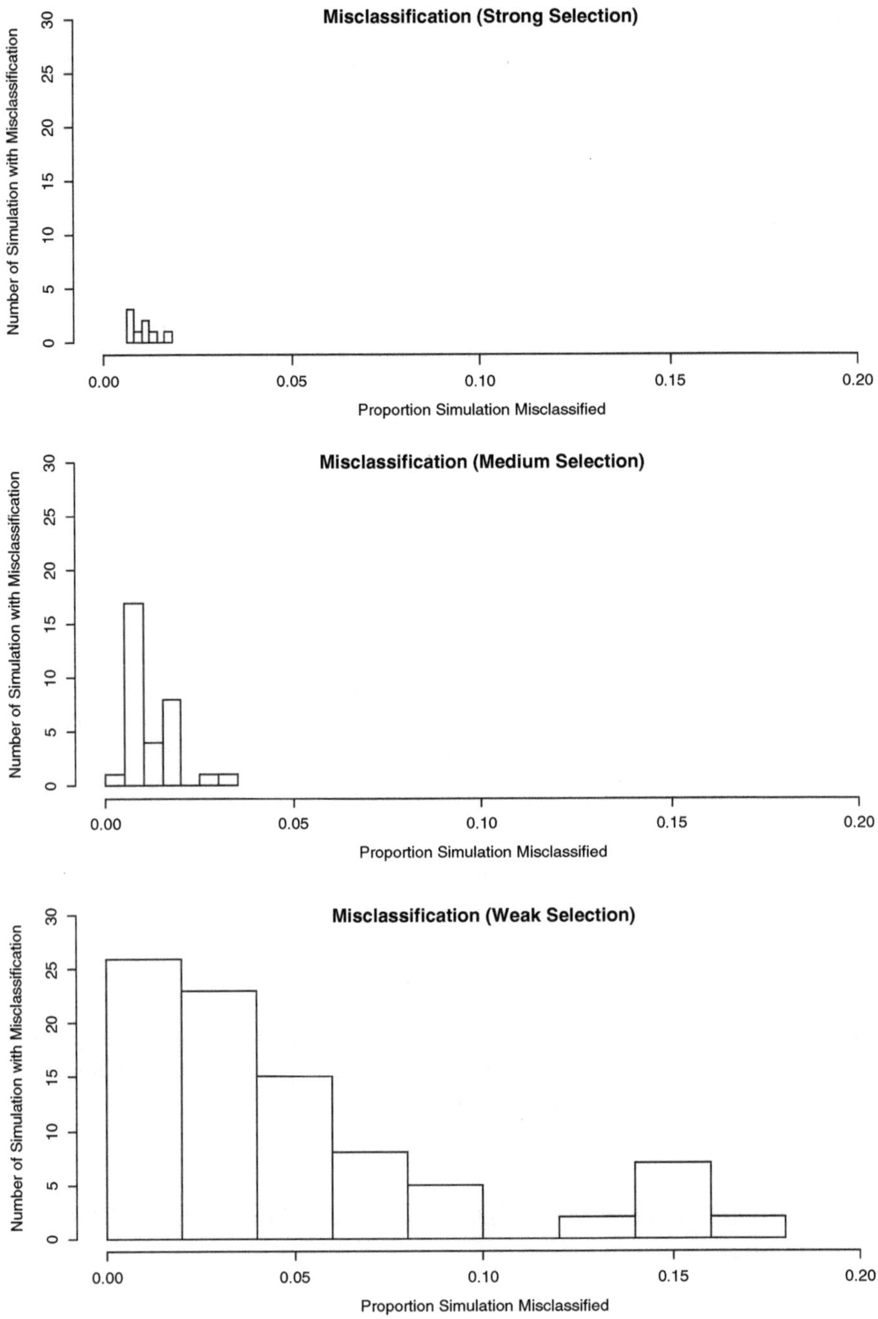

Fig. 1 Misclassification frequency distributions with strong, medium, and weak selection mechanisms in estimated Selection Class 1

et al. (2016). As such, in this section, the discussion will focus on the estimation of average treatment effects using within-cluster matching (WC), and across-cluster within-class matching (ACWC) and when to use the two techniques. Among other findings, several important results of the simulations are summarized below:

High Within-Cluster Overlap versus Low Within-Cluster Overlap When the treatment selection mechanism in the first selection class was weak (i.e., with high within-cluster overlap), from the top part (row number "1") of Table 7, one may observe there was only minimal bias when using the within-cluster matching technique. In contrast, large bias was observed using within-cluster matching when the selection mechanism was strong with little within-cluster overlap. In our simulation, cases lacking overlap were not deleted.

In addition, when the treatment selection mechanism was weak, while there was minimal bias observed for known selection classes (rows 3–7 of Table 7), gross bias was observed for estimated selection classes (rows 8–12 of Table 7). This was expected due to the higher proportion of selection-class misclassification from using

Table 7 Bias of estimated treatment effect (known versus estimated selection classification)

	The Strength of Selection in Sel.Class 1[a]		
	Strong	Medium	Weak
Known selection classification			
Within-cluster (WC)			
1. PS.Adj[b]: known SelClass1	41.834	11.187	−0.149
2. PS.Adj: known SelClass2	−0.272	−0.321	−0.321
Across-cluster within-class (ACWC)—selection only			
3. PS.Adj: known SelClass1	3.033	0.166	−1.118
4. PS.Adj: known SelClass2	−0.217	−0.271	−0.271
Across-cluster within-class (ACWC)—selection and outcome			
5. Cov.Adj[c]: known SelClass1, estimated OutClass1	−0.299	−0.263	0.097
6. Cov.Adj: known SelClass1, estimated OutClass2	−0.263	0.164	−0.532
7. Cov.Adj: known SelClass2, estimated OutClass3	−0.281	−0.325	−0.325
Estimated selection classification			
Across-cluster within-class (ACWC)—selection only			
8. PS.Adj: estimated SelClass1	4.018	1.972	−4.180
9. PS.Adj: estimated SelClass2	0.011	0.002	−18.995
Across-cluster within-class (ACWC)—selection and outcome			
10. Cov.Adj: estimated SelClass1, estimated OutClass1	−0.295	0.330	−2.705
11. Cov.Adj: estimated SelClass1, estimated OutClass2	−0.086	−0.157	−0.455
12. Cov.Adj: estimated SelClass2, estimated OutClass3	−0.046	−0.049	−20.029

Note:
[a] Selection Class 2 had random treatment assignment for all three selection conditions used in Selection Class 1
[b]PS.Adj: propensity score adjustment (using inverse propensity score weighting)
[c]Cov.Adj: covariate adjustment (using multilevel latent class linear regression)

multilevel latent class logistic regression when the selection mechanism in Selection Class 1 was weak and similar to the random selection mechanism in Selection Class 2.

Known Selection Class versus Estimated Selection Class In general, one may observe from Table 7 that for the strong selection and medium selection mechanisms in Selection Class 1, bias was minimal and similar for both known selection class and estimated selection class, since there was little misclassification. However, when there is higher misclassification from the weak selection mechanism in Selection Class 1, there is potential for gross bias resulting from misclassification using the estimated selection classes.

Propensity Score Adjustment versus Covariate Adjustment From Table 7, in most instances, the bias resulting from using propensity score adjustment and covariate adjustment was actually similar when comparing within known selection classes and estimated selection classes, respectively (see rows 4 versus 7, and rows 9 versus 12). However, one may observe from rows 3 and 8 of Tables 8 and 9, respectively,

Table 8 Mean square error (MSE) of estimated treatment effect (known versus estimated selection classification)

	The Strength of Selection in Sel.Class 1[a]		
	Strong	Medium	Weak
Known selection classification			
Within-cluster (WC)			
1. PS.Adj[b]: known SelClass1	1,787.741	147.841	3.884
2. PS.Adj: known SelClass2	5.780	5.796	5.796
Across-cluster within-class (ACWC)—selection only			
3. PS.Adj: known SelClass1	1,017.212	97.648	8.411
4. PS.Adj: known SelClass2	7.423	7.430	7.430
Across-cluster within-class (ACWC)—selection and outcome			
5. Cov.Adj[c]: known SelClass1, estimated OutClass1	10.304	4.499	3.385
6. Cov.Adj: known SelClass1, estimated OutClass2	47.700	11.603	6.305
7. Cov.Adj: known SelClass2, estimated OutClass3	5.549	5.576	5.576
Estimated selection classification			
Across-cluster within-class (ACWC)—selection only			
8. PS.Adj: estimated SelClass1	1,056.474	123.272	50.953
9. PS.Adj: estimated SelClass2	7.743	8.867	607.690
Across-cluster within-class (ACWC)—selection and outcome			
10. Cov.Adj: estimated SelClass1, estimated OutClass1	10.413	5.228	27.315
11. Cov.Adj: estimated SelClass1, estimated OutClass2	45.358	29.523	7.658
12. Cov.Adj: estimated SelClass2, estimated OutClass3	6.097	7.034	654.455

Note:
[a]Selection Class 2 had random treatment assignment for all three selection conditions used in Selection Class 1
[b]PS.Adj: propensity score adjustment (using inverse propensity score weighting)
[c]Cov.Adj: covariate adjustment (using multilevel latent class linear regression)

Table 9 Standard deviation of estimated treatment effect (known versus estimated selection classification)

	The Strength of Selection in Sel.Class 1[a]		
	Strong	Medium	Weak
Known selection classification			
Within-cluster (WC)			
1. PS.Adj[b]: known SelClass1	6.165	4.786	1.975
2. PS.Adj: known SelClass2	2.401	2.398	2.398
Across-cluster within-class (ACWC)—selection only			
3. PS.Adj: known SelClass1	31.909	9.930	2.690
4. PS.Adj: known SelClass2	2.729	2.726	2.726
Across-cluster within-class (ACWC)—selection and outcome			
5. Cov.Adj[c]: known SelClass1, estimated OutClass1	3.212	2.115	1.847
6. Cov.Adj: known SelClass1, estimated OutClass2	6.936	3.419	2.466
7. Cov.Adj: known SelClass2, estimated OutClass3	2.351	2.351	2.351
Estimated selection classification			
Across-cluster within-class (ACWC)—selection only			
8. PS.Adj: estimated SelClass1	32.417	10.981	5.815
9. PS.Adj: estimated SelClass2	2.797	2.993	15.791
Across-cluster within-class (ACWC)—selection and outcome			
10. Cov.Adj: estimated SelClass1, estimated OutClass1	3.230	2.274	4.495
11. Cov.Adj: estimated SelClass1, estimated OutClass2	6.768	5.459	2.743
12. Cov.Adj: estimated SelClass2, estimated OutClass3	2.481	2.665	15.996

Note:
[a] Selection Class 2 had random treatment assignment for all three selection conditions used in Selection Class 1
[b] PS.Adj: propensity score adjustment (using inverse propensity score weighting)
[c] Cov.Adj: covariate adjustment (using multilevel latent class linear regression)

that the mean square error and standard deviation for the estimated treatment effect in Selection Class 1 are rather large when using propensity score adjustment, indicating that estimation of the treatment effect was not very efficient. This was a direct result from the use of the inverse propensity score weighting approach in the simulations, where the strong selection mechanism led to propensity scores very close to 0 or 1, thus resulting in very large inverse propensity weights. Therefore, when the selection mechanism is strong and selection class membership is unknown, covariate adjustment can be a viable alternative if one is willing to make the required functional form and distributional assumptions.

Checking for Outcome Heterogeneity Directly Without Checking for Selection Heterogeneity The excellent classification recovery for known selection classes as discussed in Sect. 4.1 suggests that multilevel latent class linear regression was able to detect outcome heterogeneity more reliably than multilevel latent class logistic regression was able to detect selection heterogeneity and required a smaller sample size when doing so. As such, a small separate simulation study was done using the same data generation as in the original simulation study, but performing only multilevel latent class linear regression, with 20 simulations for each selection strength condition in the first selection class. All 60 simulations recovered the three-outcome classes perfectly and, as a result, had minimal bias in estimation of treatment effects. This indicates the viability of using covariate adjustment when one does not have information about selection heterogeneity. This approach can be used as a sensitivity analysis in comparing the treatment effects recovered from estimated selection classes to those obtained from multilevel latent class logistic regression.

5 Summary and Recommendations

There is a large and increasing number of observational datasets collected in social science research, and many are multilevel in structure. Naturally, there has also been increasing interest in understanding how to make causal inference involving multilevel observational data. The goal of this study is to estimate valid treatment effects with multilevel observational data by removing selection bias and also accounting for potential heterogeneity in selection and outcome processes.

The simulation study in this paper suggests that when there exists strong or moderate selection and the overlap between treatment and comparison groups are not sufficient within clusters, the joint use of selection and outcome latent class models outperforms the use of either alone. It also works substantially better than more traditional multilevel matching approaches such as within-cluster or across-cluster matching with respect to efficiency and bias reduction.

When the differences in selection processes are weak, however, it is difficult to classify units correctly. The proportions of selection class misclassification are likely to be greater than those of outcome class misspecification, as selection processes require nonlinear models for binary treatment assignment. In this case, the simulation results suggest the examination of outcome heterogeneity regardless of selection classes (i.e., fitting outcome latent class models for the entire data) and a comparison of the results to the outcome class models within each of the selection classes.

Although it is not common in practice, it is conceivable that the sample size will be large and the overlap sufficient for each of the clusters. In that case, we recommend using within-cluster matching, as located at one extreme of the multilevel matching continuum, and summarizing the cluster-specific treatment effects. When the number of latent classes is found to be one for both selection and outcome models, the one-class selection and outcome multilevel model are identical to the standard multilevel models for across-cluster matching at the other extreme of the multilevel matching continuum. Therefore, the across-cluster within-class multilevel matching continuum presented in this study provides a flexible and unifying framework for making causal inference with observational multilevel data.

Acknowledgements This research was supported by the Institute of Education Sciences, U.S. Department of Education, through Grant R305D120005. The opinions expressed are those of the authors and do not represent views of the Institute or the U.S. Department of Education.

References

B. Arpino, M. Cannas, Propensity score matching with clustered data. An application to the estimation of the impact of caesarean section on the Apgar score. Stat. Med. **35**(12), 2074–2091 (2016)

R.A. Berk, *Statistical Learning from a Regression Perspective* (Springer, New York, 2008)

H. Cooper, L.V. Hedges, J.C. Valentine, *The Handbook of Research Synthesis and Meta-Analysis*, 2nd edn. (Russell Sage Foundation, New York, 2009)

B. Grün, F. Leisch, Flexmix version 2: finite mixtures with concomitant variables and varying and constant parameters. J. Stat. Softw. **28**, 1–35 (2008)

S. Guo, M.W. Fraser, *Propensity Score Analysis: Statistical Methods and Applications*, 2nd edn. (Sage, Thousand Oaks, 2015)

P.W. Holland, Statistics and causal inference. J. Am. Stat. Assoc. **81**, 945–970 (1986)

G. Hong, S.W. Raudenbush, Evaluating kindergarten retention policy: a case study of causal inference for multilevel observational data. J. Am. Stat. Assoc. **101**, 901–910 (2006)

G.W. Imbens, D.B. Rubin, *Causal Inference for Statistics, Social and Biomedical Sciences – An Introduction* (Cambridge University Press, New York, 2015)

L. Keele, J.R. Zubizarreta, Optimal multilevel matching in clustered observational studies: a case study of the effectiveness of private schools under a large-scale voucher system (2014). ArXiv e-prints

B.M. Kelcey, Improving and assessing propensity score based causal inferences in multilevel and nonlinear settings. Unpublished doctoral dissertation, University of Michigan (2009)

B. Keller, J.-S. Kim, P.M. Steiner, Neural networks for propensity score estimation: simulation results and recommendations, Chap. 20, in *Quantitative Psychology Research*, ed. by L.A. van der Ark, D.M. Bolt, S.-M. Chow, J.A. Douglas, W.-C. Wang (Springer, New York, 2015), pp. 279–291

J. Kim, M. Seltzer, Causal inference in multilevel settings in which selection process vary across schools. Working paper 708 (Center for the Study of Evaluation (CSE), Los Angeles, 2007)

J.-S. Kim, P.M. Steiner, Multilevel propensity score methods for estimating causal effects: a latent class modeling strategy, Chap. 21, in *Quantitative Psychology Research*, ed. by L.A. van der Ark, D.M. Bolt, S.-M. Chow, J.A. Douglas, W.-C. Wang (Springer, New York, 2015), pp. 293–306

J.-S. Kim, P.M. Steiner, W.C. Lim, Mixture modeling methods for causal inference with multilevel data, in *Advances in Multilevel Modeling for Educational Research*, ed. by J.R. Harring, L.M. Stapleton, S.N. Beretvas (Information Age Publishing, Charlotte, 2016), pp. 335–359

W.-C. Lim, Selection and outcome mechanism heterogeneity in causal inference with observational multilevel data. Master's thesis, University of Wisconsin-Madison (2016)

D.F. McCaffrey, G. Ridgeway, A.R. Morral, Propensity score estimation with boosted regression for evaluating causal effects in observational studies. Psychol. Methods **9**, 403–425 (2004)

S.W. Raudenbush, A.S. Bryk, *Hierarchical Linear Models: Applications and Data Analysis Methods*, 2nd edn. (Sage, Newbury Park, 2002)

P.R. Rosenbaum, D.B. Rubin, The central role of the propensity score in observational studies for causal effects. Biometrika **70**, 41–55 (1983)

D.B. Rubin, Estimating causal effects of treatments in randomized and nonrandomized studies. J. Educ. Psychol. **66**, 688–701 (1974)

D.B. Rubin, Bayesian inference for causal effects: the role of randomization. Ann. Stat. **6**, 34–58 (1978)

W.R. Shadish, T.D. Cook, D.T. Campbell, *Experimental and Quasi–Experimental Designs for Generalized Causal Inference* (Houghton Mifflin, Boston, 2002)

P.M. Steiner, J.-S. Kim, F. Thoemmes, Matching strategies for observational multilevel data, in *Joint Statistical Meeting Proceedings, Social Statistics Section* (2012), pp. 5020–5032

E.A. Stuart, Estimating causal effects using school-level datasets. Educ. Res. **36**, 187–198 (2007)

F. Thoemmes, S.G. West, The use of propensity scores for nonrandomized designs with clustered data. Multivar. Behav. Res. **46**, 514–543 (2011)

J.K. Vermunt, J. Magidson, Bayesian propensity score estimators: incorporating uncertainties in propensity scores into causal inference. Sociol. Methodol. **40**, 151–189 (2010)

Nonequivalent Groups with Covariates Design Using Propensity Scores for Kernel Equating

Gabriel Wallin and Marie Wiberg

Abstract In test score equating, the nonequivalent groups with covariates (NEC) design use covariates with high correlation to the test scores as a substitute for an anchor test when the latter is lacking. However, as the number of covariates increases, the number of observations for each covariate combination decreases. We suggest to use propensity scores instead, which we include in the kernel equating framework using both post-stratification and chained equating. The two approaches are illustrated with data from a large-scale assessment, and the results show an increased precision in comparison with the equivalent groups design and great similarities in comparison with the results when using an anchor test.

Keywords Collateral information • Nonequivalent groups • NEC design

1 Introduction

The goal of observed-score equating is to enable the comparison between test scores from different test versions (González and Wiberg 2017). This can be done in many ways, and it depends on how the data is collected and the nature of the test forms. If we can assume that the two groups of test takers who take the different test forms are only randomly different with respect to ability, we can use the equivalent groups (EG) design. If we have the opportunity to distribute common items (i.e., an anchor test) to the test groups, we can adjust for group differences in ability by using the nonequivalent groups with anchor test (NEAT) design. The NEAT design is preferable in many situations as test groups cannot always be considered equivalent, especially if the test is given over a longer period of time (Lyrén and Hambleton 2011). A problem is that although we might have nonequivalent groups, there is not always an anchor test administered. A way to handle these situations is to use information from background covariates to improve the equating through a nonequivalent groups with covariates (NEC) design (Wiberg and Bränberg 2015). Instead of using an anchor test, categorized covariates that correlate highly with the

G. Wallin (✉) • M. Wiberg
Department of Statistics, USBE, Umeå University, Umeå, Sweden
e-mail: gabriel.wallin@umu.se; marie.wiberg@umu.se

L.A. van der Ark et al. (eds.), *Quantitative Psychology*, Springer Proceedings in Mathematics & Statistics 196, DOI 10.1007/978-3-319-56294-0_27

test scores are used. A problem with this approach is that the number of covariate categories increases rapidly with the number of covariates used, thus reducing the number of observations within each category. To avoid this problem, one might be able to use propensity scores instead in the NEC design. The aim of this paper is to propose the use of propensity scores in the NEC design with both a post-stratification equating (PSE) approach and a chained equating (CE) approach within the kernel equating framework. The proposed approaches are compared with kernel equating under the EG design and the NEAT design using data from a college admission test.

Covariates have been suggested and used before in equating: as an extra information in a matching procedure in order to take care of possible differences between (sub)populations (e.g., Kolen 1990), as a surrogate for an anchor test (Liou et al. 2001) to improve linear equating (Bränberg and Wiberg 2011), and as collateral information in equating in comparison to item response theory (IRT) true-score equating (Hsu et al. 2009). This paper differs from these previous studies as we suggest propensity scores as a way to use covariates in order to improve the precision of the equating.

Propensity scores (Rosenbaum and Rubin 1983, 1984) have been proposed before in the test equating context (Livingston et al. 1990) and have been used for matching samples (e.g., Paek et al. 2006), and also to improve the equating results by reducing the bias of the traditional PSE method (Sungworn 2009). Furthermore, propensity scores have been used to combine anchor test scores for use in PSE (Moses et al. 2010), CE, frequency estimation, IRT true-score, and IRT observed-score equating (Powers 2010). Recently, Haberman (2015) used propensity scores to make the data resemble data from an EG design, and Longford (2015) adopted a causal analysis approach for equating using both inverse proportional weighting and matched pairs based on propensity scores. This study differs from the abovementioned studies as it focuses on kernel equating instead of IRT or traditional equating methods.

2 Propensity Scores Within the NEC Design

Under the NEC design, one sample from population P has size n_P, and one sample from population Q has size n_Q. The sample from population P took test form X, and the sample from population Q took test form Y, and the scores from each respective test form are denoted X and Y. A set of common covariates collected in $\mathbf{D}$ that are correlated with X and Y are measured on both samples. A propensity score (Rosenbaum and Rubin 1983) is a scalar function of the covariates and is used to separate the test groups with respect to ability. Formally, it is the conditional probability for test taker i, $i = 1, \ldots, n_P + n_Q$, of being assigned a well-defined active treatment, given the covariate vector $\mathbf{D} = (D_1, \ldots, D_m)^t$, t denoting the transpose. Let Z_i be a binary random treatment variable which equals 1 if test form X is given to test taker i. The propensity score is defined as

$$e(\mathbf{D}_i) = \Pr(Z_i = 1|\mathbf{D}_i) = E(Z|\mathbf{D}) \tag{1}$$

The propensity score is in general unknown and in this paper, logistic regression is used to estimate it. When the vector of propensity scores has been estimated, it is divided into subgroups based on the percentiles, in accordance with Rosenbaum and Rubin (1984). Within the same propensity score category, the test takers are viewed as equivalent with respect to ability.

3 Using Propensity Scores in Kernel Equating with the NEC Design

There are five steps within the kernel equating framework: (1) presmoothing, (2) estimation of the score probabilities, (3) continuization, (4) equating, and (5) calculating the standard error of equating (von Davier et al. 2004, pp. 45–47). In observed score equating, there is a target population T defined that specifies the population upon which the equating is performed on. The target population is defined differently depending on the data collection design: In the EG design, the target population is either P or Q depending on what population the sample was taken from. In the NEAT and NEC designs, the target population is defined as a mixture between the two populations P and Q such that $T = wP + (1-w)Q$, $0 \leq w \leq 1$.

By viewing the test takers as random samples, the test scores X and Y are regarded as random variables with realizations denoted by x_j and y_k, respectively, $j = 1, \ldots, J$ and $k = 1, \ldots, K$. The probability that a randomly selected test taker in the target population T gets a specific test score on the test forms is denoted by $r_j = Pr(X = x_j|T)$ and $s_k = Pr(Y = y_k|T)$, respectively. Let $F(x) = Pr(X \leq x|T)$ and $G(y) = Pr(Y \leq y|T)$ represent the cumulative distribution functions (CDFs) of the test scores.

Typically, the equipercentile equating transformation is used to perform the equating:

$$y = \varphi_Y(x) = G^{-1}(F(x)), \tag{2}$$

where $\varphi_Y(x)$ represents the equating transformation from test form X to test form Y. However, to be able to use this transformation, $F(\cdot)$ and $G(\cdot)$ need to be monotonically increasing, continuous functions. Since test scores usually are discrete, this problem will be addressed in Sect. 3.2.

To use propensity scores in kernel equating, we need to define how we use it. Let $e(\mathbf{D}_{Xl})$ and $e(\mathbf{D}_{Yl})$ be categorized versions of the propensity score for each test form, each containing L categories. The categorized propensity scores are used in the first step of kernel equating as described in the next section.

3.1 Presmoothing and Estimation of Score Probabilities

Log-linear models are used to model the empirical distributions for both samples. Let n_{Pjl} denote the number of test takers from P with a test score equal to x_j and an observed propensity score equal to $e(\mathbf{d}_{Xl})$, and let n_{Qkl} denote the number of test takers from Q with a test score equal to y_k and propensity score equal to $e(\mathbf{d}_{Yl})$. With the joint probabilities $p_{jl} = \Pr(X = x_j, e(\mathbf{D}_{Xl}) = e(\mathbf{d}_{Xl}|P))$ and $q_{kl} = \Pr(Y = y_k, e(\mathbf{D}_{Yl}) = e(\mathbf{d}_{Yl}|Q))$ for each respective population, we assume that the vectors $\mathbf{n}_P = (n_{P11}, \ldots, n_{PJL})^t$ and $\mathbf{n}_Q = (n_{Q11}, \ldots, n_{QKL})^t$ are independent and multinomially distributed.

Let PSE-NEC-PS indicate that propensity scores are used within the NEC design using a PSE approach. Using PSE-NEC-PS, the following probabilities are needed: $r_{Pj} = \Pr(X = x_j|P)$, $r_{Qj} = \Pr(X = x_j|Q)$, $s_{Pk} = \Pr(Y = y_k|P)$, $s_{Qk} = \mathrm{P}(Y = y_k|Q)$. From the definition of the target population, it follows that the test score distributions in population T are defined as

$$r_j = \Pr\left(X = x_j \mid T\right) = w r_{Pj} + (1 - w)\, r_{Qj}, \text{ and} \tag{3}$$

$$s_k = \Pr\left(Y = y_k \mid T\right) = w s_{Pk} + (1 - w)\, s_{Qk}. \tag{4}$$

r_{Pj} and s_{Qk} are estimated by means of

$$\widehat{r}_{Pj} = \sum_l \widehat{p}_{jl} \text{ and } \widehat{s}_{Qk} = \sum_l \widehat{q}_{kl}, \tag{5}$$

where $\widehat{p}_{jl}$ is the sample proportion from P with $X = x_j$ and $e(\mathbf{D}_{Xl}) = e(\mathbf{d}_{Xl})$ and $\widehat{q}_{kl}$ is the corresponding sample proportion from Q. It is however impossible to estimate r_{Qj} and s_{Pk} using only the observed data since, by design, the test takers from population Q only have scores on test form Y and the test takers from population P only have scores on test form X. It is therefore assumed that the conditional distribution of X given $e(\mathbf{D})$ and Y given $e(\mathbf{D})$ does not differ in the P and Q population. This assumption is closely related to the assumption made for PSE under the NEAT design, where the conditional distributions of X and Y are assumed independent conditional on the anchor test. In a NEC design where propensity scores are used, it is also assumed that every test taker has a positive probability of receiving either test form, given its observed value on $\mathbf{D}$. As such, the propensity score in the NEC design replaces the anchor test score in the NEAT design so that r_{Qj} and s_{Pk} can be estimated by

$$\widehat{r}_{Qj} = \sum_l \left(\frac{\widehat{p}_{jl}}{\sum_j \widehat{p}_{jl}} \cdot \sum_k \widehat{q}_{kl} \right) \text{ and } \widehat{s}_{Pk} = \sum_l \left(\frac{\widehat{q}_{kl}}{\sum_k \widehat{q}_{kl}} \cdot \sum_j \widehat{p}_{jl} \right), \tag{6}$$

where the expressions in Eq. (6), given the assumption made, follow directly from the law of total probability.

Let CE-NEC-PS indicate a NEC design in which propensity scores with a CE approach are used. The equated scores are obtained by going from the original score distributions through the propensity score distributions onto the target distributions in T. In addition to r_{Pj} and s_{Qk}, the following probabilities are required to obtain the equated scores: $t_{Pl} = \Pr(e(\mathbf{D}_{Xl}) = e(\mathbf{d}_{Xl}) \mid P) = \sum_j p_{jl}$ and $t_{Ql} = \Pr(e(\mathbf{D}_{Yl}) = e(\mathbf{d}_{Yl}) \mid Q) = \sum_k q_{kl}$

3.2 *Continuization and Equating*

A Gaussian kernel is used to approximate the estimated discrete CDFs by a smooth, continuous distribution (von Davier et al. 2004, pp. 56–61). Let $\Phi(\cdot)$ denote the distribution function of the standard normal distribution, $\mu_X = \sum_j x_j r_j$, $a_X = \sqrt{\sigma_X^2 / \left(\sigma_X^2 + h_X^2\right)}$ where σ_X^2 denotes the X score variance in T, and let $h_X > 0$ denote the bandwidth. The score CDF approximation for X and likewise for Y is defined as

$$F_{h_X}(x) = P\left(X\left(h_X\right) \leq x\right) = \sum_j r_j \Phi\left(\frac{x - a_X x_j - \left(1 - a_X\right)\mu_X}{a_X h_X}\right). \tag{7}$$

Although several options to select the bandwidths h_X and h_Y exist (e.g., double smoothing (Häggström and Wiberg 2014)), the most common was used here, which is to minimize the penalty function given in von Davier et al. (2004, pp. 61–64).

For PSE-NEC-PS, the estimates of $F_{h_X}(x)$ and $G_{h_Y}(y)$ are used to construct the equating transformation using the equipercentile transformation. Thus, Eq. (2) becomes

$$\widehat{\varphi}_{Y(PSE)}(x) = \widehat{G}_{h_Y}^{-1}\left(\widehat{F}_{h_X}(x)\right) \tag{8}$$

The transformations needed to equate X onto Y in the target population T using CE-NEC-PS are

$$\widehat{\varphi}_{e(\mathbf{d}_{Xl})}\left(x; \boldsymbol{r}_P, \boldsymbol{t}_P\right) = \widehat{H}_{Ph_{e(\mathbf{d}_{Xl})P}}^{-1}\left(\widehat{F}_{Ph_{XP}}\left(x; \boldsymbol{r}_P\right); \boldsymbol{t}_P\right), \text{and} \tag{9}$$

$$\widehat{\varphi}_Y\left(e\left(\mathbf{d}_{Yl}\right); \boldsymbol{t}_Q, \boldsymbol{s}_Q\right) = \widehat{G}_{Qh_{YQ}}^{-1}\left(\widehat{H}_{Qh_{e(\mathbf{d}_{Yl})Q}}\left(e\left(\mathbf{d}_{Yl}\right); \boldsymbol{t}_Q\right); \boldsymbol{s}_Q\right), \tag{10}$$

where $\widehat{H}_{Ph_{e(\mathbf{d}_{Xl})P}}$ and $\widehat{H}_{Qh_{e(\mathbf{d}_{Yl})Q}}$ are the estimated continuized CDFs of the categorized propensity score for population P and Q, respectively, and $\widehat{F}_{Ph_{XP}}$ and $\widehat{G}_{Qh_{YQ}}$ are the estimated continuized CDFs of X in P and Y in Q, respectively.

$\boldsymbol{r}_P$ and $\boldsymbol{s}_Q$ denote the vectors of score probabilities for X in P and Y in Q, respectively, and $\boldsymbol{t}_P$ and $\boldsymbol{t}_Q$ are the vectors of propensity score probabilities for each respective population. The equating transformation using the CE-NEC-PS approach is then defined as

$$\begin{aligned}\widehat{\varphi}_{Y(CE)}\left(x;\boldsymbol{r}_P,\boldsymbol{t}_P,\boldsymbol{t}_Q,\boldsymbol{s}_Q\right) &= \widehat{\varphi}_Y\left(\widehat{\varphi}_{e(\mathbf{d}_{XI})}\left(x;\boldsymbol{r}_P,\boldsymbol{t}_P\right);\boldsymbol{t}_Q,\boldsymbol{s}_Q\right) \\ &= \widehat{G}^{-1}_{Qh_{YQ}}\left(\widehat{H}_{Qh_{e(\mathbf{d}_{YI})Q}}\left(\widehat{H}^{-1}_{Ph_{e(\mathbf{d}_{XI})P}}\left(\widehat{F}_{Ph_{XP}}(x)\right)\right)\right).\end{aligned} \tag{11}$$

3.3 *The Standard Error of Equating*

In kernel equating, the standard error of equating (SEE) is the square root of the asymptotical variance of the estimated equating transformation $\widehat{\varphi}$ (von Davier et al. 2004, p. 71). It is assumed that the estimator of the score probabilities is asymptotically normally distributed, making large-sample approximations using the delta method possible to calculate the variance of $\widehat{\varphi}$. The SEE is formed by the Jacobian of the equated score, $\widehat{\mathbf{J}}_{\varphi_Y}$, the Jacobian of the design function (DF), $\widehat{\mathbf{J}}_{DF}$, and a matrix $\mathbf{C}$ which is used to define the covariance between $\mathbf{P}$, the $J \times L$ matrix with p_{jl} as entries, and $\mathbf{Q}$, the $K \times L$ matrix with q_{kl} as entries. The SEE is obtained from these three components as follows:

$$\mathrm{SEE}_Y(x) = \left\| \widehat{\mathbf{J}}_{\varphi_Y} \widehat{\mathbf{J}}_{\mathrm{DF}} \mathbf{C} \right\|, \tag{12}$$

where $\|\cdot\|$ denotes the Euclidean distance. $\widehat{\mathbf{J}}_{DF}$ is used when moving from the joint probabilities in $\mathbf{P}$ and $\mathbf{Q}$ to the marginal probabilities in $\boldsymbol{r} = (r_1, \ldots, r_J)^t$ and $\boldsymbol{s} = (s_1, \ldots, s_K)^t$, and where the data collection design adapted determines the form of the DF. The SEE expressions for the PSE-NEC-PS and CE-NEC-PS can be found in Wallin and Wiberg (2017).

4 Empirical Example

To illustrate the NEC design with propensity scores in kernel equating, the Swedish Scholastic Aptitude Test (SweSAT) was used. It is a college admission test with 160 binary-scored multiple-choice items. This paper and pencil test is given twice a year and contains a quantitative section with 80 items and a verbal section with 80 items which are equated separately. SweSAT has a history of using covariates in the equating process, and only in the last five years, an anchor test has been added in order to facilitate the equating. More details about equating methods for the SweSAT can be found in Lyrén and Hambleton (2011).

We equated the test scores from two consecutive administrations from the quantitative section of the SweSAT using the R package (R Core Development Team 2016) *kequate* (Andersson et al. 2013). The R code used can be obtained upon request from the corresponding author. The approaches being compared were PSE-NEC-PS, CE-NEC-PS, PSE and CE using a NEAT design (abbreviated PSE NEAT and CE NEAT, respectively), and equating using an EG design. A comparison was also made against the equating transformation suggested in Wiberg and Bränberg (2015), which here is abbreviated PSE-NEC-RAW-COV. An alternative to PSE-NEC-PS, abbreviated ANCHOR-NEC-PS, is to include the anchor test scores as a covariate in the estimation model of the propensity score. This is thus the same as the PSE-NEC-PS method although anchor items are included in the propensity scores.

To facilitate a comparison with the results when using a NEC design with covariates directly, the same raw test score data was used as in the study by Wiberg and Bränberg (2015). There were in total 14,644 test takers who took both administrations. We divided this group into two halves, allowing for differences in the covariate distribution in the groups, so that samples of 7,322 test takers for each test form were used. A 24-item anchor test was constructed through the selection of 12 items from both test administrations. Note, slightly different samples than in Wiberg and Bränberg (2015) were used.

The covariates used were the same as in Wiberg and Bränberg (2015), that is, criterion-referenced grades (Grade, range 0–320) and the test takers' verbal test scores (Vtest, range 0–80) which are known to correlate with the quantitative section. The Grade covariate was categorized into the following categories: Grade1 = 0–225, Grade2 = 226–255, Grade3 = 256–290, and Grade4 = 291–320, and Vtest was categorized into the following categories: Vtest1 = 0–30, Vtest2 = 31–40, Vtest3 = 41–50, and Vtest4 = 51–80.

The propensity score model was fitted using the uncategorized versions of the covariates, and then different categorizations of the propensity score were investigated. The final categorization yielded twenty categories based on equally spaced percentiles. The estimated, categorized propensity scores showed a correlation to the test scores almost as strong as that between the anchor test scores and the test scores (0.71 and 0.81, respectively), with equated scores that were stable to a slight change in the number of propensity score categories.

Log-linear models were fitted to the empirical score distributions of the two test forms. For all scenarios, the best log-linear model was chosen using the Akaike Information Criterion, taking on a parsimonious model selection approach. For PSE-NEC-PS, Eqs. (5) and (6) were used to estimate the score probabilities. The weight w was set to 0.5 in Eqs. (3) and (4). For CE-NEC-PS, the probabilities needed, besides those given in Eq. (5), were estimated by marginalizing the test scores in the estimated joint distributions of the propensity scores and the test scores in the two populations. A Gaussian kernel was used to transform the discrete score distributions of X and Y into continuous score distributions, as described by Eq. (7).

4.1 Results

In Fig. 1, the difference between the equated score and the raw score is plotted for every design in the study. The equated scores using PSE-NEC-PS and ANCHOR-NEC-PS get close to the equated scores using PSE NEAT and CE NEAT. This is also true for the equated scores using PSE-NEC-RAW-COV, but these results are not as close to the NEAT design results as is the equated scores from PSE-NEC-PS and ANCHOR-NEC-PS. Previous studies have shown that the assumption underlying the EG design is problematic for the SweSAT (Lyrén and Hambleton 2011), making a NEAT approach more suitable. The results of Fig. 1 thus display how covariates successfully replace the anchor scores when the latter is lacking. CE-NEC-PS does not show this similarity to the results of the two NEAT designs, and the EG approach deviates substantially from the others.

In Fig. 2, the same approaches are compared but with respect to the SEE. PSE-NEC-PS again shows a close resemblance with the results of the two NEAT designs, with a smaller SEE for a majority of the scores in comparison with the SEE of the EG design. The SEE of PSE-NEC-PS is slightly higher for almost every score in comparison with ANCHOR-NEC-PS, but close to equivalent to the results of PSE-NEC-RAW-COV. The SEE of CE-NEC-PS is in general higher in comparison with the other methods.

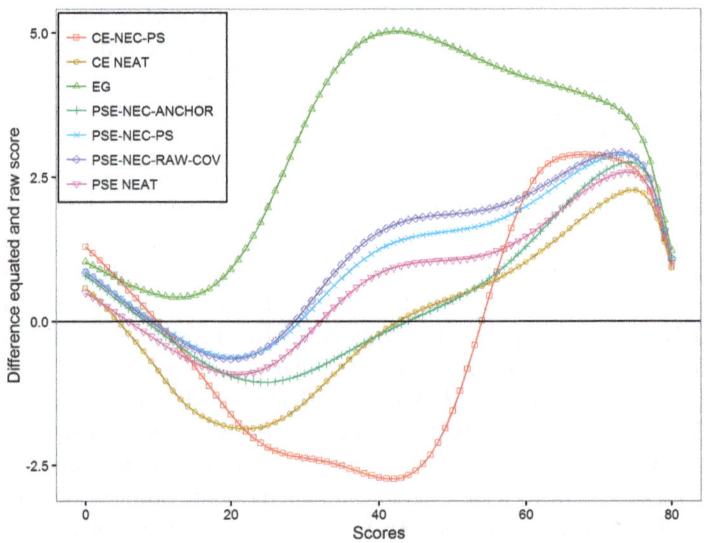

Fig. 1 The difference between the equated score and the raw score for the EG, CE NEAT, PSE NEAT, PSE-NEC-RAW-COV, CE-NEC-PS, PSE-NEC-PS, and ANCHOR-NEC-PS approach

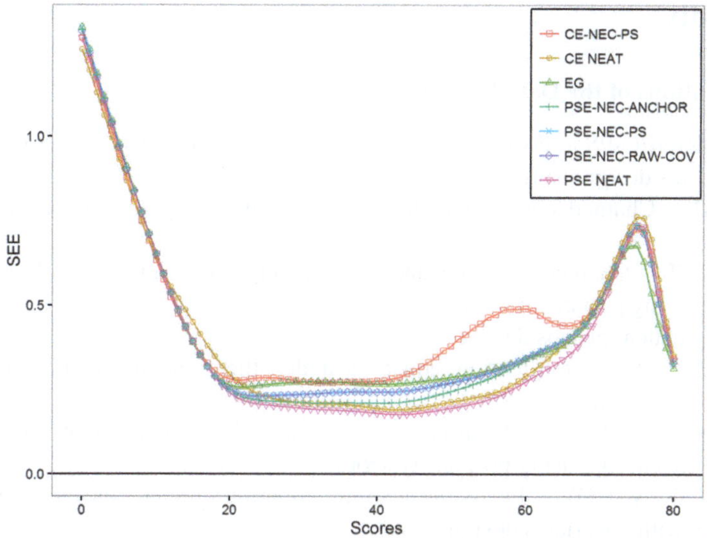

Fig. 2 The SEE for the EG, CE NEAT, PSE NEAT, PSE-NEC-RAW-COV, CE-NEC-PS, PSE-NEC-PS, and ANCHOR-NEC-PS approach

5 Concluding Remarks

Propensity scores were incorporated within the kernel equating framework, and differences were examined with this approach of handling nonequivalent groups compared to using anchor test scores. From the results of the empirical example, we concluded that the PSE-NEC-PS approach overall seems to be a more attractive choice in comparison with the EG design when the test groups are nonequivalent. The CE-NEC-PS approach was not as successful, with results relatively far from the NEAT design results. In comparison with the NEAT designs, the PSE-NEC-PS results get closer than the PSE-NEC-RAW-COV results in terms of equated scores and with close to equivalent results in terms of SEE. This indicates that the PSE-NEC-PS is a very strong alternative when there is no common items or test takers available. The PSE-NEC-PS approach is further strengthened since the incentive to use propensity scores increases with an increasing number of covariates, and we only used two. Furthermore, propensity scores facilitate the use of continuous covariates in comparison with equating using the covariates directly, meaning that the covariates are very easy to implement in a propensity score.

Acknowledgment The research in this paper was funded by the Swedish Research Council grant: 2014-578.

Appendix

Abbreviations of the Data Designs

ANCHOR-NEC-PS Post-stratification under the nonequivalent groups with covariates design using propensity scores
CE NEAT Chained equating under the nonequivalent groups with anchor test design
CE-NEC-PS Chained equating under the nonequivalent groups with covariates design using propensity scores
EG Equivalent groups design
PSE NEAT Post-stratification equating under the nonequivalent groups with anchor test design
PSE-NEC-PS Post-stratification equating under the nonequivalent groups with covariates design using propensity scores
PSE-NEC-RAW-COV Post-stratification equating under the nonequivalent groups with covariates design

References

B. Andersson, K. Bränberg, M. Wiberg, Performing the kernel method of test equating using the R package kequate. J. Stat. Softw. **55**(6), 1–25 (2013)

K. Bränberg, M. Wiberg, Observed score linear equating with covariates. J. Educ. Meas. **48**, 419–440 (2011)

J. González, M. Wiberg, *Applying Test Equating Methods Using R* (Springer, Berlin, 2017). doi:10.1007/978-3-319-51824-4

S.J. Haberman, Pseudo-equivalent groups and linking. J. Educ. Behav. Stat. **40**, 254–273 (2015)

J. Häggström, M. Wiberg, Optimal bandwidth selection in kernel equating. J. Educ. Meas. **51**, 201–211 (2014)

T. Hsu, K. Wu, J. Yu, M. Lee, Exploring the feasibility of collateral information test equating. Int. J. Test. **2**, 1–14 (2009)

M.J. Kolen, Does matching in equating work? A discussion. Appl. Meas. Educ. **3**, 23–39 (1990)

M. Liou, P.E. Cheng, M. Li, Estimating comparable scores using surrogate variables. Appl. Meas. Educ. **25**, 197–207 (2001)

S.A. Livingston, N.J. Dorans, N.K. Wright, What combination of sampling and equating methods works best? Appl. Meas. Educ. **3**, 73–95 (1990)

N.T. Longford, Equating without an anchor for nonequivalent groups of examinees. J. Educ. Behav. Stat. **40**, 227–253 (2015)

P.-E. Lyrén, R.K. Hambleton, Consequences of violated the equating assumptions under the equivalent group design. Int. J. Test. **36**, 308–323 (2011)

T. Moses, W. Deng, Y.-L. Zhang, The use of two anchors in the nonequivalent groups with anchor test (NEAT) equating. ETS Research Report RR-10-23, 2010

I. Paek, J. Liu, H.J. Oh, *Investigation of Propensity Score Matching on Linear/Nonlinear Equating Method for the P/N/NMSQT (Report SR-2006-55)* (ETS, Princeton, NJ, 2006)

S.J. Powers, Impact of matched samples equating methods on equating accuracy and the adequacy of equating assumptions, Ph.D. Dissertation, University of Iowa, 2010., http://ir.uiowa.edu/etd/875

R Core Development Team, *R: A Language and Environment for Statistical Computing* (R Foundation for statistical computing, Vienna, 2016.) http://www.R-project.org/
P.R. Rosenbaum, D.B. Rubin, The central role of the propensity score in observational studies for causal effects. Biometrika **70**, 41–55 (1983)
P.R. Rosenbaum, D.B. Rubin, Reducing bias in observational studies using subclassification on the propensity score. J. Am. Stat. Assoc. **79**, 516–524 (1984)
N. Sungworn, An investigation of using collateral information to reduce equating biases of the post-stratification equating method, Ph.D. Thesis, Michigan State University, 2009
A.A. von Davier, P.W. Holland, D.T. Thayer, *The Kernel Method of Test Equating* (Springer, New York, NY, 2004)
G. Wallin, M. Wiberg, Propensity scores in kernel equating under the non-equivalent groups with covariates design, Manuscript submitted for publication, 2017
M. Wiberg, K. Bränberg, Kernel equating under the non-equivalent groups with covariates design. Appl. Psychol. Meas. **39**, 349–361 (2015)

A Mixture Partial Credit Model Analysis Using Language-Based Covariates

Seohyun Kim, Minho Kwak, and Allan S. Cohen

Abstract A mixture partial credit model (MixPCM) can be used to classify examinees into discrete latent classes based on their performance on items scored in multiple ordered categories. Characterizing the latent classes, however, is not always straightforward, particularly when analyzing text from constructed responses. This is because there may be information in the constructed responses that is not captured by the scores. Latent Dirichlet allocation (LDA) is a statistical model that has been used to detect latent topics in textual data. The topics can be used to characterize documents, such as answers on a constructed-response test, as mixtures of the topics. In this study, we used one of the topics from the LDA as a covariate in a MixPCM to help characterize the different latent classes detected by the MixPCM.

Keywords Latent Dirichlet allocation • Mixture partial credit model • Text analysis

1 Introduction

Mixture IRT (MixIRT) models classify examinees into a number of discrete latent classes based on their performance on a test (Mislevy and Verhelst 1990). Latent classes are different from manifest groups, such as groups classified by gender or ethnicity, in that they cannot be observed directly. MixIRT models have been studied in a variety of applications. Mislevy and Verhelst (1990) used the mixture Rasch model to capture students' use of strategies on mathematics problems. Rost (1990) used the mixture Rasch model for dichotomous data and a mixture partial credit

S. Kim (✉)
The University of Georgia, 125P Aderhold Hall, 110 Carlton street, Athens, GA 30602, USA
e-mail: seohyun@uga.edu

M. Kwak
The University of Georgia, 126C Aderhold Hall, 110 Carlton street, Athens, GA 30602, USA
e-mail: minho.kwak25@uga.edu

A.S. Cohen
The University of Georgia, 125M Aderhold Hall, 110 Carlton street, Athens, GA 30602, USA
e-mail: acohen@uga.edu

L.A. van der Ark et al. (eds.), *Quantitative Psychology*, Springer Proceedings in Mathematics & Statistics 196, DOI 10.1007/978-3-319-56294-0_28

model for polytomous data (Rost 1991) to find subgroups of examinees that share the same item difficulties. Cohen and Bolt (2005) used a mixture 3PL model to detect differential item functioning.

Characterizing the latent classes, however, is not always straightforward. Modeling latent class membership with covariates can be helpful in explaining why examinees were classified in different latent classes (Cho et al. 2013; Smit et al. 1999). Dayton and Macready (1988) incorporated sex and the math score of a standard achievement test as covariates to model latent class membership. Smit et al. (1999) explored the use of covariates in a mixture Rasch model. Results suggested that the accuracy of class membership assignment and the standard errors of parameter estimates were improved by using covariates that were strongly related to latent class membership. Choi et al. (2015) used internet access as a covariate to explain students' latent class membership on the Trends in International Mathematics and Science Study (TIMSS, Mullis et al. 2005).

In this study, we used a text-based covariate to explain latent class membership in a mixture partial credit model (MixPCM) for constructed-response (CR) item score data. This information was obtained from the words students used in their written response to the CR items. Students' written response data were first analyzed using latent Dirichlet allocation (LDA; Blei et al. 2003) to extract clusters of words, called topics, that characterized students' written responses. These topics are probability distributions over words. In the present study, we first extracted topics based on students' written responses to the CR items and then used one of these topics as a covariate in a MixPCM that was estimated from the scores students received for their responses to the CR items.

2 Theoretical Framework

This paper has two parts, the LDA analysis and the analysis of mixture partial credit modeling with a covariate (MixPCM-cov). First, latent topics from students' written response to the CR items were detected using LDA. Then, students' use of words from one of the topics was incorporated into a MixPCM with a covariate (MixPCM-cov) model to help characterize the classifications into latent classes. The number of latent classes for the MixPCM-cov was determined using two model fit indexes, Akaike's information criterion (AIC) and Bayesian information criterion (BIC).

2.1 *The Partial Credit Model*

The partial credit model is an IRT model that can handle polytomous data scored in ordered categories (Masters 1982). The model is described as follows: Let the score on item i be x, where x is one of $l_i + 1$ categories, $0, 1, \ldots, l_i$, then the conditional response probability for category $x = 0, \ldots, l_i$ is

$$P(X_{ij} = x|\theta_j, b_i) = \frac{exp \sum_{k=0}^{x}[\theta_j - b_{ik}]}{\sum_{y=0}^{l_i} exp \sum_{k=0}^{y}[\theta_j - b_{ik}]}, \tag{1}$$

where θ_j is an ability parameter for person j and b_{ik} represents the ith item step parameter for category k.

2.2 The Mixture Partial Credit Model

The MixPCM is based on the partial credit model and assumes that the partial credit model holds for each latent class, but that each class may have a different set of model parameters (Rost 1991). That is, given θ_j, the response patterns can be different between latent classes. The probability for category $x = 0, \ldots, l_i$ in the MixPCM is described as follows:

$$P(X_{ij} = x|\theta_j, b_i) = \sum_{g=1}^{G} \pi_g \frac{exp \sum_{k=0}^{x}[\theta_{jg} - b_{ikg}]}{\sum_{y=0}^{l_i} exp \sum_{k=0}^{y}[\theta_{jg} - b_{ikg}]}, \tag{2}$$

where g is an index for the latent class and π_g is a mixing proportion that represents the proportion of examinees in class g. The other parameters, θ_{jg} and b_{ikg}, indicate the same parameters as in Eq. (1) but they are all unique to latent class g. In this model, the latent classes are mutually exclusive and exhaustive.

2.3 The Mixture Partial Credit Model with Covariates

Including covariates for modeling the probability of latent class membership is useful in that it helps explaining why individuals are classified into one of the G different latent classes (Cho et al. 2013; Smit et al. 1999). A MixPCM can be extended to a MixPCM-cov by incorporating covariates for modeling the mixing proportion parameter, π_g for each person. In this case, the probability of person j belonging in class g is modeled as follows:

$$\pi_{jg|\mathbf{W_j}} = \frac{exp(\beta_{0g} + \sum_{p=1}^{P} \beta_{pg} W_{jp})}{\sum_{g=1}^{G} exp(\beta_{0g} + \sum_{p=1}^{P} \beta_{pg} W_{jp})}, \tag{3}$$

where $\mathbf{W_j} = (W_{j1}, W_{j2}, \ldots, W_{jP})$ is a vector of P covariates, and $\beta_{pg}(p = 0, 1, \ldots, P)$ is the class-specific effect of covariates on group membership. This equation is a multinomial logistic regression with the covariates W_{jp}s.

2.4 Latent Dirichlet Allocation (LDA)

The objective of LDA is to summarize a text corpus into latent topics using statistical modeling. Texts have been analyzed in various fields using statistical modeling. For example, Griffiths and Steyvers (2004) analyzed papers in a journal to discover topics that were latent issues underlying research activity (e.g., hot topics) compared to topics that were no longer actively being pursued (i.e., cold topics), Phan et al. (2008) analyzed medical texts to verify hidden topics underlying the texts, and Paul and Dredze (2011) analyzed Twitter messages to extract health-related issues using an applied LDA model. To date, however, little research has been reported on the detection of latent topics in responses to CR items. Extracting latent topics underlying students' responses has the potential to help understand the major content structures underlying students' answers.

In LDA, it is assumed that each document in a corpus of students' written responses is generated by a mixture of several topics. Each word in a document is generated from a single topic. A topic in this framework is represented as a multinomial distribution over the words in a corpus. Topics are not mutually exclusive, so it is possible that the same word in a corpus can be generated from different topics. The generative process of LDA is as follows (Heinrich 2009):

1. Choose $\gamma_k \sim \textit{Dirichlet}(\beta)$: γ_k is an $V \times 1$ vector, and V is the total number of unique words in corpus C. Each element of the vector $\gamma_k = (\gamma_{k1}, \ldots, \gamma_{kV})$ indicates the probability that the corresponding word appears in a document under topic k, and $\sum_{v=1}^{V} \gamma_{kv} = 1$.

Each document d in corpus C is modeled as follows:

2. Choose $\eta_d \sim \textit{Dirichlet}(\alpha)$: η_d is a $K \times 1$ vector, and K is the number of topics in the corpus. Each element of the vector $\eta_d = (\eta_{d1}, \eta_{d2}, \ldots, \eta_{dK})$ indicates the corresponding topic proportion for document d, and $\sum_{k=1}^{K} \eta_{dk} = 1$.
3. For each word ($w_{d,n}$) in the document d:

   (a) Choose a topic $z_{d,n} \sim \textit{Multinomial}(\eta_d)$ and
   (b) Choose a word $w_{d,n} \sim \textit{Multinomial}(\gamma_{z_{d,n}=k})$.

α is the parameter of the Dirichlet distribution of the topic proportions; β is the parameter of the Dirichlet distribution of word probabilities. In the LDA analysis, these parameters are used for the priors, respectively, in the MCMC estimation of these two distributions. $z_{d,n}$ indicates the topic for the nth word in a document $d(= 1, 2, \ldots, D)$, and $\gamma_{z_{d,n}=k}$ is the word distribution of topic $z_{d,n} = k$.

3 Method

3.1 *Data*

The data for this study were taken from a larger NSF-funded host study, Language-Rich Inquiry Science for middle grades English Language Learners (LISELL). The LISELL project focused on teaching the use of academic language for understanding science inquiry practices.

3.1.1 Measures

The science assessment in the host study was designed to measure students' use of academic language and understanding of science practices. The assessment consisted of six question scenarios each from two to four CR items and designed to measure the use of independent and dependent variables, cause and effect, and construction of hypotheses. There were 27 CR items considered on the test. Students' responses were written in separate answer documents handed out with the test booklets. These responses were scored on four elements (science inquiry practices, use of everyday language, use of academic language, and science content) and were scored from two to three points. For the LDA analysis, the answer documents for the students in the sample described below were transcribed into text files.

3.1.2 Sample

The sample consisted of 138 middle school students. In the sample, 52 (37.68%) students were in the 7th grade, and 86 (62.32%) students were in the 8th grade. There were 79 (57.25%) female and 59 (42.75%) male students. The students used 404 unique words with an average document length of 98 words after removing stop words. Stop words refer to types of functional words that do not provide any information such as the, of, and etc. Words that were used less than three times were also removed.

3.2 *Estimation of LDA*

The LDA model was fit to the CR data using the *lda* package in R (Chang 2015). The *lda* package uses collapsed Gibbs sampling (Griffiths and Steyvers 2004; Heinrich 2009), a Monte Carlo Markov chain (MCMC) method for estimating model parameters. In this study, 60,000 iterations were conducted; the first 40,000 iterations were discarded as burn-in.

The hyperparameters (α and β) were assumed to be known, an assumption made in previous studies (Griffiths and Steyvers 2004; Chang 2010). However, there is no theoretically guaranteed approach for selecting optimal values for these parameters (Chang 2010; Thomas et al. 2014). For this study, we used $\alpha = \frac{1}{K}$ (Chang 2010), where K is the number of topics. The smaller the α value is, the sparser the topic proportions are. For example, in a testing situation, a small value for α assumes that students tended to use words from a small number of topics rather than using words distributed evenly among all topics to construct their answers. With regard to β, we used $\frac{1}{V}$, where V is the number of unique words. Many earlier studies (Arun et al. 2010; Bíró et al. 2009; Canini et al. 2009; Griffiths and Steyvers 2004; Griffiths et al. 2007; Lu and Wolfram 2012; Porteous et al. 2008; Rosen-Zvi et al. 2010) used either 0.01 or 0.1; however, in this study, $\frac{1}{V}$ appeared to provide better topic results than either 0.01 or 0.1. Similar to α, a smaller β leads to distinct topics (Griffiths and Steyvers 2004; Heinrich 2009).

3.3 *Estimation of the MixPCM-cov Model*

Bayesian estimates of the MixPCM-cov model were obtained using MCMC as implemented in the computer software WinBUGS (Lunn et al. 2000). This method uses the Gibbs sampling technique, in which a sample of a parameter is drawn from the parameter's full conditional distribution up to that point over a large number of iterations. In each iteration, the parameter estimates are sampled from the corresponding posterior distribution, and the means of the parameter estimates over the sampling iterations are used as the Bayesian posterior estimates of the parameters. In order to obtain the Bayesian estimates, prior distributions of the parameters need to be specified. The following prior distributions were used for obtaining the Bayesian estimates of the MixPCM-cov.

$$\begin{aligned} b_{ikg} &\sim Normal(0,1), \quad i = 1,\ldots,I \\ \theta_{jg} &\sim Normal(\mu_g,1), \quad j = 1,\ldots,J \\ \mu_1 = 0, &\quad \mu_2 \sim Normal(0,1) \\ \beta_{pg} &\sim Normal(0,10), \quad p = 0,\ldots,P, \end{aligned}$$

where I is the number of items, J is the number of students, and P is the number of covariates. In this study, 20,000 iterations were run. The first 10,000 iterations were discarded as burn-in. Convergence was determined by Heidelberger and Welch (1981) convergence diagnostics using the R package, CODA (Plummer et al. 2006).

4 Results

4.1 LDA

4.1.1 The Number of Topics in the LDA Model

The number of topics was determined by comparing log-likelihoods of candidate LDA models having different numbers of topics. In addition, topics from the models were examined with regard to interpretability. LDA models with two-, three-, four-, and five-topics were considered. Table 1 shows the mean of the full log-likelihoods over 20,000 post-burn-in iterations. The log-likelihood of the LDA model with three topics was higher than that of the LDA models with two-, four-, and five-topics. Interpretation of the topics also indicated that three topics were interpretable given the instructional intervention. The characteristics of the topics are described below.

4.1.2 Topic Characteristics

Table 2 shows the 15 words that had the highest probabilities of occurring in each topic. The three topics were named as follows based on the characteristics of the words in each topic: preponderance of everyday language (Topic 1), preponderance of general academic language (Topic 2), and preponderance of discipline specific language (Topic 3).

4.1.3 Distribution of Students over the Three Topics

Figure 1 illustrates where students fell with respect to the three topics. The values on the X and Y axes represent the values of η_d for Topic 1 (η_{d1}) and Topic 3 (η_{d3}), respectively, for each student in the sample. η_{d2} for Topic 2 is $\eta_{d2} = 1 - \eta_{d1} - \eta_{d3}$. The upper left, lower left, and lower right corners of the plot represent degree of strength in the use of the words from the corresponding topic. As an example, consider the following hypothetical situations. Student A is located in the upper left corner ($\eta_d = (0, 0, 1)$); this location means that this student almost always used the words from Topic 3 when answering the CR items. Student B is located in the lower right corner ($\eta_d = (1, 0, 0)$), meaning that the student almost always used the words from Topic 1. Finally, Student C is located in the lower left corner ($\eta_d = (0, 1, 0)$)

Table 1 The log-likelihoods of LDA models

The number of topics	Log-likelihood
2	−72,751.20
3	−72,047.77
4	−72,112.40
5	−72,161.76

Table 2 The 15 words having the highest probability of occurring in each topic

Rank order	Topic 1: Preponderance of everyday language		Topic 2: Preponderance of general academic language		Topic 3: Preponderance of discipline specific language	
1	Water	0.069	Change	0.081	Energy	0.069
2	Salt	0.053	Variable	0.079	Fish	0.061
3	Fish	0.048	Move	0.052	Kinetic	0.036
4	Weight	0.040	Think	0.052	Weight	0.035
5	Because	0.038	Independent	0.038	Increase	0.035
6	More	0.035	Cause	0.035	Potential	0.033
7	Eat	0.028	Bigger	0.033	Amount	0.032
8	Boil	0.028	Dependent	0.031	Small	0.032
9	If	0.027	Effect	0.030	Population	0.029
10	Lift	0.025	Down	0.026	Decrease	0.029
11	Hold	0.023	Wide	0.026	Time	0.028
12	Small	0.022	Need	0.025	Temperature	0.027
13	Algae	0.020	Sun	0.023	Person	0.024
14	Shadow	0.020	Might	0.020	Disease	0.023
15	Bottle b	0.017	Fly	0.019	Same	0.021

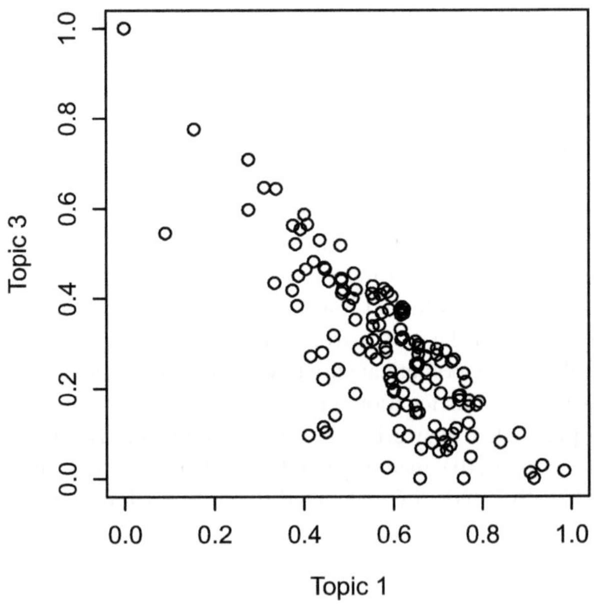

Fig. 1 Posterior distribution of students' membership in the three topics. The *X* and *Y* axes represent the proportions of words from Topic 1 (η_{d1}) and Topic 3 (η_{d3}) in a document, respectively. The proportion of words from Topic 2 is $\eta_{d2} = 1 - \eta_{d1} - \eta_{d3}$

indicating that the student almost always used the words from Topic 2. As can be seen in Fig. 1, Topics 1 and 3 for most students were located between the two topics (along the diagonal of the plot) and indicate that these students tended to use words primarily from Topic 1 and Topic 3. There are fewer dots along the X axis, meaning that fewer students wrote their answers to the CR items using words from Topics 1 and 2 only.

4.1.4 Topics and Deciding on a Covariate for the MixPCM

The *lda* package provides the number of words from each topic used by each student based on his or her responses. Topic 1 appeared to have a positive but weak relationship with total score ($r = 0.24$), Topic 2 had a negative relationship with total score ($r = -0.22$), and Topic 3 had a moderate positive relationship with total score ($r = 0.70$). There also was a stronger relationship between latent class membership, as detected by the MixPCM and the number of words from Topic 3 than from the other two topics. Therefore, the use of words from Topic 3 was selected as the covariate to help explain latent class membership in the MixPCM.

4.1.5 MixPCM-cov

Model Selection

The number of latent classes for the MixPCM was determined by comparing AIC and BIC, for one-, two-, three-, and four-latent-class solutions. Table 3 shows AIC and BIC values for each solution. Smaller AIC and BIC values indicate better model fit. As can be seen in the table, AIC selected the three-class solution, and BIC selected the two-latent-class solution. When the two indexes do not match, Li et al. (2009) suggest the BIC results for determining the number of latent classes for dichotomous variables. Kang et al. (2009) suggest BIC for the use in model selection for polytomous IRT models. There does not yet appear to be research on model selection specifically for mixture polytomous IRT models. In this paper, we used BIC for model selection for the MixPCM. Thus, two latent classes were assumed in the CR score data.

The covariate was created as the percentile of the number of words from Topic 3 and coded as follows: 1 = less than the 25th percentile, 2 = between the 25th

Table 3 Model comparison for mixture partial credit model solutions

Number of latent classes	AIC	BIC
1	5381	5507
2	5109	5367
3	5063	5450
4	5069	5584

Table 4 Posterior means of coefficient estimates for the mixing proportion in MixPCM-cov

Class	β_0 (SE)	β_1 (SE)	β_2 (SE)	β_3 (SE)
1	0	0	0	0
2	1.776 (0.749)	−2.229 (0.788)	−3.210 (0.921)	−3.022 (0.858)

Table 5 Descriptive statistics for latent classes from MixPCM-cov

Class	Number of students	Topic 3 (mean frequency)	Ability estimate (mean)
Class 1	80	35	0.04
Class 2	58	19	−1.31
All	138	28	−0.53

and 50th percentiles, 3 = between the 50th and 75th percentiles, and 4 = above the 75th percentile. Each level of the new variable was dummy coded, and the baseline was taken to be the first level (less than the 25th percentile). The coefficients for the dummy variables are as follows: β_{0g} indicates the log odds of falling into Class 2 rather than Class 1 for those students who used words from Topic 3 less than the 25th percentile (level 1). β_{1g}, β_{2g}, and β_{3g} indicate the amount of change in the log odds for students in levels 2, 3, and 4, respectively, compared to the baseline students (level 1). The coefficients for Class 1 were set to 0 for identification (Cho et al. 2013; Choi et al. 2015).

Table 4 shows the posterior mean estimates for $\beta_2 = (\beta_{02}, \beta_{12}, \beta_{22}, \beta_{32})$. If students were in the first level, the log odds of falling into Class 2 rather than Class 1 increased by 1.776. If students were above the first level, the log odds decreased of being in Class 2. The log odds decreased even more if students were in the third level rather than in the second level. In essence, the more students use words from Topic 3, the less likely that the students were to be classified into Class 2.

4.1.6 Characterizing the Latent Classes

The sample sizes for the two latent classes are shown in Table 5. More students belonged to Class 1 (N = 80) than Class 2 (N = 58). Class 1 tended to include more high-performing students on the test than Class 2 in that the average posterior mean ability estimate for Class 1 was higher than that of Class 2. Also, as indicated in Table 4, students in Class 1 tended to use more words from Topic 3 than did students in Class 2.

Figure 2 illustrates the posterior item difficulty estimates from the two-class MixPCM-cov. Here the item difficulty parameter for each item was defined as the sum of all item step parameters b_{ikg} of the corresponding item in class g. This index yields a general item difficulty for the item in class g (Rost 1991). The X-axis of Fig. 2 refers to the CR subitem number (referred to in the sequel as items), and the Y-axis refers to item difficulty estimates in each class. Overall, for the item difficulty

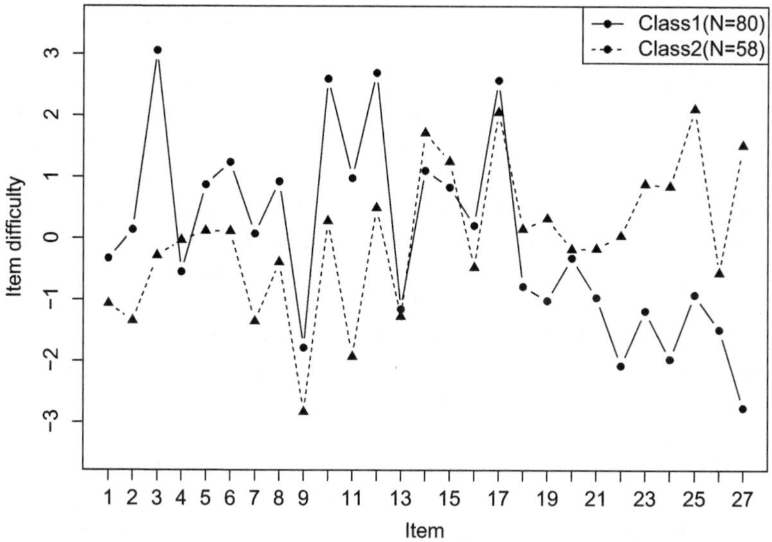

Fig. 2 Item difficulty parameter estimates of the two latent classes from MixPCM-cov

parameter estimates from Item 1 to Item 17, members of Class 1 had slightly higher item difficulty estimates than Class 2. The differences in item difficulty estimates between Classes 1 and 2 increased for the Items 18–27. This suggests that students in Class 1 and Class 2 generally responded similarly to these items. The two latent classes responded differently, however, for the remaining items. Item 1 to Item 17 and Item 18 to Item 27 measure different kinds of science inquiry practices. Items 1–17 ask about controlling variables and hypothesis, observation, and evidence. Items 18–27 ask about cause and effect relationships.

5 Discussion

Constructed-response items are becoming increasingly prominent in educational and psychological research. The objective of this study was to investigate the utility of students' use of words in their responses to CR items as a covariate for a MixPCM. This was done in order to examine whether students in different latent classes used different clusters of words to construct their answers. Results showed that students' latent class memberships were related to their use of words from Topic 3. The more a student used the words from Topic 3, the more likely the student was classified in latent Class 1.

The two models used in this study, MixPCM and LDA, used two sets of data. The MixPCM was based on polytomous item scores, and the LDA was based on the text of students' responses to the items. Thus, the two models measure different aspects of students' responses. This use of textual data as a covariate was based on the assumption that there exists additional information in students' responses that is not captured in the rubric-based scores. For example, there were two students in the sample who had similar ability estimates of 0.20 from the MixPCM but who constructed their answers differently. One student's answer for item 22 was "It will decrease." The other student's answer for the same item was "It decreases because their food source was small fish and most of them died." Both students received the same credit for the item; however, the first student used 41 words from Topic 3, and the second student used 50 words from Topic 3. One thing this shows is that the scores provided only a partial analysis of the information in students' answers in that they do not completely show how they constructed their responses. Students with the same ability constructed their answers quite differently yet received the same score.

Students' use of words from Topic 3 did not guarantee that the students used words that were directly related to discipline-specific language because some words, such as fish and small, also occupy large portions of Topic 3. The words used from Topic 3, however, do provide a clearer sense of the types of words used in constructing their answers.

References

R. Arun, V. Suresh, C.V. Madhavan, M.N. Murthy, On finding the natural number of topics with latent Dirichlet allocation: some observations, in *Advances in Knowledge Discovery and Data Mining: Vol. 21. Topic Modeling/Information Extraction*, ed. by M.J. Zaki, J.X. Yu, B. Ravindran, V. Pudi (Springer, Heidelberg, 2010), pp. 391–402

I. Bíró, D. Siklósi, J. Szabó, A.A. Benczúr, Linked latent Dirichlet allocation in web spam filtering, in *Proceedings of the 5th International Workshop on Adversarial Information Retrieval on the Web: Temporal Analysis*, ed. by D. Fetterly, Z. Gyöngyi (Association for Computing Machinery, New York, 2009), pp. 37–40

D.M. Blei, A.Y. Ng, M.I. Jordan, Latent dirichlet allocation. J. Mach. Learn. Res. **3**, 993–1022 (2003)

K.R. Canini, L. Shi, T.L. Griffiths, Online inference of topics with latent Dirichlet allocation, in *International Conference on Artificial Intelligence and Statistics: Vol. 5*, ed. by D. Dyk, M. Welling (2009), pp. 65–72

J. Chang, Not-so-latent Dirichlet allocation: collapsed Gibbs sampling using human judgments, in *Proceedings of the North American Chapter of the Association for Computational Linguistics Human Language Technologies 2010 Workshop on Creating Speech and Language Data with Amazon's Mechanical Turk*, ed. by C. Callison-Burch, M. Dredze (Association for Computational Linguistics, Stroudsburg, 2010), pp. 131–138

J. Chang, lda: collapsed Gibbs sampling methods for topic models. R package version 1.4.2 (2015)

S.J. Cho, A.S. Cohen, S.H. Kim, Markov chain Monte Carlo estimation of a mixture item response theory model. J. Stat. Comput. Simul. **83**(2), 278–306 (2013)

Y.J. Choi, N. Alexeev, A.S. Cohen, Differential item functioning analysis using a mixture 3-parameter logistic model with a covariate on the TIMSS 2007 mathematics test. Int. J. Test. **15**(3), 239–253 (2015)

A.S. Cohen, D.M. Bolt, A mixture model analysis of differential item functioning. J. Educ. Meas. **42**(2), 133–148 (2005)

C.M. Dayton, G.B. Macready, Concomitant-variable latent-class models. J. Am. Stat. Assoc. **83**(401), 173–178 (1988)

T.L. Griffiths, M. Steyvers, Finding scientific topics. Proc. Natl. Acad. Sci. **101**(suppl 1), 5228–5235 (2004)

T.L. Griffiths, M. Steyvers, J.B. Tenenbaum, Topics in semantic representation. Psychol. Rev. **114**, 211 (2007)

P. Heidelberger, P.D. Welch, A spectral method for confidence interval generation and run length control in simulations. Commun. ACM **24**(4), 233–245 (1981)

G. Heinrich, Parameter estimation for text analysis. University of Leipzig, Technical Report (2009)

T.-H. Kang, A.S. Cohen, H.-J. Sung, IRT model selection methods for polytomous items. Appl. Psychol. Meas. **33**(7), 499–518 (2009)

F. Li, A.S. Cohen, S.H. Kim, S.J. Cho, Model selection methods for mixture dichotomous IRT models. Appl. Psychol. Meas. **33**(5), 353–373 (2009)

K. Lu, D. Wolfram, Measuring author research relatedness: a comparison of word-based, topic-based, and author co-citation approaches. J. Am. Soc. Inf. Sci. Technol. **63**, 1973–1986 (2012)

D.J. Lunn, A. Thomas, N. Best, D. Spiegelhalter, WinBUGS-a Bayesian modelling framework: concepts, structure, and extensibility. Stat. Comput. **10**(4), 325–337 (2000)

G.N. Masters, A Rasch model for partial credit scoring. Psychometrika **47**(2), 149–174 (1982)

R.J. Mislevy, N. Verhelst, Modeling item responses when different subjects employ different solution strategies. Psychometrika **55**(2), 195–215 (1990)

I.V.S. Mullis, M.O. Martin, G.J. Ruddock, C.Y. O'Sullivan, A. Arora, E. Erberber, *TIMSS 2007 Assessment Frameworks* (Boston College, Chestnut Hill, 2005)

M.J. Paul, M. Dredze, You are what you Tweet: analyzing Twitter for public health, in Proceedings of the 5th International AAAI Conference on Weblogs and Social Media (2011), pp. 265–272

X.H. Phan, L.M. Nguyen, S. Horiguchi, Learning to classify short and sparse text & web with hidden topics from large-scale data collections, in *Proceedings of the 17th International Conference on World Wide Web* (ACM, New York, 2008), pp. 91–100

M. Plummer, N. Best, K. Cowles, K. Vines, CODA: convergence diagnosis and output analysis for MCMC. R News **6**(1), 7–11 (2006)

I. Porteous, D. Newman, A. Ihler, A. Asuncion, P. Smyth, M. Welling, Fast collapsed Gibbs sampling for latent Dirichlet allocation, in *Proceedings of the 14th Association for Computing Machinery SIG-Knowledge Discovery and Data Mining International Conference on Knowledge Discovery and Data mining: Vol. 1. Research Papers*, ed. by Y. Li, B. Liu, S. Sarawagi (Association for Computing Machinery, New York, 2008), pp. 569–577

M. Rosen-Zvi, C. Chemudugunta, T. Griffiths, P. Smyth, M. Steyvers, Learning author-topic models from text corpora. Assoc. Comput. Mach. Trans. Inf. Syst. **28**, 4 (2010)

J. Rost, Rasch models in latent classes: an integration of two approaches to item analysis. Appl. Psychol. Meas. **14**(3), 271–282 (1990)

J. Rost, A logistic mixture distribution model for polychotomous item responses. Br. J. Math. Stat. Psychol. **44**(1), 75–92 (1991)

A. Smit, H. Kelderman, H. van der Flier, Collateral information and mixed Rasch models. Methods Psychol. Res. Online **4**(3), 19–32 (1999)

S.W. Thomas, B. Adams, A.E. Hassan, D. Blostein, Studying software evolution using topic models. Sci. Comput. Program. **80**, 457–479 (2014)

C.M. Dayton, G.B. Macready, Concomitant-variable latent-class models. J. Am. Stat. Assoc. 83(401), 173–178 (1988)

[illegible]

Investigating Constraint-Weighted Item Selection Procedures in Unfolding CAT

Ya-Hui Su

Abstract Computerized adaptive testing (CAT) cannot only enable efficient and precise ability estimation but also increase security of testing materials. To meet a large number of statistical and nonstatistical constraints in CAT, the maximum priority index approaches can be used to handle these constraints simultaneously and efficiently for the construction of assessments. Many previous CAT studies were investigated for dominance items; however, only few CAT studies were investigated for unfolding items. In practice, an attitude measurement or personality test, such as the Minnesota Multiphasic Personality Inventory-2 or Cattell's 16 Personality Factors Test, might fit better with the unfolding models than with the dominance ones. Besides, these tests commonly have hundreds of items from complex structures. Therefore, the purpose of this study was to investigate constraint-weighted item selection procedures in unfolding CAT. It was found that the maximum priority index was implemented with the Fisher information, the interval information, and the posterior expected Kullback-Leibler information successfully in unfolding CAT. These three item information criteria had similar performance in terms of measurement precision, exposure control, and constraint management. The generalized graded unfolding model and the two-parameter logistic model had similar performance in item selection.

Keywords Item selection • Weighted • Unfolding model • Computerized adaptive

1 Introduction

In addition to statistical optimization, the construction of assessments usually involves fulfilling various statistical and nonstatistical constraints. Because items are selected sequentially, it is challenging to meet many nonstatistical constraints simultaneously in computerized adaptive testing (CAT). The maximum priority

Y.-H. Su (✉)
Department of Psychology, National Chung Cheng University, 168 University Road, Minhsiung Township, Chiayi County 62102, Taiwan
e-mail: psyyhs@ccu.edu.tw

L.A. van der Ark et al. (eds.), *Quantitative Psychology*, Springer Proceedings in Mathematics & Statistics 196, DOI 10.1007/978-3-319-56294-0_29

index approaches can be used to handle these constraints simultaneously and efficiently in unidimensional CAT (Cheng and Chang 2009; Cheng et al. 2009) and multidimensional CAT (Su 2015, 2016; Su and Huang 2015; Yao 2011, 2012, 2013).

In psychological inventory, it is common to have Likert-type items that subjects specify their level of agreement or disagreement on a symmetric agree-disagree scale for a series of statements. The generalized graded unfolding model (GGUM; Roberts et al. 1996, Roberts et al. 1998, Roberts et al. 2000) and the two-parameter logistic model (2PLM; Birnbaum 1968) are commonly employed to analyze Likert-type items. The assumptions of the GGUM and 2PLM are quite different. The 2PLM is a dominance model; that is, the probability of getting a correct answer is increased with the ability level. In contrast, the GGUM is an unfolding model; that is, there exits an idea point. A higher item score is expected when a person's ability level is close to a given item on a unidimensional latent continuum. Therefore, a higher item indicates stronger levels of agreement or attraction (Andrich 1996; Roberts 1995; Roberts et al. 1999). When a person disagrees with an attitude item because its content is either too negative or too positive relative to his/her own opinion. The GGUM is available for dichotomous or polytomous items. Many CAT studies were conducted for dominance items. However, only few CAT studies were conducted for unfolding items (Roberts et al. 2001). In practice, an attitude measurement or personality test fit better with the unfolding models than the dominance models. Therefore, many issues in unfolding CAT still need further attention.

Besides, items with high discrimination parameters are the most informative and useful in CATs. When ability estimation is not considerable certainty, these items are not needed in the early stages. In a-stratified design, item selection begins with low discriminating items and saves high discriminating ones to later stages of testing. In this way, measurement efficiency and accuracy should be improved. Hence, when a-stratification (Chang et al. 2001; Chang and van der Linden 2003; Chang and Ying 1999) is implemented in the present study, it can obtain better precise ability estimation and achieve better item usage in some degree. However, the constraint-weighted item selection method hasn't been implemented with a-stratification in unfolding CAT. Therefore, the purpose of the study is to investigate the performance of the constraint-weighted item selection procedures for dominance and unfolding CAT through simulations.

1.1 The Maximum Priority Index (MPI) Method

Cheng and Chang (2009) proposed the maximum priority index (MPI) method to monitor several statistical and nonstatistical constraints simultaneously. K is the total number of constraints. $c_{ik} = 1$ represents constraint k relevant to item i and $c_{ik} = 0$ otherwise. Each constraint k is given a weight w_k to match its importance. The priority index (PI) of item i can be computed as

$$\mathrm{PI}_i = I_i \prod_{k=1}^{K} (w_k f_k)^{c_{ik}}, \tag{1}$$

where I_i is the Fisher information of item i evaluated at the current $\widehat{\theta}$, which is the estimated ability. In fact, the Fisher information can be replaced with other item information criteria, such as interval information (Veerkamp and Berger 1997) or Kullback-Leibler information (Chang and Ying 1996). For a content constraint k, the priority index can be considered in a certain content area. If X_k is the number of items required from the content area, after x_k items have been selected, f_k is defined as

$$f_k = \frac{(X_k - x_k)}{X_k}. \tag{2}$$

For item exposure control constraint k, f_k can be defined as

$$f_k = \frac{1}{r_{\max}}\left(r_{\max} - \frac{n}{N}\right), \tag{3}$$

where $r_{\max}$ is the maximum item exposure rate, N is the number of examinees who have taken the CAT, and n is the number of examinees have seen item i. An item with the largest priority index will be administered. When flexible content balancing constraints are considered, l_k and u_k are lower and upper bounds of content area k, respectively. Let μ_k is the number of items to be selected from content area k. Then,

$$l_k \leq \mu_k \leq u_k, \tag{4}$$

and

$$\sum_{k=1}^{K} \mu_k = L, \tag{5}$$

where L is test length. To incorporate both upper and lower bounds for a one-phase item selection strategy, Su and Huang (2015) suggested that f_k can be replaced with $f_{1k}f_{2k}$, which f_{1k} and f_{2k} are defined as

$$f_{1k} = \frac{1}{u_k}(u_k - x_k), \tag{6}$$

and

$$f_{2k} - \frac{(L - l_k) - (t - x_k)}{L - l_k}, \tag{7}$$

respectively. f_{1k} represents the closeness to the upper bound whereas f_{2k} represents the closeness to the lower bound. t is the number of items that have already been administered and $t = \sum_{k=1}^{K} x_k$. When f_{2k} is equal to 0, the sum of items from other constraints has reached its maximum; $f_{1k}f_{2k}$ is defined as 1 to ensure that items from constraint k can be still included for item selection. Cheng et al. (2009) indicated that item selection in a-stratification should be considered on the basis of matching item difficulty parameter b to the current $\widehat{\theta}$, rather than matching the largest Fisher information to the $\widehat{\theta}$. They modified the PI for one-phase and two-phase item selection as

$$\mathrm{PI}_i = \frac{1}{\left|b_i - \widehat{\theta}\right|} \prod_{k=1}^{K} (f_{1k} f_{2k})^{c_{ik}}, \tag{8}$$

and

$$\mathrm{PI}_i = \frac{1}{\left|b_i - \widehat{\theta}\right|} \prod_{k=1}^{K} (f_k)^{c_{ik}}, \tag{9}$$

respectively. This version of a-stratification allows for inclusion of many constraints on item type and format as well as constrains to ensure balanced item exposure. It was found the weighted mechanism successfully addresses the constraints. This method not only helps to a great extent balancing item exposure rates but also improves measurement precision.

2 Method

2.1 Simulation Study

Both the GGUM and 2PLM are commonly used to analyze Likert-type scale, such as attitude measurement or personality test. In the GGUM, the probability of obtaining a score y to attitude statement i is defined as

$$\mathrm{P}\left(\mathrm{Y}_i = y|\theta_s\right) = \frac{\exp\left\{a_{\mathrm{i}}\left[y\left(\theta_s - b_i\right) - \sum_{k=0}^{y} \tau_{ik}\right]\right\}}{\sum_{w=0}^{M}\left\{\exp\left\{a_{\mathrm{i}}\left[y\left(\theta_s - b_i\right) - \sum_{k=0}^{w} \tau_{ik}\right]\right\}\right\}}, \tag{10}$$

where $\sum_{k=0}^{M} \tau_{ik} = 0$. θ_s is the location of person s on the attitude continuum. b_i is the location of attitude statement i on the attitude continuum. a_{i} is the discrimination of attitude statement i. τ_{ik} is the location of kth threshold on the attitude continuum relative to the location of the item i. In this study, the GGUM and the 2PLM were investigated the performance of the constraint-weighted item selection procedures in unfolding and dominance CAT through simulations, respectively.

A simulated item pool was constructed to mimic a real item pool consisting of six content areas: 25%, 15%, 15%, 15%, 20%, and 10%. Item pool size is 500 items. The fixed-length stopping rule was used. The test length was set at 30 items with content areas distributed as follows: 7-9, 4-6, 3-5, 3-5, 4-7, and 2-4 items. Thus, the distribution of the six content areas in a 30-item test was similar to that in the item

Table 1 The weights, upper bounds, and lower bounds of the constraints in unfolding CAT

Constraints	Weight	Lower bound	Upper bound
Content 1	1	7	9
Content 2	1	4	6
Content 3	1	3	5
Content 4	1	3	5
Content 5	1	4	7
Content 6	1	2	4
Item exposure rate	1		0.2
Item information	1		

pool. In the study, eight constraints were considered, including content balancing, exposure control, and item information. The corresponding weights, upper bounds, and lower bounds of the constraints in CAT list in Table 1. All 5000 simulated examinees were drawn from a standard normal distribution. The expected a posterior (EAP) estimate was used to estimate examinees' ability.

Three independent variables were manipulated in this study: models (two levels), item selection procedures (two levels), and item information criteria (three levels). The GGUM and 2PLM were considered in this study. Two item selection procedures were the MPI and a-stratified MPI. Three item information criteria were Fisher information (FI; Dodd et al. 1995), interval information (II; van Rijn et al. 2002), and posterior expected Kullback-Leibler information (PEKLI; Veldkamp and van der Linden 2002). When the FI criterion was used, Fisher information function for a single item i was defined as

$$I_i(\theta) = \mathrm{a}_\mathrm{i}^2 \left[\sum_{k=1}^{m} k^2 P_{ik}(\theta) - \left(\sum_{k=1}^{m} k P_{ik}(\theta) \right)^2 \right], \tag{11}$$

where a_i is the discrimination in Eq. (10). When the II criterion was used, the interval information function (van Rijn et al. 2002) for a single item was defined as

$$\text{Interval Information Function} = \int_{\widehat{\theta}-\delta}^{\widehat{\theta}+\delta} I_i(\theta)\, d\theta, \tag{12}$$

where δ was a small constant defining the width of the interval. When Kullback-Leibler information criterion was used, Kullback-Leibler information for a single item was defined as

$$K_i(\theta, \theta_0) \equiv \sum_{k=1}^{m} P_{ik}(\theta_0) \ln \left(\frac{P_{ik}(\theta_0)}{P_{ik}(\theta)} \right). \tag{13}$$

Because the true ability of the examinee, θ_0, was unknown, the posterior expected information of ability was used (van der Linden 1998). After $(t-1)$ items were administered, the PEKLI criterion (Veldkamp and van der Linden 2002) was defined as

$$KL_i\left(\widehat{\theta}^{t-1}\right) \equiv \int_{\theta} K_i\left(\theta, \widehat{\theta}^{t-1}\right) f\left(\theta | u_{i1}, \ldots, u_{i_{t-1}}\right) d\theta, \tag{14}$$

where $(u_{i1}, \ldots, u_{i_{t-1}})$ was response vector after (t−1) items were administered.

When the MPI item selection procedure was used, one of the item information criteria would be used to integrate with the MPI item selection procedure in Eq. (1). When a-stratified MPI item selection procedure was used, item selection should be considered on the basis on matching item difficulty parameter b to the current $\widehat{\theta}$, rather than matching the largest Fisher information to the $\widehat{\theta}$ (Cheng et al. 2009). That is, Eq. (8) would be used for item selection. Since there were more than one item difficulty parameters for the unfolding items, item selection would be considered on the basis on matching the averaged item difficulty parameters of item i to the current $\widehat{\theta}$.

2.2 Evaluation Criteria

The results of the simulation study were analyzed and discussed based on the following criteria: measurement precision, exposure control, and constraint management. With respect to measurement precision, latent trait recovery was evaluated with the bias and root mean squared error of estimation (RMSE). The formulas for bias and RMSE were given as follows:

$$\text{bias} = \frac{1}{N}\sum_{n=1}^{N}\left(\widehat{\theta}_n - \theta_n\right), \tag{15}$$

and

$$\text{RMSE} = \sqrt{\frac{1}{N}\sum_{n=1}^{N}\left(\widehat{\theta}_n - \theta_n\right)^2}, \tag{16}$$

where $\widehat{\theta}_n$ and θ_n are the estimated and true abilities, respectively. With respect to exposure control, for each item information criterion, the maximum item exposure rate and the number of unused items were reported. To measure the skewness of item exposure rate distribution (Chang and Ying 1999), the χ^2 statistic was defined as

$$\chi^2 = \frac{1}{L/Z}\sum_{i=1}^{Z}\left(r_i - L/Z\right)^2, \tag{17}$$

where r_i is the exposure rate of item i. L is the test length and Z is the number of items in the pool. The higher the χ^2 statistic, the worse the item exposure control.

With respect to constraint management, the number of violated constraints in each test was obtained. For each item information criterion, the averaged numbers of violated constraints were calculated over all examinees.

3 Results

The results of the simulations were summarized according to measurement precision, exposure control, and constraint management in Tables 2, 3, and 4, respectively. With respect to measurement precision, the bias, RMSE, and relative efficiency for different item selection methods list in Table 2. The MPI item selection method with the FI criterion was considered as the baseline while the GGUM or the 2PLM was used. In general, the 2PLM yielded slightly better performance than the GGUM in measurement precision. Three different item information criteria obtained similar performance in terms of bias, RMSE, and relative efficiency. The MPI item selection method yielded slightly better performance than the a-stratified MPI method.

With respect to exposure control, the actual item exposure rates of each item were recorded. The maximum item exposure rate, the number of overexposed items, the number of unused items, and the chi-square statistic measuring the skewness of the item exposure rate distribution were calculated. The results of exposure control for different item selection methods list in Table 3. In general, two different IRT models obtained similar performance in exposure control. Three different item information

Table 2 Measurement precision results for the item selection methods

Item selection methods		Bias	RMSE	Relative efficiency
Unfolding-GGUM	*MPI*			
	FI	0.019	0.322	1.000
	II	0.021	0.356	0.904
	PEKLI	0.018	0.301	1.070
	a-stratified MPI			
	FI	0.031	0.437	0.737
	II	0.029	0.444	0.725
	PEKLI	0.027	0.412	0.782
Dominance-2PLM	*MPI*			
	FI	0.002	0.213	1.000
	II	0.003	0.215	0.991
	PEKLI	0.009	0.198	1.076
	a-stratified MPI			
	FI	0.011	0.225	0.947
	II	0.013	0.231	0.922
	PEKLI	0.012	0.221	0.964

Table 3 Exposure control results for the item selection methods

Item selection methods		Maximum exposure rate	Overexposured items	Unused items	Chi-square
Unfolding-GGUM	*MPI*				
	FI	0.158	0	40	21.531
	II	0.161	0	56	22.133
	PEKLI	0.163	0	38	14.146
	a-stratified MPI				
	FI	0.075	0	6	0.195
	II	0.089	0	8	0.215
	PEKLI	0.112	0	12	0.201
Dominance-2PLM	*MPI*				
	FI	0.141	0	39	20.098
	II	0.112	0	50	19.563
	PEKLI	0.139	0	42	18.114
	a-stratified MPI				
	FI	0.081	0	7	0.278
	II	0.075	0	7	0.232
	PEKLI	0.086	0	11	0.155

Table 4 Constraint management results for the item selection methods

Item selection methods		Number of violation							
		0	1	2	3	4	5	6	Averaged
Unfolding-GGUM	*MPI*								
	FI	1.00	0.00	0.00	0.00	0.00	0.00	0.00	0.00
	II	1.00	0.00	0.00	0.00	0.00	0.00	0.00	0.00
	PEKLI	1.00	0.00	0.00	0.00	0.00	0.00	0.00	0.00
	a-stratified MPI								
	FI	0.17	0.4	0.35	0.08	0.00	0.00	0.00	2.68
	II	0.16	0.43	0.34	0.07	0.00	0.00	0.00	2.64
	PEKLI	0.32	0.42	0.2	0.05	0.01	0.00	0.00	2.01
Dominance-2PLM	*MPI*								
	FI	1.00	0.00	0.00	0.00	0.00	0.00	0.00	0.00
	II	1.00	0.00	0.00	0.00	0.00	0.00	0.00	0.00
	PEKLI	1.00	0.00	0.00	0.00	0.00	0.00	0.00	0.00
	a-stratified MPI								
	FI	0.96	0.04	0.00	0.00	0.00	0.00	0.00	0.10
	II	0.97	0.03	0.00	0.00	0.00	0.00	0.00	0.06
	PEKLI	0.95	0.05	0.00	0.00	0.00	0.00	0.00	0.08

criteria also obtained similar performance in exposure control. The *a*-stratified MPI item selection method yielded much better performance than the MPI method in terms of less unused items and smaller chi-square statistics.

Since the violation was considered at each examinee level, only the first six constraints in Table 1 were included to evaluate the efficiency of the constraint management. The proportions of assembled tests violating a certain number of constraints and the average number of violated constraints for different item selection methods list in Table 4. In general, the MPI item selection method with two different IRT models obtained similar performance in constraint management. However, the *a*-stratified MPI item selection with the 2PLM obtained slightly better performance than that with the GGUM in managing constraints.

4 Discussion

Two item response models (unfolding-GGUM and dominance-2PLM), two item selection procedures (MPI and *a*-stratified MPI) and three item information criteria (FI, II, and PEKLI) were considered in this study. The algorithms of the MPI item selection procedure with different item information criteria and *a*-stratification were derived successfully in CAT. The *a*-stratified MPI item selection procedure yielded better performance than the MPI method in controlling item exposure; however, the *a*-stratified MPI item selection procedure yielded slightly worse performance than the MPI method in measurement precision and constraint management. This is because *a*-stratified MPI method begins with low discriminating items and saves high discriminating ones to later stages of testing. Therefore, there is still a trade-off between measurement precision and item exposure. Three different item information criteria obtained similar performance. The 2PLM obtained slightly better performance than the GGUM.

Today one of the main challenges in educational and psychological measurement is to develop theories and methods for the new mode of large-scale implementation of computerized assessment, especially in developing item selection methods for CATs. The MPI method has great potential in operational CATs. The stopping rule is used to stop a cyclical item selection process in CATs (Reckase 2009; Wainer 2000). The stopping rule can be when a fixed number of test items have been administered or when a desired precision level has been achieved. In this study, when a stopping rule of fixed length is applied, the precisions are different at different examinee levels. It results in a high misclassification rate, which might be costly. To achieve the same level of precision, some examinees may need to take more items and some may need to take fewer items. However, some research questions need to be investigated when a stopping rule of precision is used. The administered tests for certain examinees may be undesirably lengthy or short because the required precision cannot be met or few items have improved the precision significantly. Under the CAT framework, some research has been done on using different stopping rules (Dodd et al. 1993), such as the minimum standard error stopping rule, the minimum information stopping rule (Dodd et al. 1989), and the predicted standard error reduction stopping rule (Choi et al. 2011). It is important to investigate the MPI method in variable-length condition for unfolding items in the future.

It is of great value to develop item selection procedures in the unfolding CATs that facilitates efficient control over nonpsychometric constraints, item exposure, and content balance. It is also important to develop quality control procedures for integration of the item selection to identify potential problems in the item pool structure design. This research has important implications for educational intervention research. Research findings from this study would not only advance our knowledge about unfolding CAT but also has great potential to be applied to educational and psychological intervention research. Assessments for testing particular research interventions can be delivered via computers. It improves measurement accuracy and increases assessment security. As a consequence, this kind of precise measurement has the potential for contributing to strengthening the empirical base for evidence-based educational policy-making.

References

D. Andrich, A general hyperbolic cosine latent trait model for unfolding polytomous responses: reconciling the Thurstone and Likert methodologies. Br. J. Math. Stat. Psychol. **49**, 347–365 (1996)

A. Birnbaum, in *Statistical theories of mental test scores*, ed. by F. M. Lord, M. R. Novick. Some latent trait models and their use in inferring an examinees' ability (Addison-Wesley, Reading, MA, 1968), pp. 397–479

H.-H. Chang, J. Qian, Z. Ying, a-stratified multistage CAT with b-blocking. Appl. Psychol. Meas. **25**, 333–341 (2001)

H.-H. Chang, W.J. van der Linden, Optimal stratification of item pools in alpha-stratified computerized adaptive testing. Appl. Psychol. Meas. **27**, 262–274 (2003)

H.-H. Chang, Z. Ying, A global information approach to computerized adaptive testing. Appl. Psychol. Meas. **20**, 213–229 (1996)

H.-H. Chang, Z. Ying, a-stratified multistage computerized adaptive testing. Appl. Psychol. Meas. **23**, 211–222 (1999)

Y. Cheng, H.-H. Chang, The maximum priority index method for severely constrained item selection in computerized adaptive testing. Br. J. Math. Stat. Psychol. **62**, 369–383 (2009)

Y. Cheng, H.-H. Chang, J. Douglas, F. Guo, Constraint-weighted a-stratification for computerized adaptive testing with nonstatistical constraints: balancing measurement efficiency and exposure control. Educ. Psychol. Meas. **69**, 35–49 (2009)

S.W. Choi, M. Grady, B.G. Dodd, A new stopping rule for computerized adaptive testing. Educ. Psychol. Meas. **71**, 37–73 (2011)

B.G. Dodd, R.J. De Ayala, W.R. Koch, Computerized adaptive testing with polytomous items. Appl. Psychol. Meas. **19**, 5–22 (1995)

B.G. Dodd, W.R. Koch, R.J. De Ayala, Operational characteristics of adaptive testing procedures using the graded response model. Appl. Psychol. Meas. **13**, 129–143 (1989)

B.G. Dodd, W.R. Koch, R.J. De Ayala, Computerized adaptive testing using the partial credit model: effects of item pool characteristics and different stopping rules. Educ. Psychol. Meas. **53**, 61–77 (1993)

M.D. Reckase, *Multidimensional Item Response Theory* (Springer, New York, NY, 2009)

J.S. Roberts, Item response theory approaches to attitude measurement (Doctoral dissertation, University of South Carolina, Columbia, 1995). Dissertation Abstracts International, 56, 7089B, 1995

J.S. Roberts, J.R. Donoghue, J.E. Laughlin, A generalized item response model for unfolding responses from a graded scale. Paper presented at the 61st annual meeting of the Psychometric Society, Banff, Alberta, Canada, 1996

J.S. Roberts, J.R. Donoghue, J.E. Laughlin, *The Generalized Graded Unfolding Model: A General Parametric Item Response Model for Unfolding Graded Responses (Research Report No. RR-98-32)* (Educational Testing Service, Princeton NJ, 1998)

J.S. Roberts, J.R. Donoghue, J.E. Laughlin, A general model for unfolding unidimensional polytomous responses using item response theory. Appl. Psychol. Meas. **24**, 3–32 (2000)

J.S. Roberts, J.E. Laughlin, D.H. Wedell, Validity issues in the Likert and Thurstone approaches to attitude measurement. Educ. Psychol. Meas. **59**, 211–233 (1999)

J.S. Roberts, Y. Lin, J.E. Laughlin, Computerized adaptive testing with the generalized graded unfolding model. Appl. Psychol. Meas. **25**, 177–196 (2001)

Y.-H. Su, in *Quantitative Psychology Research*, ed. by L. A. van der Ark, D. M. Bolt, W.-C. Wang, J. A. Douglas, S.-M. Chow. The performance of the modified multidimensional priority index for item selection in variable-length MCAT, vol 140 (Springer, Cham, 2015), pp. 89–97

Y.-H. Su, A comparison of constrained item selection methods in multidimensional computerized adaptive testing. Appl. Psychol. Meas. **40**(5), 346–360 (2016). doi:10.1177/0146621616639305

Y.-H. Su, Y.-L. Huang, in *Quantitative Psychology Research*, ed. by R. E. Millsap, D. M. Bolt, L. A. van der Ark, W.-C. Wang. Using a modified multidimensional priority index for item selection under within-item multidimensional computerized adaptive testing, vol 89 (Springer, Cham, 2015), pp. 227–242

W.J. van der Linden, Bayesian item selection criteria for adaptive testing. Psychometrika **63**, 201–216 (1998)

P.W. van Rijn, T.J.H.M. Eggn, B.T. Hemker, P.F. Sanders, Evaluation of selection procedures for computerized adaptive testing with polytomous items. Appl. Psychol. Meas. **26**, 393–411 (2002)

W.J.J. Veerkamp, M.P.F. Berger, Some new item selection criteria for adaptive testing. J. Educ. Behav. Stat. **22**, 203–226 (1997)

B.P. Veldkamp, W.J. van der Linden, Multidimensional adaptive testing with constraints on test content. Psychometrika **67**, 575–588 (2002)

H. Wainer (ed.), *Computerized Adaptive Testing: A Primer*, 2nd edn. (Erlbaum, Mahwah, NJ, 2000)

L. Yao, Multidimensional CAT item selection procedures with item exposure control and content constraints. Paper presented at the International Association of Computer Adaptive Testing (IACAT) Conference, Pacific Grove, CA, 2011

L. Yao, Multidimensional CAT item selection methods for domain scores and composite scores: theory and applications. Psychometrika **77**, 495–523 (2012)

L. Yao, Comparing the performance of five multidimensional CAT selection procedures with different stopping rules. Appl. Psychol. Meas. **37**, 3–23 (2013)

Rating Scale Format and Item Sensitivity to Response Style in Large-Scale Assessments

Sien Deng and Daniel M. Bolt

Abstract This study examines the relationship between rating scale format and instrument sensitivity to response style (RS). A multidimensional nominal response model (MNRM) that allows item discrimination to vary across items for both substantive and RS dimensions can be used for this purpose. We first conduct a set of simulations to examine the recovery of item slopes under the model. Then we apply the model to PISA and PIRLS survey items to investigate the item sensitivity to RS in relation to different rating scale types. Simulation results indicate good recovery of item slopes on both substantive and RS traits, and different rating scale formats (agreement-type versus frequency-type scales) from PISA and PIRLS are found to have varying sensitivities to RS, such that frequency-type scales are less affected.

Keywords Response style • Multidimensional nominal response model • Rating scale types

1 Introduction

Self-report rating scales are widely used in behavioral and psychological research. One disadvantage of this approach is that rating scales are often susceptible to response styles (RS), stylistic tendencies in how a respondent uses a rating scale that is independent of the item content (Baumgartner and Steenkamp 2001). Extreme response style (ERS) is one frequently cited example and is the focus of this study. ERS refers to a tendency to overuse the endpoints of a rating scale (e.g., over select 1 or 7 on a 7-point Likert scale). Other various RS types have also been discussed in the literature (see Van Vaerenbergh and Thomas 2013 for a review). As response styles tend to be stable within respondents across different constructs and over time (Weijters et al. 2010), their biasing effects reflect a likely source of systematic measurement error.

S. Deng (✉) • D.M. Bolt
Department of Educational Psychology, University of Wisconsin-Madison, 1025 West Johnson Street, Madison, WI 53706, USA
e-mail: sdeng7@wisc.edu; dmbolt@wisc.edu

L.A. van der Ark et al. (eds.), *Quantitative Psychology*, Springer Proceedings in Mathematics & Statistics 196, DOI 10.1007/978-3-319-56294-0_30

The effects of RS can be consequential not only in introducing bias to individual trait estimates or scale scores but also in estimates of the relationship between scale scores and other variables (Moors 2012). ERS, for example, has been found to associate with respondent characteristics such as gender (De Jong et al. 2008; Weijters et al. 2010), and education (Meisenberg and Williams 2008), and also to vary across cultures (Lu and Bolt 2015). However, the magnitude and implications of the biasing effects have been the subject of some debate. Some studies suggested that ERS effects are typically negligible (Wetzel et al. 2016; Plieninger 2016), while others found that ERS is more substantially biases means, variances, and correlations based on scale scores (Weijters et al. 2010).

The complex biasing effects of RS and inconsistencies across studies may be due to RS not only correlating with person characteristics but also other aspects of the instrument design (see Fowler 1995 for more details). For example, rating scales and items can have varying numbers of rating anchors (e.g., 5-point, 7-point), rating types (e.g., agreement- versus frequency- or likelihood-based ratings), and wording strategies (e.g., positive, negative, or mixed value types). The sensitivity of items to RS may vary in relation to these and other item characteristics, potentially explaining why the results of RS are often not consistent.

Approaches to measuring RS based on item response theory (IRT) provide the possibility to simultaneously estimate how both items and persons are affected by RS. Among a variety of IRT-based models developed to measure and/or control for RS, the multidimensional nominal response models (MNRM) and variants based on the Bock (1972) nominal response model can be appealing, as separate traits are introduced to model both the substantive constructs and RS traits (Bolt and Johnson 2009; Falk and Cai 2015).

The current work aims to use one such model to better understand how to design rating scale instruments so as to minimize their susceptibility to ERS effects. Specifically, a recently extended MNRM proposed by Falk and Cai (2015) includes item-level discrimination parameters on both substantive and RS traits, allowing items to be differentially influenced by RS, making it potentially ideal for this purpose. We further study this model in two ways: (1) by conducting simulations to examine the recovery of item slopes and (2) by applying the model to real survey instruments to investigate the varying sensitivity of items to ERS in relation to rating scale types.

In the next section, the extended MNRM is presented. Next, the recovery of item slopes particularly on the ERS dimension is examined by a set of simulations. The subsequent study considers application to large-scale assessment data including agreement-type and frequency-type rating scales. Conclusions and future directions are discussed in the final section.

2 An Extended MNRM to the Study of Extreme Response Style

While prior applications of the MNRM have been used to measure and control for RS (Bolt and Newton 2011; Johnson and Bolt 2010), a simplifying assumption was that RS had the same impact for all items (i.e., RS loadings are equal across items). However, this may not be true, and it may be useful to test this assumption. Another limitation of the model is that it can only model a limited number of latent traits due to heavy computational demands. These models were usually estimated using maximum likelihood estimation techniques based on numerical integration using either Gaussian or Bayesian methods (Falk and Cai 2015). To overcome these limitations, Falk and Cai (2015) proposed an extended MNRM with a novel parameterization and efficient estimation via the Metropolis-Hastings Robbins-Monro (MH-RM) algorithm (Cai 2010a,b).

Compared to previous applications of the MNRM, the Falk and Cai (2015) model demonstrates flexibility in three main respects. First, the overall item slopes for each latent trait (i.e., substantive and RS traits) are separated from the slopes for individual categories, which allows for investigation of whether the effect of RS is constant or varying across items. Second, the model allows users to accommodate various types of RS, corresponding to over-selection of different categories. Third, the MH-RM algorithm allows for larger numbers of continuous and correlated latent traits to be estimated and is more efficient compared to the traditional algorithms mentioned above (Cai 2010a,b).

Suppose there are $i = 1, \ldots, N$ independent subjects who respond to $j = 1, \ldots, J$ items, and assume $k = 1, \ldots, K$ are response options for each item, where $Y_{ij} = k$ represents selection of category k for subject i on item j, m is an index for item category, and $\boldsymbol{\theta}_i$ is a $D \times 1$ vector representing subject i's latent trait scores on $d = 1, \ldots, D$ latent dimensions, which are assumed multivariate normal, i.e., $\boldsymbol{\theta}_i \sim N(\boldsymbol{\mu}, \Sigma)$. Then the extended MNRM can be statistically formulated as:

$$P(Y_i = k|\boldsymbol{\theta}_i, \mathbf{a}, \mathbf{S}, \mathbf{c}) = \frac{\exp([\mathbf{a} \circ \mathbf{s_k}]'\boldsymbol{\theta}_i + c_k)}{\sum\limits_{m=1}^{K} \exp([\mathbf{a} \circ \mathbf{s_m}]'\boldsymbol{\theta}_i + c_m)}. \tag{1}$$

Note that the item subscript is dropped in (1), and item parameters and latent trait scores may vary across items. Specifically, $\mathbf{a}$ is a vector of item slope (discrimination) parameters of length D, and $\mathbf{c}$ is a vector of item intercept parameters of length K, where the element c_k represents the intercept corresponding to category k. The constraint $\sum\limits_{k=1}^{K} c_k = 0$ is applied for identification purposes within each item. Item categories with positive c parameters tend to be more frequently selected at $\theta = 0$; items with negative c parameters are less frequently. $\mathbf{S}$ is a $D \times K$ matrix where each column corresponds to a score category, and each row corresponds to a different latent dimension. The scoring function values $\mathbf{s_k}$ determine the kth column of **S**. The symbol $\circ$ denotes the entrywise product. In this model, the scoring function **S** is separated from the overall item slopes **a**. This constraint allows the order of categories for the dimension to be fixed, while the overall item slope parameters for that dimension are estimated and may vary across items. Thus, we can investigate the potential varying effects of RS across items in relation to certain item characteristics, such as the type of rating scale as considered in this study. In the current analysis, we focus on ERS, so one of the latent traits is defined to represent an ERS trait by specifying the item category slopes. The ERS trait is defined by a propensity toward selection of the rating scale endpoints.

3 Simulation Study

Before applying this model to real data, to examine the effectiveness of the model in recovering item slope parameters for the ERS trait, we generated response data from the three-dimensional NRM in (1), including two intended-to-be-measured traits and one ERS trait. We consider designs of two sample sizes ($N = 500$ or 2000) and two test lengths ($n = 10$ or 20 items for each substantive latent trait dimension). We chose these values both because they seemed to reflect realistic test conditions and are also different enough to allow us to see the effects of the manipulated factor. In each simulation, person parameters are generated from a multivariate normal distribution with moderate positive correlations among traits, i.e., $\theta_i \sim N(\mu, \Sigma)$, where $\Sigma = \begin{bmatrix} 1 & 0.5 & 0.3 \\ 0.5 & 1 & 0.3 \\ 0.3 & 0.3 & 1 \end{bmatrix}$, implying that the correlation is 0.5 between the two substantive traits and 0.3 between each substantive trait and the ERS trait. For item parameters, a four-point rating scale is assumed, as in our real data analyses described shortly. The item slopes **a** and category intercepts **c** are separately generated from uniform distributions, i.e., $a_j \sim Unif(0.5, 2)$ and $c_{jk} \sim Unif(-2.5, 2.5)$ with constraint $\sum_{k=1}^{K} c_k = 0$. To define the latent trait, the scoring functions for the substantive and ERS traits are fixed at prespecified values as detailed below, while all a_j and c_{jk} are estimated. Specifically, we fix the scoring function values of the two substantive traits at $\begin{bmatrix} 1 & 2 & 3 & 4 \end{bmatrix}$ and the ERS trait at $\begin{bmatrix} 1 & 0 & 0 & 1 \end{bmatrix}$. The use of equal interval category slopes for the substantive traits is consistent with the use of an equal interval rating scale and is a common constraint that leads to the frequently used partial credit or generalized partial credit models (Thissen and Steinberg 1986). With this fixed scoring function matrix, higher values of θ_i for substantive traits imply a greater likelihood of higher scores on the rating scales. Similarly, a more positive θ_{ERS} implies a greater likelihood of picking categories 1 or 4 and a more negative θ_{ERS} implies avoiding 1 and 4. To examine the recovery of parameters, the data-generating model is fit to the generated response data using the "mirt" R package (Chalmers 2012) with the implementation of the

Table 1 Mean and standard deviation of RMSE for item slope parameters across 15 replication runs, simulation analyses

Mean RMSE (SD)						
	Slope of dimension 1		Slope of dimension 2		Slope of dimension ERS	
	$n = 10$	$n = 20$	$n = 10$	$n = 20$	$n = 10$	$n = 20$
$N = 500$	0.161 (0.056)	0.154 (0.032)	0.156 (0.065)	0.132 (0.035)	0.218 (0.056)	0.187 (0.033)
$N = 2000$	0.074 (0.028)	0.067 (0.015)	0.074 (0.019)	0.075 (0.013)	0.103 (0.022)	0.097 (0.012)

MH-RM algorithm. The recovery of parameters is evaluated by root mean square error (RMSE), where lower RMSE indicates better recovery. Recovery results are averaged over 15 replications under each simulation condition. Here we mainly focus on the recovery of item discrimination parameters.

Table 1 displays the mean and standard deviation of the RMSEs over the 15 replications for the estimates of item slope parameters. From the table it can be seen that the overall recovery of item slopes for both substantive and ERS traits is quite good for both sample size and both test length conditions. Moreover, as expected, increasing the sample size and the number of items helps reduce the RMSE. Specifically, when the sample size is increased to 2000, the RMSEs are found to be reduced to half and also yield a smaller standard deviation. RMSEs of slope estimates for the ERS trait are not recovered as well as the item slopes for the two substantive traits. This is as anticipated and still reflects overall good recovery (0.218).

4 Empirical Data Application

To illustrate the extended MNRM (Falk and Cai 2015) and to study the variability of item slope parameters on the ERS trait in relation to rating scale types, we consider data from the Program for International Student Assessment (PISA) and Progress in International Reading Literacy Study (PIRLS) student questionnaires from different years, each assessment including both agreement-type and frequency-type rating scales. All items are scored using four-point rating scales. For each assessment, four scales measuring potentially correlated constructs were used. From each assessment, we analyzed data from two agreement-type (e.g., 1 = strongly disagree to 4 = strongly agree) and two frequency-type (e.g., 1 = very often to 4 = never or hardly ever) rating scales. Item examples are given in Table 2 from both the agreement type and frequency-type rating scales.

Table 3 displays the assessment, scale type, construct, and test length for each of the assessments considered. For each assessment, item responses from four constructs are simultaneously analyzed using a five-dimensional extended MNRM in (1), with one dimension for each of the substantive constructs, and a single ERS factor measured by all of the items. This allows for a comparison of the estimates of

Table 2 Examples of items from agreement- and frequency-type rating scales, 2006 PISA and 2006 PIRLS

Agreement-type
Item 1(2006 PISA): I generally have fun when I am learning <broad science> topics
1 = strongly agree 2 = agree 3 = disagree 4 = strongly disagree
Item 2 (2006 PIRLS): Like being in school
1 = agree a lot 2 = agree a little 3 = disagree a little 4 = disagree a lot
Frequency-type
Item 1(2006 PISA): Watch TV programmes about <broad science>
1 = very often 2 = regularly 3 = sometimes 4 = never or hardly ever
Item 2 (2006 PIRLS): Use internet/music
1 = every data or almost every day 2 = once or twice a week 3 = once or twice a month
4 = never or almost never

Table 3 Rating scales in PISA and PIRLS analyses

Assessment	Scale type	Trait name	Trait description	Number of items
PISA 2006	Agreement-type	ENJ	Science enjoyment	5
		USE	Science usefulness	4
	Frequency-type	SCI	Science activity	5
		ENV	Environment activity	5
PISA 2012	Agreement-type	CPT	Math self-concept	5
		AXT	Math anxiety	5
	Frequency-type	BHV	Math behavior	5
		COG	Math cognitive	8
PIRLS 2006	Agreement-type	REN	Reading enjoyment	5
		SCH	School feeling	5
	Frequency-type	ACTR	Activity after reading	4
		INTF	Internet function	5
PIRLS 2011	Agreement-type	RFUN	Reading fun	4
		RUSE	Reading usefulness	5
	Frequency-type	PARC	Parent caring	4
		REDF	Reading function	3

ERS slopes for the different rating scale types. We calculate the mean and standard deviation (SD) of ERS slope estimates according to constructs and rating scale types (shown in Tables 4 and 5) and compare the mean differences in ERS slopes between rating scale types using *t*-tests.

Figure 1 presents the ERS slope estimates for each item. Items with higher ERS slopes are more affected by ERS, while those with lower ERS slopes are less affected. From Fig. 1 we see most items on the agreement-type scale have larger ERS slope estimates than those on the frequency-type scale. However, the ERS slope estimates within a construct are also varying, and the variances also appear

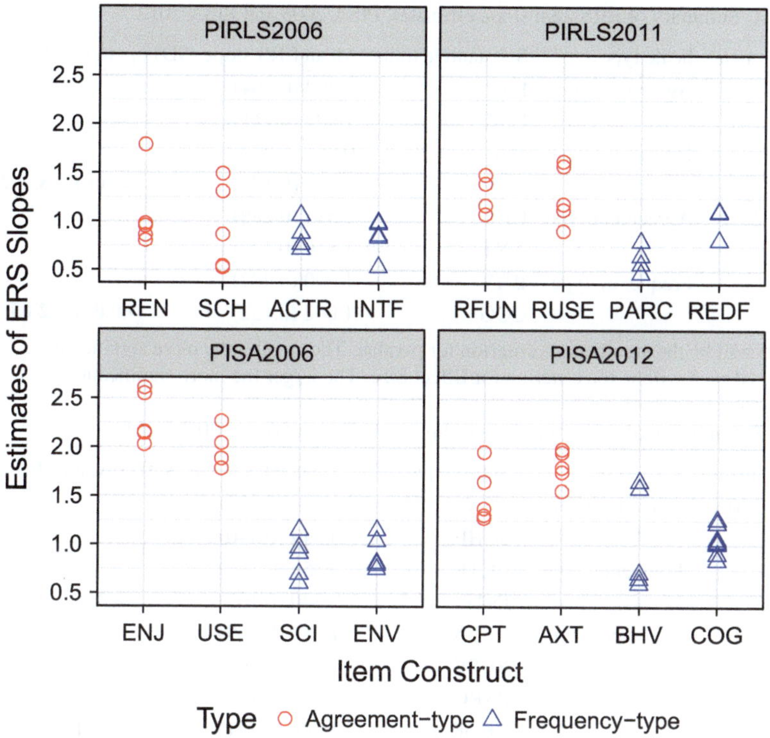

Fig. 1 Estimates of ERS slopes by rating scales for PISA and PIRLS

to be heterogeneous across rating scale types. The variance is not surprising as the sensitivity of items to ERS may also be impacted by other design factors such as ambiguous wording and content.

Tables 4 and 5 show the summary statistics (mean, SD) of the item discrimination estimates for ERS for each construct, the resulting p-value of associated *t*-tests, and effect size (Cohen's *d*) estimates. As seen in these two tables, the mean ERS slopes of the agreement-type rating scales are all greater than those of the frequency-type rating scales across assessments. In addition, most comparisons show statistically significant differences between the two rating scale types. Most effect sizes are greater than 0.8 (Cohen 1988) suggesting nontrivial differences between the two rating scale types (agreement-type versus frequency-type) in terms of ERS slope estimates. Such significant differences suggest that ERS has less effect on items of frequency-type rating scales compared to those of agreement-type rating scales. In other words, items rated on the agreement-type scales seem more prone to leading ERS respondents to select the 1 or 4 categories than the frequency-type scales.

Table 4 Summary of ERS item slope estimates, PISA 2006 and PISA 2012

Assessment	Scale type	Substantive traits	Mean ERS slope (SD)	*p*-value	Effect size
PISA 2006	Agreement-type	ENJ	2.299 (0.264)		
		USE	1.996 (0.211)		
	Frequency-type	SCI	0.868 (0.217)		
		ENV	0.908 (0.174)	**<0.001**	**5.392**
PISA 2012	Agreement-type	CPT	1.514 (0.288)		
		AXT	1.810 (0.172)		
	Frequency-type	BHV	1.090 (0.487)		
		COG	1.042 (0.132)	**<0.001**	**2.056**

$p = .05$ will be the significance criterion for p-value. The smaller the more significant
Cohen's $d = .8$ will be the criterion for Effect Size. The larger the more significant

Table 5 Summary of ERS item slope estimates, PIRLS 2006 and PIRLS 2011

Assessment	Scale type	Substantive traits	Mean ERS slope (SD)	*p*-value	Effect size
PIRLS 2006	Agreement-type	RENJ	1.078 (0.403)		
		SCHF	0.949 (0.440)		
	Frequency-type	ACTR	0.859 (0.156)		
		INTF	0.840 (0.185)	**0.256**	**0.537**
PIRLS 2011	Agreement-type	RFUN	1.284 (0.189)		
		RUSE	1.285 (0.307)		
	Frequency-type	PARC	0.641 (0.139)		
		REDF	1.006 (0.171)	**0.001**	**2.017**

$p = .05$ will be the significance criterion for p-value. The smaller the more significant
Cohen's $d = .8$ will be the criterion for Effect Size. The larger the more significant

5 Conclusion and Discussion

Item slope parameters for both the substantive and ERS traits appear to be recovered well in the Falk and Cai (2015) extended MNRM for the measurement of response style. Such findings suggest that item-level RS discrimination parameters have the potential to inform the development of rating scales so as to reduce susceptibility to RS effects. From the real data analysis, we find that frequency-type rating scale items appear to yield less susceptibility to ERS than agreement-type items, potentially resulting in more objective responses that are less subject to idiosyncratic response style tendencies. Thus, rating scale type might be carefully considered in terms of potentially varying sensitivities to RS when designing a new instrument. This study also indicates that the ERS biasing effect may not be constant across items, with varying item slopes of ERS traits potentially affecting the measurement of ERS effects. Further work may focus on those aspects of item stems that might be related to response style susceptibility. One theory is that more ambiguously worded statements might be more susceptible to RS. Prior work (Lu 2012) has suggested a link between external ratings of item ambiguity and the influence of response style. Indeed, this theory also provides a possible explanation for the current finding regarding rating scale types. The use of frequency-based anchors may imply less subjectivity than agreement-based anchors.

One limitation of this study is that other types of RS such as midpoint response style, acquiescence, or social desirability have not been investigated. The correlations between traits may also affect the recovery of item-level discrimination parameters. The relationship between the number of rating scale anchor points and RS effects is also worthy of study in the future. Another limitation concerns our ability to attribute the effects we observe to rating scales specifically. It is conceivable that the differences we observed are related to other aspects of the survey content that is reflected by the types of rating scale used, but not directly attributable to the rating scale type itself. The ideal test of a rating scale effect might seek to administer the same scale using the two different rating scale formats and compare results using the same analytic approach. Thus, there is much potential for future work in this area.

References

H. Baumgartner, J.B.E. Steenkamp, Response styles in marketing research: a cross-national investigation. J. Mark. Res. **38**(2), 143–156 (2001)

R.D. Bock, Estimating item parameters and latent ability when responses are scored in two or more nominal categories. Psychometrika **37**(1), 29–51 (1972)

D.M. Bolt, T.R. Johnson, Applications of a MIRT model to self-report measures: addressing score bias and DIF due to individual differences in response style. Appl. Psychol. Meas. **33**(5), 335–352 (2009)

D.M. Bolt, J.R. Newton, Multiscale measurement of extreme response style. Educ. Psychol. Meas. **71**(5), 814–833 (2011)

L. Cai, High-dimensional exploratory item factor analysis by a Metropolis–Hastings Robbins–Monro algorithm. Psychometrika **75**(1), 33–57 (2010a)

L. Cai, Metropolis-Hastings Robbins-Monro algorithm for confirmatory item factor analysis. J. Educ. Behav. Stat. **35**(3), 307–335 (2010b)

R.P. Chalmers, mirt: a multidimensional item response theory package for the R environment. J. Stat. Softw. **48**(6), 1–29 (2012)

J. Cohen, *Statistical Power Analysis for the Behavior Science*, 2nd edn. (Lawrence Erlbaum, Hillsdale, 1988)

M.G. De Jong, J.B.E. Steenkamp, J.P. Fox, H. Baumgartner, Using item response theory to measure extreme response style in marketing research: a global investigation. J. Mark. Res. **45**(1), 104–115 (2008)

C.F. Falk, L. Cai, A flexible full-information approach to the modeling of response styles. Pyschol. Methods **21**(3), 328–347 (2015)

F.J. Fowler, *Improving Survey Questions. Design and Evaluation*. Applied Social Research Methods Series, vol. 38 (Sage, Thousand Oaks, 1995)

T.R. Johnson, D.M. Bolt, On the use of factor-analytic multinomial logit item response models to account for individual differences in response style. J. Educ. Behav. Stat. **35**(1), 92–114 (2010)

Y. Lu, A multilevel multidimensional item response theory model to address the role of response style on measurement of attitudes in PISA 2006. PhD thesis, University of Wisconsin–Madison, Madison WI (2012)

Y. Lu, D.M. Bolt, Examining the attitude-achievement paradox in PISA using a multilevel multidimensional IRT model for extreme response style. Large Scale Assess. Educ. **3**(1), 1–18 (2015)
G. Meisenberg, A. Williams, Are acquiescent and extreme response styles related to low intelligence and education? Personal. Individ. Differ. **44**(7), 1539–1550 (2008)
G. Moors, The effect of response style bias on the measurement of transformational, transactional, and laissez-faire leadership. Eur. J. Work Organ. Psychol. **21**(2), 271–298 (2012)
H. Plieninger, Mountain or molehill? A simulation study on the impact of response styles. Educ. Psychol. Meas. **77**(1), 32–53 (2016)
D. Thissen, L. Steinberg, A taxonomy of item response models. Psychometrika **51**(4), 567–577 (1986)
Y. Van Vaerenbergh, T.D. Thomas, Response styles in survey research: a literature review of antecedents, consequences, and remedies. Int. J. Publ. Opin. Res. **25**(2), 195–217 (2013)
B. Weijters, M. Geuens, N. Schillewaert, The stability of individual response styles. Psychol. Methods **15**(1), 96–110 (2010)
E. Wetzel, J.R. Böhnke, N. Rose, A simulation study on methods of correcting for the effects of extreme response style. Educ. Psychol. Meas. **76**(2), 304–324 (2016)

Mode Comparability Studies for a High-Stakes Testing Program

Dongmei Li, Qing Yi, and Deborah J. Harris

Abstract Mode comparability between paper and online versions of a test cannot be simply assumed. This paper presents the research designs, statistical analyses, and major findings from a series of special studies intended to ensure score comparability for a high-stakes testing program, including an online timing study and two mode comparability studies as well as a general framework that guided the design of these studies. The framework views score comparability as a matter of degree and the evaluation of score comparability as a matter of score validation. The high-stakes uses of the test scores required stringent score comparability which was obtained by applying test-equating methodologies under a randomly equivalent groups design. Meanwhile, score equivalency and construct equivalency were examined through statistical analyses of test results and responses to survey questions. The comparability framework and results from these studies may provide guidance for other testing programs transitioning from paper to online or when evaluating score comparability in general.

Keywords Mode comparability • Online testing • Score comparability

1 Introduction

When test scores are used to inform decision making, one important requirement is that the scores are *comparable*. However, the test scores could be based on different test questions, obtained under different administration conditions, or derived based on different scoring methodologies. Therefore, score comparability cannot be simply assumed but should be carefully examined before such claims are made, especially for high-stakes decisions. Mroch et al. (2015) suggested a general framework for the evaluation of score comparability.

D. Li (✉) • Q. Yi • D.J. Harris
ACT, Inc., 500 ACT Drive, Iowa City, IA 52243, USA
e-mail: dongmei.li@act.org

L.A. van der Ark et al. (eds.), *Quantitative Psychology*, Springer Proceedings in Mathematics & Statistics 196, DOI 10.1007/978-3-319-56294-0_31

1.1 A General Framework for Score Comparability

The framework was developed based on a literature review of comparability studies, earlier comparability frameworks (e.g., Kolen 1999; Lottridge et al. 2008; Wang and Kolen 2001), and contemporary perspectives on score validation (e.g., Kane 2006). It proposes that, for scores to be comparable, the validity of test score interpretation to be compared should be similar. Therefore, gathering evidence for score comparability should follow a process similar to gathering validity evidence, and the sufficiency of evidence is dictated by the particular interpretation and uses of the test scores. Based on this overarching point of view, the framework suggested some guiding principles in the evaluation of score comparability. A slightly modified version of Fig. 1 from Mroch et al. (2015) is presented in Fig. 1 in this paper to illustrate these principles.

The first principle, as illustrated at the bottom section of Fig. 1, emphasizes that score comparability is a matter of degree, with the required stringency determined by the intended purpose of score use. Higher-stakes uses of test scores require more stringent score comparability, and lower-stakes uses require less stringent score comparability. The second principle, as illustrated in the top left section of the figure, suggests that score incomparability may be impacted by all relevant factors throughout the process of testing, from test development, test administration, to test scoring, and the linking/equating process, if applicable. Evaluation of score comparability should examine and document differences in each step of the process so that sources of score incomparability can be identified. Finally, the third principle, as typified by the top right section of the figure, suggests that evaluation of score comparability implies examination of both construct equivalency and metric equivalency, with construct equivalency focusing on what is being measured and metric equivalency focusing on how (e.g., scoring methods and score scales) and how well (e.g., precision and consistency) it is being measured. These two aspects may be intertwined with no clear boundary between them.

1.2 Mode Comparability in High-Stakes Testing

With the advances in technology, transitioning from paper test administration to online administration is something that has been or will be considered for almost all testing programs. Transferring test items from paper booklets to computer for online delivery is more complicated than it might appear (Leeson 2006; Mutler 1996; Parshall et al. 2002; Pommerich 2004; Schroeders and Wilhelm 2011). If score equivalence is sought between online and paper versions of a test, careful decisions need to be made not only to optimize the presentation of items but also to minimize mode effects, especially in situations where paper and online tests are both administered. To best achieve both maximum comparability to the paper version and optimal online interface and delivery, an iterative process may be needed. This paper describes a series of studies that were conducted for a high-stakes testing program to ensure the comparability of scores between paper and online administrations.

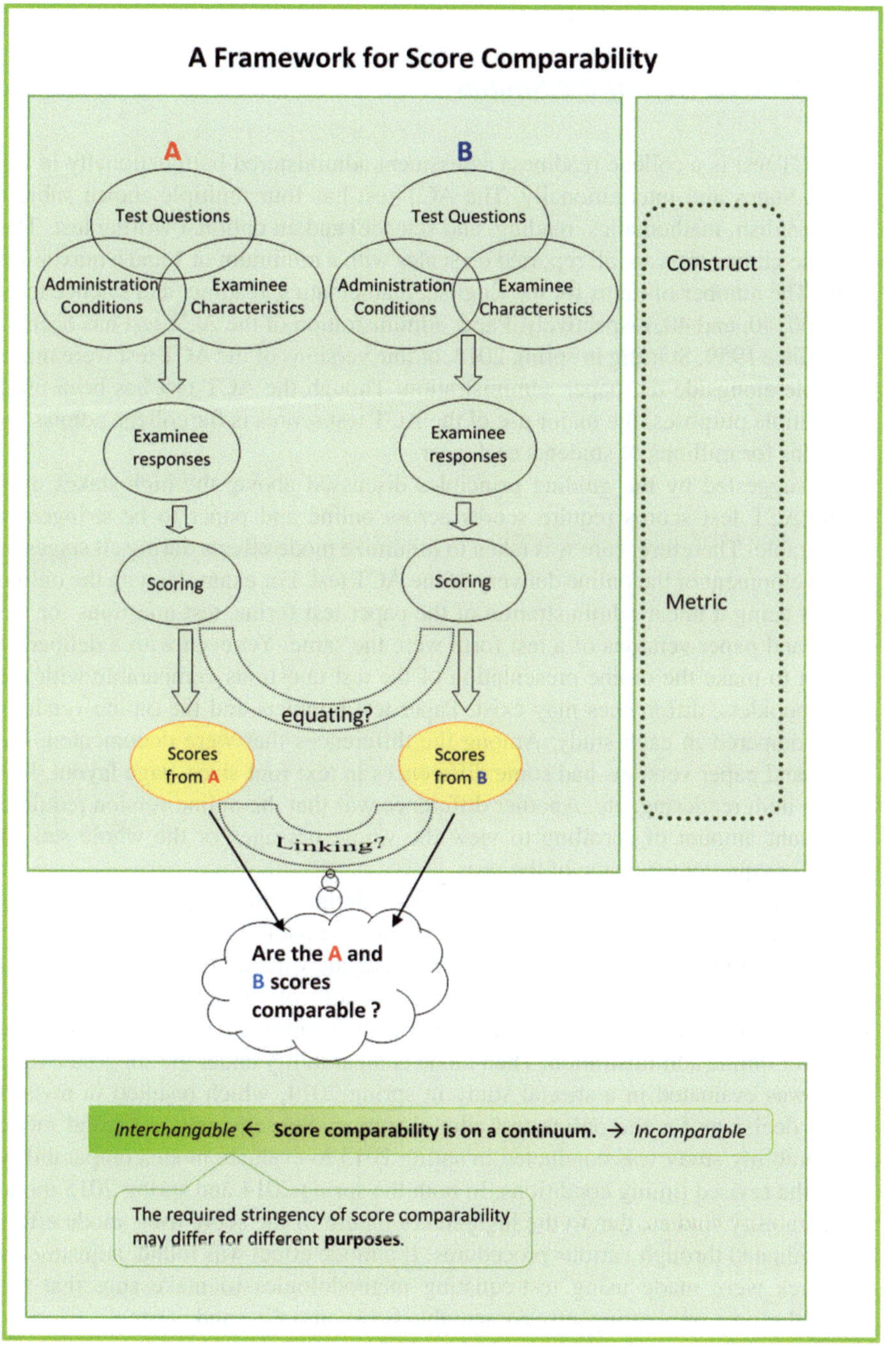

Fig. 1 A general framework of score comparability, a slightly modified version of Fig. 1 from Mroch et al. (2015)

2 Studies

2.1 An Overview of the Studies

The ACT test is a college readiness assessment administered both nationally in the United States and internationally. The ACT test has four multiple-choice subject tests (English, mathematics, reading, and science) and an optional writing test. The multiple-choice tests are all reported on scales with a minimum of 1 and a maximum of 36. The number of items for the English, mathematics, reading, and science tests is 75, 60, 40, and 40, respectively. Paper administration of the ACT test has been in place since 1959. Starting in spring 2015, online versions of the ACT test were made available alongside the paper administration. Though the ACT test has been used for multiple purposes, the major use of the ACT test scores is for college admission decisions for millions of students each year.

As suggested by the guiding principles discussed above, the high-stakes uses of the ACT test scores require scores across online and paper to be stringently comparable. Therefore, care was taken to minimize mode effects during all stages in the development of the online delivery of the ACT test. For example, with the online version being a linear administration of the paper test forms, test questions for the online and paper versions of a test form were the same. Yet even with a deliberate attempt to make the online presentation of the test questions comparable with the paper booklets, differences may exist. Paper test booklets and the online versions were compared in each study. Among the differences that were documented, the online and paper versions had some differences in text font sizes, page layout, line breaks, item rendering, etc. Another difference was that the online version required significant amount of scrolling to view the whole passages or the whole sets of figures or equations in some of the tests.

Prior to the official launch of the online administrations, a series of studies was conducted to evaluate and to ensure the comparability of scores between paper and online administrations. First, to investigate whether the online test would require additional time than the paper test because of the amount of scrolling, a timing study was conducted in fall 2013 to help inform decisions on the time limits for online administration. Then mode comparability under the suggested time limits was evaluated in a special study in spring 2014, which resulted in revised timing decisions for the online test administration. Subsequently, a second mode comparability study was conducted in spring 2015 to evaluate mode comparability under the revised timing conditions. In both the spring 2014 and spring 2015 mode comparability studies, due to the high-stakes nature of the score uses, mode effect was evaluated through various procedures. If a mode effect was found, adjustments to scores were made using test-equating methodologies to make sure that the reported scores were stringently comparable between online and paper.

A randomly equivalent groups design (Kolen and Brennan 2014) was used in all three studies. Students were randomly assigned to take the test under different timing conditions in the online timing study and were randomly assigned to take the paper or online test in both mode studies. As part of each study, students were provided with online tutorials so they could get familiar with the online testing system before taking the test. After the test was taken, feedback from students and test administrators was collected through survey questions. In addition to questions on the sufficiency of testing time, students were also asked various questions concerning their preparation for online testing, computer experience and typing skills, ease of navigation, and their use of various features of the online test, use of scratch paper, testing mode preference, and others. They were also asked to provide any additional comments that they might have regarding their testing experience. Test data and responses to the survey questions were analyzed to inform decisions concerning online administration time and to evaluate mode comparability.

More details about each of the studies and the major findings from each study are discussed next, focusing on results from the multiple-choice tests. The ACT writing test had its own timing and comparability studies which are not included in this paper. Below is a more detailed description of the specific data collection designs, data analyses, and results for each of the studies.

2.2 *Fall 2013 Timing Study*

2.2.1 Purpose

The standard time limits for the paper administration of the ACT test are 45, 60, 35, and 35 min for the English, mathematics, reading, and science tests, respectively. Because of the potential differences in the online and paper testing experiences, especially the amount of scrolling involved for the online testing, it was not known whether slightly different administration times should be used for online. Therefore, a special study was conducted to inform timing decisions for the online administration of the ACT test.

2.2.2 Data Collection Design

Students participating in the study were randomly assigned to take the subject tests under one of the three timing conditions: (1) the current paper time limit, (2) the current time limit plus 5 min, and (3) the current time limit plus 10 min. The tests under different timing conditions were augmented with survey questions of different lengths (separately timed), so the total administration time was the same for all participants who took the same subject test. One of the survey questions included in all timing conditions asked to what extent the students agreed/disagreed that they had enough time to finish the test.

Over 3000 examinees from 58 schools participated in the study, with each examinee responding to one subject test. Previous performance of these schools on the ACT test was examined to make sure that this group of schools was representative of the national testing population in terms of achievement level. In this timing study, students did not receive college-reportable scores, so students' motivation was likely lower than in operational testing. Therefore, data were cleaned to eliminate records with signs of low motivation or nonstandard administration, based on a review of the irregularity reports, the item response time data, the test completion rates, etc. The final data set contained 2794 students.

2.2.3 Data Analyses

Students' item and test level scores, item omission rates, item and test latency information, and student survey results were compared across the different timing conditions. Because the timing study had only online test administrations, a matched sample with similar total score distributions was also extracted from operational paper testing data of the same test form. Item mean scores and omission rates were compared between the timing study sample and the matched sample.

2.2.4 Summary of Results

Results from various analyses suggested that the online reading and science tests under the standard paper timing condition might be overly speeded. For example, under the standard paper timing condition, the percentage of students omitting 3 or more items was 16% for English, 20% for mathematics, 36% for reading, and 27% for science; the percentage of students not reaching the last item was 20% for English, 22% for mathematics, 40% for reading, and 29% for science; and the percentage of students who disagree or strongly disagree with the statement that they had enough time to complete the test was 23% for English, 21% for mathematics, 54% for reading, and 52% for science. In addition, compared with the matched operational paper sample, the average number of items omitted was higher for the timing study sample for all subject tests, under the standard paper testing timing condition. The timing study sample also had lower item *p*-values for the last few items than the matched sample, especially for reading and science.

Evidence from the timing study seemed to suggest that online scores on the reading and science tests would be more likely to be comparable to paper administration scores with an increase in testing time, given the specific online delivery system at the time. However, the findings from the timing study might have been confounded with issues of low motivation and unfamiliarity with the online testing format. For example, even though an online tutorial was provided for students to view prior to taking the tests, the post-test survey indicated that less than half of the students made use of the resource, with an even lower percentage for students who took the reading and the science tests. After evaluating results from different analyses and considerations from different perspectives, a decision was made to tentatively increase online testing time for the reading and science tests by 5 min and continue to evaluate the timing issue in the subsequent mode comparability studies.

2.3 *Spring 2014 Mode Comparability Study*

2.3.1 Purpose

As in the fall 2013 timing study, test items for the online and the paper version of a test form were the same in the mode comparability studies, though small differences in the presentation of the items may have occurred (e.g., changes in line breaks). In addition, improvements were made to the online test delivery system based on experiences and feedback from the previous study.

In the spring 2014 study, the testing time for the online and paper administrations was the same for the English and mathematics tests, but it was different for the reading and science tests. Five additional minutes were added to the online versions of the reading and science tests based on results from the fall 2013 timing study.

The purposes of the 2014 mode comparability study were to (1) investigate the comparability of the ACT scores from the online and paper testing modes, (2) obtain interchangeable scores across modes for operational score reporting, (3) reevaluate the timing decisions for the online administration of the ACT, and (4) gain additional insights about the online administration process.

2.3.2 Data Collection Design

Students participating in the spring 2014 mode comparability study were randomly assigned to take one of the three test forms (two online and one paper). The second online form was included to evaluate mode effect in light of form differences. After the administration, survey questions were sent to students who participated in the study to ask for their comments and feedback on their testing experience. This study was conducted in an operational test setting. Students took all four subject tests and received college-reportable scores.

Thousands of students from about 80 schools across the country participated in this study. Data were cleaned based on reviews of the proctor comments, phone logs, irregularity reports, item response time information, and an examination of the random assignment. Students with invalid scores and test centers with large discrepancies in form counts across modes were excluded from further analyses. The final data set contained 5593 students.

2.3.3 Method

Analyses were conducted to investigate mode comparability at two levels: metric and construct equivalency. Metric equivalency was first examined in terms of the similarity of test score distributions between the two modes, such as means, standard deviations, and relative cumulative frequency distributions. Then, the similarity of item-level information, such as the item *p*-values, item discrimination, item response distributions, and item omission rates, was compared. Test level and item-level comparisons were also conducted using item response theory (IRT). In addition, measurement precision (reliability and conditional standard errors of measurement) was compared across modes, and the item response time information for the online test items was also examined. Construct equivalency was examined by comparing the dimensionality and factor loadings, correlations among the subject tests, and by examining differential item functioning (DIF) between online and paper scores.

Because of the high-stakes uses of the ACT test scores, in addition to a thorough evaluation of mode comparability, equating methodology was used to ensure that the scores were comparable across modes. The equating methodology used for adjusting for mode differences was the same as what is used to equate the ACT test across different paper forms, that is, equipercentile equating based on a randomly equivalent group data collection design (Kolen and Brennan 2014).

2.3.4 Summary of Results

The analyses showed little mode differences in test reliability, correlations among the four subject tests, effective weights, or factor structures but some differences in score distributions for some of the subject tests. Item scores and test scores tended to be higher, and omission rates tended to be lower for the online group than for the paper group, especially for the reading and science tests.

Test Score Distributions

Table 1 presents the scale score descriptive statistics and statistical test of the differences across modes, with the online scale scores obtained by applying the same raw to scale score conversions as the paper form. The online group tended to have slightly higher mean scores than the paper group for all tests. Though not shown in the table, the descriptive statistics were also compared with those of the second online form. The magnitude of the mean differences between online and paper was larger than the mean differences between the two online forms. Except for the mathematics test, the between-mode mean differences were all statistically significant with *t*-test *p*-values smaller than 0.0001. The reading test had the largest mean difference (two scale score points) between online and paper. The plots of the relative cumulative frequency distributions of scale scores are shown in Fig. 2a, which also suggested the existence of mode differences especially for reading.

Table 1 Scale score mean differences across modes (online minus paper) for spring 2014

	Online (N = 1801)		Paper (N = 1987)		Scale score comparison		
Test	Mean	SD	Mean	SD	Mean difference	Effect size	*t*-test *p*
English	21.39	5.95	20.47	6.12	0.93	0.15	<0.0001
Mathematics	21.30	5.26	21.02	5.15	0.28	0.05	0.0942
Reading	23.56	6.43	21.47	6.43	2.09	0.32	<0.0001
Science	22.12	5.23	21.14	5.03	0.98	0.19	<0.0001

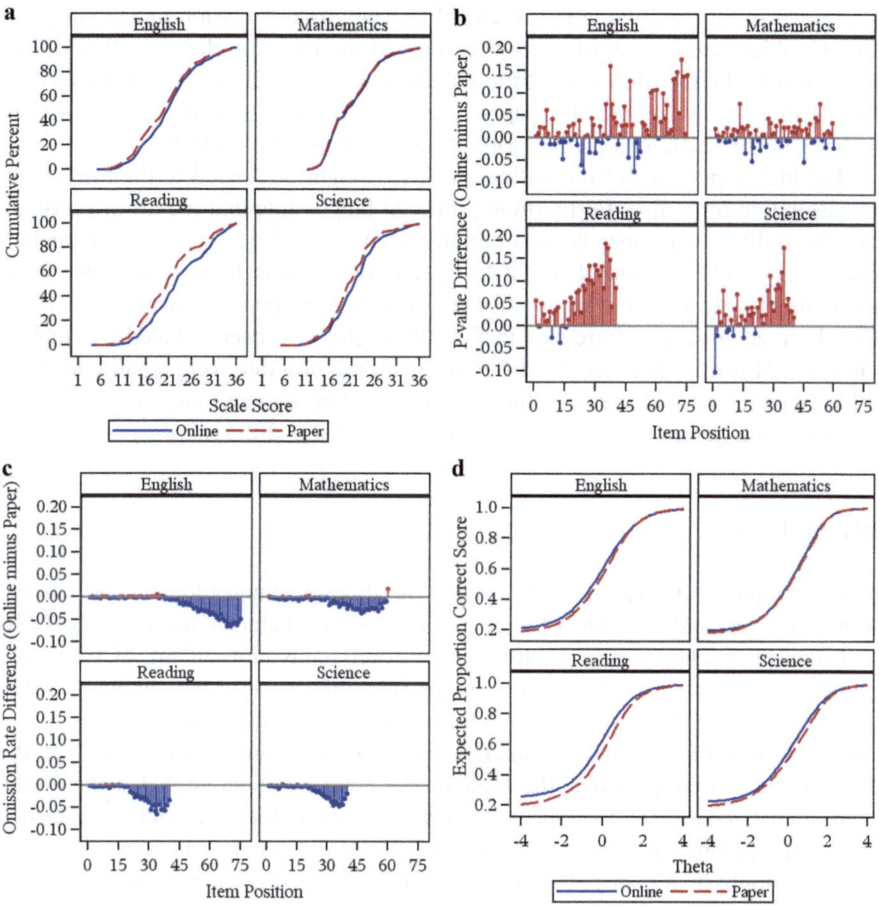

Fig. 2 Selected results of the spring 2014 mode comparability analyses: (**a**) scale score distributions, (**b**) *p*-value differences, (**c**) omission rate differences, (**d**) test characteristic curves

Item Scores and Item Omission Rates

Figure 2b presents the item *p*-value (i.e., the proportion of students answering the item correctly) differences across modes, with positive differences indicating that the item was easier for the online administration. Except for mathematics, the item *p*-values tended to be higher for online than for paper, especially for items in the latter half of the tests. For the reading test, almost all items had a higher *p*-value for online than for paper, with about one half of the items having a difference larger than 0.05 and about one fourth of the items having a difference larger than 0.10.

The omission rate (i.e., the proportion of missing responses) for each item was also compared across modes. Figure 2c presents the proportion of students omitting each item for online minus the proportion of students omitting the item for paper.

As shown in Fig. 2c, online items tended to have lower omission rates than paper for all subject tests, except for the first third of the items on the tests. This might be related to the fact that it is a little bit easier to click a choice for online than to bubble in an answer choice for paper. The differences of omission rates, however, were all below 0.10, with the majority of the differences below 0.05.

The higher *p*-values of the online test items might be related to the lower item omission rate for online. Furthermore, it might be possible that the lower omission rate for online than paper is, to a certain extent, due to a higher rate of random guessing, simply because it might be easier for students to click an answer on the computer than bubble in an answer choice on the paper answer sheet. If so, the observed *p*-value differences in Fig. 2b might have been inflated by random guessing. However, because the differences in omission rate were small compared with the differences in item *p*-values, the assumption of more random guessing for the online test cannot fully explain the observed *p*-value differences.

IRT Analyses

Mode comparability was examined using the three-parameter logistic IRT model at both the test and item level by comparing the test characteristic curves (TCCs) and item parameters across modes. Figure 2d contains plots of the TCCs across modes for each subject. The across-mode TCC difference is the smallest for the mathematics test, but largest for reading, which is consistent with the test score distributions in Fig. 2a. Item parameters were compared across modes. As with the comparison of item *p*-values, the *b*-parameter comparison showed that the online items tended to be easier than the paper items, especially for the reading and science tests. The *c* parameters tended to be higher for the online items, which indicated that guessing might be higher for online than paper, as noted when evaluating the omit rates. The *a* parameters did not show consistent mode differences.

Measurement Precision and Consistency

Score reliability was compared between paper and online, and little difference was found. For both paper and online, Cronbach's alpha reliability coefficient was 0.93, 0.92, 0.87, and 0.86, for the English, mathematics, reading, and science tests, respectively. A multivariate generalizability analysis was also conducted for paper and online, under a person-crossed-with-item design, treating the different content categories within a subject test as different variables. Reliability indices from the generalizability analyses were very similar across modes.

Factor Analysis and DIF

Exploratory factor analysis was conducted. The online and paper eigenvalue scree plots were similar for all four subject tests. A one-factor model showed sufficiently good fit for both paper and online, and the factor loadings correlated highly between modes (0.87 and 0.90). DIF was examined using the Mantel-Haenszel procedure (Camilli and Shepard 1994; Mantel and Haenszel 1959), with just a few items identified. Further investigation did not reveal any concrete sources of DIF for these items.

Equating

Equating methodology was used for all four multiple-choice tests to adjust for differences so that the college-reportable scores of students participating in the mode comparability study were comparable to national test takers, regardless of the testing mode. The adjustment resulted in no change or change within just one scale score point for the majority of the score points (around 50% to nearly 100%) for English, mathematics, and science. For reading, however, the adjustment was two scale score points or more for over half of the score points.

Online Timing Decision

Findings from the above analyses showed that students performed slightly higher for online than for paper, especially on the reading and science tests which had five more minutes than paper administration. In addition, the percentage of students who either agreed or strongly agreed that they had enough time to finish the test turned out to be higher for the online group than for the paper group. Therefore, a decision was made to eliminate the extra 5 min for the online reading and science tests in the spring 2015 mode comparability study. Refinements in the delivery of the online assessments may be one of the factors contributing to the different recommendations of online test time limits in this study compared to the previous study.

2.4 Spring 2015 Mode Comparability Study

2.4.1 Purpose

The spring 2015 mode comparability study was conducted to examine mode comparability under the revised online administration time for reading and science, i.e., the same administration time for paper and online of all subject tests. Because English and mathematics administration time did not change between the two mode studies, the spring 2015 study was a replicate of the spring 2014 study for these two subjects.

2.4.2 Design, Data, and Method

The mode comparability study in spring 2015 used the same data collection design as the spring 2014 study, with randomly equivalent groups of students taking one paper form and two online forms in an operational testing environment where students received college-reportable scores. The final data set contained 3192 students.

2.4.3 Summary of Results

With the revised online testing time, mode differences decreased for reading and science. Similar to findings from the 2014 study, mode differences were mostly observed in test and item scores, omission rates, and IRT statistics, but not in the other analyses such as reliability, factor structure, etc. Equating methodology was applied to ensure stringent score comparability. For the analyses that showed mode differences, the results are presented in Table 2 and Fig. 3.

Table 2 presents the descriptive statistics of scale scores and differences across modes. The plots of the relative cumulative frequency distributions of scale scores are shown in Fig. 3a. Figure 3b presents item difficulty differences across modes, and it shows that while later items tended to be harder compared to earlier items for each test regardless of mode, the items tended to be easier for the online administration, especially for items that appeared later in the test. Figure 3c shows the omission rate difference between online and paper for each item. Figure 3d contains plots of the TCCs across modes for each subject. The between-mode TCC difference was the smallest for the mathematics and science tests. There were some differences in TCC for the English and reading tests.

Results showed that students performed similarly across modes on the science test but still higher on the online reading test even without the extra 5 min. Equating methodology was used for all tests to ensure that the college-reportable scores of students participating in this study were comparable to other examinees.

Table 2 Scale score mean differences across modes (online minus paper) for spring 2015

	Online (N = 1092)		Paper (N = 1056)		Scale score comparison		
Test	Mean	SD	Mean	SD	Mean difference	Effect size	*t*-test *p*
English	20.79	5.98	19.79	6.03	1.00	0.17	0.0001
Mathematics	20.69	5.20	20.58	5.16	0.11	0.02	0.6199
Reading	21.99	6.24	20.91	6.08	1.08	0.18	<0.0001
Science	20.86	5.17	20.80	4.96	0.06	0.01	0.7717

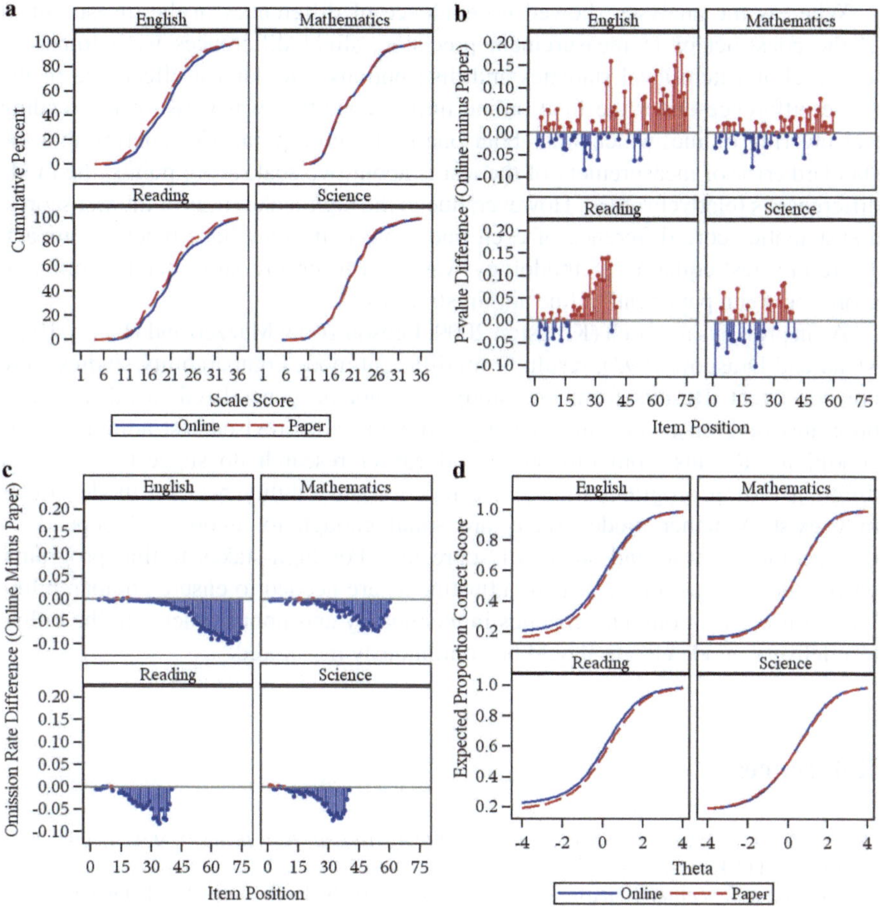

Fig. 3 Selected results of the spring 2015 mode comparability analyses: (**a**) scale score distributions, (**b**) *p*-value differences, (**c**) omission rate differences, (**d**) test characteristic curves

3 Summary and Discussion

This paper described a series of special studies conducted for a high-stakes testing program prior to the introduction of online testing as an alternative to existing paper testing. Guided by a general framework for score comparability, these special studies were intended to ensure stringent comparability of scores between online and paper testing due to the high-stakes uses of the scores, in terms of both construct comparability and score comparability. These studies all used a strong research design (Kingston 2009) involving random assignment of examinees to mode conditions. The two mode comparability studies, one with initial timing decisions and one with the final timing decisions for the online administration, were both conducted in an operational testing environment where student motivation was high.

Whereas the analyses showed no evidence of differences in the measurement of the construct or in measurement precision, slight differences were found on test level and item-level statistics and distributions. The largest effect size of the mode difference under the final online timing condition was 0.18 for the reading test favoring online, which was about one scale score point. Considering that the standard error of measurement of the test is about two scale score points, the mode difference is relatively small. However, due to the high-stakes uses of the test scores, a systematic score difference of even one score point may have practical impact. Therefore, test-equating methodology was used to ensure strict comparability of scores between paper and online administrations.

As in previous research (Kingston 2009; Leeson 2006; Mazzeo and Harvey 1988; Mead and Drasgow 1993), results from different mode comparability studies were not consistent. Results from one study may not be generalized to other testing programs or even to the same testing program under revised test administration conditions. Results from this study and earlier research do suggest a need for different testing programs to evaluate mode comparability because mode effects may exist. Whether mode effects are small enough to ignore will depend on the specific purpose and stakes of score use. For high-stakes testing programs, when mode effects are identified, adjustments are needed to ensure comparability. Furthermore, with ongoing changes in technology and improvements in the online test delivery, mode effects should be continuously monitored.

References

G. Camilli, L.A. Shepard, *Methods for Identifying Biased Test Items* (SAGE Publications, Thousand Oaks, CA, 1994)

M.T. Kane, in *Educational Measurement*, 4th edn., ed. by R. L. Brennan. Validation (American Council on Education and Praeger, Westport, CT, 2006), pp. 17–64

N.M. Kingston, Comparability of computer- and paper-administered multiple-choice tests for k-12 populations: a synthesis. Appl. Meas. Educ. **22**, 22–37 (2009)

M.J. Kolen, Threats to score comparability with applications to performance assessments and computerized adaptive tests. Educ. Assess. **6**(2), 73–96 (1999)

M. Kolen, R. Brennan, *Test Equating, Scaling, and Linking: Methods and Practices*, 3rd edn. (Springer, New York, NY, 2014)

H.V. Leeson, The mode effect: a literature review of human and technological issues in computerized testing. Int. J. Test. **6**(1), 1–24 (2006)

S. Lottridge, A. Nicewander, M. Schulz, H. Mitzel, Comparability of paper-based and computer-based tests: a review of the methodology. Paper submitted to the CCSSO Technical Issues in Large Scale Assessment Comparability Research Group, 2008. Retrieved from https://docs.google.com/viewer?a=v&pid=sites&srcid=ZGVmYXVsdGRvbWFpbnxwYXBlcnZlcnN1c3NjcmVlbnxneDoxNTU0NmE0NDY0NTQ4MzA4

N. Mantel, W. Haenszel, Statistical aspects of the analysis of data from retrospective studies of disease. J. Natl. Cancer Inst. **22**, 719–748 (1959)

J. Mazzeo, A.L. Harvey, *The Equivalence of Scores from Automated and Conventional Educational and Psychological Tests: A Review of the Literature (College Board Report No. 88–8; ETS RR No. 88–21)* (College Entrance Examination Board, New York, NY, 1988)

A.D. Mead, F. Drasgow, Equivalence of computerized and paper-and-pencil cognitive ability tests: a meta-analysis. Psychol. Bull. **114**, 449–458 (1993)

A. Mroch, D. Li, T. Thompson, A framework for evaluating score comparability. Paper presented at the Annual Meeting of the National Council on Measurement in Education, Chicago, IL, 2015

P. Mutler, in *Cognitive Aspects of Electronic Next Processing*, ed. by H. van Oostendorp, S. de Mul. Interface design and optimization of reading of continuous text (Ablex, Norwood, NJ, 1996), pp. 161–180

C.G. Parshall, J.A. Spray, J.C. Kalohn, T. Davey, *Practical Considerations in Computer-Based Testing* (Springer-Verlag, New York, NY, 2002)

M. Pommerich, Developing computerized versions of paper-and-pencil tests: mode effects for passage-based tests. J. Technol. Learn. Assess. **2**(6) (2004.) Available from http://ejournals.bc.edu/ojs/index.php/jtla/article/view/1666/1508

U. Schroeders, O. Wilhelm, Equivalence of reading and listening comprehension across test media. Educ. Psychol. Meas. **71**(5), 849–869 (2011)

T. Wang, M.J. Kolen, Evaluating comparability in computerized adaptive testing: issues, criteria and an example. J. Educ. Meas. **38**, 19–49 (2001)

Power Analysis for *t*-Test with Non-normal Data and Unequal Variances

Han Du, Zhiyong Zhang, and Ke-Hai Yuan

Abstract A Monte Carlo-based power analysis is proposed for *t*-test to deal with non-normality and heterogeneity in real data. The step-by-step procedure of the proposed method is introduced in the paper. For comparing the performance of the Monte Carlo-based power analysis to that of conventional pooled-variance *t*-test, a simulation study was conducted. The results indicate the Monte Carlo-based power analysis provided well-controlled empirical Type I error rate, whereas the conventional pooled-variance *t*-test failed to yield nominal-level Type I error rate. Both an R package and its corresponding online interface are provided to implement the proposed method.

Keywords Power analysis • Monte Carlo simulation • Non-normality • Heterogeneity

Power analysis is widely used for sample size determination (e.g., Cohen 1988). With appropriate power analysis, an adequate but not "too large" sample size is determined to detect an existing effect. The conventional method for power analysis for the *t*-test is limited by two strict assumptions: normality and homogeneity (two-sample pooled-variance *t*-test). The two-sample separated-variance *t*-test (also known as the Welch's *t*-test; Welch 1947) tolerates heterogeneity but still assumes normally distributed data. Thus, the corresponding exact power solution for the separated - variance *t*-test assumes normality with either numerical integration of noncentral density function or approximation (Moser et al. 1989; Disantostefano and Muller 1995).

Practical data in social, behavioral, and education research are rarely normal or homogeneous (Blanca et al. 2013; Micceri 1989). This poses challenges on statistical power analysis for the *t*-test (Cain et al. in press). To deal with the

H. Du (✉) • K.-H. Yuan
Department of Psychology, University of Notre Dame, 123A Hagger Hall, Notre Dame, IN 46556, USA
e-mail: hdu1@nd.edu

Z. Zhang
University of Notre Dame, 118 Haggar Hall, Notre Dame, IN 46556, USA

L.A. van der Ark et al. (eds.), *Quantitative Psychology*, Springer Proceedings in Mathematics & Statistics 196, DOI 10.1007/978-3-319-56294-0_32

problems, we develop a general method to conduct power analysis for t-test through Monte Carlo simulation. The method can flexibly take into account non-normality in one-sample t-test, two-sample t-test, and paired t-test and unequal variances in two-sample t-test. We provide an R package as well as an online interface for implementing the proposed Monte Carlo-based power analysis procedure.

1 One-Sample *t*-Test

The one-sample t-test concerns whether the population mean μ is different from a specific target value μ_0 (usually $\mu_0 = 0$). Thus, the null hypothesis is

$$H_0 : \mu = \mu_0.$$

The alternative hypothesis can be either two-sided (H_{a1}) or one-sided (H_{a2} or H_{a3}):

$$H_{a1} : \mu \neq \mu_0,$$
$$H_{a2} : \mu > \mu_0,$$
$$\text{or } H_{a3} : \mu < \mu_0.$$

The statistic given sample size n, $t = \frac{\bar{y}-\mu_0}{s\sqrt{\frac{1}{n}}}$, follows a t distribution with degrees of freedom $n-1$ under the normality assumption, where s is the sample standard deviation. When the normality assumption is violated, the t statistic does not follow a t distribution any more. When sample size increases, the statistic approximately follows a normal distribution. However, power analysis is less meaningful with a huge sample size because the power would be always 1.

Non-normality can take many forms. In this study, we focus on continuous variables with skewness and kurtosis different from a normal distribution (e.g., Cain et al. in press). With such non-normal data, it is extremely difficult to use an analytical formula to calculate power as in traditional power analysis. Instead, a Monte Carlo simulation method can be conveniently used (e.g., Muthén and Muthén 2002; Zhang 2014). The basic procedure of the Monte Carlo method is to first simulate the empirical null distribution of a chosen test statistic with the first four moments under the null distribution to get the critical value for null hypothesis testing and then simulate the distribution of the test statistic under the alternative hypothesis. Finally, the power can be estimated using the empirical distribution under the alternative hypothesis and the empirical critical value.

To use the Monte Carlo method, information regarding the first four moments is needed. Specifically, we need the population mean (μ) and standard deviation (σ). In addition, we need the population skewness $\gamma_1 = E\left[\left(\frac{x-\mu}{\sigma}\right)^3\right] = \frac{\mu_3}{\sigma^3}$ and kurtosis $\gamma_2 = E\left[\left(\frac{x-\mu}{\sigma}\right)^4\right] = \frac{\mu_4}{\sigma^4}$. For testing the population mean, the means under the null and alternative hypotheses should be different, denoted by μ_0 and μ_1, respectively. However, we assume that the shapes of distributions under the null and alternative are the same with the same standard deviation, skewness, and kurtosis in this study although they can be different. In practice, the population statistics are unknown, but they can be decided based on meta-analysis or literature review (e.g., Schmidt and Hunter 2014).

For the one-sample test, the following step-by-step procedure can be used to obtain the power for a given sample size n for testing:

$$H_0 : \mu = \mu_0 \text{vs.} H_1 : \mu = \mu_1.$$

1. Given the mean (μ_0), standard deviation (σ), skewness (γ_1), and kurtosis (γ_2), generate R_0 sets of non-normal data, each with the sample size n. R_0 should be sufficiently large, and we recommend a minimum value 100,000.
2. Calculate the mean and variance for each of the R_0 datasets denoted as $\bar{y}_{0j}$ and $s_{0j}^2, j = 1, \ldots, R_0$. Calculate the statistics $t_{0j}^* = \frac{\bar{y}_{0j} - \mu_0}{s_{0j}^2 \sqrt{\frac{1}{n}}}$. Obtain the critical value c_α according to the prespecified Type I error rate α, typically 0.05, and the alternative hypothesis. For example, if the alternative hypothesis is H_{a2}, c_α is the 100(1 − α)th percentile of $t_{0j}^{*\prime}s$.
3. Generate R_1 sets of non-normal data, each with the sample size (n), mean (μ_1), standard deviation (σ), skewness (γ_1), and kurtosis (γ_2). We recommend a minimum value 1000 for R_1.
4. Calculate the mean and variance for each dataset in Step (3) and denote them as $\bar{y}_{ai}$ and $s_{ai}^2, i = 1, \ldots, R_1$, and calculate the corresponding statistic $t_{ai}^* = \frac{\bar{y}_{ai} - \mu_0}{s_{ai}^2 \sqrt{\frac{1}{n}}}$ statistic.
5. The power is estimated as the proportion that t_{ai}^* is greater than the critical value c_α: $\pi = \# \left(t_{ai}^* > c_\alpha \right) / R_1$.

The Monte Carlo procedure works equally for the normal data, in which the data in Steps (1) and (3) can be generated from normal distributions. The procedure above also works for the paired samples where the population mean, standard deviation, skewness, and kurtosis of the difference scores are used.

2 Two-Sample t-Test

The two-sample t-test is used to test whether two independent population means are equal. The null hypothesis is

$$H_0 : \mu_1 - \mu_2 = 0.$$

The alternative hypothesis can be either two-sided or one-sided:

$$H_{a1} : \mu_1 - \mu_2 \neq 0,$$

$$H_{a2}: \mu_1 - \mu_2 > 0,$$

$$\text{or } H_{a3}: \mu_1 - \mu_2 < 0.$$

The pooled-variance t-test where the statistic $t_{pooled} = \frac{\bar{y}_1 - \bar{y}_2}{\sqrt{\frac{(n_1-1)s_1^2+(n_2-1)s_2^2}{n_1+n_2}}\sqrt{\frac{1}{n_1}+\frac{1}{n_2}}}$ follows a t distribution with degrees of freedom $n_1 + n_2 - 2$, where n_1 and n_2 are sample sizes for the two independent samples. $\bar{y}_1$ and $\bar{y}_2$ are the sample means and s_1^2 and s_2^2 are the sample variances of the two groups, respectively. The pooled t-test assumes homogeneity and normality. When the variances of the two groups are not the same, the separated-variance t-test should be used where the test statistic $t = \frac{\bar{y}_1 - \bar{y}_2}{\sqrt{\frac{s_1^2}{n_1}+\frac{s_2^2}{n_2}}}$ follows a t distribution with the degrees of freedom $\frac{\left(s_1^2/n_1+s_2^2/n_2\right)^2}{\left(s_1^2/n_1\right)^2/(n_1-1)+\left(s_2^2/n_2\right)^2/(n_2-1)}$. As for the one-sample t-test, when the normality assumption is violated, the distribution of the statistic is not a t distribution. Therefore, the Monte Carlo-based method could be used for power analysis.

As in one-sample t-test, we assume that the shapes of the data distribution for each group under the null and alternative are the same with the same standard deviation, skewness, and kurtosis, which can be estimated from meta-analysis or based on literature review. The step-by-step procedure for the two-sample t-test power calculation with given sample sizes n_1 and n_2 for the two groups is given below:

1. Let μ_{10} and μ_{20} be the means of the two groups under the null hypothesis, typically, $\mu_{10} - \mu_{20} = 0$. Given the population means (μ_{10} and μ_{20}), standard deviations (σ_1 and σ_2), skewness values (γ_{11} and γ_{12}), and kurtosis values for two groups (γ_{21} and γ_{22}), generate R_0 sets of non-normal data, one with sample size n_1 and another with sample size n_2. We recommend a minimum value 100,000 for R_1.
2. For the R_0 sets of data from previously simulated data pool, calculate the mean and variance of each group for each dataset denoted as $\bar{y}_{01j}$, $\bar{y}_{02j}$, s_{01j}^2, and s_{02j}^2, $j = 1, \ldots, R_0$. Calculate the separated-variance test statistics $t_{0j}^* = \frac{\bar{y}_{01j} - \bar{y}_{02j}}{\sqrt{\frac{s_{01j}^2}{n_1}+\frac{s_{02j}^2}{n_2}}}$. Obtain the critical value c_α according to the prespecified Type I error rate α and the alternative hypothesis.
3. Let μ_{11} and μ_{21} be the means of the two groups under the alternative hypothesis. Generate R_1 sets of non-normal data, each with the sample sizes (n_1 and n_2), means (μ_{11} and μ_{21}), standard deviations (σ_1 and σ_2), skewness values (γ_{11} and γ_{12}), and kurtosis values (γ_{21} and γ_{22}) for the two groups separately. We recommend a minimum value 1000 for R_1.

4. Calculate the means and variances for each group in each dataset in Step (3) and denote them as $\bar{y}_{a1j}$, $\bar{y}_{a2j}$, s^2_{a1j}, and s^2_{a2j}, $i = 1, \ldots, R_1$, and calculate the corresponding $t^*_{ai} = \frac{\bar{y}_{a1j} - \bar{y}_{a2j}}{\sqrt{\frac{s^2_{a1j}}{n_1} + \frac{s^2_{a2j}}{n_2}}}$ statistic.
5. The power is estimated as the proportion that t^*_{ai} is greater than the critical value c_α: $\pi = \#\left(t^*_{ai} > c_\alpha\right)/R_1$.

3 Implementation

The Monte Carlo procedure for power analysis for the one-sample, paired-sample, and two-sample analysis is implemented in an R package WebPower. Specifically, the function `wp.mc.t()` is utilized. The basic usage of the function `wp.mc.t()` has the following form:

```
wp.mc.t(n, R0, R1, mu0, mu1, sd, skewness, kurtosis,
alpha, type, alternative).
```

In the function, `n` is the sample size; `mu0`, `mu1`, `sd`, `skewness`, and `kurtosis` are the mean under the null hypothesis, mean under the alternative hypothesis, standard deviation, skewness, and kurtosis, with the default values 0, 0, 1, 0, and 3, respectively. `R0` and `R1` specify the total number of replications under null and alternative hypotheses with the default value 100,000 and 1000, respectively. `alpha` is the significance level with the default value 0.05. `type` specifies the type of analysis such as one-sample test or two-sample test, and `alternative` specifies the direction of the alternative hypothesis.

We briefly illustrate the application of the `wp.mc.t` function via three examples. First, in a one-sample t-test, we are interested in whether the population mean is equal to 0 with a two-sided alternative hypothesis. The population distribution follows a normal distribution with mean equal to 0.5 and standard deviation equal to 1. To calculate the power with sample size equal to 20, the R input is as follows:

```
wp.mc.t(n=20 , mu0=0, mu1=0.5, sd=1, skewness=0,
kurtosis=3, type = c("one.sample"), alternative =
c("two.sided")).
```

The power is 0.557 in this example.

Second, in a paired t-test, we plan to test whether the matched pairs have equal means with one-sided alternative hypothesis ($H_a : \mu_D > 0$). The mean, standard deviation, skewness, and kurtosis of the difference scores are 0.3, 1, 1, and 6, respectively. To calculate the power with sample size equal to 40, the specification of the R function is as follows:

```
wp.mc.t(n=40 , mu0=0, mu1=0.3, sd=1, skewness=1,
kurtosis=6, type = c("paired"), alternative =
c("larger")).
```

The power is 0.657 in this example.

Third, in a two-sample independent t-test, we plan to examine whether two independent population means are equal with one-sided alternative hypothesis ($H_a: \mu_1 - \mu_2 < 0$). The means for two groups are 0.2 and 0.5, standard deviations for two groups are 0.2 and 0.5, skewnesses for two groups are 1 and 2, and kurtoses for two groups are 4 and 6, respectively. To calculate the power with sample size equal to 15 per group, the specification of the R function is as follows:

```
wp.mc.t(n=c(15, 15), mu1=c(0.2, 0.5), sd=c(0.2, 0.5),
skewness=c(1, 2), kurtosis=c(4, 6), type = c("two.
sample"), alternative = c("less")).
```

The power is 0.879 in this example.

For those who are not familiar with R, an online application is also created to conduct the same power analysis using a simple interface on this webpage: http://w.psychstat.org/tnonnormal.

4 A Simulation Study

We conducted a simulation study to examine the performance of the Monte Carlo-based power analysis for the two-sample analysis under the null hypothesis $H_0: \mu_1 - \mu_2 = 0$. This is to investigate whether the Type I error can be well controlled. The performance of the Monte Carlo method (MC) is also compared with conventional pooled-variance t-test (CP).

We varied the following four factors in the simulation: normality of data (either normal or non-normal), ratio of variance of group 1 to that of group 2 with $\sigma_2^2 = 50$ ($\frac{\sigma_1^2}{\sigma_2^2}$ =0.2, 1, 2, and 5), ratio of sample size of group 1 to that of group 2 ($\frac{n_1}{n_2}$ =0.2, 1, and 2), and sample size of group 1 (n_1=10, 50, and 100). The non-normal data are generated from a Gamma distribution. Overall, a total of 72 conditions ($2 \times 4 \times 3 \times 3$) are evaluated.

The empirical Type I error rates are listed in Table 1. Clearly, the Monte Carlo-based power analysis controlled the Type I error rates well around the nominal level ($\alpha = 0.05$) regardless of the shape of distribution, the level of heterogeneity ($\frac{\sigma_1^2}{\sigma_2^2}$), the ratio of sample size of group 1 to that of group 2 ($\frac{n_1}{n_2}$), and the sample size of group 1 (n_1). The conventional pooled-variance t-test only controlled the Type I error rates at the nominal level under homogeneity and/or equal-sample-size situations as expected. When two groups have different variance and sample sizes, the conventional pooled-variance t-test yielded either too small rejection rate (e.g., 0.002) or too large rejection rate (e.g., 0.242). Given that practical data are often non-normal and heterogeneous, the Monte Carlo-based power analysis is therefore recommended.

Table 1 The empirical Type I error in Monte Carlo-based power analysis (MC) and conventional pooled-variance t-test (CP) under the null hypothesis

		$\frac{\sigma_1^2}{\sigma_2^2} = 0.2$		$\frac{\sigma_1^2}{\sigma_2^2} = 1$		$\frac{\sigma_1^2}{\sigma_2^2} = 2$		$\frac{\sigma_1^2}{\sigma_2^2} = 5$	
$\frac{n_1}{n_2}$	n_1	MC	CP	MC	CP	MC	CP	MC	CP
Normal data									
0.2	10	0.048	0.003	0.051	0.049	0.049	0.117	0.049	0.227
0.2	50	0.050	0.001	0.047	0.048	0.056	0.120	0.047	0.219
0.2	100	0.052	0.002	0.047	0.048	0.051	0.116	0.050	0.225
1	10	0.053	0.057	0.052	0.051	0.048	0.050	0.048	0.055
1	50	0.049	0.051	0.052	0.050	0.051	0.051	0.047	0.050
1	100	0.053	0.054	0.052	0.053	0.048	0.049	0.048	0.048
2	10	0.050	0.131	0.052	0.050	0.047	0.028	0.052	0.020
2	50	0.046	0.116	0.051	0.051	0.048	0.028	0.048	0.015
2	100	0.049	0.121	0.052	0.054	0.052	0.029	0.050	0.015
Non-normal data									
0.2	10	0.050	0.005	0.047	0.048	0.049	0.109	0.050	0.234
0.2	50	0.051	0.003	0.046	0.046	0.053	0.119	0.051	0.242
0.2	100	0.050	0.002	0.047	0.050	0.049	0.119	0.047	0.224
1	10	0.050	0.065	0.052	0.047	0.053	0.056	0.052	0.103
1	50	0.047	0.055	0.049	0.049	0.049	0.049	0.049	0.067
1	100	0.052	0.052	0.048	0.048	0.044	0.048	0.048	0.062
2	10	0.047	0.131	0.053	0.047	0.048	0.038	0.045	0.072
2	50	0.049	0.122	0.049	0.048	0.051	0.032	0.050	0.034
2	100	0.050	0.120	0.050	0.050	0.046	0.029	0.050	0.027

5 Conclusion

To flexibly deal with non-normality and unequal variances in the real data, we proposed a Monte Carlo-based power analysis procedure for one-sample t-test, two-sample t-test, and paired t-test. Simulation results showed that the Monte Carlo-based method achieved well-controlled Type I rate even when the assumptions for the conventional power analysis do not hold. In contrast, when homogeneity assumption does not hold and/or two groups have unequal sample size, the conventional pooled-variance t-test could be either too liberal or too conservative. Both an R package WebPower and an online application are provided for researchers to easily carry out the Monte Carlo-based power analysis. The Monte Carlo-based method can be generalized to power analysis for ANOVA, regression, structural equation modeling, and multilevel modeling to handle non-normal data. Missing data can also be considered in the Monte Carlo method.

References

J. Cohen, *Statistical Power Analysis for the Behavioral Sciences*, 2nd edn. (Lawrence Erlbaum, Hillsdale, NJ, 1988)

B.L. Welch, The generalization of student's' problem when several different population variances are involved. Biometrika **34**(1/2), 28–35 (1947)

B.K. Moser, G.R. Stevens, C.L. Watts, The two-sample t test versus Satterthwaite's approximate F test. Commun. Stat. Theory Methods **18**(11), 3963–3975 (1989)

R.L. Disantostefano, K.E. Muller, A comparison of power approximations for Satterthwaite's test. Commun. Stat. Simul. Comput. **24**(3), 583–593 (1995)

M.J. Blanca, J. Arnau, D. López-Montiel, R. Bono, R. Bendayan, Skewness and kurtosis in real data samples. Methodology **9**(2), 78–84 (2013)

T. Micceri, The unicorn, the normal curve, and other improbable creatures. Psychol. Bull. **105**(1), 156 (1989)

M. Cain, Z. Zhang, K. Yuan, Univariate and multivariate skewness and kurtosis for measuring nonnormality: prevalence, Influence and estimation. Behav. Res. Methods (in press)

L.K. Muthén, B.O. Muthén, How to use a Monte Carlo study to decide on sample size and determine power. Struct. Equation Model. **9**(4), 599–620 (2002)

Z. Zhang, Monte Carlo based statistical power analysis for mediation models: methods and software. Behav. Res. Methods **46**(4), 1184–1198 (2014)

F.L. Schmidt, J.E. Hunter, *Methods of Meta-Analysis: Correcting Error and Bias in Research Findings* (Sage, Newbury Park, CA, 2014)

Statistical Power Analysis for Comparing Means with Binary or Count Data Based on Analogous ANOVA

Yujiao Mai and Zhiyong Zhang

Abstract Comparison of population means is essential in quantitative research. For comparing means of three or more groups, analysis of variance (ANOVA) is the most frequently used statistical approach. Typically, ANOVA is used for continuous data, but discrete data are also common in practice. To compare means of binary or count data, the classical ANOVA and the corresponding power analysis are problematic, because the assumption of normality is violated. To address the issue, this study introduces an analogous ANOVA approach for binary or count data, as well as the corresponding methods for statistical power analysis. We first introduce an analogous ANOVA table and a likelihood ratio test statistic for comparing means with binary or count data. With the test statistic, we then define an effect size and propose a method to calculate statistical power. Finally, we develop and show software to conduct the proposed power analysis for both binary and count data.

Keywords Statistical power • Analogous ANOVA • Binary data • Count data

1 Introduction

Comparison of population means is one of the essential statistical analyses in quantitative research (Moore et al. 2013). For comparing means of three or more groups, analysis of variance (ANOVA) is the most frequently used statistical approach in psychological research (Howell 2012). Typically, it is used for continuous data and produces an F-statistic as the ratio of the between-group variance to the within-group variance that follows an F-distribution. To use the F-test for ANOVA, three assumptions must be met. The first is the independence of observations, which assumes that all samples are drawn independently of each other. The second is the normality assumption that requires the distribution of the residuals to be normal. The third is the equality of variances, which assumes that the variance of the data in all groups should be the same. In practice, studies with even continuous data cannot

Y. Mai (✉) • Z. Zhang (✉)
University of Notre Dame, 118 Haggar Hall, Notre Dame, IN 46556, USA
e-mail: ymai@nd.edu; zzhang4@nd.edu

L.A. van der Ark et al. (eds.), *Quantitative Psychology*, Springer Proceedings in Mathematics & Statistics 196, DOI 10.1007/978-3-319-56294-0_33

always meet all three assumptions. For binary or count data, the assumption of normality is apparently violated. Therefore, it is unreliable to use classical ANOVA to compare means of binary or count data. Furthermore, the corresponding power analysis is expected to be problematic.

Since discrete data are very common in practice, there have been discussions on the statistical methods for mean comparison with binary or count data. The existing approaches include $k \times 2$ contingency tables and logistic regression to analyze the mean (proportion) difference among groups of binary data (Cox and Snell 1989; Collett 1991). Contingency tables are commonly used along with Pearson's chi-squared test (Pearson 1947; Larntz 1978), likelihood ratio test (Birch 1963; Grove 1984; Williams 1976), Freeman-Tukey chi-squared statistic (Bishop et al. 1975; Freeman and Tukey 1950), and Fisher's exact test (Fisher 1922; Agresti 1992). Pearson's chi-squared test is related to Goodman and Kruskal's τ (Goodman and Kruskal 1954; Efron 1978). This test is less accurate with small sample size (less than 10 for each cell) and is unreliable if more than 20% of cells have expected values less than 5 (Yates et al. 1999). For likelihood ratio test and Freeman-Tukey chi-squared test, simulation studies found that the Type I error rates became very high when the sample size was small and there were cells with small observed means and moderate expected values (Larntz 1978). Fisher's exact test is related to Goodman and Kruskal's λ (Turek and Suich 1989; Efron 1978). This test is more accurate than the chi-squared tests with small sample size, but it becomes difficult to calculate with large samples or unbalanced tables (Mehta et al. 1984). Although none of these tests is perfect, in general, the likelihood ratio test is preferred by many statisticians (Larntz 1978; Collett 1991), because it is based on the exact Bernoulli distribution for binary data, and researches (Hoeffding 1965; Bahadur 1967) suggested that it has some asymptotically optimal properties.

Researchers have also used logistic regression to estimate and compare the group means of binary data (Cox and Snell 1989; Collett 1991). This method utilizes the likelihood ratio test, which performs well when there are enough observations to justify the assumptions of the asymptotic chi-squared tests. However, the models and procedures might be more complicated than necessary. First, the procedure requires creating dummy variables since regression models are used with categorical predictors. These dummy variables not only increase the complexity of the model itself but also make the interpretation of the model more difficult for applied researchers. Second, the procedure using logistic regression is more complex with the current software. Third, researchers are interested in whether the groups are from populations with different means using ANOVA, while logistic regression is more efficient for parameter estimation (Cox and Snell 1989) and prediction of proportions (Collett 1991). The meaning of parameters in logistic regression is not easy to interpret for the purpose of mean comparison.

Although contingency tables and logistic regression are two different approaches, it is not difficult to show that contingency table and logistic regression lead to the same conclusions when using likelihood ratio tests. Then, is it possible to provide the equivalent results for binary data by applying likelihood ratio test to ANOVA? In fact, as suggested by Efron (1978), log-likelihood can be used as a general measure of variation. From the perspective of variation decomposition, Efron (1978) constructed an ANOVA-like table for binary data with emphasis on descriptive statistics. Based on the work by Efron (1978), we will introduce an analogous ANOVA table with a closed-form likelihood ratio test statistic for comparing means with binary data. Then we will define an effect size and provide a corresponding power analysis method. Software to conduct the power analysis will also be developed. After that, we will extend the method for binary data to count data.

The rest of the chapter is organized as follows. Section 2 is a review of one-way ANOVA with continuous data. Section 3 proposes the method for binary data. Section 4 discusses the method for count data. Section 5 illustrates the developed software through examples. Section 6 summarizes and concludes this study.

2 One-Way ANOVA with Continuous Data

Analysis of variance (ANOVA) is a collection of statistical models used to analyze the differences among group means through variance decomposition (Maxwell and Delaney 2004; Fisher 1921). The current study focuses on the use of one-way ANOVA. We first review the basics of one-way ANOVA with continuous data. Let Y be the outcome variable and A be a categorical variable of k levels; with A as the grouping variable, we divide the population of Y into k groups. The null hypothesis H_0 states that different groups have equal population means, while the alternative hypothesis H_1 supposes that at least two groups have different population means. Let μ_j be the population mean of the jth group, $j = 1, 2, \cdots, k$ and μ_0 be the grand population mean. The null and alternative hypotheses can be specified as follows:

$$H_0: \quad \mu_1 = \mu_2 = \ldots = \mu_k = \mu_0,$$
$$H_1: \quad \exists\ \mu_g \neq \mu_j, \quad \text{where} \quad g \neq j \quad \text{and} \quad g,\ j \in [1, 2, \cdots, k].$$

Consider the corresponding models with H_0 and H_1. The null model M_0 is

$$\mathrm{E}\{Y|A = j\} = \mu_0, \tag{1}$$

where $Y|(A = j) \sim N(\mu_0, \sigma_0^2)$. The alternative model M_1 is

$$\mathrm{E}\{Y|A = j\} = \mu_j, \tag{2}$$

where $Y|(A = j) \sim N(\mu_j, \sigma_1^2)$.

In one-way ANOVA, the observed variance in the outcome variable is partitioned into between-group variance and within-group variance. If the between-group variance is greater than the within-group variance, the group means are considered to be different. For continuous data, "squared error" is deployed as a measure of variation between an observed data point and corresponding expectation ("explanatory point", see Efron 1978). Its function is defined as

$$S(y, \mu) = (y - \mu)^2 \tag{3}$$

Table 1 ANOVA table for continuous data

Source	Sum of squares	Degree of freedom	Test statistic	P-value
Between-group	$SS_B = \sum_{j=1}^{k}(\bar{y}_j - \bar{y})^2$	$k-1$	$F = \frac{SS_B/(k-1)}{SS_W/(n-k)}$	$\Pr\{\mathrm{F}(k-1, n-k) \geq F\}$
Within-group	$SS_W = \sum_{j=1}^{k}\sum_{i=1}^{n_j}(y_{ij} - \bar{y}_j)^2$	$n-k$		
Total	$SS_T = \sum_{j=1}^{k}\sum_{i=1}^{n_j}(y_{ij} - \bar{y})^2$	$n-1$		

Note: $\mathrm{F}(k-1, n-k)$ is the F-distribution with $df_1 = k-1$ and $df_2 = n-k$

with y denoting a data point and μ denoting the expectation. Given a sample of data $\mathbf{Y} = (\mathbf{y}_j) = \{y_{ij}\}$, $i = 1, 2, \cdots, n_j, j = 1, 2, \cdots, k$, with n_j denoting the sample size of the jth group, the test statistic is equal to the ratio of between-group sample variance and within-group sample variance and follows an F-distribution under H_0:

$$F = \hat{\sigma}^2_{between}/\hat{\sigma}^2_{within} \sim \mathrm{F}(k-1, n-k), \tag{4}$$

where $\hat{\sigma}^2_{between} = \sum_{j=1}^{k}(\bar{y}_j - \bar{y})^2/(k-1)$, and $\hat{\sigma}^2_{within} = \sum_{j=1}^{k}\sum_{i=1}^{n_j}(y_{ij} - \bar{y}_j)^2/(n-k)$ with $\bar{y}_j$ denoting the sample mean of the jth group and $\bar{y}$ denoting the grand mean of data. ANOVA is often conducted by constructing the source of variance table shown in Table 1.

3 One-Way Analogous ANOVA with Binary Data

3.1 Model and Test Statistic for Binary Data

For comparison of group means, often called proportions, for binary data, the hypotheses are the same as one-way ANOVA with continuous data. But the models are different since the distribution of the outcome variable is not normal.

Let Y be a zero-one outcome variable and A be a categorical variable of k levels, with A as the grouping variable we can divide the population of Y into k groups. Let μ_0 denote the grand probability of the outcome 1, and μ_j denote the jth group probability of observing $1, j = 1, 2, \cdots, k$. Then the null and alternative hypotheses are

$$\begin{aligned} H_0: &\quad \mu_1 = \mu_2 = \ldots = \mu_k = \mu_0, \\ H_1: &\quad \exists\ \mu_g \neq \mu_j, \quad \text{where} \quad g \neq j \quad \text{and} \quad g,\ j \in [1, 2, \cdots, k]. \end{aligned}$$

The null model M_0 is

$$\mathrm{E}\{Y|A = j\} = \mu_0, \tag{5}$$

where $Y|(A=j) \sim Bernoulli(\mu_0)$, and the alternative model M_1 is

$$\mathrm{E}\{Y|A=j\} = \mu_j, \tag{6}$$

where $Y|(A=j) \sim Bernoulli(\mu_j)$. Given a sample of data $\mathbf{Y} = (\mathbf{y}_j) = \{y_{ij}\}$, $i = 1, 2, \cdots, n_j, j = 1, 2, \cdots, k$, with n_j denoting the sample size of the jth group, we define minus twice the log-likelihood ratio of M_0 to M_1 as a statistic:

$$\begin{aligned} D &= -2\ln\frac{\mathscr{L}(\mu_0|\mathbf{Y})}{\mathscr{L}(\mu_1, \mu_2, \ldots, \mu_k|\mathbf{Y})} \\ &= -2[\ell(\mu_0|\mathbf{Y}) - \ell(\mu_1, \mu_2, \ldots, \mu_k|\mathbf{Y})] \\ &= -2(\ell_{M_0} - \ell_{M_1}), \end{aligned} \tag{7}$$

where $\mathscr{L}(\boldsymbol{\theta}|\mathbf{Y})$ denotes the likelihood function of $\boldsymbol{\theta}$ given data $\mathbf{Y}$. Under the null hypothesis H_0, this statistic follows a chi-squared distribution $D \sim \chi^2(df)$ with the degrees of freedom $df = k-1$ if the sample size tends to infinity (Wilks 1938).

Let the observed grand mean $\bar{y} = \sum_j^k \sum_i^{n_j} y_{ij}/n$ be the estimate of μ_0 and the observed group mean $\bar{y}_j = \sum_i^{n_j} y_{ij}/n_j$ be the estimate of μ_j. For a given sample of data, we calculate the test statistic as $\tilde{D}$ as follows. We first calculate minus twice the log-likelihood for M_0 and M_1:

$$\begin{aligned} -2\hat{\ell}_{M_0} &= -2\ell(\bar{y}|\mathbf{Y}) \\ &= -2\sum_{j=1}^{k}\sum_{i=1}^{n_j}\left[y_{ij}\ln\bar{y} + (1-y_{ij})\ln(1-\bar{y})\right], \end{aligned} \tag{8}$$

$$\begin{aligned} -2\hat{\ell}_{M_1} &= -2\ell(\bar{y}_1, \bar{y}_2, \ldots, \bar{y}_k|\mathbf{Y}) \\ &= -2\sum_{j=1}^{k}\sum_{i=1}^{n_j}\left[y_{ij}\ln\bar{y}_j + (1-y_{ij})\ln(1-\bar{y}_j)\right]; \end{aligned} \tag{9}$$

and then $\tilde{D}$ as their difference:

$$\begin{aligned} \tilde{D} &= -2(\hat{\ell}_{M_0} - \hat{\ell}_{M_1}) \\ &= -2\sum_{j=1}^{k} n_j\left\{\bar{y}_j(\ln\bar{y} - \ln\bar{y}_j) + (1-\bar{y}_j)\left[\ln(1-\bar{y}) - \ln(1-\bar{y}_j)\right]\right\}. \end{aligned} \tag{10}$$

It can be proven that the observed grand mean $\bar{y}$ is the maximum likelihood estimate of μ_0, and the observed group mean $\bar{y}_j$ is the estimate of μ_j (Efron 1978). Other than the "squared error" used by standard analysis of variance, if we use minus twice the

Table 2 Analogous ANOVA table for binary data

Source	Sum of variance	Degree of freedom	Test statistic	*P*-value
Between-group	$SS_B = -2(\hat{\ell}_{M_0} - \hat{\ell}_{M_1})$	$k-1$	$\tilde{D} = -2(\hat{\ell}_{M_0} - \hat{\ell}_{M_1})$	$\Pr\{\chi^2(k-1) \geq \tilde{D}\}$
Within-group	$SS_W = -2\hat{\ell}_{M_1}$	$n-k$		
Total	$SS_T = -2\hat{\ell}_{M_0}$	$n-1$		

Note: $\chi^2(k-1)$ is the chi-squared distribution with $df = (k-1)$

log-likelihood as a measure of variation, the variation function for binary data is as follows:

$$S_1(y, \mu) = \begin{cases} -2\ln(\mu) & \text{if} \quad y = 1 \\ -2\ln(1-\mu) & \text{if} \quad y = 0 \end{cases} \tag{11}$$

with y denoting a data point and μ the expectation. Then the sum of variation $SS_1 = \sum S_1(\mathbf{Y}, \mu) = -2\ln\mathscr{L}(\mu|\mathbf{Y})$ with $\mathbf{Y}$ denoting a sample of data (Efron 1978). With these functions, we can obtain the analogous total variance, within-group variance, and between-group variance as follows:

$$\begin{aligned} SS_T &= -2\hat{\ell}_{M_0} \\ SS_W &= -2\hat{\ell}_{M_1} \\ SS_B &= -2(\hat{\ell}_{M_0} - \hat{\ell}_{M_1}). \end{aligned} \tag{12}$$

Now with these statistics, we can create an analogous ANOVA table in Table 2 for binary data similar to that for continuous data.

From the analogous ANOVA table, we see that the likelihood ratio test statistic here equals the between-group variation. The ratio of between-group variation to total variation is exactly the R^2 coefficient for model M_1 (see Efron 1978), which is also used by Goodman (1971) for contingency tables.

3.2 *Measure of Effect Size for Binary Data*

Standardized effect-size measures facilitate comparison of findings across studies and disciplines, while unstandardized effect-size measures (simple effect size) with "immediate meanings" may be preferable for reporting purposes (Ellis 2010; Baguley 2009). The r-family and the d-family effect-size measures are standardized (Rosenthal 1994), while R^2-family effect-size measures such as f^2 and η^2 are unstandardized and immediately meaningful (Cameron and Windmeijer 1997). Both types of effect-size measures could be defined. But not all types of effect-size measures can be used for power analysis with a specific test statistic. For the purpose of power analysis, in this study, we use a standardized effect-size measure like Cramer's V, which is a member of the r family (Ellis 2010). It is also an adjusted version of phi coefficient ϕ that is frequently reported as the measure of effect size for a chi-squared test (Cohen 1988; Ellis 2010; Fleiss 1994). It can be viewed as the association between two variables as a percentage of their maximum possible variation. In the case of one-way analogous ANOVA, the two variables are the outcome variable and the grouping variable.

For one-way analogous ANOVA with binary data, we define the effect size V:

$$V = \sqrt{-2\sum_{j=1}^{k} w_j \left\{\mu_j(\ln\mu_0 - \ln\mu_j) + (1-\mu_j)\left[\ln(1-\mu_0) - \ln(1-\mu_j)\right]\right\} / (k-1)}, \quad (13)$$

where $w_j = n_j/n$ is the weight of the jth group, and $n = \sum_j^k n_j$ is the total sample size. The small, medium, and large effect size can be defined as 0.10, 0.30, and 0.50, borrowed from Cohen's effect-size benchmarks (Cohen 1988; Ellis 2010).

For a given sample of data, the sample effect size can be calculated as

$$\begin{aligned}\hat{V} &= \sqrt{\tilde{D}/n(k-1)} \\ &= \sqrt{-2\sum_{j=1}^{k} w_j \left\{\bar{y}_j(\ln\bar{y} - \ln\bar{y}_j) + (1-\bar{y}_j)\left[\ln(1-\bar{y}) - \ln(1-\bar{y}_j)\right]\right\} / (k-1)}.\end{aligned} \quad (14)$$

3.3 Statistical Power Analysis with Binary Data

Power analysis is often applied in the context of ANOVA in order to assess the probability of successfully rejecting the null hypothesis if we assume a certain ANOVA design, effect size in the population, sample size, and significance level. Power analysis can assist in study design by determining what sample size would be required in order to have a reasonable chance of rejecting the null hypothesis when the alternative hypothesis is true (Strickland 2014).

For one-way analogous ANOVA with binary data, when the null hypothesis H_0 is true, the test statistic D follows a central chi-squared distribution $\chi^2(df)$, where $df = k - 1$ is the degree of freedom. If $\tilde{D}$ is larger than the critical value $C = \chi^2_{1-\alpha}(df)$, one would reject the null hypothesis H_0. When the alternative hypothesis H_1 is true, the test statistic D follows a noncentral chi-squared distribution $\chi^2(df, \lambda)$, where $df = k-1$ is the degree of freedom and $\lambda = D = n(k-1)V^2$ is the noncentral parameter. Let $\Phi_{\chi^2(df,\lambda)}(x)$ be the cumulative distribution function of the noncentral chi-squared distribution; then the statistical power of the test is

$$\begin{aligned} power &= \Pr\{D \geq C|H_1\} \\ &= \Pr\{\chi^2(df, \lambda) \geq C\} \\ &= 1 - \Phi_{\chi^2(df,\lambda)}(C) \\ &= 1 - \Phi_{\chi^2[k-1,n(k-1)V^2]}\left[\chi^2_{1-\alpha}(k-1)\right]. \end{aligned} \tag{15}$$

With this formula, the power, minimum detectable effect size V, minimum required sample size n, or significance level α can be calculated given the other parameters.

4 One-Way Analogous ANOVA with Count Data

For comparison of group means with count data, the statistical inference is similar to that for binary data. The main difference lies in that the distribution of the outcome variable in the model is Poisson instead of Bernoulli.

4.1 Model and Test Statistic for Count Data

To construct the models for count data, let Y be the outcome variable, which can take only the nonnegative integer values, and A be a categorical variable of k levels. The null and alternative hypotheses are

$$\begin{aligned} H_0 &: \quad \mu_1 = \mu_2 = \ldots = \mu_k = \mu_0, \\ H_1 &: \quad \exists\ \mu_g \neq \mu_j, \quad \text{where} \quad g \neq j \quad \text{and} \quad g,\ j \in [1, 2, \cdots, k]. \end{aligned}$$

The null model M_0 is

$$\text{E}\{Y|A = j\} = \mu_0, \tag{16}$$

where $Y|(A = j) \sim Poisson(\mu_0)$, and the alternative model M_1 is

$$\text{E}\{Y|A = j\} = \mu_j, \tag{17}$$

where $Y|(A = j) \sim Poisson(\mu_j)$, $j = 1, 2, \cdots, k$. Given a sample of data, $\mathbf{Y} = (\mathbf{y}_j) = \{y_{ij}\}$, $i = 1, 2, \cdots, n_j, j = 1, 2, \cdots, k$, with n_j denoting the sample size of the jth group, minus twice the log-likelihood ratio of model M_0 to M_1 is

$$\begin{aligned} D &= -2\ln\frac{\mathscr{L}(\mu_0|\mathbf{Y})}{\mathscr{L}(\mu_1, \mu_2, \ldots, \mu_k|\mathbf{Y})} \\ &= -2[\ell(\mu_0|\mathbf{Y}) - \ell(\mu_1, \mu_2, \ldots, \mu_k|\mathbf{Y})] \\ &= -2(\ell_{M_0} - \ell_{M_1}) \end{aligned} \tag{18}$$

Under null hypothesis H_0, this statistic follows a chi-squared distribution $D \sim \chi^2(df)$ with the degrees of freedom $df = k - 1$ if the sample size tends to infinity (Wilks 1938). Let the grant mean $\bar{y} = \sum_j^k \sum_i^{n_j} y_{ij}/n$ be the estimate of μ_0, and the group mean $\bar{y}_j = \sum_i^{n_j} y_{ij}/n_j$ be the estimate of μ_j. For a given sample of data, we can calculate the test statistic as $\tilde{D}$ as follows. We first calculate minus twice the log-likelihood for M_0 and M_1:

$$SS_T = -2\hat{\ell}_{M_0} = -2\sum_{j=1}^{k}\sum_{i=1}^{n_j}\left(\bar{y}_{ij}\ln\bar{y} - \bar{y}\right), \tag{19}$$

$$SS_W = -2\hat{\ell}_{M_1} = -2\sum_{j=1}^{k}\sum_{i=1}^{n_j}\left(\bar{y}_{ij}\ln\bar{y}_j - \bar{y}_j\right); \tag{20}$$

and then $\tilde{D}$ as their difference:

$$SS_B = \tilde{D} = -2(\hat{\ell}_{M_0} - \hat{\ell}_{M_1}) = -2\sum_{j=1}^{k} n_j\left[\bar{y}_j(\ln\bar{y} - \ln\bar{y}_j) - (\bar{y} - \bar{y}_j)\right]. \tag{21}$$

For count data, we can also create an analogous ANOVA table like Table 2.

4.2 *Effect Size and Power Analysis for Count Data*

For one-way analogous ANOVA with count data, the effect size is also defined as $V = \sqrt{D/n(k-1)}$. The sample effect size can be calculated as

$$\begin{aligned}\hat{V} &= \sqrt{\tilde{D}/n(k-1)}\\ &= \sqrt{-2\sum_{j=1}^{k} w_j\left[\bar{y}_j(\ln\bar{y} - \ln\bar{y}_j) + (\bar{y}_j - \bar{y})\right]/(k-1)},\end{aligned} \tag{22}$$

where $w_j = n_j/n$ is the weight of the jth group, and $n = \sum_j^k n_j$ is the total sample size. The power analysis of one-way analogous ANOVA with count data is the same as that with binary data.

5 Software

To carry out the power analysis for analogous ANOVA with binary or count data, we have developed online applications that can be used within a Web browser. The link for the binary analogous ANOVA is http://psychstat.org/anovabinary and for the count analogous ANOVA is http://psychstat.org/anovacount. The software interface of power analysis for analogous ANOVA with binary data is shown in Fig. 1. Among *number of groups*, *sample size*, *effect size*, *significance level*, and *power*, any of them can be calculated given the rest of the information. The following examples illustrate the usage of the interface.

Suppose a student researcher hypothesizes that freshman, sophomore, junior, and senior college students have different rates of passing a reading exam. Based on his prior knowledge, he expects that the effect size is about 0.15. Based on the information, he wants to know (1) the power for him to find the significant difference among the four groups if he plans to collect data from 25 students in each of the four groups and (2) the minimum required sample size for him to find the significant difference among the four groups with power 0.8.

For the calculation of power, the number of group $k = 4$, the total sample size $n = 25\times 4 = 100$, and the effect size $V = 0.15$. Let the significance level $\alpha = 0.05$, then we can use formula (15) to calculate the power as

$$\begin{aligned} power &= 1 - \Phi_{\chi^2[k-1,n(k-1)V^2]}\left[\chi^2_{1-\alpha}(k-1)\right] \\ &= 1 - \Phi_{\chi^2(3,100\times 3\times 0.15^2)}\left[\chi^2_{0.95}(3)\right] \\ &= 1 - \Phi_{\chi^2(3,6.75)}(7.8147) \\ &= 0.572. \end{aligned}$$

We can also use the online interface to estimate the power (see Fig. 1a). Given four groups, sample size 100, effect size 0.15, and significance level 0.05, the output indicates the power for this design is again 0.572.

With the required $power = 0.8$, $k = 4$, $V = 0.15$, and $\alpha = 0.05$, we solve the following equation:

$$\begin{aligned} power &= 1 - \Phi_{\chi^2[k-1,n(k-1)V^2]}\left[\chi^2_{1-\alpha}(k-1)\right] \\ 0.8 &= 1 - \Phi_{\chi^2(3,n\times 3\times 0.15^2)}\left[\chi^2_{0.95}(3)\right] \\ 0.8 &= 1 - \Phi_{\chi^2(3,n\times 0.0675)}(7.8147) \\ n &= 161.520. \end{aligned}$$

So, the minimum required sample size is 162.

Figure 1b shows how to use the interface to calculate the minimum required sample size. Given four groups, effect size 0.15, significance level 0.05, and the desired power 0.8, the output showed that a sample size 162, the near integer of

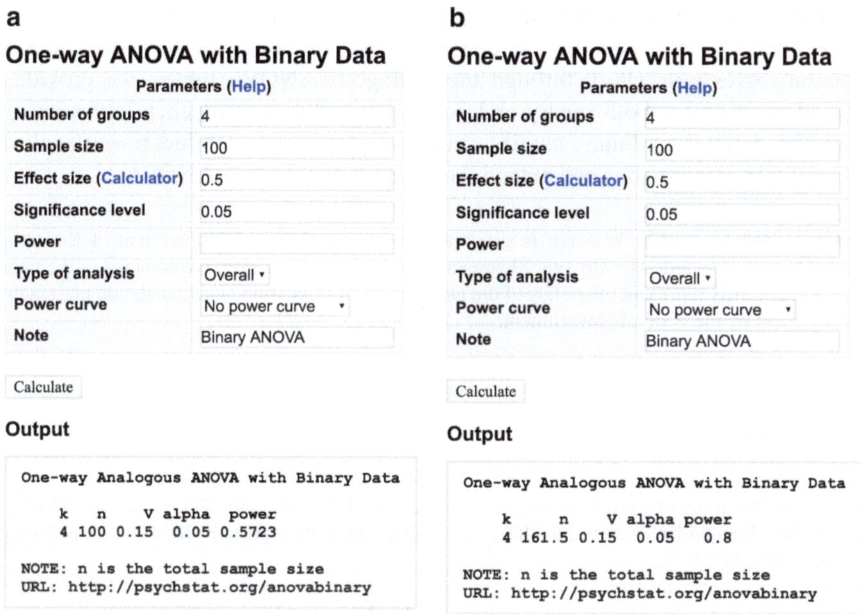

Fig. 1 Examples of power analysis for analogous ANOVA with binary data. (**a**) Given sample size, calculate power. (**b**) Given power, calculate sample size

161.5, is needed. A power curve can also be plotted by providing multiple sample sizes in the *Sample size* field. The interface for analogous ANOVA with count data is the same.

6 Discussion

In this chapter, an analogous ANOVA table and the closed-form likelihood ratio test statistic were introduced for comparing mean differences among groups of binary and count data, respectively. Based on the analogous ANOVA table and test statistic, the effect size V statistic, an adjusted phi coefficient, was defined. The power analysis involved four parameters, number of groups, total sample size, statistical significance level, and effect size. In addition, corresponding free online software were developed.

We recommend the application of these methods in binary and count data analysis. First, these methods are analogous to procedures in classical ANOVA as they decompose variation in observed outcomes for binary and count data. Specifically, the analogous ANOVA tables can help the researchers intuitively understand the exact meanings of the likelihood ratio test statistics used to compare means of binary or count data. Second, by using raw data and closed-form statistics, these methods are easier to use and more efficient than logistic regression or Poisson regression. Third, through the analogous ANOVA tables, we provide a unified solution for both binary and count data, while contingency tables cannot deal with count data. Future studies can investigate how to conduct power analysis for multiple comparisons and extend the methods to two-way ANOVA.

Acknowledgements This research is supported by a grant from the Department of Education (R305D140037) awarded to Zhiyong Zhang and Ke-Hai Yuan. However, the contents of the paper do not necessarily represent the policy of the Department of Education, and you should not assume endorsement by the Federal Government.

References

A. Agresti, A survey of exact inference for contingency tables. Stat. Sci. **7**(1), 131–153 (1992)

T. Baguley, Standardized or simple effect size: What should be reported? Br. J. Psychol. **100**(3), 603–617 (2009)

R. Bahadur, An optimal property of the likelihood ratio statistic, in *Proceedings of Fifth Berkeley Symp. Math. Statist. Probab*, vol. 1 (1967), pp. 13–26

M. Birch, Maximum likelihood in three-way contingency tables. J. R. Stat. Soc. Ser. B Methodol. **25**(1), 220–233 (1963)

Y.M. Bishop, S.E. Fienberg, P.W. Holland, *Discrete Multivariate Analysis: Theory and Practice* (MIT Press, Cambridge, 1975)

A.C. Cameron, F.A. Windmeijer, An r-squared measure of goodness of fit for some common nonlinear regression models. J. Econ. **77**(2), 329–342 (1997)

J. Cohen, *Statistical Power Analysis for the Behavioral Sciences*, 2nd edn. (Lawrence Erlbaum Associates, Hillsdale, 1988)

D. Collett, *Modelling Binary Data* (Chapman & Hall, New York, 1991)

D.R. Cox, E.J. Snell, *Analysis of Binary Data*, 2nd edn. (Chapman & Hall, New York, 1989)

B. Efron, Regression and anova with zero-one data: measures of residual variation. J. Am. Stat. Assoc. **73**(361), 113–121 (1978)

P.D. Ellis, *The Essential Guide to Effect Sizes: Statistical Power, Meta-Analysis, and the Interpretation of Research Results* (Cambridge University Press, New York, 2010)

R.A. Fisher, On the "probable error" of a coefficient of correlation deduced from a small sample. Metron **1**, 3–32 (1921)

R.A. Fisher, On the interpretation of χ 2 from contingency tables, and the calculation of p. J. R. Stat. Soc. **85**(1), 87–94 (1922)

J.L. Fleiss, Measures of effect size for categorical data, in *The Handbook of Research Synthesis*, ed. by H. Cooper, L.V. Hedges (Russell Sage Foundation, New York, 1994), pp. 245–260

M.F. Freeman, J.W. Tukey, Transformations related to the angular and the square root. Ann. Math. Stat. **21**(4), 607–611 (1950)

L.A. Goodman, The analysis of multidimensional contingency tables: stepwise procedures and direct estimation methods for building models for multiple classifications. Technometrics **13**(1), 33–61 (1971)

L.A. Goodman, W.H. Kruskal, Measures of association for cross classifications. J. Am. Stat. Assoc. **49**(285), 732–764 (1954)

D.M. Grove, Positive association in a two-way contingency table: likelihood ratio tests. Commun. Stat. Theory Methods **13**(8), 931–945 (1984)

W. Hoeffding, Asymptotically optimal tests for multinomial distributions. Ann. Math. Stat. **36**(2), 369–401 (1965)

D.C. Howell, *Statistical Methods for Psychology* (Duxbury/Thomson Learning, Pacific Grove, 2012)

K. Larntz, Small-sample comparisons of exact levels for chi-squared goodness-of-fit statistics. J. Am. Stat. Assoc. **73**(362), 253–263 (1978)

S.E. Maxwell, H.D. Delaney, *Designing Experiments and Analyzing Data: A Model Comparison Perspective*, 2nd edn. (Lawrence Erlbaum Associates, Mahwah, 2004)

C.R. Mehta, N.R. Patel, A.A. Tsiatis, Exact significance testing to establish treatment equivalence with ordered categorical data. Biometrics **40**(3), 819–825 (1984)

D.S. Moore, W. Notz, M.A. Fligner, *Essential Statistics* (W.H. Freeman and Company, New York, 2013)

E.S. Pearson, The choice of statistical tests illustrated on the interpretation of data classed in a 2× 2 table. Biometrika **34**(1/2), 139–167 (1947)

R. Rosenthal, Parametric measures of effect size, in *The Handbook of Research Synthesis*, ed. by H. Cooper, L.V. Hedges (Russell Sage Foundation, New York, 1994), pp. 231–244

J.S. Strickland, *Predictive Analytics Using R* (2014), Lulu.com

R.J. Turek, R.C. Suich, An exact test on the goodman-kruskal λ for prediction on a dichotomy: power considerations. Comput. Stat. Data Anal. **8**(2), 171–178 (1989)

S.S. Wilks, The large-sample distribution of the likelihood ratio for testing composite hypotheses. Ann. Math. Stat. **9**(1), 60–62 (1938)

D. Williams, Improved likelihood ratio tests for complete contingency tables. Biometrika **63**(1), 33–37 (1976)

D. Yates, D. Moore, G. McCabe, *The Practice of Statistics* (W.H. Freeman and Company, New York, 1999) [illegible] W. Hoeffding, Asymptotically optimal tests for multinomial distributions. Ann. Math. Stat. 36 [illegible]

[illegible] D. C. Howell, Statistical Methods for Psychology [illegible]

[illegible]

[illegible]

Robust Bayesian Estimation in Causal Two-Stage Least Squares Modeling with Instrumental Variables

Dingjing Shi and Xin Tong

Abstract In causal randomized experiments or psychological trials, the two-stage least squares (2SLS) model with instrument variables (IVs) is a widely used approach to address the issue of treatment endogeneity. The IVs are used to estimate a part of the causal effect whose estimation is not affected by the violation of the linearity assumption in the causal model, and the causal effect of interest in the 2SLS model becomes the local average treatment effect (LATE). Because practical data usually violate the normality assumption, the LATE estimate from the traditional normal-distribution-based method may be inefficient or even biased. This study proposes a robust Bayesian estimation method using Student's t distributions to model data with heavy tails or containing outliers and compares the performance of the proposed robust method to that of the traditional normal-distribution-based method. A Monte Carlo simulation study is conducted and shows that the proposed robust method outperforms the traditional method when data are contaminated. The robust method provides more accurate and efficient LATE estimates and better model fits and thus is recommended to be used in general in the 2SLS modeling with IVs.

Keywords Robust Bayesian method • Causal two-stage least squares modeling • LATE • Instrumental variable

1 Introduction

In randomized experiments or psychological trials, the average treatment effect (ATE) is often the center of the causal research interests. Researchers measure the ATE by studying the outcome difference between participants who are assigned to the treatment and those who are assigned to the control. Regression models with ordinary least squares (OLS) estimation are commonly used to estimate the ATE.

D. Shi (✉) • X. Tong
University of Virginia, 102 Gilmer Hall, Department of Psychology, Charlottesville, VA 22904, USA
e-mail: ds4ue@virginia.edu; xtong@virginia.edu

L.A. van der Ark et al. (eds.), *Quantitative Psychology*, Springer Proceedings in Mathematics & Statistics 196, DOI 10.1007/978-3-319-56294-0_34

Traditional OLS regression in causal analysis assumes that there is no correlation between the regressors and the errors. However, in the presence of endogenous regressors that are correlated with the errors, the linearity assumption may be violated, leading to a biased estimate of the ATE (Angrist and Krueger 1990). To eliminate the bias caused by the correlation between the regressors and the errors, a commonly applied strategy is to include instrumental variables (IVs) in the causal model (Angrist and Pischke 2008, 2014).

The idea of incorporating IVs is to use some variables as instruments to estimate a part of the causal effect whose estimation is not contaminated by the violation of the linearity assumption in the causal OLS model. Specifically, when the linearity assumption is violated, exogenous factors that cause some of the variations in the treatment status and that are uncorrelated with the errors are selected and treated as IVs. The exogenous portion of variations in the treatment that has been partialled out by the IVs is used to estimate the corresponding treatment effect. For example, in the study of the effect of schooling on educational returns (e.g., earnings), because there is likely omitted variable bias such as unobserved personal ability, researchers choose the proximity to college as a candidate IV. Particularly, people whose homes are far away from the college are less likely to attend college. The college proximity has some effect on schoolings, but has no direct effect on the outcome variable earnings. It was argued that people who live a long way from a college are more likely to be in a low-wage labor market (Card 1995). Because the ATE for only a subset of observations is studied, the generalizability of the ATE (i.e., external validity) for the whole sample is traded for the improvement of the estimation performance (i.e., internal validity), and the ATE for the subset of participants is called the local average treatment effect (LATE) (Angrist et al. 1996; Imbens and Rubin 1997). Angrist and Imbens (1995) proposed a two-stage least squares (2SLS) model to estimate the LATE, and this model is widely used in causal inference. In particular, there are two modeling stages in the framework. In the first stage, IVs are used to predict the partial treatment effect that can be explained by the variations of IVs, and in the second stage, the fitted treatment values are used to predict the study outcome and to estimate the LATE. LATE is the ATE for a small group of subjects whose variations are explained by the IVs. The 2SLS model provides reliable LATE estimates as long as valid IVs are used (Angrist and Imbens 1995; Angrist et al. 1996). There are two selection criteria for valid IVs: the instruments are related to the treatment in some way, so that some variations of the outcome could be explained by the instruments; and the instruments are not correlated with other determinants, so that the instruments only affect the outcome through the treatment (Angrist et al. 1996; Imbens and Rubin 1997).

Given valid IVs, the traditional causal 2SLS modeling assumes that the measurement errors at both stages are normally distributed. However, data in psychological or behavioral research usually violate the normality assumption and may have heavy tails or contain outliers. Fitting the heavy-tailed data as if they were normally distributed can result in inflated type I error rates, biased and inefficient parameter estimates (Yuan et al. 2004; Zimmerman 1994, 1998; Zu and Yuan 2010), which may eventually lead to incorrect statistical inferences. Therefore, various robust procedures have been developed to provide more accurate and precise parameter estimates and been used in complex modeling frameworks, such as linear and generalized linear mixed-effects modeling (e.g., Pinheiro et al. 2001; Song et al. 2007), structural equation modeling (e.g., Lee and Xia 2006; Yuan and Bentler 1998), and hierarchical linear and nonlinear modeling (e.g., Rachman-Moore and Wolfe 1984; Wang et al. 2015).

In the last decades, robust methods based on Student's *t* distributions have been developed and advanced to model heavy-tailed data and outliers (e.g., Yuan and Zhang 2012). Student's *t* distributions have been applied to many complex data analyses. For example, Lee and Xia (2006) discussed the use of the *t* distributions in structural equation modeling; Shoham (2002) and Wang et al. (2004) applied the *t* distributions in robust mixture models; Seltzer and Choi (2003) conducted a sensitivity analysis using Student's *t* distributions in robust multilevel models; and in longitudinal data, Tong and Zhang (2012) made use of the *t* distributions for robust growth curve modeling. The complex data structure is not uncommon in causal inference for observational studies. Although the robust methods based on Student's *t* distributions have been studied to provide reliable parameter estimates and confidence intervals in the preceding complex data analyses, few have been examined under the causal inference modeling framework.

The purpose of this study is to propose a robust Bayesian estimation method based on Student's *t* distributions for the causal 2SLS modeling with IVs, and to evaluate the performance of the robust Bayesian method in estimating the LATE, the causal effect of interest. In the following section, the 2SLS models with IVs and the associated LATE estimate in causal inference are reviewed. The robust method based on Student's t distributions is also introduced. Next, a Monte Carlo simulation study is conducted to evaluate the performance of the robust method in the 2SLS modeling with IVs. In the end, the results are summarized and discussions are provided.

2 Robust Method for 2SLS Modeling with IVs

2.1 *Two-Stage Least Squares Modeling (2SLS) with Instrumental Variables (IVs)*

Let $\boldsymbol{D}_i$ and $\boldsymbol{y}_i$ be the treatment and the outcome for individual *i*, respectively, and $\boldsymbol{Z}_i = (Z_{i1}, \ldots, Z_{iJ})'$ be a vector of instrumental variables for individual *i* $(i = 1, \ldots, N)$. Here *N* is the sample size and *J* is the total number of instrumental variables. In the first stage of the 2SLS model, the IVs $\boldsymbol{Z}$ are used to predict the treatment $\boldsymbol{D}$. In other words, the portion of variations in the treatment $\boldsymbol{D}$ is identified and estimated by the IVs $\boldsymbol{Z}$; and then the second stage relies on the estimated exogenous portion of treatment variations in the form of the predicted treatment values to estimate the treatment effect on the outcome $\boldsymbol{y}$. A typical form of the 2SLS model with IVs can be expressed as

$$\boldsymbol{D}_i = \pi_{10} + \boldsymbol{\pi}_{11}\boldsymbol{Z}_i + e_{1i}, \tag{1}$$

$$y_i = \pi_{20} + \pi_{21}\hat{\boldsymbol{D}}_i + e_{2i}, \tag{2}$$

where π_{10} and $\boldsymbol{\pi}_{11} = (\pi_{11}, \ldots, \pi_{1J})'$ are the intercept and regression coefficients for the linear model where the treatment $\boldsymbol{D}$ is regressed on the IVs $\boldsymbol{Z}$, respectively; and π_{20} and π_{21} are the intercept and slope for the linear model where the outcome y is regressed on the predicted treatment values of $\hat{\boldsymbol{D}}$, respectively. $\boldsymbol{\pi}_{11}$ is the causal effect of the IVs $\boldsymbol{Z}$ on the treatment $\boldsymbol{D}$; and π_{21} is the treatment effect on the outcome y for a subset of participants whose treatment effect has been partialled out and explained by the IVs $\boldsymbol{Z}$. Traditional causal 2SLS model with IVs is commonly estimated using OLS methods. The measurement errors at both stages, e_{1i} and e_{2i}, are assumed to be normally distributed as $e_{1i} \sim N(0, \sigma_{e_1}^2)$ and $e_{2i} \sim N(0, \sigma_{e_2}^2)$.

2.2 *Instrumental Variables (IVs)*

There are two crucial assumptions for selecting valid IVs. One is the instrument exogeneity assumption: the instruments are related to the treatment in some way, so that some variations of the outcome could be explained by the instruments, and the assumption is expressed as $cov(\boldsymbol{D}, \boldsymbol{Z}_j) \neq 0,\ j = 1, \ldots, J$. The other is the exclusion restriction assumption: the instruments are not correlated with other determinants, so that the instruments only affect the outcome through the treatment, and the assumption is expressed as $cov(\boldsymbol{Z}_j,\ other\ determinants) = 0,\ j = 1, \ldots, J$. The instrument exogeneity assumption guarantees that the predicted value $\hat{\boldsymbol{D}}_i$ is related to the treatment $\boldsymbol{D}_i$; and the exclusion restriction assumption ensures that the predicted value $\hat{\boldsymbol{D}}_i$ is uncorrelated with the second stage error e_{2i}.

Finding proper IVs is always a practical challenge, and different approaches have been used to help find IVs. First, in general, the selection of IVs relies on the previous substantive theory. When researchers suspect there are potential important factors being omitted in a causal effect model, they may use their previous knowledge to find variables that are related with the treatment but do not affect the outcome except through the treatment and use these variables as instruments to study the partial treatment effect. Statistically, we can test whether the selected instruments are strong or weak, through the incremental F test (Staiger and Stock 1997). Second, there are noncompliance situations, where some people are treatment always-takers or never-takers regardless of whether they have been assigned to the treatment group or not, so that the errors in the causal model may correlate with the treatment. Only the treatment effect on the treated is, or the participants who have been assigned to the treatment and who have actually take the treatment are, studied. For example, Steel et al. (2010) studied the effect of using Governor's Teaching Fellowship (GTF) as an incentive to attract talented teachers to and retain them in low-performing schools. Because only teachers that meet certain criteria are eligible to apply for the GTF, it is infeasible to estimate the

ATE of GTF for all teachers. The article used the eligibility status as the instrument to study the effect of GTF on the outcome to GTF-eligible-only teachers. In all, using IVs to study the effect of treatment on the treated has been found effective and been introduced to other science fields such as epidemiology (e.g., Greenland 2000). Third, in situations where potential IVs are supported by a strong previous theory but are difficult to locate, researchers were suggested to refer to different data sources for the treatment, outcome, or instrument variables. Duncan et al. (1968) studied the causal effect of adolescents' educational aspirations on their peers. As the previous theory assumes that a person's family background affects its own educational aspirations but not its peer's, the authors used each child's family background as the instrument and found the information source different from the sample data. Similar to this idea, Angrist and Krueger (1992) and Angrist and Imbens (1995) proposed a two-sample instrumental variable technique to locate IV information from different sources. Using the two sample IV technique, Currie and Yelowitz (2000) studied the effect of a public housing voucher program on housing quality and educational attainment. Because the previous theory supports that a household having extra number of kids is entitled to a larger housing unit, whether there are extra kids in the household is chosen as the IV. In the study, the outcome data comes from the Survey of Income and Program Participation, the endogenous treatment data comes from the March Current Population Survey, and both sources contain the IV data.

2.3 Local Average Treatment Effect (LATE)

Despite the use of IVs, the treatment effect is still the causal effect of interest. IVs initiate a causal chain, where the causal effect of IVs on the treatment is first estimated, and in the second causal chain, the partial causal effect of the treatment on the outcome is estimated. When IVs are used in the presence of endogenous regressors, the causal effect of interest becomes the local average treatment effect (LATE). For example, Angrist and Krueger (1990) studied the effect of military service on the labor market earnings during the Vietnam era. Young men were drafted for military service to serve in Vietnam, and later a draft lottery was introduced because of some fairness concerns about the military conscription policy. The draft-eligible men were not necessarily drafted if they had a high lottery number (i.e., above the cutoff lottery number which means they don't need to be drafted). In this study, it was almost impossible to estimate the ATE of the military service for all the draft-eligible men. Instead, the author estimated the LATE of the military service for those who were draft-eligible and who had a low lottery number that actually participated in the service.

Under the 2SLS model, $\hat{\pi}_{21}$ is the estimated LATE, or the estimated ATE for a subset of participants whose variations are explained by $\mathbf{Z}$. In other words, the LATE is the causal effect of interest for the 2SLS model. Although LATE can be estimated through the Wald estimator method or from the reduced form equation (e.g., Angrist and Pischke 2008), there are several advantages in using 2SLS modeling to estimate LATE. First, 2SLS models provide a standard error estimate of the LATE, whereas the Wald estimator only provides a point estimate. Second, by using 2SLS models, covariates could be controlled simultaneously at both stages of the 2SLS model when the effect of $\boldsymbol{Z}$ on $\boldsymbol{D}$ and the effect of $\hat{\boldsymbol{D}}$ on $\boldsymbol{y}$ are estimated. The estimated LATE $\hat{\pi}_{21}$ in 2SLS can be derived as

$$\hat{\pi}_{21} = \frac{cov(y_i, \hat{D}_i)}{var(\hat{D}_i)} = \frac{cov(y_i, \hat{\pi}_{10} + \hat{\pi}_{11}Z_{i1} + \cdots + \hat{\pi}_{1J}Z_{iJ})}{var(\hat{\pi}_{10} + \hat{\pi}_{11}Z_{i1} + \cdots + \hat{\pi}_{1J}Z_{iJ})}$$

$$= \frac{\hat{\pi}_{11}cov(y_i, Z_{i1}) + \cdots + \hat{\pi}_{1J}cov(y_i, Z_{iJ})}{\hat{\pi}_{11}^2 var(Z_{i1}) + \cdots + \hat{\pi}_{1J}^2 var(Z_{iJ})} \tag{3}$$

Mathematically, when the instrument $\boldsymbol{Z}$ is the same as the treatment $\boldsymbol{D}$, the LATE equals the ATE. In other words, when Z has perfect predictions on $\boldsymbol{D}$, the second stage slope estimate $\hat{\pi}_{21}$ is the standard OLS regression slope estimate. From a practical perspective, the instruments $\boldsymbol{Z}$ rarely replace the treatment $\boldsymbol{D}$, as the treatment is usually the research interest of the causal inference.

2.4 Robust Method Based on Student's t Distributions

In a 2SLS model, although we assume that the measurement errors at both stages are normally distributed, practical data usually violate the normality assumption. Routine methods to accommodate nonnormality, such as data transformation or data truncation, can be problematic. For example, transformed data can be difficult to interpret when the raw scores have meaningful scales; and the exclusion of outliers may result in reduced efficiency (e.g., Yuan et al. 2002). Recently, different robust methods have been developed as alternative approaches to provide reliable parameter estimates, the associated standard errors and statistical tests in the presence of nonnormal data. The fundamental idea of the robust procedure is to either model the nonnormality using certain types of nonnormal distributions or assign a weight to each case and properly downweight the cases that are far from the center of the majority of the data (Hampel et al. 1986; Tong and Zhang 2012; Huber 1981).

Robust methods based on Student's t distributions were developed by Lange et al. (1989) and have been applied to complex models such as structural equation modeling, multilevel modeling and growth curve modeling. The shape of a t distribution is controlled by its degrees of freedom. When the degrees of freedom are small, the distribution is flatter and captures more heavy-tailed data. When the degrees of freedom are large, a t distribution approaches a normal distribution. In addition, because methods based on t distributions can be considered as a natural extension of normal-distribution-based methods for heavy-tailed data and t distributions have a parametric form, these methods are relatively straightforward to understand.

For the 2SLS modeling, the equation in the second stage is used to estimate the LATE. In the traditional model estimation, the measurement error is assumed to follow a normal distribution, as $e_{2i} \sim N(0, \sigma_{e_2}^2)$. To deal with heavy-tailed data or outliers, we use a robust method and model the measurement error with a Student's t distribution, so that $e_{2i} \sim T(0, \sigma_{e_2}^2, k)$, where k is the degrees of freedom of the t distribution, and can be set a priori or estimated from the model. Under certain conditions, the degrees of freedom have been recommended setting a priori. Lange et al. (1989) and Zhang et al. (2013) suggested fixing the value for the degrees of freedom of the t distributions when sample size is small, as small sample sizes could lead to a biased degrees of freedom estimate. Moreover, Tong and Zhang (2012) argued that by fixing the degrees of freedom, more accurate parameter estimates and credible intervals can be obtained when model specification is built on solid substantive theories. In contrast, estimating the degrees of freedom can make the model more flexible. When the degrees of freedom k are freely estimated, the Student's t distributions have an additional parameter k, compared with the normal distributions. As the degrees of freedom k increase, the Student's t distribution approaches the normal distribution, and therefore the robust 2SLS model becomes the normal 2SLS model.

3 A Monte Carlo Simulation Study

3.1 Study Design

In this section, the performance of the robust method based on Student's t distributions is evaluated and compared to that of the traditional method in estimating the LATE in the 2SLS model with an IV through a Monte Carlo simulation study. Data are generated from the general causal inference model

$$y_i = 3 + 0.5x_i + e_i,$$

where y_i is the causal outcome, x_i is the causal treatment, and e_i is the measurement error following a standard normal distribution in general. Three potential influential factors are considered. First, sample size (N) is either 200 or 600. Second, correlation between x_i and e_i (ϕ) is manipulated to be either 0.3 or 0.7, reflecting relatively weak or strong linear relationship between the treatment and the measurement error. Third, we manipulate a proportion of observations to contain outliers. For these observations, the measurement error e_i is generated from a different normal distribution with the mean being 8 standard deviations away from the mean of the original normal distribution. The proportion of outliers (OP) is considered to be 0%, 5%, or 10%. When the OP is 0%, data contain no outliers and are normally distributed.

The 2SLS model presented in Eqs. (1) and (2) is used to fit the data. We use one IV in this simulation study for illustration purposes. The IV is generated from a normal distribution and is correlated to the treatment x_i with the correlation coefficient being 0.6. In the first stage, the IV is used to predict the endogenous treatment, and the estimated treatment is then used in the second stage to estimate the LATE. From Eq. (3), the theoretical LATE estimate is $5/6$. Both the traditional 2SLS modeling and the robust 2SLS modeling are applied, and the LATEs are estimated using Bayesian methods. The bias and standard errors (SE) of the LATE estimates for both traditional and robust methods are assessed. The deviance information criterion (DIC) for each condition is also examined to study the model fit. A lower value of DIC indicates a better model fit.

3.2 *Results*

The bias and SEs of the LATE estimates, and DICs from the traditional method and the robust method when $\phi = 0.3$ are given in Table 1. When data are normally distributed (i.e., $OP = 0\%$), the traditional method and the robust method perform almost equally well as they provide similar bias, SEs, and DICs. However, in the presence of outliers, the robust method outperforms the traditional method in terms of the accuracy and efficiency of the LATE estimates and the model fits across all conditions. Overall, the robust method produces smaller bias and SE of the LATE estimate than the traditional method does. Additionally, the DIC from the robust method is much smaller, indicating a better model fit than the traditional method. For example, in the condition $N = 200$ and $OP = 10\%$, the bias and SE from the traditional method are 0.392 and 0.367, respectively, whereas the bias and SE decrease sharply to −0.069 and 0.159, respectively, in the robust method. Also the DIC from the robust method is 175 less than that from the traditional method, suggesting the strong evidence of better model fit when the robust method is applied.

As the proportions of outliers increase, the better performance of the robust method becomes more salient. For example, when $N = 600$ and $OP = 5\%$, the bias in the traditional and robust methods are 0.199 and −0.029, respectively, a 0.228

Table 1 Bias, SEs of the LATE estimates and DICs for all the conditions when $\phi = 0.3$

		Traditional method			Robust method		
N	OP	Bias	SE	DIC	Bias	SE	DIC
200	0%	0.023	0.147	1141.558	0.021	0.147	1142.157
	5%	0.213	0.288	1381.255	−0.028	0.152	1242.425
	10%	0.392	0.367	1487.053	−0.069	0.159	1312.132
600	0%	0.007	0.078	3417.444	0.004	0.078	3418.617
	5%	0.199	0.155	4123.277	−0.029	0.083	3705.855
	10%	0.384	0.199	4446.370	−0.056	0.087	3916.545

Table 2 Bias, SEs of the LATE estimates and DICs for all the conditions when $\phi = 0.7$

		Traditional method			Robust method		
N	OP	Bias	SE	DIC	Bias	SE	DIC
200	0%	0.044	0.097	1167.509	0.041	0.098	1168.137
	5%	0.509	0.226	1388.200	−0.068	0.137	1262.594
	10%	0.921	0.321	1490.574	−0.139	0.150	1334.795
600	0%	0.017	0.057	3493.816	0.013	0.057	3495.078
	5%	0.485	0.115	4151.484	−0.039	0.073	3777.674
	10%	0.872	0.160	4455.258	−0.122	0.079	3984.497

difference in magnitude; when *OP* becomes 10%, the bias in the robust method is −0.056, while the bias in the traditional method increases to 0.384, a 0.440 difference in magnitude. This suggests that the robust method produces a much less biased LATE estimate. The SE and DIC follow similar patterns, namely, when *OP* increases, the robust method is preferred to the traditional method in terms of SE and DIC. Similarly, given different sample sizes, the robust method provides less biased LATE estimates and smaller SEs and DICs. The advantage of the robust method is more apparent under small sample conditions.

Table 2 presents results from the traditional and robust methods for all the conditions when $\phi = 0.7$. Consistent with the results from previous conditions when $\phi = 0.3$, when data contain outliers, the robust method produces more accurate and efficient LATE estimates and better model fits than the traditional method does. The advantage of the robust method is clearer when sample size is small and the proportion of outliers is large.

Furthermore, comparing Tables 1 with 2, it is shown that the strength of the covariance between the endogenous treatment and the error term affects the performances of the two methods in estimating the LATE. When the endogenous correlation between the treatment and the corresponding error is high (i.e., $\phi = 0.7$), the LATE estimates are more biased. In addition, the robust method is better at providing more accurate LATE estimates when the endogenous correlation is high. For example, in the conditions where $N = 200$ and $OP = 5\%$, when $\phi = 0.3$, the robust method has 0.185 less bias than the traditional method; when $\phi = 0.7$, the robust method has 0.441 less bias.

4 Concluding Comments

In sum, the 2SLS model with IVs is a widely used approach to address the issue of treatment endogeneity in causal inference research. In 2SLS modeling with IVs, LATE becomes the causal effect of interest. Because practical data usually violate the normality assumption, the LATE estimate from the traditional method may be inefficient or even biased. This study proposed a robust Bayesian method using

Student's t distributions to model data that are contaminated. The Monte Carlo simulation study shows that the proposed robust Bayesian method outperforms the traditional method when data contain outliers. The robust method is especially preferred when sample size is small and the proportion of outliers is large as it produces more accurate and efficient LATE estimates and better model fits. When data are normally distributed, the performance of the robust methods is about the same as the traditional method. Consequently, we recommend using the robust method in general in the 2SLS modeling with IVs.

Note that we need to be cautious when using Student's t distributions to accommodate the effects of outliers because Student's t distributions are sensitive to the skewness. If data are highly skewed, some alternative robust methods might be considered, such as robust methods based on skewed-t distributions (Azzalini and Genton 2008). We also want to note that we consider the nonnormality only in the second stage of the 2SLS modeling; however, nonnormality may also come from the first stage. A future study could extend the robust procedure to model the nonnormality at both stages.

References

D. Angrist, G. Imbens, Two stage least squares estimation of average causal effects in models with variable treatment intensity. J. Am. Stat. Assoc. **90**, 431–442 (1995)

D. Angrist, A.B. Krueger, Does compulsory school attendance affect schooling and earnings? Tech. Rep. Natl. Bur. Econ. Res. **10**, 1–14 (1990)

D. Angrist, A.B. Krueger, The effect of age at school entry on educational attainment: an application of instrumental variables with moments from two samples. J. Am. Stat. Assoc. **87**, 328–336 (1992)

D. Angrist, J. Pischke, *Mostly Harmless Econometrics: An Empiricist's Companion* (Princeton University Press, Princeton, 2008)

D. Angrist, J. Pischke, *Mastering Metrics: The Path from Cause to Effect* (Princeton University Press, Princeton, 2014)

D. Angrist, G. Imbens, D. Rubin, Identification of causal effects using instrumental variables. J. Am. Stat. Assoc. **91**, 444–455 (1996)

A. Azzalini, M.G. Genton, Robust likelihood methods based on the sked-t and related distributions. Int. Stat. Rev. **76**, 106–129 (2008)

D. Card, *Using Geographical Variation in College Proximity to Estimate the Return to Schooling. Aspects of Labor Market Behavior: Essays in Honor of John Vanderkamp* (University of Toronto Press, Toronto, 1995)

J. Currie, A.Yelowitz, Are public housing projects good for kids? J. Public Econ. **75**, 99–124 (2000)

O.D. Duncan, A.O. Haller, A. Portes, Peer influence on aspirations: a reinterpretation. Am. J. Sociol. **74**, 119–137 (1968)

S. Greenland, An introduction to instrumental variables for epidemiologists. Int. J. Epidemiol. **29**, 722–729 (2000)

F.R. Hampel, E.M. Ronchetti, P.J. Rousseeuw, W.A. Stahel, *Robust Statistcs: The Approach Based on Influence Functions* (Wiley, New York, 1986)

P.J. Huber, *Robust Statistics* (Wiley, New York, 1981)

G. Imbens, D. Rubin, Bayesian inference for causal effects in randomized experiments with noncompliance. Ann. Stat. **29**, 305–327 (1997)

K.L. Lange, R.J.A. Little, J.M.G. Taylor, Robust statistical modeling using the t distribution. J. Am. Stat. Assoc. **84**(408), 881–896 (1989)

S.Y. Lee, Y.M. Xia, Maximum likelihood methods in threating outliers and symmetrically heavy-tailed distributions for nonlinear structural equation models with missing data. Psychometrika **71**, 565–585 (2006)

J.C. Pinheiro, C. Liu, Y.N. Wu, Efficient algorithms for robust estimation in linear mixed-effects models using the multivariate t distribution. J. Comput. Graph. Stat. **10**(2), 249–276 (2001)

D. Rachman-Moore, R.G. Wolfe, Robust analysis of a nonlinear model for multilevel educational survey data. J. Educ. Stat. **9**(4), 277–293 (1984)

M. Seltzer, K. Choi, Sensitivity analysis for hierarchical models: downweighting and identifying extreme cases using the t distribution. Multilevel Model. Methodol. Adv. Issues Appl. **13**, 25–52 (2003)

S. Shoham, Robust clustering by deterministic agglomeration em of mixtures of multivariate t-distributions. Pattern Recogn. **35**, 1127–1142 (2002)

P. Song, P. Zhang, A. Qu, Maximum likelihood inference in robust linear mixed-effects models using multivariate t-distribution. Stat. Sin. **17**, 929–943 (2007)

D.O. Staiger, J.H. Stock, Instrumental variables regression with weak instruments. Econometrica **65**, 557–586 (1997)

J. Steele, R. Murnane, J. Willet, Do financial incentives help low-performing schools attract and keep academically talented teachers? Evidence from California. J. Policy Anal. Manag. **29**, 451–478 (2010)

X. Tong, Z. Zhang, Diagnostics of robust growth curve modeling using student's t distribution. Multivar. Behav. Res. **47**, 493–518 (2012)

H.X. Wang, Q.B. Zhang, B. Luo, S. Wei, Robust mixture modelling using multivariate t-distributions with missing information. Pattern Recogn. Lett. **25**, 701–710 (2004)

J. Wang, Z. Lu, A.S. Cohen, The sensitivity analysis of two-level hierarchical linear models to outliers, in *Quantitative Psychology Research*. Springer Proccedings in Mathematics and Statistics (Springer, Cham, 2015), pp. 307–320

K.-H. Yuan, P.M. Bentler, Structural equation modeling with robust covariances. Sociol. Methodol. **28**, 363–396 (1998)

K.-H. Yuan, Z. Zhang, Structural equation modeling diagnostics using R package semdiag and EQS. Struct. Equ. Model. **19**, 683–702 (2012)

K.-H. Yuan, L.L. Marshall, P.M. Bentler, A unified approach to exploratory factor analysis with missing data, nonnormal data, and in the presence of outliers. Psychometrika **67**, 95–111 (2002)

K.-H. Yuan, P.L. Lambert, R.T. Fouladi, Mardia's multivariate kurtosis with missing data. Multivar. Behav. Res. **39**, 413–437 (2004)

Z. Zhang, K. Lai, Z. Lu, X. Tong, Bayesian inference and application of robust growth curve models using student's t distribution. Struct. Equ. Model. **20**, 47–78 (2013)

D. Zimmerman, A note on the influence of outliers on parametric and nonparametric tests. J. Gen. Psychol. **121**, 391–401 (1994)

D. Zimmerman, Invalidation of parametric and nonparametric statistical tests by concurrent violation of two assumptions. J. Exp. Educ. **67**, 55–68 (1998)

J. Zu, K.-H.Yuan, Local influence and robust procedures for mediation analysis. Multivar. Behav. Res. **45**, 1–44 (2010)

Measuring Grit Among First-Generation College Students: A Psychometric Analysis

Brooke Midkiff, Michelle Langer, Cynthia Demetriou, and A. T. Panter

Abstract The concept of grit is of interest in the field of education, particularly as it pertains to persistence to a 4-year college degree. This study offers an IRT analysis of the Grit Scale when used among first-generation college students (FGCSs) as well as recent first-generation college graduates and non-FGCS recent graduates. The Grit Scale was included in surveys administered as part of an array of other research projects within The Finish Line Project—a US Department of Education First in the World grant-funded project that seeks to improve FGCS access to, persistence in, and completion of postsecondary education through rigorous research into various programs and supports for FGCSs. The reliability and validity of the Grit Scale have not yet been analyzed for use with FGCS or overall with students at large, research universities. By comparing enrolled students and recent graduates, the psychometric analysis in this study offers insight into the measurement of student grit for use in program development and policy-making to improve student retention. Item response theory (IRT) analyses, analysis of differential item functioning (DIF), reliability analyses, convergent and discriminant validity analyses, and known groups validity analyses were used to examine the Grit Scale.

Keywords Grit • First-generation college students • Item response theory • Differential item functioning

1 Factor Structure and Uses of the Grit Scale

The latent construct of grit is reported to be comprised of two elements—perseverance of effort and consistency of interest in the original research into grit (Duckworth et al. 2007). Grit has been shown to be an effective predictor of success and retention in a variety of contexts such as the national spelling bee, military,

B. Midkiff (✉) • C. Demetriou • A.T. Panter
University of North Carolina at Chapel Hill, Chapel Hill, NC 27599, USA
e-mail: bmidkiff@email.unc.edu; cyndem@email.unc.edu; panter@email.unc.edu

M. Langer
American Institutes for Research, Washington, DC 20007, USA
e-mail: mlanger@air.org

L.A. van der Ark et al. (eds.), *Quantitative Psychology*, Springer Proceedings in Mathematics & Statistics 196, DOI 10.1007/978-3-319-56294-0_35

workplace, school, and marriage (Duckworth et al. 2011; Eskreis-Winkler et al. 2014; Strayhorn 2014). At the same time, first-generation college students (FGCSs) experience lower likelihoods of completing a 4-year degree when compared to their continuing-generation peers (Chen and Carroll 2005; D'Amico and Dika 2013; Engle and Tinto 2008; Vaughan et al. 2014). Therefore, the measurement of grit among this population has the potential to uncover underlying issues that impact FCGS retention, which could lead to improved interventions and supports. Additionally, within the extant literature on grit, a recent meta-analytic study states that "the grit literature may benefit from a refinement of the Grit Scale using methods based on Item Response Theory (IRT)" (Credé et al. 2016). In this chapter, we examine the Grit Scale (Duckworth et al. 2007; Duckworth and Quinn 2009) when used among FGCSs, a population as yet unstudied in conjunction with grit, compared to non-FGCSs, and offer an IRT analysis useful for the emergent area of research into refinements of the Grit Scale.

1.1 Predictive Validity of Grit in the Extant Literature

Previous research has provided evidence of the predictive validity of the Grit Scale on metacognition (Arslan et al. 2013), the retention of first-year military cadets (Maddi et al. 2012), educational attainment among adults, and grade point average (GPA) among Ivy League undergraduates (studied by Duckworth et al. 2007). However, these populations and scenarios differ in important ways from FGCSs pursuing a baccalaureate degree at a large, public research university. For example, the distribution and variance of undergraduate GPAs among students at Ivy League universities are likely to differ from those of FGCSs at a public university, as students in the research conducted by Duckworth et al. (2007) are positioned in the most advantageous university settings, with student populations, largely derived from the most advantaged high school students in the United States. Similarly, research done by Arslan et al. (2013) studied grit and metacognition among college students, yet is contextually different in that it focuses on Turkish university students—a group likely to have nontrivially different cultural norms than the average FGCS at a public university in the United States. Other studies of grit, such as those around the retention of military cadets and spelling bee champions, present radically different contexts from that of FGCSs completing a baccalaureate degree. Further, while the previous studies mentioned here demonstrate successful use of the Grit Scale, none offer an IRT analysis that can more deeply examine the psychometric properties of the scale.

1.2 Factor Structure of the Grit Scale

In line with research by Credé et al. (2016) which calls for IRT analyses of grit, the research presented here offers insight into the existing scholarly disagreement over grit as a latent construct—whether it is substantively different from conscientiousness, its incremental validity, and even its factor structure. For example, Duckworth and Quinn (2009) used confirmatory factor analysis (CFA) to determine that grit has a higher-order structure with two first-order factors and one second-order factor. Yet this finding is tempered by the fact that, using CFA, a higher-order model would exhibit identical fit to a model using two correlated first-order factors and no higher-order factor, making the analysis of limited utility (Credé et al. 2016). In fact, such a factor structure was examined by Duckworth et al. (2007) and found to have poor fit based on comparative fit index (0.83) and root mean square error of approximation (RMSEA) (0.11). Credé et al. (2016) suggest that a more meaningful way to assess the factor structure would be to examine the correlation between the two theoretical components of grit—perseverance of effort and consistency of interest. However, empirical estimates of this correlation summarized in the meta-analysis by Credé et al. (2016) find that the strength of the correlation has wide variation, with the correlation dropping as low as zero in some empirical studies (Chang 2014; Datu et al. 2016; Jordan et al. 2015). IRT analysis offers insights into the factor structure of the scale to shed light on the contradictory findings from previous CFA analyses.

1.3 Grit and FGCSs

Despite substantial debate over grit as an important noncognitive factor related to student success, grit has become an important buzzword in education as both an explanation for student achievement and as an intervention (Anderson et al. 2016). As the role of grit in education gains popular attention, it is important to understand if the measurement of grit is both reliable and valid among populations whose retention has historically been at greater risk. This research seeks to assess the validity and reliability of the Grit Scale among one such group—FGCSs—and to answer the overarching research question: What are the psychometric properties of the Grit Scale when used among FGCSs? To answer this question, we examine the reliability and factor structure of the scale and test for local dependence and differential item functioning.

2 Methods

IRT analysis was used to examine the psychometric properties of the items on the Grit Scale. IRT also allows for the assessment of local dependence between items. Analysis of DIF through logistic regression was used to test the validity of the scale for FGCSs and non-FGCSs, as well as by race/ethnicity and gender.

2.1 *Data*

A total of 648 participants completed a version of the Grit Scale. The sample consisted of 190 undergraduates who completed a survey at the end of their first year of college that contained the 12-item Grit Scale (Duckworth et al. 2007) in Spring 2015. An additional 458 recent graduates completed a survey in Fall 2015 that contained 9 items taken from the 12-item Grit Scale.

2.1.1 Differences in Scale Administration

The 9-item scale is a subset of the 12 items from the published 12-item scale (Duckworth et al. 2007). However, the 9-item scale used response options "very much like me" (1) to "not like me at all" (5), while the 12-item scale used response options "strongly disagree" (1) to "strongly agree" (5). In addition to the different response wording, responses also differed in direction. The 12-item scale contained a neutral option "neither disagree nor agree," although the 9-item scale did not. Lastly, different prompts were used between the two administrations. The 9-item scale was preceded by the following prompt: "Please indicate how true the following statements are for you. Rate each statement." The 12-item scale was preceded by the following prompt: "Here are a number of statements that may or may not apply to you. For the most accurate score, when responding think of how you compare to most people—not just the people you know well, but most people in the world. There are no right or wrong answers, so just answer honestly!" Relevant items were reverse coded so that higher scores reflect more grit across both administrations.

2.1.2 Descriptive Statistics

Of those who answered at least one Grit Scale item, <1% (four students) skipped one or more items. For the IRT analyses, only participants who did not respond to any Grit Scale item were removed from the analyses. For DIF analyses, participants with any missingness on the Grit Scale were listwise deleted. The scale items, along with associated sample size and mean scores, are given in Table 1. Differences in sample sizes are due to the combination of different surveys as explained previously.

2.1.3 Demographics

Of the sample who completed at least one Grit Scale item, 155 FGCSs completed the 12-item scale, 182 FGCS recent graduates completed the 9-item scale, and 254 non-FGCS recent graduates completed the 9-item scale. Participants included 189 men and 402 women. There were 361 non-Hispanic, White participants compared to 234 other races and ethnicities. When disaggregated by FGCS status, there were 333 FGCSs compared to 268 non-FGCSs.

Table 1 Combined scale sample item means

Item	N	Mean	Std. Dev.	Minimum	Maximum
I have overcome setbacks to conquer an important challenge	609	4.05	0.90	1	5
New ideas and projects sometimes distract me from previous ones	610	2.74	0.98	1	5
My interests change from year to year	174	2.73	1.16	1	5
Setbacks do not discourage me	609	3.15	1.01	1	5
I have been obsessed with a certain idea or project for a short time but later lost interest	610	2.98	1.03	1	5
I am a hard worker	608	4.45	0.75	1	5
I often set a goal but later choose to pursue a different one	610	3.23	0.96	1	5
I have difficulty maintaining my focus on projects that take more than a few months to complete	174	3.01	1.11	1	5
I finish whatever I begin	610	3.82	0.89	1	5
I have achieved a goal that took years of work	609	4.10	1.03	1	5
I become interested in new pursuits every few months	609	2.85	1.01	1	5
I am diligent	174	4.25	0.77	1	5

2.2 Analytic Strategy

2.2.1 IRT

Structural validity was evaluated with a factor analytic approach using IRT, implemented using the software IRTPRO (Cai et al. 2011). Each subscale of the Grit Scale was examined in a unidimensional confirmatory factor analysis (CFA) with the graded response IRT model (Samejima 2010) as well as within a bifactor graded response IRT model to examine if one underlying factor explained most of the variability in the two subscales. For the bifactor model, the explained common variance (ECV) of the general factor was examined and unidimensionality considered for values greater than 0.85. In addition to inspection of item content to assess local independence, the LD χ^2 proposed by Chen and Thissen (1997) wherein values larger than 10 are considered evidence of local dependence and values between 5 and 10 may indicate either local dependence or sparseness in the underlying table of frequencies was used as the statistical criteria for determining local dependence.

Item fit was assessed based on the SS-χ^2 fit statistics proposed by Orlando and Thissen (2000, 2003), for which a nonsignificant result ($p > 0.05$, adjusted for multiple comparisons) was an indicator of adequate model fit. Model fit was determined piecemeal through item fit, as well as the M_2 statistic proposed by

Maydeu-Olivares and Joe (2005, 2006) and its associated RMSEA. Additionally, the −2 log likelihood, the Akaike information criteria (AIC) (Akaike 1974), and the Bayesian information criterion (BIC) (Schwarz 1978) were also examined.

2.2.2 DIF

Analysis of potential DIF by FGCS status, gender, and race/ethnicity was conducted using ordinal logistic regression (OLR) of summed scores. For each item within a domain, an OLR model was used to examine whether item responses were significantly associated with group membership after controlling for students' summed score on the measure. Uniform DIF was detected by a likelihood ratio test comparing an OLR model with one predictor, summed score, to an OLR model with an additional predictor, group membership, representing a shift in the use of the response options due to group membership. Nonuniform DIF was detected by a likelihood ratio test comparing the OLR model with two predictors, summed score and group membership, to an OLR model with an additional interaction term, representing a difference in how strongly the item is related to the underlying construct due to group membership. With each paired-group analysis, an initial OLR model was run to identify a clean anchor group of items without DIF. For each sequential OLR model, any items previously identified as having DIF were removed from the summed score computation. The final OLR model used a summed score computed with only the DIF-free anchor items to test for DIF. Subsequent OLR models used a summed score computed with only the clean anchor items to test for DIF. The Benjamini-Hochberg procedure was used to make inferential decisions in the context of the multiple comparisons. In addition to examining the significance ($p < 0.05$), magnitude of DIF was further evaluated by examining the expected item scores and estimating the effect sizes ($\Delta R2 > 0.02$ indicative of salient DIF).

2.2.3 Reliability

Internal consistency was evaluated by Cronbach's α for both versions of the scale, as well as for both subscales within each version using the software Mplus (Muthén and Muthén 1998). Alpha values of 0.70 or greater are an acceptable minimum for group-level assessment.

2.2.4 Convergent and Discriminant Validity

Participants who completed the Grit Scale also completed the Growth Mindset Scale (Dweck 2008) and the 5-item Guilt Proneness Scale (Cohen et al. 2011; Cohen et al. 2014). To assess convergent validity, for both grit subscales, the correlation between the mean item score and the mean item score of the Growth Mindset Scale was computed. To assess divergent validity, for both grit subscales, the correlation between the mean item score and the mean item score of the Guilt Proneness Scale was computed. Growth Mindset response items ranged from "strongly disagree" (1) to "strongly agree" (5) and were scored such that higher mean scores reflect more growth mindset. Guilt Proneness response items ranged from "extremely unlikely" (1) to "extremely likely" (5) and were scored such that higher mean scores reflect more guilt proneness.

2.2.5 Known Groups Validity

The validity of the Grit Scale was examined by assessing the extent to which it could discriminate between several known groups that should, in theory, differ. These groups included FGCSs who were current students, FGCS recent graduates, non-FGCS recent graduates, race and ethnicity, race and ethnicity interacted with gender, and participants grouped by their reported use of university resources. The use of university resources was measured by a list of resources and the frequency with which students used them while in college, with responses ranging from "never" (1) to "ten or more times" (6). We compared means across all groups using a one-way analysis of analysis of variance (ANOVA). Statistical significance was defined at the 0.05 alpha level for evaluation of convergent, discriminant, and known groups validity.

3 Results

3.1 Item Response Theory

Two items on the perseverance of effort subscale had low cell counts in the extreme categories, and thus for those two items, categories were collapsed for IRT analysis. The factor structure was further examined by estimating the graded response model (Samejima 2010), on the item scores for each grit subscale, and then modeling all scores using a bifactor model. Table 2 shows the fit of the graded response IRT models fit to the Grit Scale. The bifactor model fit well, although one item had a high factor loading (0.73) on the overall factor and a low loading (0.25) on the perseverance of effort subscale. The ECVs for this model were 0.45 for consistency of interest and 0.63 for perseverance of effort, suggesting these two subscales do not support one underlying factor. Therefore, unidimensional IRT models were fit to each of the Grit Scale subscales, including both a 5-item and 4-item (dropping the problematic item identified in the bifactor model) version of the perseverance of effort subscale. As Table 2 indicates, these models also fit the data well.

The IRT parameters of the bifactor model as well as the final two unidimensional models for the Grit Scale subscales are presented in Table 3. For the consistency of interest subscale, all items have high IRT a parameters; the item "I have been

Table 2 Comparison of IRT graded response models

Model	Consistency of interest subscale	Perseverance of effort subscale	Perseverance of effort subscale	Bifactor model
Number of items	4	4	5	9
Item fit				
Local dependence	None	None	None	None
SS-$\chi 2$ item fit	All nonsignificant*	All nonsignificant*	One item does not fit well*	All nonsignificant*
Model fit				
M_2	231.25	165.58	277.25	1081.81
RMSEA	0.05	0.05	0.05	0.05
−2 log likelihood	6287.33	5596.55	6974.20	13121.35
AIC	6327.33	5632.55	7018.20	13223.35
BIC	6415.60	5712.00	7115.29	13448.43

Notes: (*) indicates the use of Benjamini-Hochberg to control multiple comparisons

Table 3 IRT parameters

Item	Bifactor		Consistency of interest subscale		Perseverance of effort subscale									
			a parameters				*b* parameters							
	a	*s.e.*	*a*	*s.e.*	*a*	*s.e.*	b_1	*s.e.*	b_2	*s.e.*	b_3	*s.e.*	b_4	*s.e.*
New ideas and projects sometimes distract me from previous ones	0.78	0.14	1.29	0.13	–	–	−2.14	0.19	−0.35	0.09	1.21	0.12	3.44	0.34
I have been obsessed with a certain idea or project for a short time but later lost interest	1.25	0.19	2.22	0.25	–	–	−1.76	0.13	−0.54	0.07	0.49	0.07	2.26	0.17
I often set a goal but later choose to pursue a different one	1.46	0.20	1.72	0.17	–	–	−2.4	0.19	−1.05	0.1	0.21	0.07	2.35	0.19
I become interested in new pursuits every few months	0.15	0.15	1.10	0.12	–	–	−2.35	0.23	−0.61	0.11	0.94	0.12	4.08	0.45
Setbacks don't discourage me	0.68	0.12	–	–	2.09	0.24	−2.77	0.24	−1.74	0.13	−0.91	0.08	0.12	0.07
I have achieved a goal that took years of work	1.20	0.19	–	–	2.25	0.29	−2.63	0.23	−1.57	0.12	−0.25	0.06	–	–
I have overcome setbacks to conquer an important challenge	0.32	0.17	–	–	0.66	0.10	−5.16	0.81	−1.64	0.27	0.81	0.18	3.68	0.56
I am a hard worker	1.56	0.27	–	–	1.30	0.14	−2.74	0.26	−1.07	0.11	0.49	0.09	–	–
I finish whatever I begin	1.97	0.30	–	–	–	–	–	–	–	–	–	–	–	–

obsessed with a certain idea or project for a short time but later lost interest" has the strongest relationship with the underlying construct. For the perseverance of effort subscale, the items also have strong IRT a parameters. However, the item "I am a hard worker" has a slightly weaker relationship to the underlying construct, although it also measures the lower end of the Grit Scale ($b_1 = -5.16$).

3.1.1 Overall Test Information Curves

Figure 1 shows the overall test information curve and standard error for the consistency of interest subscale. This figure indicates that test information is high across the range of consistency of interest, providing adequate measurement from −3 to 3 SDs below and above the mean. Figure 2 shows the same information for the 4-item version of the perseverance of effort subscale, after dropping the poorly fitting item "I finish whatever I begin." However, test information is only high for the perseverance of effort subscale at the lower end of the scale, indicating that the Grit Scale measures perseverance of effort well for those with levels at 1 SD above the mean and below.

3.2 Differential Item Functioning

Analysis of potential DIF resulted in only one item being flagged. The item, "I have overcome setbacks to conquer an important challenge," had significant uniform DIF (p.03) with a large effect size for FGCSs ($n = 333$) and non-FGCSs ($n = 268$). Figure 3 displays the item means by summed score for each of these groups. FGCSs have slightly higher item means across the score range, indicating they are slightly more likely to endorse this item than non-FGCSs, resulting in a higher overall score on the Grit Scale.

3.3 Reliability

Cronbach's alphas are adequate for both administered versions of the scale: 0.72 for the 12-item scale and 0.65 for the 9-item scale. The perseverance of effort subscale had lower alphas (0.60, 0.57) in both versions of the scale administered than did the consistency of interest subscale (0.69 in both versions).

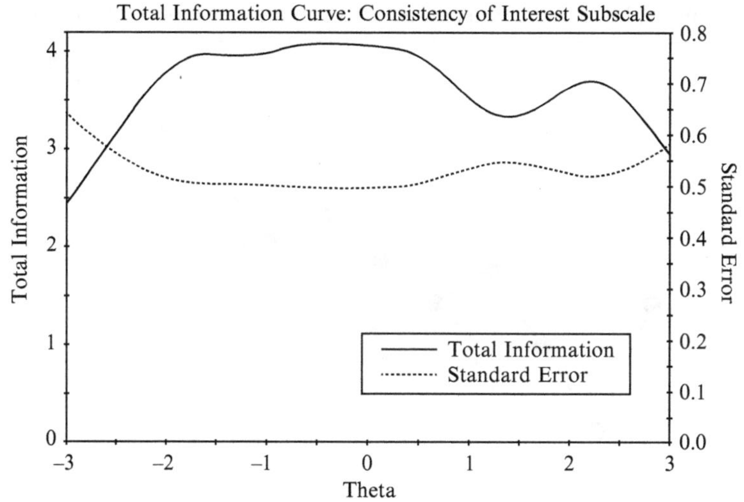

Fig. 1 Total information curve for consistency of interest subscale

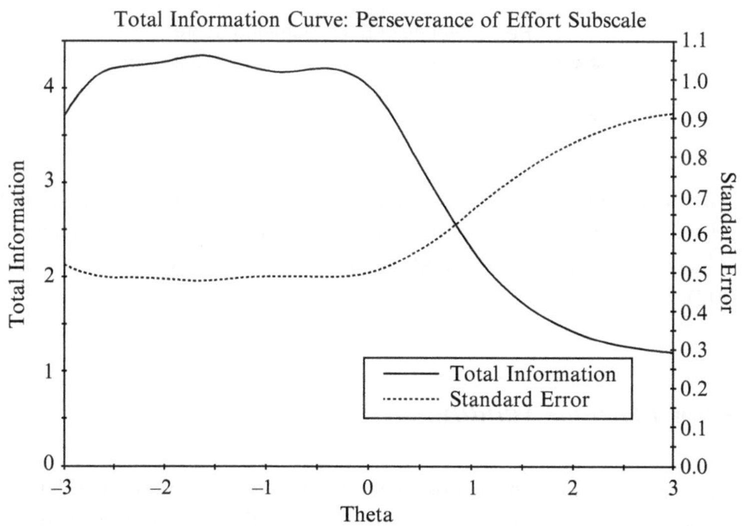

Fig. 2 Total information curve for perseverance of effort subscale

3.4 Convergent and Discriminant Validity

The mean item score on growth mindset was 3.55 (s.d. 0.84). It was significantly correlated with perseverance of effort ($r = 0.20$, $p = 0.001$), but not with consistency of interest ($r = 0.01, p = 0.73$). This indicates grit and growth mindset are moderately related; however, it also provides evidence of the difference between

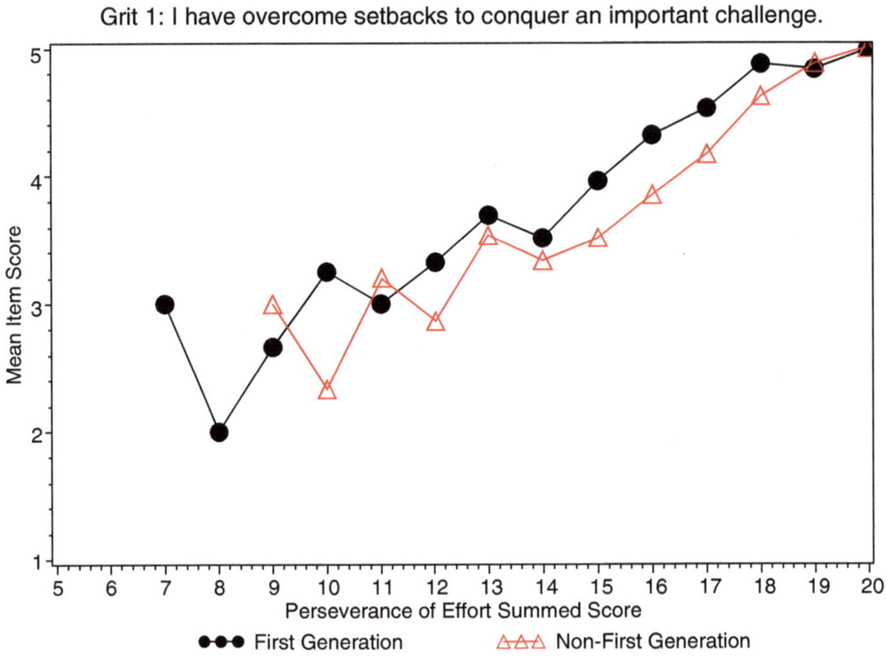

Fig. 3 DIF between FGCSs and non-FGCSs

the two Grit Scale subscales. The mean item score on guilt proneness was 4.12 (s.d. 0.74) and was significantly correlated with both subscales. Perseverance of effort correlated at $r = 0.19$ ($p < 0.001$) and consistency of interest correlated at $r = 0.12$ ($p = 0.002$).

3.5 *Known Groups Validity*

The results from the known groups analysis are given in Table 4. Of the groups analyzed (see Sect. 2.2.5), only two were statistically significantly different from one another. On the perseverance of effort subscale, men showed less perseverance of effort than females (mean = 3.85 vs. 3.97). In consistency of interest, the interaction of gender, race/ethnicity, and FGCS status resulted in statistically significant differences. Women who were White, non-Hispanic, and not FGCSs scored the highest in consistency of interest (mean = 3.13), while men who were non-White and FGCSs scored the lowest (mean = 2.80). Across all gender, race/ethnicity, and FGCS groupings, non-FGCSs (means range from 3.13 to 2.94) scored the highest in consistency of interest than FGCS (means range from 2.92 to 2.80).

Table 4 ANOVA results of known groups

Groups	DF	Sum of squares	Mean square	F	Pr > F
Perseverance of effort subscale					
Male, White, non-Hispanic, FGCS	7	8.25	1.18	2.29	0.0261
Consistency of interest subscale					
Gender	1	1.64	1.64	4.16	0.0418

Notes: Benjamini-Hochberg procedure used to control for Type I error rate in multiple comparisons. Only statistically significant differences after correction are provided; no other group differences were significant after the use of Benjamini-Hochberg procedure

4 Discussion

The IRT analysis suggests that the 4-item consistency of interest subscale fits well with no local dependence, as does the 4-item perseverance of effort subscale, after dropping the poorly fitting item "I finish whatever I begin." The item "I have overcome setbacks to conquer an important challenge" exhibits uniform DIF among FGCS and non-FGCS with a large effect size. Consistent with previous literature (see Credé et al. 2016 for a comprehensive overview), our findings indicate that the higher-order factor structure suggested by Duckworth and Quinn (2009) is not supported. IRT analysis provided here demonstrates that two unidimensional subscales fit better than the bifactor model. Factor loadings from the IRT analysis suggest that there is little evidence of a higher-order construct, when using the data analyzed in this research. Given the discrepancies in previous research about the factor structure of the Grit Scale, along with the recent notation that IRT analysis is needed, this research contributes the important finding that IRT analysis, conducted using these data, does not support a factor structure wherein consistency of interest and perseverance of effort load onto the higher-order construct grit.

References

H. Akaike, A new look at the statistical model identification. IEEE Trans. Autom. Control **19**(6), 716–723 (1974)

C. Anderson, A.C. Turner, R.D. Heath, C.M. Payne, On the meaning of grit... and hope... and fate control... and alienation... and locus of control... and... self-efficacy... and... effort optimism... and.... Urban Rev. **48**(2), 198–219 (2016). doi:10.1007/s11256-016-0351-3

S. Arslan, A. Akin, N. Çitemel, The predictive role of grit on metacognition in Turkish university students. Stud. Psychol. **55**(4), 311 (2013)

L. Cai, D. Thissen, S.H.C. DuToit, *ITPRO 3 (Version 3) [Windows]* (Scientific Software International, Lincolnwood, IL, 2011)

W. Chang, Grit and academic performance: is being grittier better?, Ph.D. Dissertation, University of Miami, Miami, FL, 2014

X. Chen, C.D. Carroll, First-generation students in postsecondary education: a look at their college transcripts. Postsecondary Education Descriptive Analysis Report. NCES 2005-171. National Center for Education Statistics (2005)

W.-H. Chen, D. Thissen, Local dependence indexes for item pairs using item response theory. J. Educ. Behav. Stat. **22**(3), 265–289 (1997)
T.R. Cohen, S.T. Wolf, A.T. Panter, C.A. Insko, Introducing the GASP scale: a new measure of guilt and shame proneness. J. Pers. Soc. Psychol. **100**(5), 947–966 (2011). doi:10.1037/a0022641
T.R. Cohen, Y. Kim, A.T. Panter, *The Five-Item Guilt Proneness Scale (GP-5)* (Carnegie Mellon University, Pittsburgh, PA, 2014), p. 1
M. Credé, M.C. Tynan, P.D. Harms, Much ado about grit: a meta-analytic synthesis of the grit literature. J. Pers. Soc. Psychol. (2016). doi:10.1037/pspp0000102
M.M. D'Amico, S.L. Dika, Using data known at the time of admission to predict first-generation college student success. J. Coll. Stud. Retent. Res. Theory Pract. **15**(2), 173–192 (2013)
J.A.D. Datu, J.P.M. Valdez, R.B. King, Perseverance counts but consistency does not! Validating the short grit scale in a collectivist setting. Curr. Psychol. **35**(1), 121–130 (2016)
A.L. Duckworth, P.D. Quinn, Development and validation of the short grit scale (Grit–S). J. Pers. Assess. **91**(2), 166–174 (2009). doi:10.1080/00223890802634290
A.L. Duckworth, C. Peterson, M.D. Matthews, D.R. Kelly, Grit: perseverance and passion for long-term goals. J. Pers. Soc. Psychol. **92**(6), 1087 (2007)
A.L. Duckworth, T.A. Kirby, E. Tsukayama, H. Berstein, K.A. Ericsson, Deliberate practice spells success why grittier competitors triumph at the national spelling bee. Soc. Psychol. Personal. Sci. **2**(2), 174–181 (2011)
C.S. Dweck, *Mindset: The New Psychology of Success*, Ballantine Books Trade Pbk edn. (Ballantine Books, New York, NY, 2008)
J. Engle, V. Tinto, Moving beyond access: college success for low-income, first-generation students. Pell Institute for the Study of Opportunity in Higher Education (2008)
L. Eskreis-Winkler, E.P. Shulman, S.A. Beal, A.L. Duckworth, The grit effect: predicting retention in the military, the workplace, school and marriage. Front. Psych. **5** (2014). doi:10.3389/fpsyg.2014.00036
M.H. Jordan, T. Gabriel, R. Teasley, W.J. Walker, M. Schraeder, An integrative approach to identifying factors related to long-term career commitments: a military example. Career Dev. Int. **20**(2), 163–178 (2015)
S.R. Maddi, M.D. Matthews, D.R. Kelly, B. Villarreal, M. White, The role of hardiness and grit in predicting performance and retention of USMA cadets. Mil. Psychol. **24**(1), 19 (2012)
A. Maydeu-Olivares, H. Joe, Limited-and full-information estimation and goodness-of-fit testing in 2 n contingency tables: a unified framework. J. Am. Stat. Assoc. **100**(471), 1009–1020 (2005)
A. Maydeu-Olivares, H. Joe, Limited information goodness-of-fit testing in multidimensional contingency tables. Psychometrika **71**(4), 713–732 (2006)
L.K. Muthén, B.O. Muthén, *Mplus User's Guide*, 7th edn. (OECD, Los Angeles, CA, 1998)
M. Orlando, D. Thissen, Likelihood-based item-fit indices for dichotomous item response theory models. Appl. Psychol. Meas. **24**(1), 50–64 (2000)
M. Orlando, D. Thissen, Further investigation of the performance of S-X2: an item fit index for use with dichotomous item response theory models. Appl. Psychol. Meas. **27**(4), 289–298 (2003)
F. Samejima, in The general graded response model, ed. by M.L. Nering, R. Ostini. Handbook of Polytomous Item Response Theory Models, (Routledge, New York, NY, 2010), pp. 77–108
G. Schwarz, Estimating the dimension of a model. Ann. Stat. **6**(2), 461–464 (1978)
T.L. Strayhorn, What role does grit play in the academic success of black male collegians at predominantly White institutions? J. Afr. Am. Stud. **18**(1), 1–10 (2014)
A. Vaughan, J. Parra, T. Lalonde, First-generation college student achievement and the first-year seminar: a quasi-experimental design. J. First-Year Exp. Stud. Trans. **26**(2), 51–67 (2014)

A Comparison of Item Parameter and Standard Error Recovery Across Different R Packages for Popular Unidimensional IRT Models

Taeyoung Kim and Insu Paek

Abstract With the advent of the free statistical language R, several item response theory (IRT) programs have been introduced as psychometric packages in R. These R programs have an advantage of a free open source over commercial software. However, in research and practical settings, the quality of results produced by free programs may be called into questions. The aim of this study is to provide information regarding the performance of those free R IRT software for the recovery item parameters and their standard errors. The study conducts a series of comparisons via simulations for popular unidimensional IRT models: the Rasch, 2-parameter logistic, 3-parameter logistic, generalized partial credit, and graded response models. The R IRT programs included in the present study are "eRm," "ltm," "mirt," "sirt," and "TAM." This study also reports convergence rates reported by both "eRm" and "ltm" and the elapsed times for the estimation of the models under different simulation conditions.

Keywords R IRT packages • eRm • ltm • mirt • TAM • sirt • Item parameter recovery

Many item response theory (IRT) estimation programs have been developed for the past years. Some commercial IRT programs are very widely used. For instance, PARSCALE (Muraki and Bock 1997) and MULTILOG (Thissen 1991) have frequently been used for research and in practice (Tao et al. 2014). Most notably, a free statistical language, R (R Core Team 2015), has provided several packages which have enabled researchers to conduct psychometric analyses. Rusch et al. (2013) outline the ongoing development of R packages in psychometrics, particularly in terms of breadth and depth in IRT.

T. Kim (✉)
State University of New York at Buffalo, Buffalo, NY 14228, USA
e-mail: tkim33@buffalo.edu

I. Paek
Florida State University, Tallahassee, FL 32306, USA
e-mail: ipaek@fsu.edu

L.A. van der Ark et al. (eds.), *Quantitative Psychology*, Springer Proceedings in Mathematics & Statistics 196, DOI 10.1007/978-3-319-56294-0_36

As several IRT software have been introduced, comparisons among them for various model estimations have been studied. However, most studies have been limited to comparisons among commercial IRT packages. The earliest of these studies (e.g., Ree 1979) compared PARSCALE and MULTILOG under different population distributions for binary items. Later studies (e.g., DeMars 2002) encompassed a broad range of evaluations of these programs to polytomous items with Samejima's (1969) graded response model (GRM) and Masters' (1982) partial credit model (PCM).

Though previous studies have compared commercial and free IRT software (e.g., Pan and Zhang 2014), the IRT programs in R have not been rigorously evaluated in a systematic manner. Furthermore, most of the software evaluation studies have only investigated the recovery of item parameters in a variety of settings, and not of standard errors of item parameters. In this study, a comparison study was conducted via a series of simulations with popular unidimensional IRT models using five IRT programs in R, which are the Rasch model, the 2-parameter logistic (2-PL) and the 3-parameter logistic (3-PL) models, Muraki's (1992) generalized partial credit model (GPCM), and GRM, with respect to the recovery of item parameters and their standard errors.

The R IRT programs, at the time of the study, included the most updated versions[1] of "eRm" (extended Rasch modeling; Mair et al. 2015), "ltm" (latent trait models under IRT; Rizopoulos 2006), "mirt" (multidimensional item response theory; Chalmers 2012), "sirt" (supplementary item response theory models; Robitzsch 2015), and "TAM" (Test Analysis Modules; Kiefer et al. 2015). Except "ltm," the rest of the IRT programs in R were recently released.

1 Method

1.1 Conditions

We evaluated item parameter and standard error (SE) recovery under the following conditions for the dichotomous item response models: 2 (test forms) × 2 (sample sizes). Four conditions for the Rasch and 2-PL models were constructed by two test forms (test lengths of 25 and 50) and two different sample sizes (500 and 1000 examinees). For 3-PL model, the two test lengths were kept the same as those in the Rasch and 2-PL models, but sample sizes were increased to 2000 and 4000 based on preliminary analyses which have suggested a large sample size to avoid non-convergence issues. For the two polytomous models (GPCM and GRM), a single condition was considered: a large sample size of 5000 and a test length of six with each item having five categories. The purpose of using the large sample size was to avoid the zero frequency in some of the option(s), which presents challenges

[1]Note that this study used the latest version of each package available at the time of study: "eRm" (0.15–6; November 12, 2015), "ltm" (1.0–0; December 20, 2013), "TAM" (1.15–0; December 15, 2015), "sirt" (1.8–9; June 28, 2015), and "mirt" (1.15; January 21, 2016).

Table 1 Simulation design

Models	Test length (n)	Number of examinees (p)	Package(s) used
Rasch	25, 50	500, 1000	eRm, ltm
2-PL	25, 50	500, 1000	ltm, sirt, TAM, mirt
3-PL	25, 50	2000, 4000	ltm
GPCM/GRM	6	5000	ltm, mirt

in terms of evaluating the recovery of item parameters in reference to the true item parameters in the polytomous item response modeling. Also, the currently employed R IRT polytomous item response models do not provide a procedure to deal with this problem. While one package ("ltm") was evaluated for the 3-PL model, two packages ("eRm" and "ltm") and four packages ("ltm," "sirt," "TAM," and "mirt") were assessed for the Rasch model and the 2-PL model, respectively. For GPCM and GRM, "ltm" and "mirt" were evaluated. Table 1 encapsulates the simulation design in this study.

2 Data Generation

Item response data were generated following the standard IRT procedure. One thousand replications were made for each condition. Across all models, examinee ability (θ) was drawn from N(0, 1). True values of item parameters of dichotomous models were randomly drawn from logN(0, 0.5^2) for item discrimination or slope (a) parameters, N(0, 1) for item difficulty (b) parameters, and beta(5, 17) for the (pseudo) guessing (g) parameters. For the simulated tests, the true item difficulties ranged from 1.748 to 2.017 (mean = 0.088, *SD* = 1.024), the true discrimination ranged from 0.468 to 1.553 (mean = 1.000, *SD* = 1.72), while the true guessing ranged from 0.054 to 0.286 (mean = 0.185, *SD* = 0.056). For GRM, the same underlying distributions (i.e., logN(0, 0.5^2), N(0, 1)) were used again to generate true values of item discrimination parameters and step difficulty parameters, respectively. (It should be mentioned that the step difficulties (bs) were generated from N(0,1) and transformed into intercept parameters (d) by $d = ab$.) However, a simple item parameter set, which is not based on a random draw from the above distributions, was used for the GPCM data generation. This is because the current version of "mirt" does not use a popular GPCM parameterization, adopting a different parametrization from "ltm" with respect to the slope-intercept form in GPCM. (The current "mirt" GPCM parameterization is $a\theta - k^2$, where k is defined as a difference of adjacent intercept parameters, which is not conventionally used in the popular GPCM parameterization.) In this

[2]Note that "mirt" uses actually "+ intercept" but for consistency with the "ltm" expression, "–intercept" was used in this article.

regard, to make the metric transformation from "mirt" GPCM parametrization to the other usual slope-intercept form efficient, *a*s for GPCM were either 1 or 2, and *b*s were −1, −0.5, 0.5, and 1 for the "mirt" GPCM calibration. Of note is that for the polytomous models, the recovery of the $a\theta - d$ parameterization was examined, while in the dichotomous models, the recovery of the $a(\theta - b)$ parameterization was investigated.

2.1 Recovery of Item Parameters and Their Standard Errors

The recovery of item parameters and their standard errors was examined after checking convergence of the model estimation. The evaluation criteria were absolute bias and root-mean-square errors (RMSEs). For the standard error recovery, the standard deviation of the parameter estimates was used as the (approximate) true value. With respect to standard error estimation, default methods provided by R IRT packages were used. "ltm" and "mirt" clearly delineated what the default standard error estimation method was. "ltm" reported standard errors using delta method under the usual IRT parameterization (i.e., $a(\theta - b)$ form). In "mirt" package, a variety of options for standard error computations, including "crossprod" which is the default, were available.

2.2 Convergence Check and Elapsed Time

This study reported estimation run times for all packages and convergence rates for "eRm" and "ltm" which provided a convergence indicator as part of the program run. Non-convergence rates and average elapsed estimation time per one data set are summarized in Table 2. Non-convergence rates shown in Table 2 represent the percentage of replication diagnosed by the program convergence indicator. Notably, the issue of convergence was critical in 3-PL model using "ltm." For the 3-PL model with "ltm," unreasonable estimates (e.g., very large unreasonable estimates) were sometimes observed despite the program reporting that there was no flag in the converge check.

3 Results

The results of this study, which excludes non-convergence replications, are summarized in Figs. 1, 2, 3 and 4: Rasch, 2-PL, 3-PL, and GRM, respectively. The summary measures (i.e., absolute bias, RMSE) in each of the figures represent averages across items. Our results suggest that absolute bias, and RMSE of item parameter estimates and their standard errors in "eRm," and "ltm" for the Rasch model, was nearly the same (see Fig. 1). We used a metric transformation to obtain equivalent parameter

Table 2 Average running time in minutes, average running time per iteration in seconds, and percentage of analyses that did not converge

Model	Sample size	Test length	Package	Time	Time/iter	Non-conv.
Rasch	500	25	eRm	17	1.02	0
			ltm	13	0.78	0
		50	eRm	41	2.46	0
			ltm	30	1.80	0
	1000	25	eRm	27	1.62	0
			ltm	20	1.20	0
		50	eRm	70	4.20	0
			ltm	56	3.36	0
2-PL	500	25	ltm	28	1.68	0
			sirt	16	0.96	NA
			TAM	27	1.62	NA
			mirt	16	0.96	NA
		50	ltm	56	3.36	0
			sirt	34	2.04	NA
			TAM	25	1.50	NA
			mirt	20	1.20	NA
	1000	25	ltm	36	2.16	0
			sirt	19	1.14	NA
			TAM	28	1.68	NA
			mirt	14	0.84	NA
		50	ltm	119	7.14	0
			sirt	41	2.46	NA
			TAM	44	2.64	NA
			mirt	33	1.98	NA
3-PL	2000	25	ltm	192	11.52	18.8
		50		365	21.90	19.3
	4000	25		480	28.80	22.8
		50		990	59.40	33.7

Note: Time = Average running time in minutes; Time/iter = Average running time per iteration in seconds; Non-conv. = Percentage of analyses that did not converge; and NA in Non-conv. represents convergence flag that was not available for those packages

estimates for the Rasch model, as "eRm" is based upon Rasch framework and uses sum-to-zero constraints for item difficulty estimates while this is not the case for "ltm" where a common item discrimination parameter is estimated.

Unlike the Rasch model, the 2-PL parameter recovery showed different performances across "ltm," "TAM," "sirt," and "mirt." Specifically, "TAM" showed relatively poor performance on point estimate recovery compared to other programs. For the SE recovery, "sirt" indicated poor performance as compared to the other programs. In general, "ltm" and "mirt" provided better results than the other two

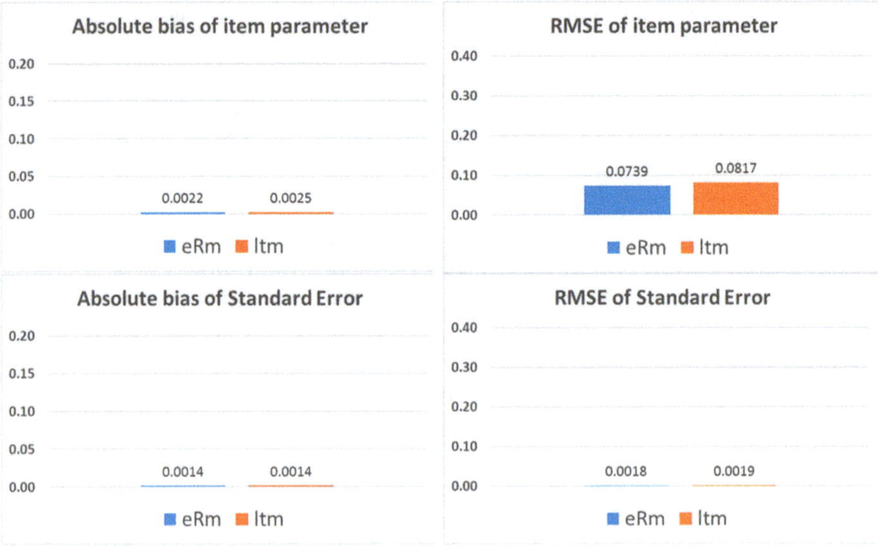

Fig. 1 Rasch result in case of n = 1000, p = 50

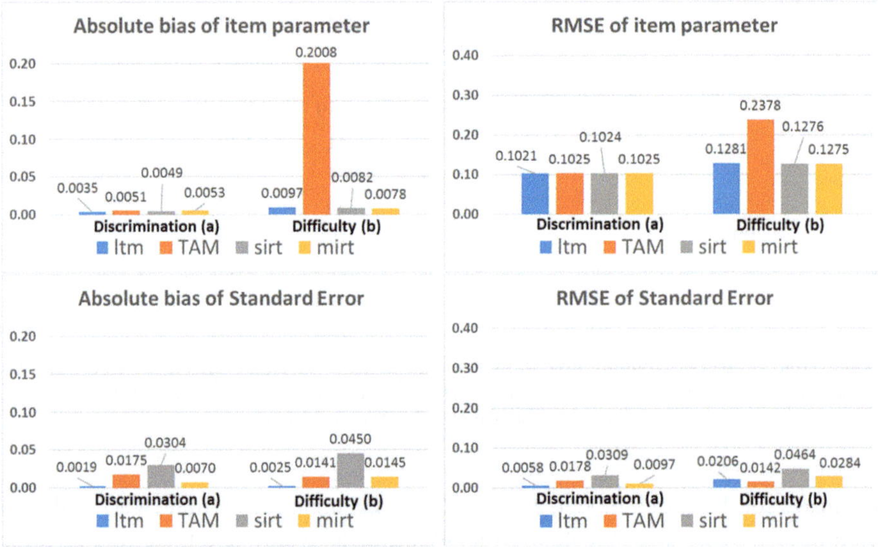

Fig. 2 2-PLM result in case of n=1000, p = 50

packages (see Fig. 2). For example, while the average RMSE of "ltm," "TAM," "sirt," and "mirt" for discrimination parameter was 0.1021, 0.1025, 0.1024, and 0.1025, those for difficulty parameter were 0.1281, 0.2378, 0.1276, and 0.1275, respectively. "TAM" exhibited about twice average RMSE than the others. As well,

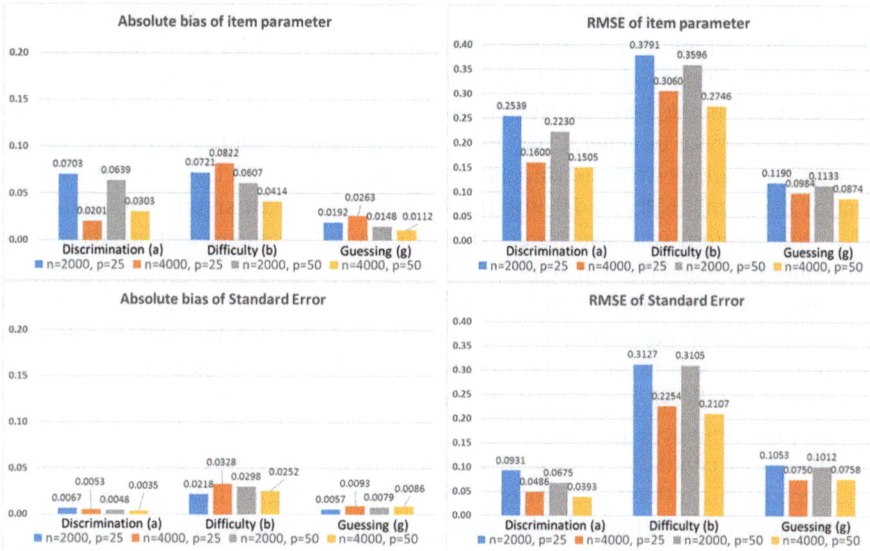

Fig. 3 3-PLM result

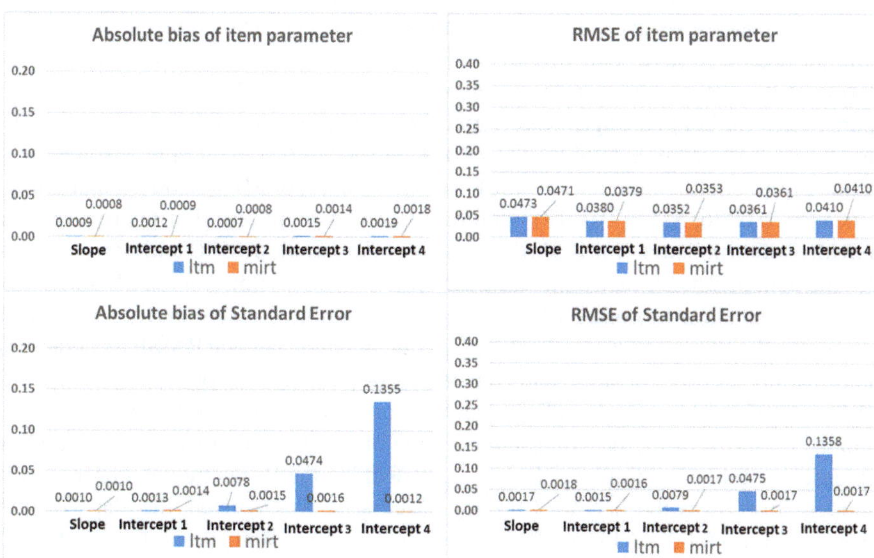

Fig. 4 GRM result

the average RMSE of "sirt" for the standard error of difficulty parameter was 0.0464, which was higher than those of "ltm," "TAM," and "mirt" (0.0206, 0.0142, and 0.0284, respectively).

In terms of the 3-PL model, only one package, "ltm," was used. As mentioned previously, the non-convergence rate was high in the estimation of the 3-PL model by "ltm," which seems to be due to the lack of no item prior provision, especially for the low asymptote in the current "ltm" program. In addition, we observed that the convergence rate did not increase as the sample size increased (see Table 2). The RMSE values in the 3-PL model estimated by "ltm" were relatively high compared to the 2-PL model in general (see Fig. 3). In particular, while the average RMSE across four packages for discrimination parameter in the 2-PL model was 0.1024, that of the "ltm" 3-PL model was 0.1968 across different simulation conditions. This same pattern was also observed for difficulty parameter and standard error estimations of *a* and *b* parameters.

Both "ltm" and "mirt" provided either slope-intercept (i.e., $a\theta - d$ form) or conventional IRT parametrization (i.e., $a(\theta - b)$ form) for the polytomous item response models. However, as previously indicated, the intercept parameter in both programs for GPCM was not defined in the same manner. In "ltm," the intercept is the usual intercept parameter itself (again, d in $a\theta - d$), while in "mirt," it is defined sequentially (k in $a\theta - k$, which is the difference between adjacent intercepts). This different parameterization in both program made the comparison of SE challenging, although one may use a delta method. The current "mirt" program does not provide built-in standard error computation for $a\theta - d$ or $a(\theta - b)$. For this reason, only the evaluation of item parameter recovery was attempted in GPCM, and this study had more emphasis on GRM in terms of comparison of parameter and SE recovery for a polytomous model. The detailed results for GPCM are not presented here, but, overall, both "ltm" and "mirt" performed similarly in terms of the recovery of item parameters of GPCM. The RMSE values of all item parameters for both packages were very comparable (mean = 0.0425, $SD = 0.0049$ for "ltm", and mean = 0.0421, $SD = 0.0053$ for "mirt"), while the absolute bias values were slightly smaller for "mirt" (mean = 0.0016, $SD = 0.0005$) than "ltm" (mean = 0.0036, SD = 0.0018), with respect to absolute bias. For the recovery of GRM, both "ltm" and "mirt" showed, again, comparable RMSE and absolute bias for the item parameter recovery, while the recovery of SEs noticeably differed across the two packages. In contrast to "ltm," "mirt" exhibited stable performance with respect to the SE recovery. As illustrated in Fig. 4, while average RMSE of SE across item parameters (i.e., a slope and four intercept parameters) for "ltm" was 0.1945 ($SD = 0.057$), the corresponding quantity for "mirt" was 0.0017 ($SD < 0.001$). Finally, in terms of the program running time of the dichotomous response models (please see Table 2), "ltm" was faster than "eRm" in the Rasch model. For the 2-PL model, "mirt" was the fastest of the four packages. As expected, the elapsed time per replication for the 3-PL model by "ltm" was longest. With a sample size of 4000 and a test length of 50, it took nearly a minute for a single replication.

4 Discussion

This study evaluated the performance of free IRT programs in R regarding item parameter and its SE recovery. Because the programs are free, practitioners and researchers may consider those programs for classroom instruction, research, or other practical uses. In this regard, the results of this study provide a substantial amount of insight into the performance of five R IRT programs for popular unidimensional IRT models.

The ongoing continued development/update of some IRT programs in R and several limitations in this study warrant further research. The inclusion of currently popular commercial IRT software in the comparisons of these R IRT programs could provide even more insights, which would allow researchers and practitioners to recognize availability and potential utility of these R IRT packages. Of the current R IRT programs, the 3-PL model estimation by "mirt" requires further investigation. The high non-convergence rate and relatively weak performance of the 3-PL model estimation may be improved by employing item prior distributions, which are available in the "mirt" package. Finally, we suggest that users pay attention to the model parameterization used by each program, especially for the polytomous item response models. From the point of view of the consumer, R IRT program developers might consider providing more commonly used IRT parameterizations, as well as align the SEs of those parameters with common IRT parametrizations. This would prevent users being left to calculate SEs of those parameters manually (e.g., using the delta method by users).

References

P. Chalmers, mirt: a multidimensional item response theory package for the R environment. J. Stat. Softw. **48**(6), 1–29 (2012)

C. DeMars, Recovery of graded response and partial credit parameters in MULTILOG and PARSCALE. Paper presented at the annual meeting of American Educational Research Association, Chicago, IL, 2002

T. Kiefer, A. Robitzsch, M. Wu, TAM: Test Analysis Modules. R package version 1.15-0, 2015., http://CRAN.R-project.org/package=TAM

P. Mair, R. Hatzinger, M.J. Maier, eRm: Extended Rasch Modeling. R package version 0.15-6, 2015., http://CRAN.R-project.org/package=eRm

G.N. Masters, A Rasch model for partial credit scoring. Psychometrika **47**, 149–174 (1982)

E. Muraki, A generalized partial credit model: application of an EM algorithm. Appl. Psychol. Meas. **16**, 159–176 (1992)

E. Muraki, R.D. Bock, *PARSCALE 3: IRT Based Test Scoring and Item Analysis for Graded Items and Rating Scales [Computer Software]* (Scientific Software, Chicago, IL, 1997)

T. Pan, O. Zhang, A comparison of parameter recovery using different computer programs and the latent trait models R-Packages in estimating the graded response model. Paper Presented at the annual meeting of American Education Research Association, Philadelphia, PA, 2014

R Core Team, R: a language and environment for statistical computing. R Foundation for Statistical Computing, Vienna, Austria, 2015. ISBN: 3-900051-07-0., http://www.R-project.org/

M.J. Ree, Estimating item characteristic curves. Appl. Psychol. Meas. **3**, 371–385 (1979)

D. Rizopoulos, ltm: an R package for latent variable modeling and item response theory analyses. J. Stat. Softw. **17**(5), 1–25 (2006)

A. Robitzsch, sirt: supplementary item response theory models. R package version 1.8-9, 2015., http://CRAN.R-project.org/package=sirt

T. Rusch, P. Mair, R. Hatzinger, Psychometrics with R: a review of CRAN packages for item response theory. Discussion Paper Series/Center for Empirical Research Methods, 2013/2. WU Vienna University of Economics and Business, Vienna, 2013

F. Samejima, Estimation of ability using a response pattern of graded scores. Psychometrika Monograph, No. 17, 1969

S. Tao, B. Sorenson, M. Simons, Y. Du, Item parameter recovery accuracy: Comparing PARSCALE, MULTILOG and flexMIRT. Paper presented at the 2014 National Council of Measurement in Education Annual Meeting, Philadelphia, PA, 2014

D. Thissen, *MULTILOG: Multiple Category Item Analysis and Test Scoring Using Item Response Theory [Computer Software]* (Scientific Software, Chicago, IL, 1991)

MIX
Papier aus verantwortungsvollen Quellen
Paper from responsible sources
FSC® C105338

If you have any concerns about our products,
you can contact us on
ProductSafety@springernature.com

In case Publisher is established outside the EU,
the EU authorized representative is:
Springer Nature Customer Service Center GmbH
Europaplatz 3, 69115 Heidelberg, Germany

Printed by Libri Plureos GmbH
in Hamburg, Germany